Introduction to PHYCOLOGY

G. ROBIN SOUTH
BSc PhD
Director
Huntsman Marine Science Centre
St Andrews
New Brunswick
Canada
and Professor of Biology
Memorial University of Newfoundland
St John's, Newfoundland
Canada

ALAN WHITTICK
BSc MSc PhD
Associate Professor of Biology
Memorial University of Newfoundland
St John's, Newfoundland
Canada

BLACKWELL SCIENTIFIC PUBLICATIONS
OXFORD LONDON EDINBURGH
BOSTON PALO ALTO MELBOURNE

Editorial offices:
Osney Mead, Oxford, OX2 0EL
(*Orders:* Tel. 0865 240201)
8 John Street, London, WC1N 2ES
23 Ainslie Place, Edinburgh, EH3 6AJ
52 Beacon Street, Boston
Massachusetts 02108, USA
667 Lytton Avenue, Palo Alto
California 94301, USA
107 Barry Street, Carlton
Victoria 3053, Australia

First published 1987

Set by Setrite Typesetters, Hong Kong, and printed and bound in Great Britain by Wm Clowes, Beccles, Suffolk.

DISTRIBUTORS

USA and Canada
Blackwell Scientific Publications Inc
P O Box 50009, Palo Alto
California 94303
(*Orders:* Tel. (415) 965-4081)

Australia
Blackwell Scientific Publications
(Australia) Pty Ltd
107 Barry Street,
Carlton, Victoria 3053
(*Orders*: Tel. (03) 347-0300)

British Library
Cataloguing in Publication Data

South, G. Robin
Introduction to Phycology.
1. Algae
I. Title II. Whittick, Alan
589.3 QK566

ISBN 0-632-01769-4
ISBN 0-632-01726-0 Pbk

Library of Congress
Cataloguing in Publication Data

South, G. Robin (Graham Robin)
Introduction to Phycology.

Bibliography: p.
Includes index.
1. Algology. 2. Algae. I. Whittick, Alan.
II. Title. III Title: Phycology.
QK566.S68 1987 589.3 87-7146
ISBN 0-632-01769-4
ISBN 0-632-01726-0 (pbk.)

Contents

Preface

In this age of scientific overload it is becoming increasingly difficult to provide a synoptic account of any scientific subdiscipline within the confines of a single undergraduate textbook. The study of algae is certainly no exception, and even a casual perusal of the biological literature of the past decade reveals the enormous increase in our knowledge of this group of plants. Their evolutionary diversity, their importance as primary producers, their usefulness in physiological and ecological experiments, their value as indicators of environmental quality, and their direct economic importance have attracted the interest of scientists and engineers who would not regard themselves as phycologists in the classical sense. Treatises continue to be produced on smaller and smaller segments of the subject, and from this standpoint phycology, as a scientific subdiscipline of biology, has certainly come of age. As educators, however, we are still required to present a course in the subject as a whole, often within the confines of a single semester or term.

Introduction to Phycology is our attempt to provide an integrated overview of phycology for the undergraduate student. It is a departure from the phyletic approach to the teaching of phycology as reflected in the majority of currently available textbooks, in which the various algal divisions are individually presented and discussed. We hope that this approach will place the study of the algae in a proper, holistic perspective, and that students will perceive a subject at the forefront of the biological sciences, rather than one whose practitioners are preoccupied with taxonomy and life histories.

In *Introduction to Phycology* we begin by providing an overview of the classification of the algae to provide a conceptual framework for the chapters which follow. Levels of organization are then treated from the subcellular, cellular, and morphological standpoints, followed by reviews of life cycles and reproduction, physiology and biochemistry, ecology, phylogeny and evolution, and, finally, the role of algae in human society.

We have included the blue-green algae/bacteria (Cyanophycota) in this book. We regard them, together with the Prochlorophycota, as prokaryotic algae and thus a legitimate part of the science of phycology, even though we do not dispute their affinities with the bacteria. Any comparative account of the algae must of necessity include reference to them; a comprehensive view of algal structure, physiology, ecology, or evolution is otherwise incomplete.

We believe that students should be introduced to the primary literature of a subject, and to this end have provided extensive references to key topics. In addition, we have also included references to recent review articles. The literature cited reflects our personal biases and is not intended to be an exhaustive survey of phycology. In areas where there is great current activity (such as in the classification of the green algae *sensu lato*) we have taken a conservative approach, in the full realization that future work will inevitably alter the picture that we have attempted to present. Indeed, in the short time since the manuscript of this book was completed, such events have already occurred.

We recognize that in this work there will be shortcomings, particularly those of omission; we hope, however, that these will be outweighed by the positive aspects. While we take full responsibility for this book, we gratefully acknowledge the many friends and colleagues who assisted us in its completion. In particular we acknowledge those who read all or portions of the text during its preparation, Eric Henry and Robert Hooper. Roy Ficken provided expert photographic services and Sue Meades redrew a number of the text figures. Marc Favreau and Ralph Kuhlenkamp provided assistance in checking our French and German citations, and the staff of the Memorial University of Newfoundland Queen Elizabeth II Library were helpful in locating a number of references. We are especially grateful to our many colleagues who willingly provided copies of original photographs or who gave permission for the use of their illustrations. The editorial staff of Blackwell Scientific Publications

Ltd were particularly supportive in their help and encouragement to complete what at times seemed an impossible task. Last, but certainly not least, our students, graduate and undergraduate, past and present, have been a continuous source of ideas and stimulation.

G. Robin South
Alan Whittick

CHAPTER 1

Introduction

Phycology and the algae

Phycology (Greek: *phykos* = seaweed) is the study of algae (singular, alga). Algae belong to the plant kingdom and comprise a very diverse group of organisms which has, since the earliest times, defied precise definition. Here they are classified in six divisions (or phyla), although in some treatments 12 or 13 divisions may be recognized; a number of the smaller groups are still incompletely known. Both prokaryotic and eukaryotic forms are included. From their long fossil record it is evident that the prokaryotic algae were the first photosynthetic cellular plants; it is now generally agreed that algae are the group from which all subsequent cryptogamic groups of plants, and ultimately the flowering plants (or Spermatophyta), arose.

The algae are simply constructed; they range from single-celled forms to aggregations of cells, filaments, or parenchymatous thalli. Many of the unicellular forms are motile, and may intergrade confusingly with the Protozoa. Even the most complex multicellular forms show a low level of differentiation compared with other groups of plants, with only the most advanced possessing elementary conducting tissues. The range of morphology is, however, extremely diverse, and a number of the brown algae (Phaeophycota) may attain a size comparable with that of a small tree. Yet the relative simplicity of the algae is misleading, because even the smallest may exhibit, at the cellular level, a high degree of complexity.

The algae demonstrate as much variation in reproduction as they do in morphology. Vegetative, asexual, and sexual processes are involved, and in many forms an alternation of generations occurs. In the majority, motile zoospores or gametes are formed; sexual processes have evolved independently in various divisions and have included the development of oogamy. Sexual reproduction is absent, however, from the prokaryote divisions (Cyanophycota; Prochlorophycota), and from a number of the eukaryote divisions (e.g. Euglenophycota). A feature which distinguishes algae from other cryptogams is the lack of a multicellular wall around the sporangia or gametangia (with the exception of the antheridium in the Charophyceae). The sexual reproduction of the Rhodopohyceae may be the most complex in the plant kingdom.

Biochemically and physiologically the algae are similar in many respects to other plants. They possess the same basic biochemical pathways; all possess chlorophyll *a* and have carbohydrate and protein end-products comparable to those of higher plants. Therefore many algae are ideal experimental organisms in that, due to their small size and easy manipulation in liquid media, they can be studied under controlled conditions in the laboratory.

The algae are ubiquitous, occurring in practically every habitable environment on earth, ranging from hot and cold deserts to the soil, permanent ice and snow fields, and every kind of aquatic habitat. They are the major primary producers of organic compounds, and they play a central role as the base of the food chain in aquatic systems. They are a source of human food, and chemical extracts from algae are used in the manufacture of food and many other products. They are also 'nuisance' organisms in municipal water supply systems, and in aquatic resources subject to eutrophication. As the causative organisms of 'red tides', some Dinophyceae are a source of poisoning. Only rarely, however, are algae the cause of serious human disease. It is likely that, in the future, uses of algae will proliferate in the fields of energy and food production. It is also likely that algae will be among the first organisms to be put to work by man in space.

The development of phycology

The development of phycology as a science has its roots in western culture, especially European (*see* Taylor, 1969), and can be assigned to four phases. The first began with the writings of the Greeks (Theophrastus; Dioscorides), and concluded at the end of the eighteenth century, the second spanned the period 1800 to about 1880, the third from about 1880 to the early 1950s (*see* Prescott, 1951) and the fourth, or modern phase, from the early 1950s to the present.

The long early history of phycology is entwined with the origins of botany itself. Although the first written references to the algae are to be found in the ancient Chinese classics (Porterfield, 1922), it was the Greeks who laid the foundations of the subject. The Roman word *fucus*, derived from the Greek *phykos*, persisted as a name for a poorly defined group of simple plants in early botanical writings until Linnaeus (1753) described the genus *Fucus*, thus marking the beginnings of the formal nomenclature of algae (Chapter 2). Up to the end of the eighteenth century, however, the algae were crudely classified, and were often grouped with the fungi and lichens. Von Zalusian (1592) included them with fungi, lichens, and seaweeds as Musci, under plants '*Ruda et Confusa*', while Bauhin (1620) listed under algae *Muscus, Fucus, Conferva* and *Equisetum* (= *Chara*). The flagellate forms were frequently classified as 'zoophytes'. Dillenius (1741) wrote on the algae but, as pointed out by Prescott (1951), it is difficult to know which plants he was describing. At the end of this period, the majority of the macroscopic algae were designated under a few genera, notably *Fucus*, *Conferva, Ulva* and *Corallina*, and in modern terms many diverse and unrelated species were grouped arbitrarily together.

The crude taxonomy of this early phase, coupled with a lack of understanding about sexual processes, a lack of adequate microscopes, and prevalent biases about fundamental biological principles, served to hinder the emergence of phycology as a science. There was, for example, a universal belief among the early botanists that the algae lacked sexuality, even though De Reaumur had sparked a lively debate on the subject following his publication in 1711 on the sex organs of *Fucus*. As late as 1768 Gmelin claimed that algae develop parthenogenetically.

It is J. Stackhouse who perhaps deserves the most credit for beginning the transformation of phycology into a true science, thus heralding the second phase of the development of the subject. He was probably the first to study zygote germination in *Fucus* and to describe (1801) the process of fertilization; a consequence of his work was a realization of the inadequacy of the prevailing classification of *Fucus* as defined by Linnaeus, and the need to rearrange the many diverse species included within it.

The advent of improved microscopy, coupled with the stimulus provided by Stackhouse and others, heralded an explosive growth in taxonomy. There were revolutionary advances in the systematics of cryptogams from the beginning of the nineteenth century to about 1880, by which time many of the presently known algal genera had been described. Pioneers included Dawson Turner (1802), Vaucher (1803), Lamouroux (1813), Lyngbye (1819), Greville (1830), and William Henry Harvey (1846–51, 1852–8). Most remarkable, however, were the contributions of the Swedish phycologist C.A. Agardh and his son J.G. Agardh, whose extensive publications spanned three-quarters of a century and remain a cornerstone of algal taxonomy and nomenclature, and the extensive work of the German phycologist F.T. Kützing (1843, 1849). For a more complete review of this important period of phycology, in which the foundations of modern taxonomy and systematics were laid, reference should be made to the accounts of Prescott (1951) and Taylor (1969).

Once the majority of the algae had been described the development of modern systematics became possible, marking the beginning of the third phase of phycology. At the start of this period De Toni (1889) began the monumental task of summarizing, in his *Sylloge Algarum*, the taxonomy and nomenclature of the algae, a work not completed until 35 years later (1924). The critical studies of Schmitz (1883, 1889; Rhodophycota), Kuckuck (1912) and Kjellman (1897; Phaeophycota) and Wille (1897–1911; Chlorophycota), among others, established new taxonomic treatments. Concurrently, important pioneering studies (e.g. Williams 1897, 1898; Sauvageau, 1899, 1915; Yamanouchi, 1906; Svedelius, 1908; Kylin, 1914) advanced the understanding of algal life histories. The accumulation of this new information led to the development of improved classification of the algae, and to new speculation on the origins and interrelationships among the algae.

During this third phase many floras, monographs, and general taxonomic treatments were published covering the microscopic and macroscopic groups, both freshwater and marine. Landmarks in the emergence of phycology, however, were the classical compendia of Oltmanns (1904,

1922) and Fritsch (1935, 1945); although not the only general works of their period, they brought together the many diverse aspects of the subject in the most comprehensive manner. It is indicative of the enormous growth of phycology since the time of Fritsch that it would now be impossible for any single individual to amass the sum of our knowledge of the algae within the confines of two volumes.

The beginning of the fourth, modern, phase of phycology is perhaps best marked by the publication of Smith's (1951) *Manual of Phycology*. Containing no reference to the electron microscope, it marks the end of the pre-EM phase of phycology yet, through its multiauthor approach in recognition of the ever-widening complexity of the subject, it set the trend for more modern treatments of phycology.

Phycology is being revolutionized in this, its modern, phase of development. The widespread availability of the electron microscope, both scanning and transmission, the development of highly sophisticated aids to investigations in ecology, physiology, biochemistry, and molecular biology, and the greatly expanded accessibility of algae afforded through the invention of scuba have combined to bring about this revolution. The vistas opened up would have been undreamed of by the classical workers such as F.E. Fritsch. It is likely that future historians will conclude that these changes are even more profound in their effects on phycology than, for example, those brought about by the invention of the light microscope.

These modern tools have greatly advanced our knowledge of the structure and functioning of algal cells and their diverse organelles, including an understanding of their micromorphology, cytology, physiology, molecular organization, and genetics. Contributors to modern phycology are legion; reference to recent reviews mentioned in this text is recommended for those interested in specific aspects.

Consequent upon the examination of many algae under the electron microscope, and upon a much better understanding of algal biochemical characteristics, numerous additional taxonomically important criteria have become available to phycologists. A result has been the proposal of many new systems of classification of the algae which take these features into account. As yet no concensus has emerged as to an acceptable classification, and for the specialist, let alone the student new to the subject, it is a frustrating time. The debate about the relationships among and between green algae and other green plants is particularly intense. As more algae are examined using modern techniques, many of the current discrepancies will hopefully be resolved.

The ranks of phycologists have grown in keeping with the growth of the subject, and at the beginning of the 'modern' phase new societies were founded, such as the Phycological Society of America, the British Phycological Society, La Société Phycologique de France, the Japanese Phycological Society, and the International Phycological Society. The coming of age of phycology was also marked by the founding of journals devoted exclusively to the algae: *Phycologia*, the *Journal of Phycology*, the *British Phycological Bulletin* (now *Journal*), *Phykos*, the *Japanese Journal of Phycology* and the *Bulletin de la Société Phycologique de France*. Recently, journals devoted to specific groups of algae (such *Bacillaria*, devoted to the Bacillariophycota), or to reviewing the current state of knowledge of the subject (e.g. *Advances in Phycology*), have been introduced. At an annual publication rate of several thousand articles on phycological topics, the subject can indeed be said to have matured as a science.

The scope of modern phycology

The scope of modern phycology is indeed enormous, as is apparent from an examination of titles of abstracts of papers presented at phycological meetings. Included are the fields of cytology (including electronmicroscopy), physiology, biochemistry, genetics, molecular biology, ecology, taxonomy and systematics, applied biology (e.g. aquaculture, agriculture, industrial extraction, and uses of phycocolloids), pharmacology, medicine, space biology, planktology, and others. While specialists in many of these areas might not refer to themselves as phycologists, they are using algae in their work; the fundamental importance of algae is evident from their eminent suitability as experimental organisms.

Whether one's interests are in strictly laboratory studies, in field work, or in any combination, there are endless problems available for study. Among significant advances during the past 10–15 years are the discovery of two distinct modes of cytokinesis (the phragmoplast/phycoplast), the analysis and mapping of chloroplast genomes, the realization of the importance of microtrabecular structural lattices as the basis of cytoplasmic ultrastructure, and extensive investigations of eyespots, cell membranes, flagella, and cyanophycean chloroplasts. To single out any area as the 'hottest' would be presumptuous, although few would deny the excitement of current studies with high-resolution electron microscopes, or the enormous unexplored prospects still awaiting molecular biologists. Yet the classical origins of phycology are intact, for the taxonomist remains central to modern studies: how can the molecular biologist publish his work if he does not know the name of the organism with which he is working? Many species still remain to be described, especially among the smaller flagellates; and in those groups that are relatively well known, new methods are resulting in a re-examination of old ideas.

Phycology, then, is a subject catering to all tastes and interests. It has the additional bonus that its subjects are among the most beautiful of organisms when viewed at close quarters under the microscope. It has been through research with the algae that many fundamental physiological problems have been resolved, and it is among the algae there are likely hidden many of the answers to the mysteries of the origins of life, the origins of the eukaryotic cell (and hence ourselves), and the origins of land plants.

We presume here to introduce the subject of phycology in the knowledge that we can provide no more than a superficial glance at this expanding subject. Hopefully many will be stimulated to proceed further: there are many challenges open to the neophyte phycologist!

CHAPTER 2

Classification of the algae

The classification of plants and animals has occupied man since the times of Aristotle. The Swedish botanist Linnaeus (1754) was the first to apply the name Algae to a group of plants; however, his group consisted of a mixture of algae and Hepaticae. The difficulty of defining an alga still persists; as Lewin and Gibbs (1982) aptly state: '... organisms currently regarded as algae have virtually no common features which would indicate phylogenetic homogeneity.'

Most modern phycologists would accept the spirit if not the letter of Fritsch's (1935) definition: 'Unless purely artificial limits are drawn, the designation of an alga must involve all holophytic organisms (as well as their numerous colourless derivatives) that fail to reach the level of differentiation of archegoniate plants.' The algae thus constitute an heterogeneous assemblage of oxygen-producing, photosynthetic, non-vascular organisms with unprotected reproductive structures. The definition bridges both pro- and eukaryotic forms. With the exception of the Charophyceae (division Chlorophycota) the algae are distinguished from the bryophytes by their lack of multicellular sex organs contained within sterile jackets of cells and by their lack of retention of the sporophyte within the female organ (Silva, 1982).

The classification of the algae which we employ here is adapted from Parker (1982) (Table 2.1). The algae belong in the kingdom Protoctista (Whittaker, 1959, 1969; Whittaker & Margulis, 1978; Margulis & Schwartz, 1982). The scheme of Parker (1982) represents the collective view of a number of leading phycologists. It recognizes a separation between prokaryotic forms, which lack membrane-bounded organelles and which include the bacteria, the Cyanophycota (or Cyanobacteria, the 'blue-green algae') and the recently proposed Prochlorophycota (Lewin 1976, 1977), and eukaryotic forms which include algae and all other plants.

Subdivisions of the algae have long been the subject of debate. Harvey (1836) was the first to recognize four major divisions based primarily on colour (brown, red, and green algae, and diatoms) and his scheme has proved to be the basis of most modern classifications of algae into divisions. While recent treatments recognize more than four divisions, the importance of the major photosynthetic pigments and the accompanying biochemical and structural similarities as a means of segregating algae into divisions has prevailed. There remain, nonetheless, wide differences of opinion among phycologists as to the details of classification. A continuous stream of new information on the fine structure and biochemistry of algae is stimulating new approaches to algal classification, and any scheme proposed at the present time is necessarily tentative. Factors which contribute to difficulties of classification include the evident polyphyletic origins of the algae, the seemingly poor fossil record for most groups, and the tendency to make wide-ranging conclusions on the basis of an examination of a relatively small number of species.

Silva (1982) recognizes 16 major phyletic lines, these being assigned to the taxonomic rank of class. The classes are distinguished principally on differences in pigmentation, storage products, cell wall characteristics, and the fine structure of organelles such as flagella, the nucleus, chloroplasts, pyrenoids, and eyespots. These differences are summarized in Table 2.2. The grouping of the classes into divisions is variously treated in modern algal texts, the variation being largely the result of the different stresses that authors place on the importance of the distinguishing characteristics. In this treatment six divisions are recognized: two of them prokaryotic (Cyanophycota and Prochlorophycota) and four of them eukaryotic (Rhodophycota, Chromophycota, Euglenophycota, and Chlorophycota). For some variations on the scheme employed here, refer to Chapman and Chapman (1973), Bold and Wynne (1978, 1985), Trainor (1978), Round (1973), Hoek and Jahns (1978), Lee (1980), Christensen (1980), and Scagel *et al.* (1982).

Classification below the level of class (or subclass in some, e.g. Rhodophyceae) progresses, in descending rank, to order, family, genus and species. Similarities between the taxa become

Table 2.1 The general classification of the algae. After Parker (1982).

Kingdom		Division	Class	Subclass
Plantae	Prokaryota	Cyanophycota		
		Prochlorophycota		
	Eukaryota	Rhodophycota	Rhodophyceae	Bangiophycidae Florideophycidae
		Chromophycota	Chrysophyceae	Chrysophycidae Dictyochophycidae
			Prymnesiophyceae	
			Xanthophyceae	
			Eustigmatophyceae	
			Bacillariophyceae	
			Dinophyceae	Dinophycidae Ebriophycidae Ellobiophycidae Syndiniophycidae
			Phaeophyceae	
			Raphidophyceae	
			Cryptophyceae	
		Euglenophycota	Euglenophyceae	
		Chlorophycota	Chlorophyceae	
			Charophyceae	
			Prasinophyceae	

greater with progression to the lower levels of classification. The endings of the names of the taxonomic ranks are consistent, and follow rules that are strictly applied according to the International Code of Botanical Nomenclature (Voss, 1983). They can be summarized as follows for the brown alga *Fucus vesiculosus*.

Division	Chromophycota (-phycota)
Class	Phaeophyceae (-phyceae)
Order	Fucales (-ales)
Family	Fucaceae (-aceae)
Genus	*Fucus*
Species	*vesiculosus*

It is customary, in formal accounts, to indicate the name of the author who described a genus and species. The name *Fucus vesiculosus* would therefore be given as *Fucus vesiculosus* Linnaeus (or abbreviated as *Fucus vesiculosus* L.). While many algae may have a common or vernacular name (such as 'bladder wrack' for *F. vesiculosus*), such a name has little value in scientific terms; the advantage of the formal system is that it is international and is universally understood.

The genus and species names together comprise the binomial, and are unique for each species. The binomial system of nomenclature began for most algae with Linnaeus' *Species Plantarum* (1st edition 1753). Lower (subspecific) taxa may be described and may include, in decreasing order, subspecies, variety, and form.

Table 2.2 Summary of the principal characteristics of the algal classes.

Division/class	Principal pigments	Food reserves	Chloroplast features	Cell wall	Flagella
Prokaryota					
CYANOPHYCOTA					
Cyanophyceae	Chlorophyll *a* β-Carotene c-Phycoerythrin Allophycocyanin c-Phycocyanin	Phycocyanin granules (arginine and aspartic acid)	No chloroplast Thylakoids free in cytoplasm, unstacked Phycobilisomes	Four-layered peptidoglycan (murein) principal component	Absent
PROCHLOROPHYCOTA	Chlorophyll *a* Chlorophyll *b* β-Carotene Zeaxanthin Cryptoxanthin		No Chloroplast Thylakoids free in cytoplasm, some stacking in pairs or more	As for Cyanophycota	Absent
Eukaryota					
RHODOPHYCOTA					
Rhodophyceae	Chlorophyll *a* Chlorophyll (*d*) r-Phycocyanin Allophycocyanin c-Phycoerythrin α and β-carotene	Floridean starch	Chloroplast envelope two-layered, chloroplast ER absent Thylakoids unstacked	Cellulose (xylan in some Bangiophycidae) polysaccharides; some calcified with $CaCO_3$	Absent
CHROMOPHYCOTA					
Chrysophyceae	Chlorophyll *a* Chlorophylls c_1 and c_2 Fucoxanthin	Chrysolaminarin (β-1, 3-glucopyranoside)	Two additional membranes of chloroplast ER, the outer continuous with outer membrane of nuclear envelope Thylakoids in stacks of three	Naked; scales or lorica in some	Normally two, unequal: smooth posterior shorter; hairy anterior longer
Prymnesiophyceae (Haptophyceae)	Chlorophyll *a* Chlorophyll c_1, c_2 β-Carotene	Chrysolaminarin	Two additional membranes of chloroplast ER, the outer continuous with outer membrane of the nuclear envelope Thylakoids in stacks of three	Cell covering of organic scales in one to several layers	Usually two, unequal, smooth Haptonema
Xanthophyceae	Chlorophyll *a* (chlorophyll *c* in *Vaucheria*) β-Carotene Diatoxanthin Diadinoxanthin Heteroxanthin Vaucheriaxanthin ester	β-1, 3-linked glucan Fats and oils	Two additional membranes of chloroplast ER, the outer continuous with outer membrane of the nuclear envelope Thylakoids in stacks of three	Cellulose, glucose, and uronic acids Many walls are bipartite	Usually two, unequal: the anterior longer, hairy; the posterior shorter, smooth
Eustigmatophyceae	Chlorophylls *a*, *c* β-Carotene Violaxanthin Vaucheriaxanthin	Oils	Two additional membranes of chloroplast ER, neither continuous with the nuclear envelope Thylakoids in stacks of three	Naked	Single anterior hairy flagellum, and second basal body

Table 2.2 *continued.*

Division/class	Principal pigments	Food reserves	Chloroplast features	Cell wall	Flagella
Bacillariophyceae	Chlorophyll *a* Chlorophyll *c* β-Carotene Fucoxanthin Diatoxanthin Diadinoxanthin	Chrysolaminarin Lipids	Two additional membranes of chloroplast ER, the outer continuous with outer membrane of nuclear envelope Thylakoids in stacks of three	Silica frustule	Male gametes possess single hairy flagellum which lacks central microtubules
Dinophyceae	Chlorophyll *a* Chlorophyll c_2 Peridinin Neoperidinin β-Carotene	Starch	Two additional membranes of chloroplast ER, not continuous with nuclear envelope Thylakoids in stacks of three	Cellulosic theca Body scales (rare)	Usually two, transverse and longitudinal, heterokontic, highly distinctive: transverse, longer, helical, positioned in a groove or girdle; the shorter, longitudinal directed posteriorly
Phaeophyceae	Chlorophyll *a* Chlorophyll c_1, c_2 β-Carotene Fucoxanthin	Laminarin Mannitol	Two additional membranes of chloroplast ER, the outer continuous with outer membrane of nuclear envelope Thylakoids in stacks of 2–6	Two-layered, the inner cellulose, the outer of alginic acid and fucoidan	Usually two, unequal: the longer anterior, hairy; the shorter posterior, smooth
Raphidophyceae	Chlorophyll *a* Chlorophyll *c* Diadinoxanthin Dinoxanthin Heteroxanthin ?Fucoxanthin		Two additional membranes of chloroplast ER Thylakoids in stacks of three	Naked	Usually two, unequal: the anterior longer, hairy; the posterior shorter, smooth
Cryptophyceae	Chlorophyll *a* Chlorophyll c_2 α-Carotene Diatoxanthin Phycoerythrin Phycocyanin	Starch-like	Two additional membranes of chloroplast ER Thylakoids in pairs	Periplast	Two apical or lateral, equal, hairy.
EUGLENOPHYCOTA					
Euglenophyceae	Chlorophyll *a* Chlorophyll *b* β-Carotene Astaxanthin Antheraxanthin Diadinoxanthin Neoxanthin	Paramylon (β-1, 3-glucan)	One additional chloroplast ER membrane, not continuous with nuclear membrane Thylakoids in stacks of three, sometimes more	Proteinaceous pellicle	Usually two, unequal, hairy, one often not emergent
CHLOROPHYCOTA					
Chlorophyceae	Chlorophyll *a* Chlorophyll *b* α, β and γ-carotene Lutein Siphonoxanthin Siphonein	Starch (amylose, amylopectin)	Chloroplast ER absent Thylakoids in stacks of 2–6 or more Grana may be present	Naked (some) Cellulose, hydroxyproline, glycosides, xylans, mannans Some calcified	Normally two (four) equal, smooth (may be tomentose) apically inserted

Table 2.2 *continued.*

Division/class	Principal pigments	Food reserves	Chloroplast features	Cell wall	Flagella
Charophyceae	Chlorophyll *a* Chlorophyll *b* α, β and γ-carotene Various xanthophylls	Starch (α-1, 4 glucan)	Chloroplast ER absent Thylakoids in stacks of 2–6 or more Grana present	Cellulose Many calcified	Male gamete only; two unequal, subterminally inserted; scaly
Prasinophyceae	Chlorophyll *a* Chlorophyll *b* β-Carotene Siphonein Siphonoxanthin	Starch Mannitol	Chloroplast ER absent Thylakoids in stacks of 2–6	Naked, usually scaly	One or two, unequal, or four, more or less equal; scaly

For every described species of alga, there should be a type specimen, which is then permanently associated with the given binomial. This may be in the form of a pressed specimen housed in an herbarium, in the form of a prepared microscope slide, or in the form of a liquid-preserved specimen. In some algae, especially microscopic forms such as the desmids (Chlorophyceae), an illustration (the iconotype) may serve as the type. Original descriptions of each species must have been formally published and, currently, a Latin description (known as a diagnosis) must accompany the description. For those interested in nomenclature, the International Code of Botanical Nomenclature (Voss, 1983) should be consulted. Silva (1980) has published an authoritative catalogue of the names of classes and families of living algae, while Farr *et al.* (1979) have compiled an *Index Nominum Genericorum (Plantarum)* which includes details of the publication and typification of the genera of algae.

In the following synopsis the characteristics of the algal divisions and classes are summarized. Following each section is a listing of the orders. In subsequent chapters overviews of cellular organization, morphology, growth, physiology, and ecology are provided.

General characteristics of the algae

PROKARYOTA

The Prokaryota comprise the Schizomycophyta (bacteria), the Cyanophycota, and the Archaebacteria. In the prokaryotes the nuclear material is not enclosed in a nuclear envelope, and a nucleolus and organized chromosomes are lacking. Also lacking are plastids, mitochondria, Golgi bodies, and other unit-membrane organelles. All prokaryotes possessing chlorophyll *a* and carrying out aerobic photosynthesis are classified in the Cyanophycota; those possessing chlorophylls *a* and *b* and carrying out aerobic photosynthesis are classified in the Prochlorophycota.

CYANOPHYCOTA

Cyanophyceae: blue-green algae (Figs 2.1 & 2.2)

While exhibiting an algal morphology, the Cyanophycota possess an admixture of bacterial features (Friedmann, 1982). The name Cyanophycota (= Cyanochloronta; Bold, 1973) has been generally replaced by Cyanobacteria (Stanier *et al.*, 1978; cf. Lang & Walsby, 1982), but for conformity we have retained the Cyanophyceae *sensu* Sachs (1874), as in Parker (1982).

The Cyanophycota lack a nuclear envelope and mitochondria; their photosynthetic lamellae (thylakoids) are single or unstacked and distributed peripherally in the cytoplasm and not within a membrane-bounded chloroplast. They possess chlorophyll *a* and phycobiliproteins (phycocyanin, allophycocyanin and phycoerythrin) with polyglucan granules (carbohydrate) and cyanophycin (combined nitrogen) as storage products. The cell wall is four-layered, with the major, structural layer of peptidoglycan (as in bacteria). Morphology ranges from unicellular to filamentous; many possess extensive mucilaginous sheaths. Specialized features include the ability to fix atmospheric nitrogen (which under aerobic conditions is associated with specialized

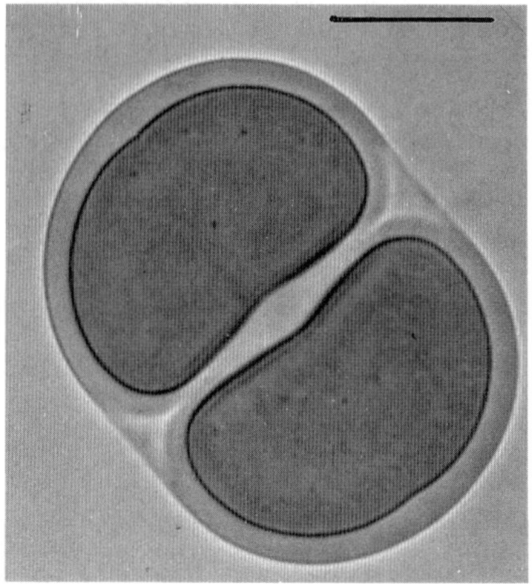

Fig. 2.1 *Chroococcus* sp. (Cyanophyceae, Chroococcales). A planktonic, freshwater species with the cells in pairs, surrounded by a thin, colourless common sheath. Scale = 10 μm.

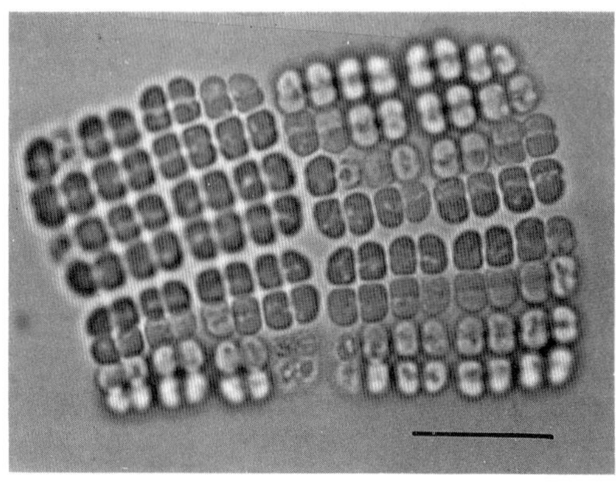

Fig. 2.2 *Merismopedia* sp. (Cyanophyceae, Chroococcales). The numerous, small cells are arranged in a somewhat curved sheet, which enlarges by cell division in two directions. Scale = 10 μm.

cells, the heterocysts), possession of akinetes, and the trichome, a feature of some filamentous forms in which the entire row of cells is sheathed by a common outer wall layer. Asexual reproduction predominates, involving binary fission, multiple fission resulting in endospores (baeocytes) and hormogonia (short, gliding trichomes). True sexual reproduction is lacking, although there are reports of a low-frequency occurrence of genetic recombination by transformation in some species (Devilly & Houghton, 1977; Stevens & Porter, 1980). Flagella are lacking, although characteristic gliding occurs in many species.

The Cyanophycota are undoubtedly among the most primitive autotrophs, and are represented in the fossil record of the Precambrian period (3000 million years BP) (Schopf, 1976). Living species are ubiquitous, and may occur in extreme habitats such as hot springs and in desert rocks. There is a wide range of planktonic, benthic and free-living forms in marine, freshwater, and soil habitats; many form symbiotic associations. Noxious species can cause toxic blooms in eutrophic environments, resulting in health hazards to humans and livestock. Nitrogen-fixing forms are important in all habitats where they occur, and especially in tropical soils, such as rice fields.

The taxonomy and nomenclature of the Cyanophycota are in a state of disorder, and there is presently no consensus among experts. Divergent opinions exist as to whether to apply bacterial (Stanier *et al.*, 1978) or conventional (Drouet, 1968, 1973; Drouet & Daily, 1956) algal taxonomic and nomenclatural principles to the Cyanophycota. The species concept remains in a highly confused state. For a review of current opinions, *see* Lang and Walsby (1982) and Friedmann (1982).

ORDERS

Chroococcales; Pleurocapsales; Stigonematales; Nostocales.

PROCHLOROPHYCOTA

Prochlorophyceae

This division was recently proposed by Lewin (1976, 1977) and contains the single genus *Prochloron* (Lewin, 1981, 1984; Chapman & Trench, 1982). The algae are prokaryotic, unicellular, obligate symbionts of marine didemnid ascidians. To date only one species has been described with certainty; Thinh (1979) believes that different ascidians harbour the same species of *Prochloron*; *see also* Stackebrandt *et al.* (1982).

Because of their obvious phyletic interest, the Prochlorophycota have attracted considerable attention. While essentially prokaryotic in nature, the cells contain chlorophylls *a* and *b* (characteristic for the Chlorophycota), β-carotene, zeaxanthin and cryptoxanthin, but lack the phycobiliprotein pigments of the Cyanophycota.

Unlike the Cyanophycota, their thylakoids are arranged in pairs or stacks; thylakoid architecture resembles that of green algae and higher plants (Giddings *et al.*, 1980; Cox & Dwarte, 1981). Wall structure is similar to that of the Cyanochloronta, with the presence of muramic acid confirmed (Moriarty, 1979). Growth is by binary fission.

Lewin (1976, 1977) claimed that these primitive algae represent a link between the prokaryotes and the Chlorophycota. Chadefaud (1978) opposed the idea that the Prochlorophycota might represent a direct line of evolution to the green algae, on the grounds that chlorophylls *a* and *b* have arisen several times during the course of evolution. Gibbs (1981) has postulated that a *Prochloron*-like ancestor gave rise endosymbiotically to the chloroplasts of Chlorophyceae, while red algal chloroplasts may have been derived from a coccoid cyanobacterium (blue-green alga). Antia (1977) and others (Stanier & Cohen-Bazire, 1977; Ragan & Chapman, 1978; Trench, 1982) have argued against establishment of a separate division. The evolutionary significance of this intriguing group is discussed in greater detail in Chapter 8.

The recent success in culturing *Prochloron* in the laboratory (Patterson & Withers, 1982), and the discovery of a free-living prochlorophyte (Burger-Wiersma *et al.*, 1986), will doubtless lead to new avenues of research in which the biochemistry and physiology of these algae will be clarified.

EUKARYOTA

All the remaining algae, together with all other living organisms, are included in the Eukaryota. The cells possess a nuclear envelope, chromosomes, and membrane-bounded organelles such as mitochondria, plastids, and Golgi bodies.

RHODOPHYCOTYA: red algae (Figs 2.3–2.18)

Rhodophyceae

The Rhodophyceae are probably the closest of the eukaryotic algae to the Cyanophycota in their possession of simple plastids with unstacked thylakoids, and phycobiliproteins, and in their lack of chloroplast ER and flagella (Dixon, 1982; Norris & Kugrens, 1982). They are also reported to possess, in common with the Cyanophycota, a different unsaturated fatty acid biosynthetic pathway from that of green algae (Erwin & Bloch, 1963).

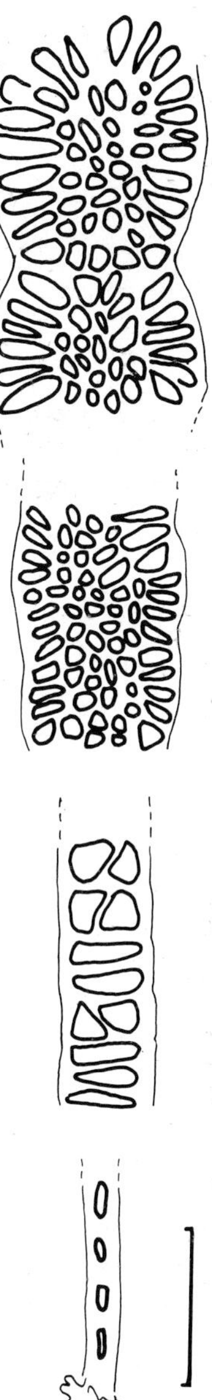

Fig. 2.3 *Bangia atropurpurea* (Rhodophyceae, Bangiales). Portions of the unbranched, parenchymatous thallus. Plants grow in dense stands in the upper intertidal zone of the cooler regions of the world. Scale = 50 μm.

Fig. 2.4 *Porphyra umbilicalis* (Rhodophyceae, Bangiales). The thallus is parenchymatous, sheet-like, with two cell layers. Scale = 2.0 cm.

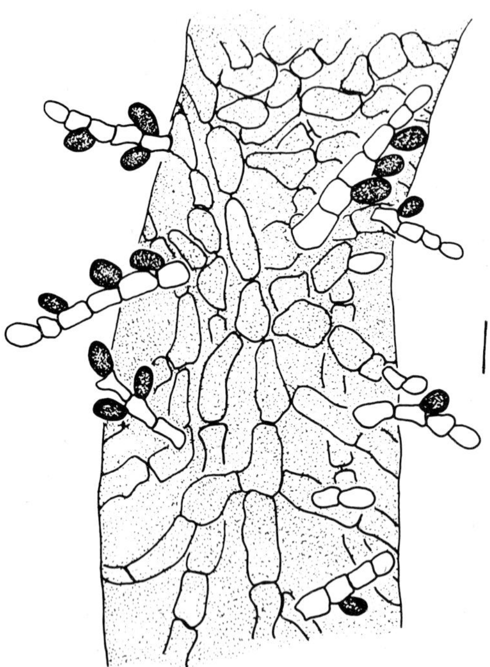

Fig. 2.5 *Audouinella infestans* (Rhodophyceae, Nemaliales). Plants grow as creeping filaments in the surface of their hydroid host. Short upright branches bear lateral monospores. Scale = 20 μm. From Woelkerling (1983) with permission.

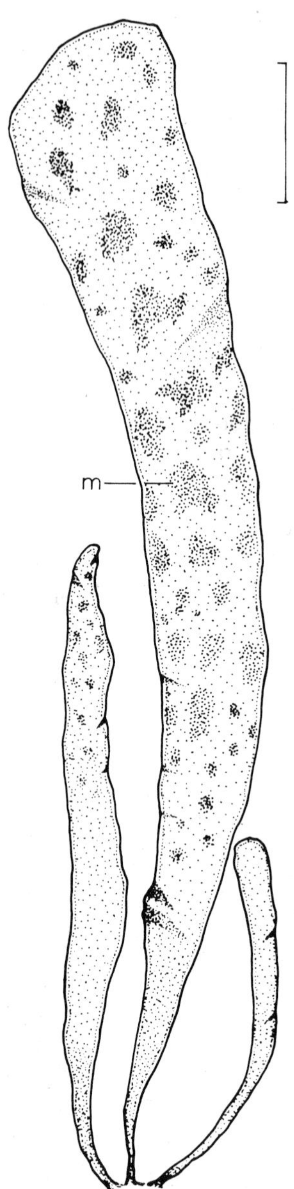

Fig. 2.6 *Erythrotrichia foliiformis* (Rhodophyceae, Bangiales). The small, leafy plants grow in tufts on the brown fucoid alga *Marginariella boryana*. Shown is a group of the simple, parenchymatous blades with patchily distributed monosporangial sori (m). Scale = 500 μm. From South & Adams (1976) with permission.

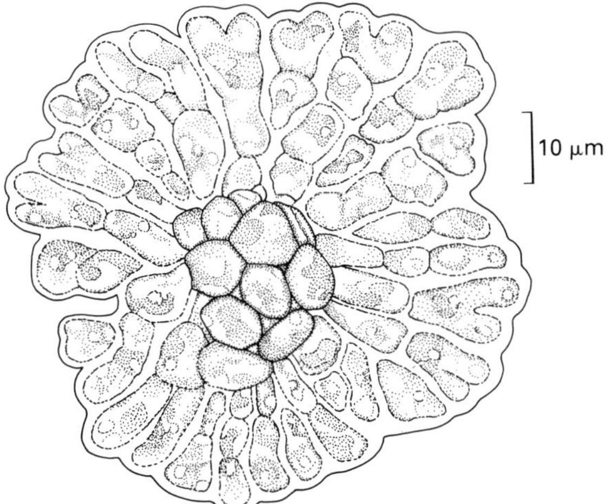

Fig. 2.7 *Erythrocladia* sp. (Rhodophyceae, Bangiales). The mature thallus is minute and discoid, with a monostromatic margin and polystromatic centre. From Nichols & Lissant (1967) with permission.

Fig. 2.8 *Beckerella scalaramosa* (Rhodophyceae, Nemaliales). Whole plant, showing bipinnate, distichous branching. Scale = 10 cm. From Kraft (1976) with permission.

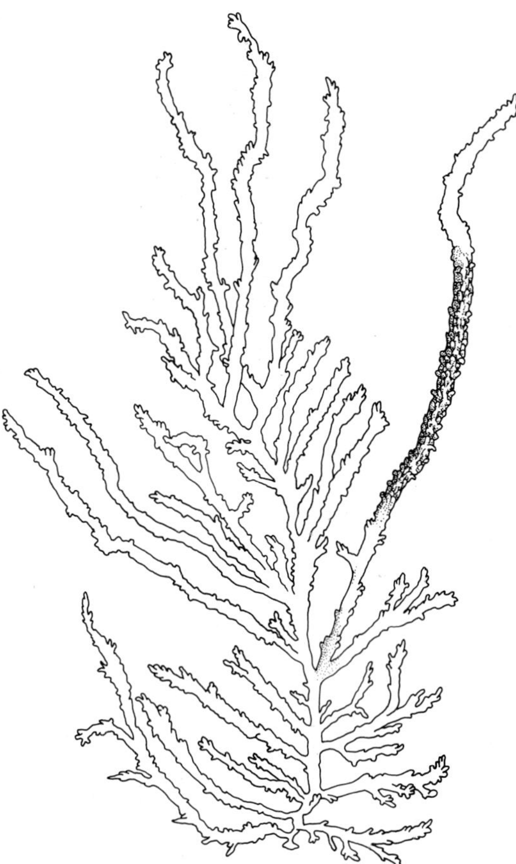

Fig. 2.9 *Liagora tanakai* (Rhodophyceae, Nemaliales). The thallus is multiaxial, with the axial filaments aggregated centrally; short determinate branches form the outer layer of the thallus. Not to scale. From Abbott (1967) with permission.

The Rhodophycota possess chlorophyll *a* (the presence of chlorophyll *d* has been reported only once; O'hEocha, 1971) in addition to phycobiliproteins (allophycocyanin; phycocyanin; phycoerythrin) and a variety of carotenoids. The colour is often red, but may be violet, brown, black or blue. The storage product is floridean starch, which lies freely in the cytoplasm, and the cell wall is composed of cellulose and, in certain species, of xylans and galactans, specifically the commercially important agar and carrageenan. The mucilages may comprise up to 70% of the dry weight of the cell wall. Calcification of the cell wall occurs in many Cryptonemiales and some Nemaliales and *Titanophora* of the Gigartinales (*see* Kraft, 1981). A specialized feature is the pit connection (or 'pit plug'; Pueschel, 1980), which is found in all Rhodophycota except for a few orders among the Bangiophycidae (Brawley & Wetherbee, 1981). While a great deal remains to be determined as to the function of this unique rhodophyte structure, it has been linked with intercellular transport (Wetherbee, 1979), structural strengthening in filamentous and pseudoparenchymatous thalli (Pueschel, 1980), and parasite–host interaction (Goff, 1979).

Morphology ranges from unicells (which are rare) to filamentous and pseudoparenchymatous forms; there is no truly parenchymatous growth, and the majority of species grow in a highly organized apical manner, with the exception of some Bangiophycidae. Some members of more advanced orders may exhibit intercalary growth (e.g. some Cryptonemiales and Ceramiales). By far the majority of the 4100 species and 675 genera (Kraft, 1981) are marine, with only approximately 200 species (42 genera; Ott & Sommerfeld, 1982) known from freshwater habitats.

Sexual reproduction is oogamous and involves many specialized features; *in-situ* post-fertilization development usually occurs, resulting in the production of a unique diploid carposporophyte generation attached to the female gametophyte. Sexes are normally separate, and the life history usually involves a sequence of haploid gametophyte and diploid carposporophyte and tetrasporophyte generations. The gametophyte and tetrasporophyte generations are usually isomorphic, but many heteromorphic examples are known.

The subdivision of the class Rhodophyceae into two subclasses is a matter of continuing debate, especially since some of the supposedly

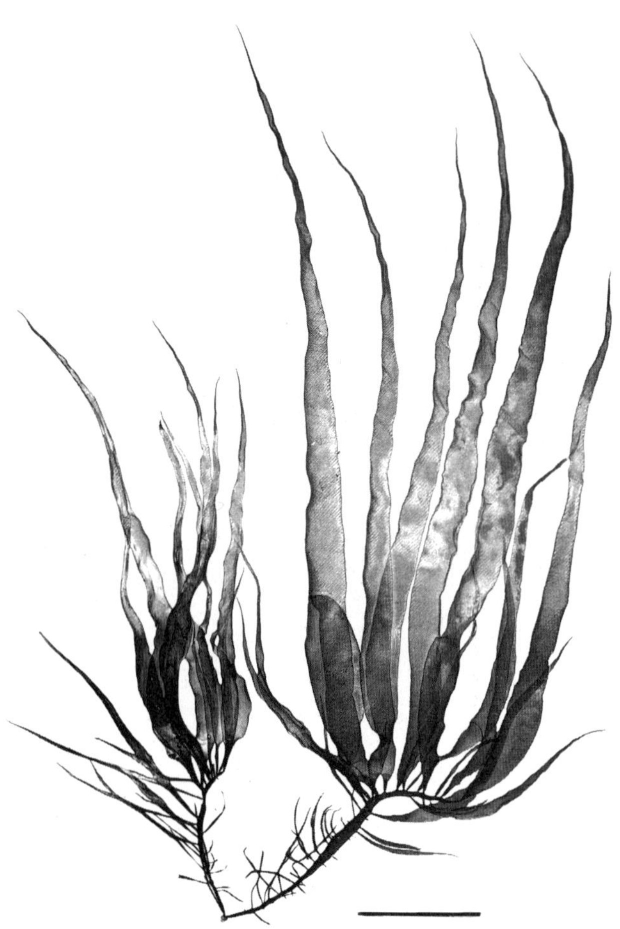

Fig. 2.10 *Grateloupia stipitata* (Rhodophyceae, Cryptonemiales). The stipitate, foliose blades are hollow and very gelatinous; they arise from a coarser, narrower, perennial stalk. The morphology is highly variable, depending on the habitat. Scale = 5 cm. (Photograph: R.C. Ficken.)

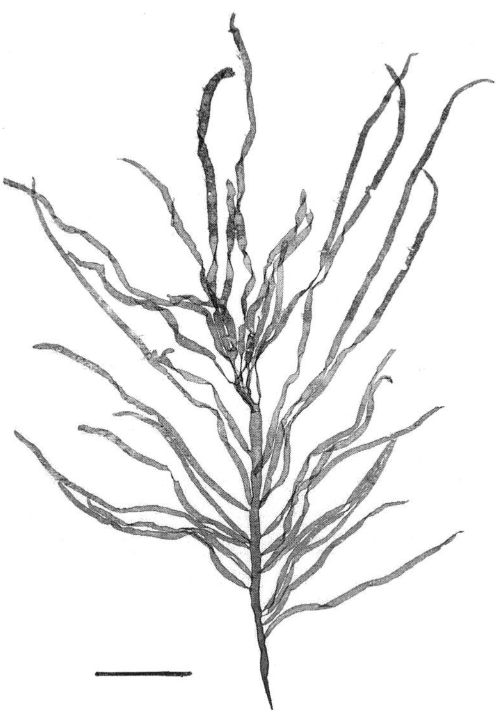

Fig. 2.11 *Dumontia incrassata* (Rhodophyceae, Cryptonemiales). The hollow, gelatinous branches are terete or somewhat flattened, or inflated. Scale = 2 cm. (Photograph: R.C. Ficken.)

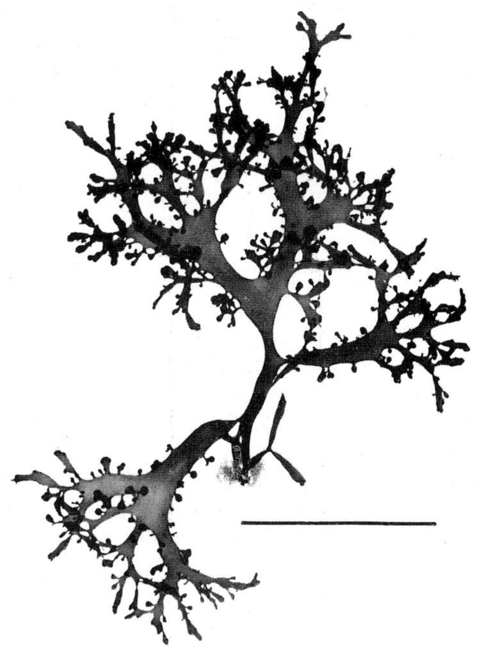

Fig. 2.12 *Mychodea marginifera* (Rhodophyceae, Gigartinales). Habit, showing the leafy thallus with spatulate proliferations arising from the blade margins. Scale = 5 cm. From Kraft (1978) with permission.

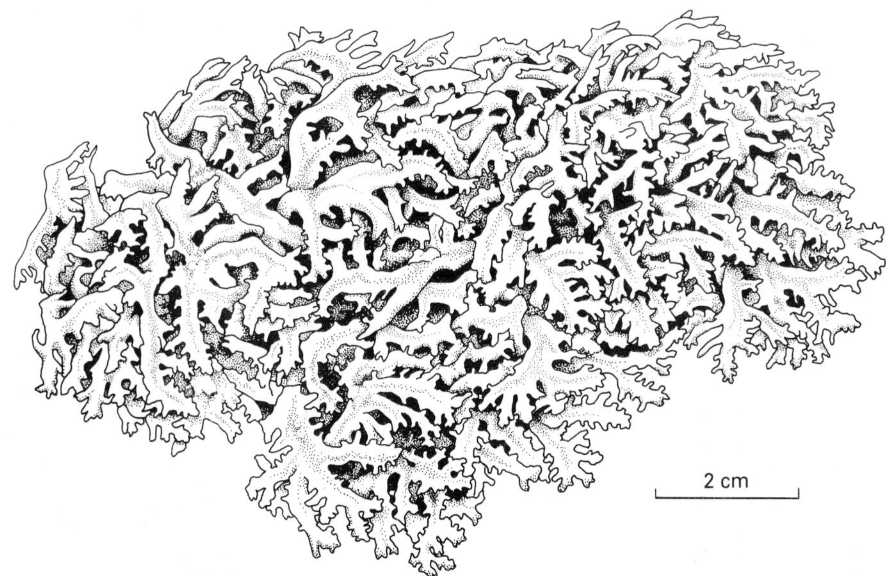

Fig. 2.13 *Eucheuma gelatinae* (Rhodophyceae, Gigartinales). Habit of whole plant. From Tseng (1981) with permission.

Fig. 2.14 *Gymnogongrus* sp. (Rhodophyceae, Gigartinales). Habit showing dense, somewhat terete, branches and dichotomous axes. Scale = 5 cm. (Photograph: R.C. Ficken.)

Fig. 2.15 *Palmaria palmata* (Rhodophyceae, Palmariales). Whole plant, showing foliose structure and pseudodichotomous branching. Scale = 5 cm. (Photograph: R.C. Ficken.)

distinguishing characters are no longer clearly segregative. Thus, the subclasses are retained here in the realization that in the future it may be preferable to discard them in favour of recognition of the unity of the division at the class level.

ORDERS

Bangiophycidae: Porphyridiales; Bangiales; Compsopogonales; Rhodochaetales.

Florideophycidae: Nemaliales; Cryptonemiales; Gigartinales; Rhodymeniales; Palmariales; Ceramiales.

CHROMOPHYCOTA

The Chromophycota include all algae possessing chlorophylls *a* and *c*, and lacking chlorophyll *b*. Accessory pigments, mainly β-carotene and various xanthophylls (the latter specific to each

Fig. 2.16 *Rhodomela confervoides* (Rhodophyceae, Ceramiales). Habit of whole plant. Plants are polysiphonous, with heavily corticated axes. Scale = 5 cm. (Photograph: R.C. Ficken.)

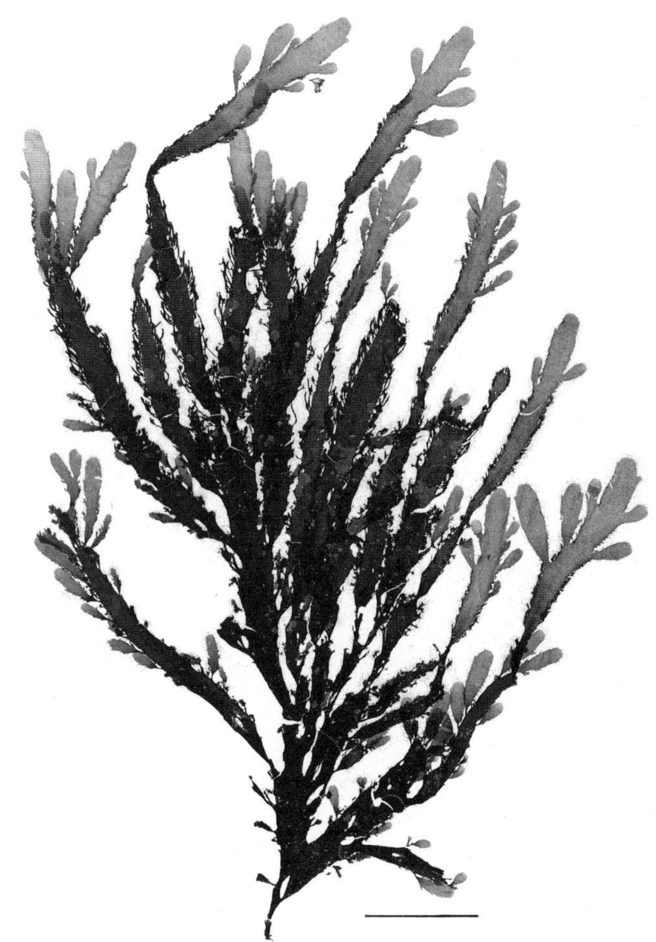

Fig. 2.17 *Cladhymenia oblongifolia* (Rhodophyceae, Ceramiales). The flattened axes bear numerous proliferous branches which also support the reproductive structures. Scale = 5 cm. (Photograph: R.C. Ficken.)

Fig. 2.18 *Phycodrys rubens* (Rhodophyceae, Ceramiales). The uniaxial thallus has a foliose structure with conspicuous mid-rib and veins resembling the leaf of a higher plant. Scale = 2 cm. (Photograph: R.C. Ficken).

class) give the chloroplasts a brownish or yellowish colour. Storage products are carbohydrates (glucans, starch) or lipids. Motile cells are widespread (except in the Bacillariophyceae) and normally possess two flagella, usually unequal in length (except in the Prymnesiophyceae, where they are equal) and heterodynamic, with one most commonly of the hairy type (two rows of hairs), the other smooth. All cells contain a periplastidial cisterna (= chloroplast ER), and the chloroplasts are rather uniformly constructed with the thylakoids in stacks of three. Cell wall composition is variable among the classes, and includes silica, pectin or cellulose. There is an extremely wide range of morphology, from unicellular flagellates to very large parenchymatous thalli, such as those of members of the order Laminariales (Phaeophyceae). The Chromophycota are widely distributed in both marine and freshwater habitats.

In this treatment the Chromophycota, including nine classes, embraces a highly heterogeneous group of organisms. In some treatments, a number of the classes are grouped into separate divisions, while others (e.g. the Xanthophyceae, Bacillariophyceae, Dinophyceae and Phaeophyceae) are given independent divisional status. In Bold and Wynne (1978) they are grouped in four divisions, the Chrysophycophyta, Pyrrhophycophyta, Phaeophycophyta and Cryptophycophyta, while in Hoek and Jahns (1978) the division Heterokontophyta includes the classes

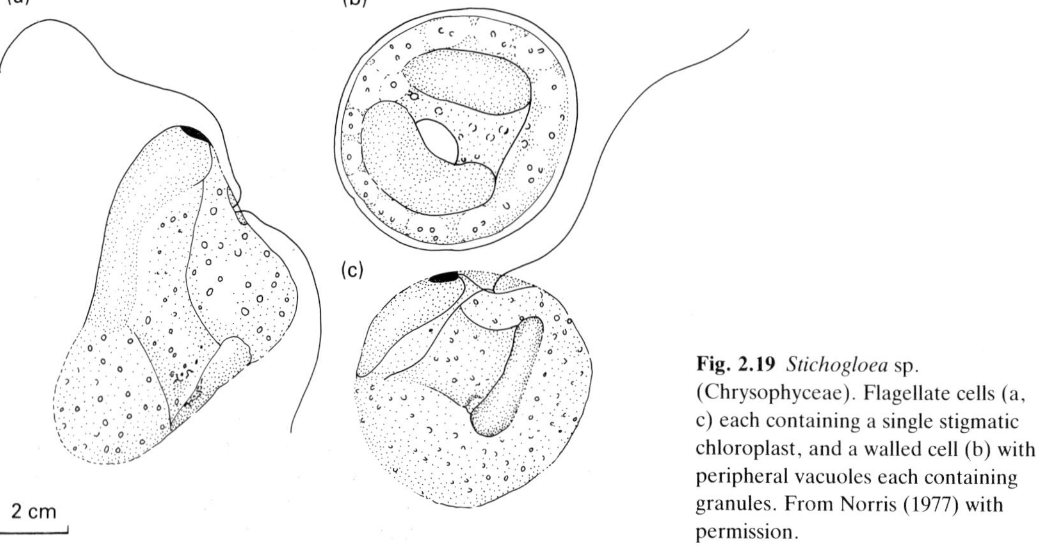

Fig. 2.19 *Stichogloea* sp. (Chrysophyceae). Flagellate cells (a, c) each containing a single stigmatic chloroplast, and a walled cell (b) with peripheral vacuoles each containing granules. From Norris (1977) with permission.

Chrysophyceae, Xanthophyceae, Bacillariophyceae, Phaeophyceae, and Chloromonadophyceae; the Prymnesiophyceae (as the Haptophyta), Eustigmatophyceae, Cryptophyceae and Dinophyceae are assigned divisional status. The treatment we are employing (following Kristiansen, 1982) can be contrasted with that of Hoek and Jahns (1978): in the former, pigmentation is considered a divisional character, whereas in the latter it is flagellar structure. In view of the heterogeneity of the Chromophycota, it is likely that there will continue to be divergent views on how the component classes should be grouped. Following are diagnostic summaries of each of the classes included.

Chrysophyceae: golden-brown algae
(Figs 2.19–2.24)

The delimitation of the Chrysophyceae is given in detail by Hibberd (1976). The characteristic group is the subclass Chrysophycidae; the subclass Dictyochophycidae is included only on the basis of common pigmentation and storage products (Kristiansen, 1982). The taxonomy of this group is difficult at the ordinal, generic, and species levels (Kristiansen & Takahashi, 1982). Cells are naked, or provided with a cell envelope or lorica; they normally possess two unequal flagella, the smooth flagellum sometimes reduced in length or internal as a basal body only. The hairy flagellum bears tripartite hairs. An eyespot is commonly found, associated with the chloroplast. Cells are uninucleate and contain one to a few chloroplasts. The predominant accessory pigment is fucoxanthin, which gives the cells their golden-brown colour. The storage product is chrysolaminarin. Specialized structures include statospores (or stomatocysts; Kristiansen, 1982) enclosed by a silicified wall and with a terminal pore. Reproduction is mainly asexual, although sexual stages are known for an increasing number of species (Kristiansen & Takahashi, 1982). Morphology ranges from unicellular flagellates (which predominate) to coccoid, rhizopodial, filamentous, and parenchymatous forms (which are rare). The majority of the Chrysophyceae are members of the freshwater phytoplankton, especially in cool environments.

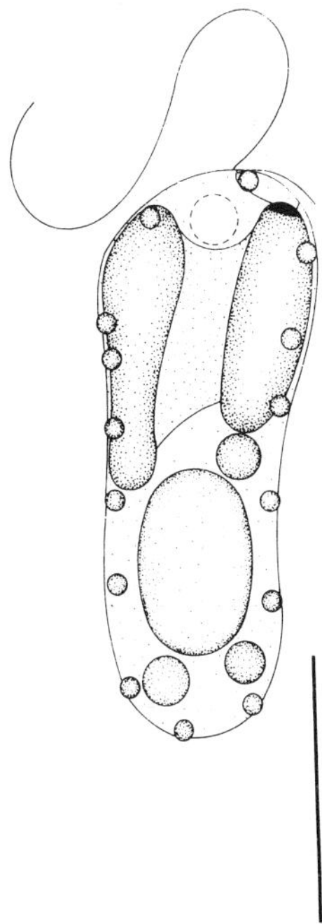

Fig. 2.20 *Ochromonas sphaerocystis* (Chrysophyceae). An elongate cell showing heterodynamic flagella, chloroplast, nucleus and peripheral discobolocysts. Scale = 10 cm. From Andersen (1982) with permission.

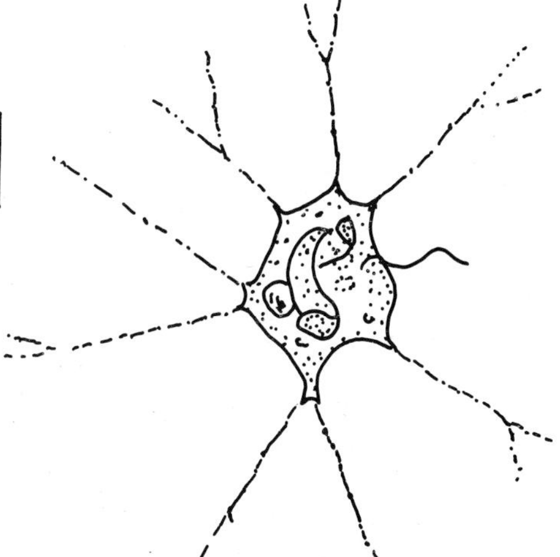

Fig. 2.21 *Chrysamoeba* sp. (Chrysophyceae). Cell showing the characteristic shape, and with beaded rhizopodia and contractile vacuoles. Scale = 10 μm. From Bourrelly (1981) with permission.

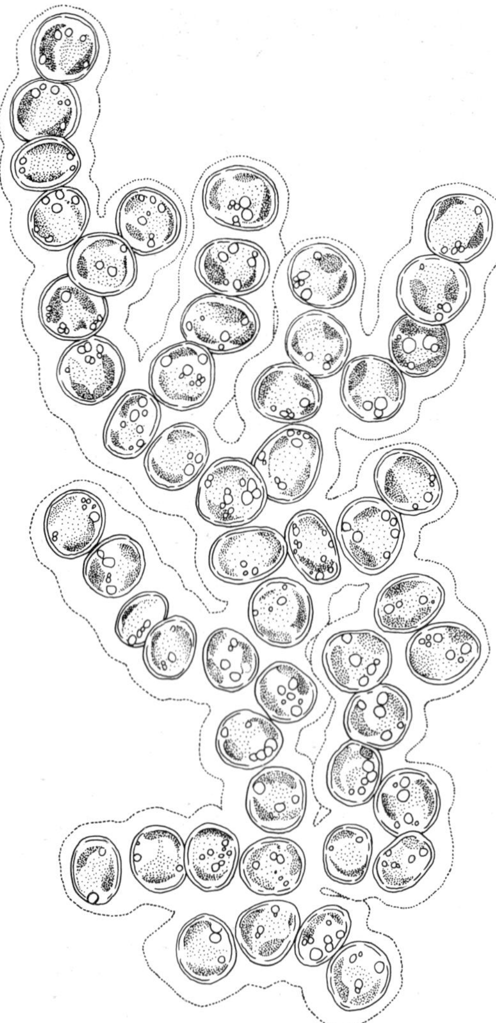

Fig. 2.22 *Sphaeridiothrix compressa* (Chrysophyceae). View of the branching, reticulate thallus (from culture). Cells contain a parietal chloroplast and several droplets of food reserve, but lack pyrenoids (× 1000). From Andrews (1970) with permission.

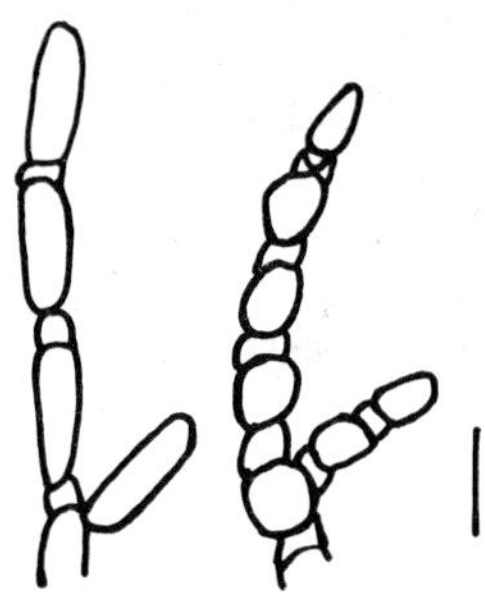

Fig. 2.23 *Phaeothamnion* sp. (Chrysophyceae). Branches showing thickened cross-walls. Scale = 10 μm. (Modified from Geitler & Schiman-Czeika, 1970.)

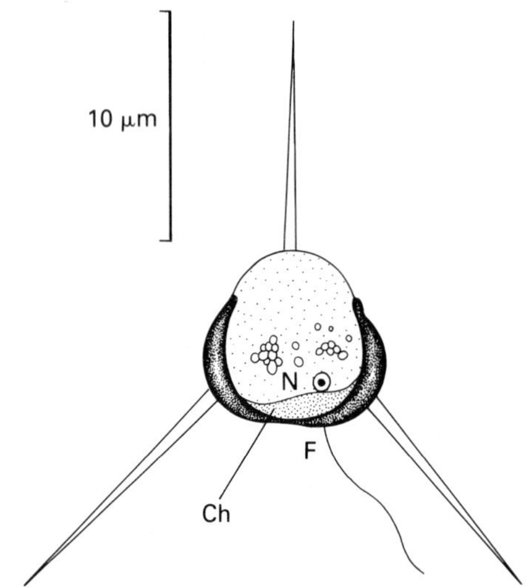

Fig. 2.24 *Chrysococcus furcatus* (Chrysophyceae). Vegetative cell showing spines, chloroplast (Ch), flagellum (F), nucleus (N) and laminarin-like storage droplets. From Nicholls (1981) with permission.

ORDERS

Chrysophycidae: Ochromonadales; Chrysamoebidales; Chrysocapsales; Chrysosphaerales; Phaeothamniales; Sarcinochrysidales.
Dictyochophycidae; Dictyochales.

Prymnesiophyceae (Haptophyceae)
(Figs 2.25–2.29)

The Prymnesiophyceae Hibberd (1976, 1980a= Haptophyceae; Christensen, 1962) are a natural group (Boney & Green, 1982). They are mostly unicellular flagellates with two equal smooth flagella (except in the order Pavlovales, where the cells possess unequal flagella, one of which is scaly; Green, 1980). A third flagellum-like appendage, the haptonema, is present in many species. Cells are uninucleate with one or two chloroplasts, each with a single pyrenoid; pigments and storage products are the same as for the Chrysophyceae. Eyespots occur in the Pavlo-

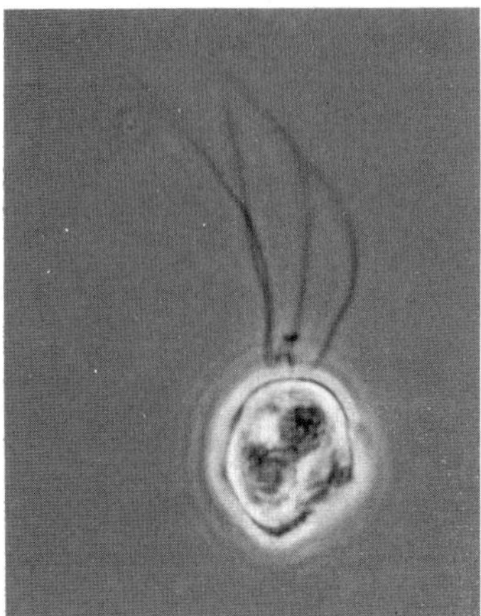

Fig. 2.25 *Chrysochromulina birgeri* (Prymnesiophyceae). Phase-contrast photomicrograph of the whole cell, fixed in Lugol's iodine (× 1000). From Hällfors & Thomsen (1979) with permission.

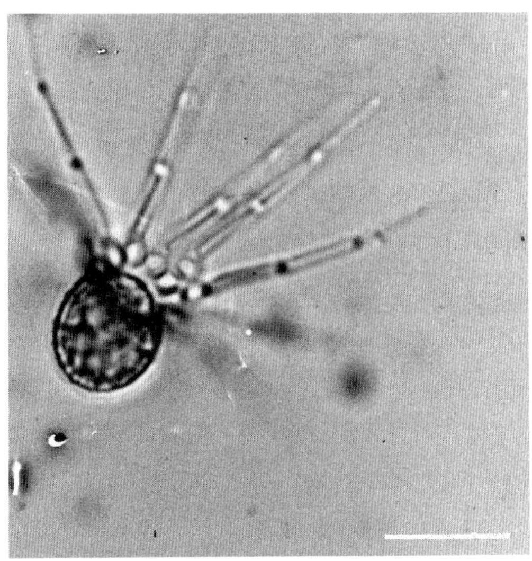

Fig. 2.26 *Michaelsarsia elegans* (Prymnesiophyceae). Light photomicrograph of whole cell. Note the whorl of distal arms. Scale = 10 μm From Heimdal & Gaarder (1981) with permission.

vales, and are contained within the chloroplast envelope. Cells are usually covered with organic (cellulosic) scales in one to several layers, the scales formed within cisternae of the Golgi apparatus. A contractile vacuole may be associated with the Golgi apparatus. Sexual reproduction occurs and the life history commonly involves alternating motile and non-motile phases. In the

Fig. 2.27 *Emiliania huxleyi* (Prymnesiophyceae). SEM of entire coccosphere, showing details of coccoliths. Scale = 1 μm. From Heimdal & Gaarder (1981) with permission.

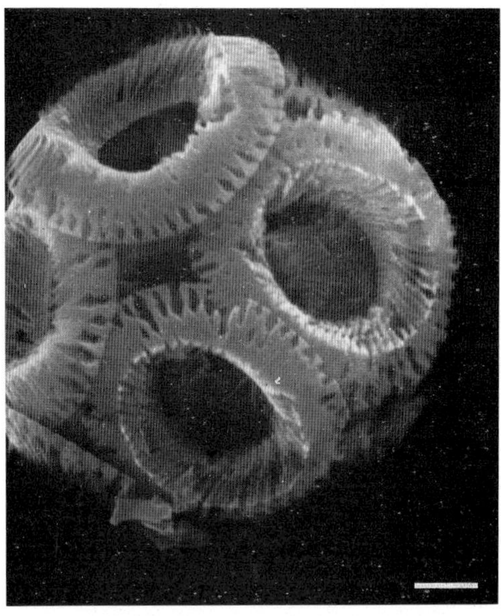

Fig. 2.28 An unidentified coccolithophorid (Prymnesiophyceae). SEM of whole coccosphere. Scale = 1 μm. From Heimdal & Gaarder (1981) with permission.

Coccosphaerales a haploid, benthic, filamentous phase may alternate with a diploid, unicellular, motile phase.

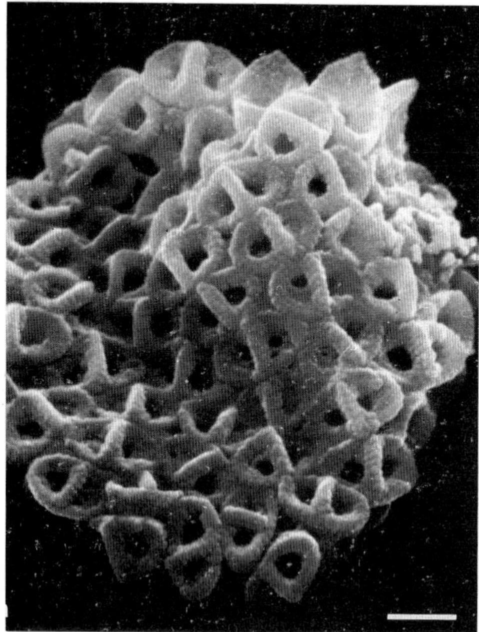

Fig. 2.29 *Corisphaera gracilis* (Prymnesiophyceae). SEM of complete coccosphere. Scale = 1 μm. From Heimidal & Gaarder (1980) with permission.

Fig. 2.30 *Vaucheria* sp. (Xanthophyceae). View of a dense growth on silt in the intertidal. The filaments frequently serve as sand and mud binders in sheltered, estuarine habitats. Approximate scale = 5 cm.

Members of the Prymnesiophyceae are generally small; there are about 75 living genera, and as many as 300 species (Norris, 1982a), most of which are marine and belong to the nanoplankton. It has been estimated that prymnesiophytes may constitute about 45% of the total phytoplankton in the middle latitudes of the South Atlantic Ocean. Calcified scales (coccoliths) of the prymnesiophytes fossilize readily and may be prominent in fossil oozes. The evidence suggests that prymnesiophytes may have been more abundant than at present, especially in the Mesozoic (the Jurassic and Cretaceous periods, 205–125 million years BP).

ORDERS

Isochrysidales; Coccosphaerales; Prymnesiales; Pavlovales.

Xanthophyceae: yellow-green algae (Figs 2.30–2.32)

The Xanthophyceae are distinguished from other 'green' algae by the absence of chlorophyll *b*, and from other Chromophycota by the lack of fucoxanthin and the predominance of diatoxanthin as the principal carotenoid (Silva, 1979; Hibberd, 1980a, 1981, 1982a; Ott, 1982). Motile cells are naked and bear two unequal flagella, the larger, hairy flagellum (which bears tubular hairs) being directed forwards, the shorter, smooth flagellum being directed backwards. The compound zoospore of *Vaucheria* bears numerous pairs of only slightly unequal flagella, each of which lacks hairs. Cell walls of coccoid forms are delicately sculptured; in filamentous and larger coccoid species, the wall is bipartite, with the two halves tapered and overlapping. Cellulose is the major constituent of the cell wall in siphonous species (e.g. *Vaucheria*), the remaining portion being composed of glucose and uronic acids.

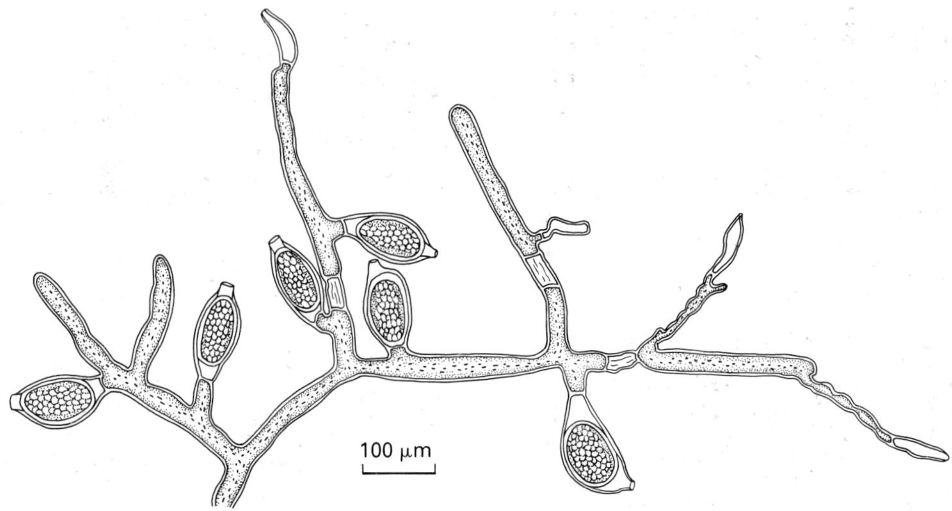

Fig. 2.31 *Vaucheria adela* (Xanthophyceae). The thallus is coenocytic; reproductive structures include the terminal antheridia borne on the male branch or androphore, and the oogonia containing the mature oospores. From Ott & Hommersand (1974) with permission.

Chloroplasts are usually discoid and surrounded by a double membrane of chloroplast ER, the outer continuous with the outer membrane of the nuclear envelope. Pyrenoids usually occur singly within the chloroplast. The eyespot, when present, lies ventrally within the chloroplast and is associated with a depression in the cell and a basal swelling of the short, posterior flagellum. Storage products are a β-1, 3-linked glucan (cf. paramylon of the Euglenophycota), fats, and oils: never starch.

Asexual reproduction is by cell division, fragmentation (in filamentous forms), or production of motile zoospores or non-motile aplanospores, akinetes, or cysts (statocysts — cf. Chrysophyceae). Sexual reproduction is known in the Vaucheriales (*Botryidium* and *Vaucheria*), where it is oogamous, and in the Tribonematales (*Tribonema*), where it is isogamous.

Classification to order is based on morphology; a morphological evolutionary series comparable to that in the Chlorophycota occurs. Members range from unicellular and flagellate to coccoid (the majority) to filamentous (*c.* 20%) and siphonous (*c.* 10%, mainly species of *Vaucheria*). Approximately 600 species have been described (Hibberd, 1982b), the majority of which occur in fresh water. Only 20 species are marine or brackish (all *Vaucheria*).

ORDERS

Chloramoebales; Rhizochloridales; Heterogloeales; Mischococcales; Tribonematales; Vaucheriales.

Eustigmatophyceae

The Eustigmatophyceae are a small class of only 12 species, distinguished from the Xanthophyceae through the unique structure of the motile cells and by the predominance of violaxanthin rather than diatoxanthin (Hibberd, 1982b, c). Zoospores are unique in the plant kingdom, with a single, anteriorly directed hairy flagellum and a second basal body (*Ellipsoidion acuminatum* and *Pseudocharaciopsis texensis,* however, possess an emergent, smooth posterior flagellum; Lee & Bold, 1973). A large orange-red eyespot occurs anteriorly, and is completely independent of the chloroplast; it is associated with a T-shaped swelling at the base of the hairy flagellum, which is closely adpressed to the plasmalemma in the region of the eyespot. Asexual reproduction is by the formation of autospores or zoospores (rare). Sexual reproduction is unknown. All are naked unicells, sometimes attached, all occurring in freshwater or soil habitats.

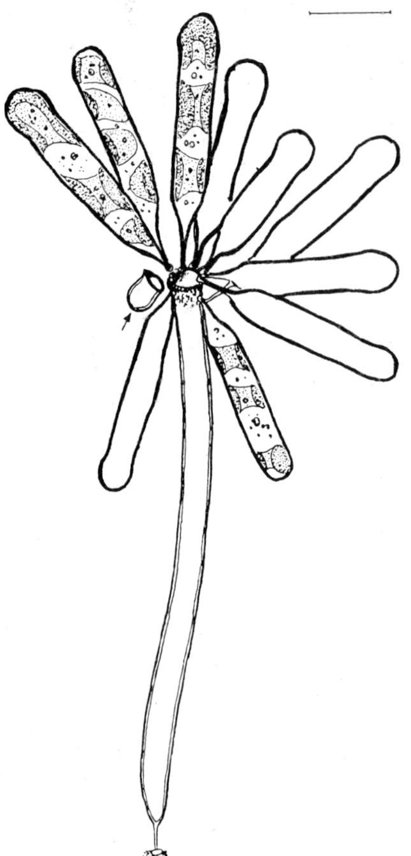

Fig. 2.32 *Ophiocytium arbuscula* (Xanthophyceae). Thallus showing 'mother' cell bearing additional cells from the apex. Cells are multinucleate and contain several parietal chloroplasts. The arrow indicates the apical cap of the 'mother' cell, which has released spores. Scale = 10 μm. From Bourrelly (1981) with permission.

(a)

(b)

Fig. 2.33 *Thalassiosira oestrupii* (Bacillariophyceae). (a) Large cell with valve having eccentric areola and two small openings near the centre of the valve (× 1920). (b) SEM of girdle region with the mature epitheca (bottom) consisting of valve, ornate valvocopula, copula with one large ring of pores, and three pleurae. The immature hypotheca (top) has valve, volvocopula, and copula visible (× 2330). From Fryxell & Hasle (1980) with permission.

Bacillariophyceae: diatoms (Figs 2.33–2.36)

The Bacillariophyceae are unicellular, sometimes colonial or pseudofilamentous, microscopic algae (Werner, 1977; Ross, 1982; Amspoker & Czarnecki, 1982). The principal accessory pigments are fucoxanthin, β-carotene, diadinoxanthin and diatoxanthin, and the principal food reserves are lipids and chrysolaminarin; in these respects the Bacillariophyceae most closely resemble the Chrysophyceae. Cells are uninucleate, and contain one or more chloroplasts. They possess a silica exoskeleton (frustule) consisting of two halves (valves), each composed of a more or less flattened plate with a connecting band attached to the edge. The two connecting bands are called the girdle. Certain pennate forms possess a slit-like raphe, which is associated with unique gliding motility. Asexual reproduction, which predominates, is normally by binary fission, the new valves and girdle formed within the parent cell; this process results in a steady decline in cell size in a percentage of the population over several generations. Sexual reproduction is iso- or anisogamous, with non-flagellate gametes, or oogamous (mostly Centrales), with a uniflagellate male gamete which is unique in that the single flagellum possesses no central microtubules. Cell size is restored following sexual reproduction, when the zygote swells to form an auxospore which regenerates the maximum size of the frustule for the species. Resting spores, with heavily silicified walls and a morphology different from that of the normal vegetative cell, are formed by most benthic and neritic diatoms as a means of persisting during conditions unfavourable for growth.

Diatoms are abundant in marine and freshwater habitats, in the benthos, and as epiphytes on large algae or higher plants. They fossilize readily, and extensive deposits (referred to as diatomaceous earth) dating back to the Cretaceous period are known. Diatomaceous earth has various commercial uses; in addition, stratigraphic analyses of deposits may indicate past environmental events in lakes (Round, 1957).

The two orders of diatoms are based on the shape of the frustules, the Biddulphiales exhibiting radial symmetry, the Bacillariales bilateral symmetry. Electron microscopy is greatly refining the systematics of this class.

ORDERS

Biddulphiales (= Centrales); Bacillariales (= Pennales).

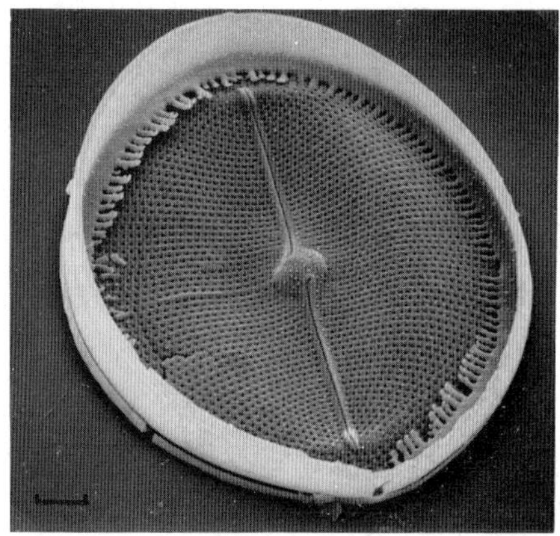

Fig. 2.34 *Cocconeis pseudomarginata* (Bacillariophyceae). SEM showing interior of valve with a fimbriate valvocopula, and a raphe. Scale = 1 μm. From Holmes *et al.* (1982) with permission.

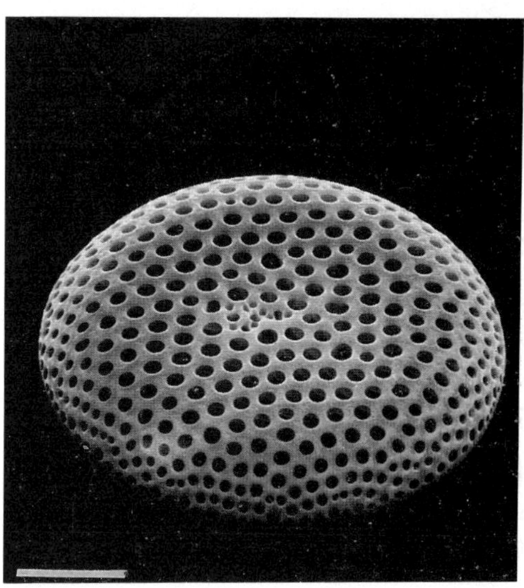

Fig. 2.35 *Thalassiosira confusa* (Bacillariophyceae). SEM showing external view of the theca, from an oblique angle. Note the open pores (foramina). Scale = 5 μm. From Fryxell & Hasle (1979) with permission.

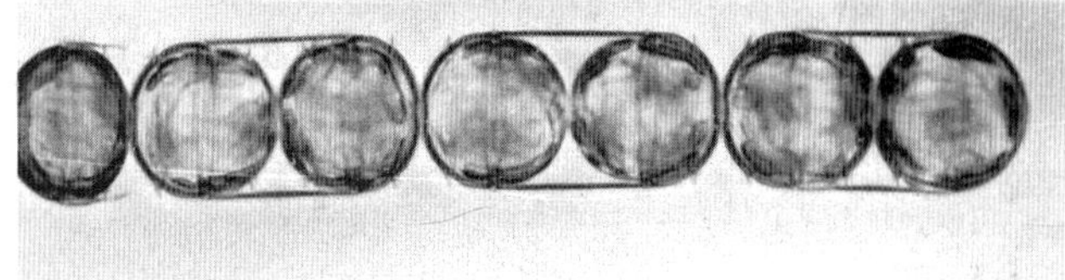

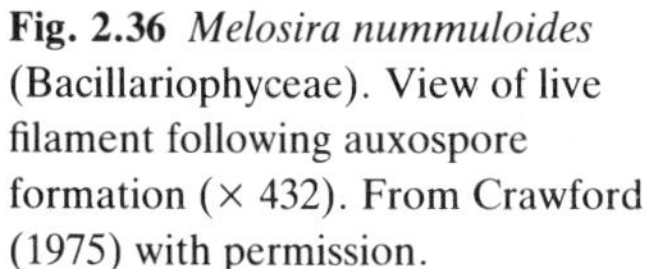

Fig. 2.36 *Melosira nummuloides* (Bacillariophyceae). View of live filament following auxospore formation (× 432). From Crawford (1975) with permission.

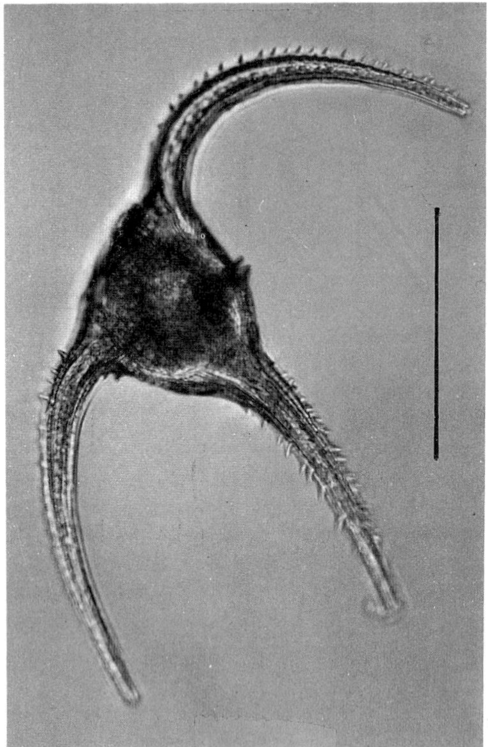

Fig. 2.37 *Ceratium* sp. (Dinophyceae). A common member of the marine phytoplankton. The heavily armoured theca bears three 'horns' which are in turn ornamented with spines. Note the conspicuous girdle. Scale = 50 μm.

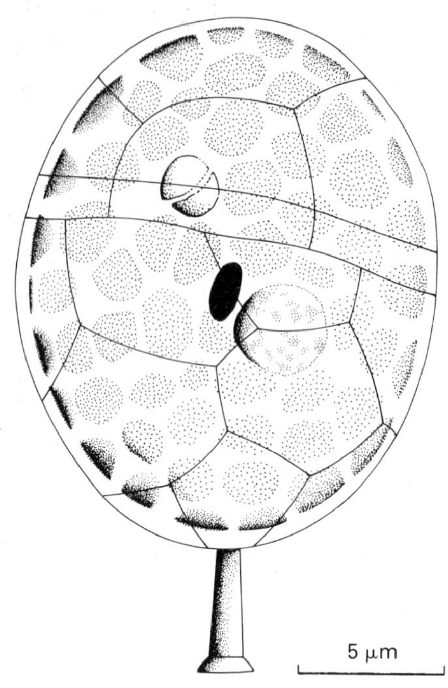

Fig. 2.38 *Stylodinium littorale* (Dinophyceae). An attached, non-motile cell. From Horiguchi & Chihara (1983) with permission.

Dinophyceae: dinoflagellates

The Dinophyceae, as exemplified by the largest subclass, the Dinophycidae, are principally planktonic, photosynthetic, biflagellate unicells of a highly distinctive structure (Dodge, 1979; Dodge & Steidinger, 1982; Loeblich, 1982). The motile cells bear the flagella ventrally; one is transverse and encircles the cell, usually in a groove-like girdle or cingulum; it is distinctive in possessing an outer axoneme and an inner, stranded band. The other, which is smooth, is directed posteriorly. Cells may be autotrophic, or colourless and heterotrophic; the latter gain their nutrition sapro- or phagotrophically, or through a parasitic mode of life. In the photosynthetic species the chloroplasts are surrounded by a single chloroplast ER membrane which is not continuous with the outer membrane of the nuclear envelope. The principal carotenoids are peridinin and neoperidinin. The storage products are oil and a starch similar to that of higher plants. The cell covering is a multilayered theca (amphiesma), and consists of an epicone and hypocone, divided into a number of thecal plates; a sulcus runs perpendicular to the girdle. Cells may contain pyrenoids and eyespots (less than 5% of the species). Projectiles known as trichocysts and cnidocysts are formed in a number of species and may have a protective or evasive function. The nucleus is unique and is described as dinokaryon or mesokaryon (with permanently condensed chromosomes). The chromosomes are membrane-attached and the nuclear envelope remains intact during mitosis. Asexual reproduction is by fission; vegetative cells are normally haploid, with the zygote the only diploid stage. Resting spores, cysts, or hypnospores are common; when fossilized they are known as hystri-

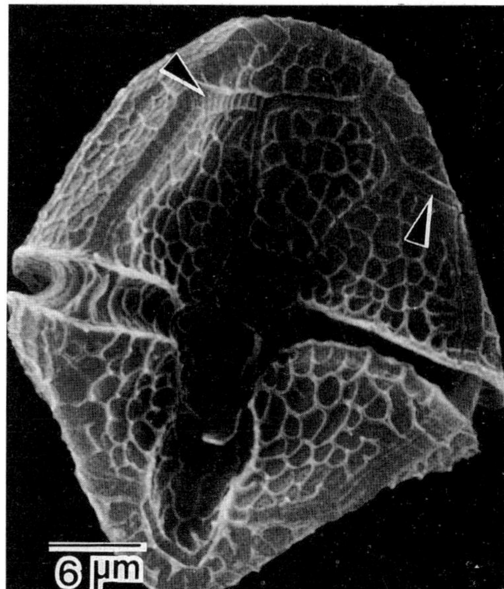

Fig. 2.39 *Peridinium cinctum* (Dinophyceae). SEM of vegetative cell. From Pfiester & Skvarla (1980) with permission.

chosphaerids, and are grouped with other cyst-like fossils as acritarchs. Dinoflagellates were probably the dominant primary producers in the Palaeozoic period (Tasch, 1973) and the oldest records are from the Silurian (435–460 million years BP). Toxic blooms of dinoflagellates are well-known phenomena and have been the subject of extensive study (Taylor & Seliger, 1979; Yentsch & Taylor, 1982).

Of the four subclasses, the Dinophycidae is the largest; the Ebriophycidae comprise only two living representatives and are largely known as fossils, the Ellobiophycidae are non-photosynthetic parasites of Arthropoda (Mysidacea) and Annelida (Polychaeta) and the Syndiniophycidae are parasites of various marine animals (Loeblich, 1982).

ORDERS

Dinophycidae: Blastodiniales; Brachydiniales; Chytriodiniales; Desmocapsales; Dinamoebales; Dinocloniales; Dinophysiales; Gloeodiniales; Gymnodiniales; Noctilucales; Peridiniales; Phytodiniales; Prorocentrales; Pyrocystales; Zooxanthellales.
Ebriophycidae: Ebriales.
Ellobiophycidae: Thalassomycetales.
Syndiniophycidae: Syndiniales.

Phaeophyceae: brown algae
(Figs 2.41–2.61)

This class of *c.* 265 genera and more than 1500 species is composed almost exclusively of benthic marine forms (Wynne, 1981, 1982a, b). The principal carotenoid is fucoxanthin which, combined with various tannins (stored in physodes or fucosan vesicles), gives the plants their characteristic brown colouration. There are typically two

Fig. 2.40 *Pyrodinium bahamense* (Dinophyceae). SEM of whole cell viewed from above the epitheca, showing plates and an apical pore. Scale = 10 μm. From Dodge & Hermes (1981) with permission.

Fig. 2.41 *Ectocarpus* sp. (Phaeophyceae, Ectocarpales). (a) Detail of uniseriate thallus showing vegetative structure and plurilocular sporangia; scale = 100 μm. (b) Habit of plant bearing unilocular sporangia; scale = 1000 μm. Based on Cardinal (1964).

membranes of chloroplast ER, the outer of which may be continuous with the outer membrane of the nuclear envelope. Cells are normally uninucleate and contain one to many chloroplasts; the storage products are laminarin and mannitol. The cell wall is two-layered, with an inner layer of cellulose and an outer layer of alginic acid and fucoidin. The alginic acid is of considerable commercial value, and is present in quantity in members of the Laminariales and Durvillaeales. Calcium carbonate is deposited in the walls of species of *Padina* (Dictyotales) in the form of aragonite crystals. Unicellular species are unknown. Zoospores and gametes normally bear two unequal flagella which are laterally inserted, the longer anterior flagellum being hairy (bearing tripartite hairs), the shorter, posterior one being smooth. (The sperm of Dictyotales, Laminariales, Desmarestiales and some members of the Fucales vary from this pattern.) Eyespots commonly occur and are similar in position and structure to those of the Chrysophyceae and Xanthophyceae.

Morphology ranges from microscopic, branched filamentous forms to macroscopic parenchymatous plants which exhibit the highest degree of anatomical differentiation in the algae (e.g. the giant kelps of the order Laminariales). Asexual reproduction is principally by means of motile zoospores, although a variety of vegetative structures are involved in some species. Sexual reproduction ranges from isogamous to anisogamous and oogamous. Meiosis occurs in unilocular sporangia, which are unicellular and form from four to many usually motile spores. Multicellular (plurilocular) structures produce asexual spores or gametes by mitotic division. Life histories typically involve an alternation of phases, which may be isomorphic or heteromorphic. They have been used as a characteristic in the classification of brown algae (Wynne & Loiseaux, 1976; Pedersen, 1981). The life histories of many species remain unknown; in some examples a 'direct' type occurs, in which there is no meiosis or syngamy, yet in which there may be a morphological alternation.

The classification of the Phaeophyceae is in a dynamic state; at the ordinal level, a combination of life history and morphological and organizational features are used. There is no clear-cut separation among some of the orders, particularly the Ectocarpales, Dictyosiphonales, Chordariales and Scytosiphonales. For a review, *see* Wynne (1981). Pedersen (1984) has proposed that the order Dictyosiphonales be merged in the Tilopteridales (*sensu lato*), since the biphasic, heteromorphic life history considered diagnostic of the Dictyosiphonales is not consistent. Henry (1984) has proposed the new monogeneric order Syringodermatales, in which the heteromorphic life history features a very reduced, two-celled gametophyte stage, and in which the vegetative structure is distinct from both the Dictyotales and the Sphacelariales.

ORDERS

Ectocarpales; Chordariales; Sporochnales; Desmarestiales; Dictyosiphonales; Scytosiphonales; Tilopteridales; Cutleriales; Sphacelariales; Dictyotales; Syringodermatales; Laminariales; Fucales; Durvillaeales; Ascoseirales.

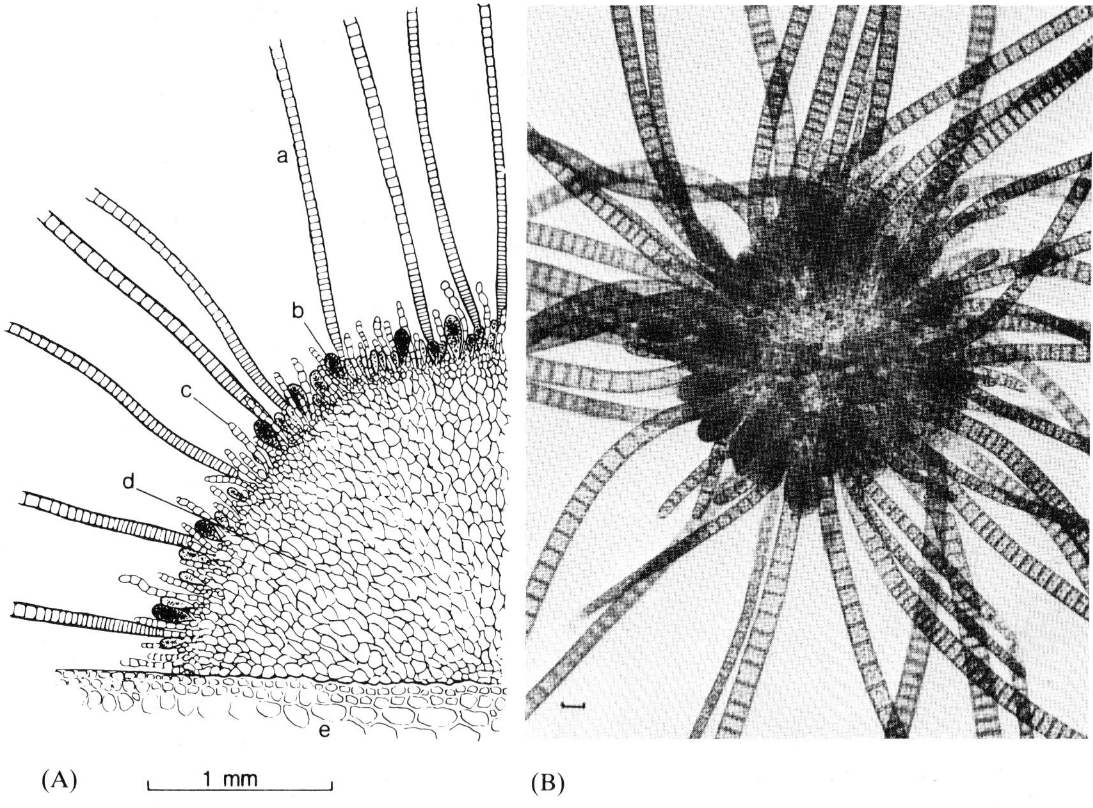

Fig. 2.42 *Elachista fucicola* (Phaeophyceae, Chordariales). (A) Sectional view of part of a mature plant showing assimilators (a), unilocular sporangia (b), paraphyses (c), inflated cells of the medulla (d) and the tissues of the *Fucus* host (e). (B) Squashed view of the thallus (grown in culture) showing basal region with short branches and conspicuous unilocular sporangia, and assimilators. Scale = 100 μm. From Koeman & Cortel-Breeman (1976) with permission.

Raphidophyceae

A very small class of freshwater flagellates sometimes included in the Xanthophyceae, with which they share similar pigments and flagellation. The name 'chloromonads' (or Chloromonadophyceae) is unacceptable, as it is based on the genus *Chloromonas* which is not a raphidophyte. There is considerable controversy as to which algae belong to this class (Heywood, 1982), although most authorities accept the genera *Chattonella, Gonyostomum, Merotricha* and *Vacuolaria*. On the basis of pigment analysis, Fiksdahl *et al.* (1984) showed that *Gonyostomum semen* and *Vacuolaria virescens* possess a pigment composition like the Xanthophyceae. The principal carotenoids are diadinoxanthin (50–60% of total carotenoids), dinoxanthin (8–17%), β-carotene (7%) and heteroxanthin (7%). By contrast, *Chattonella japonica* and *Fibrocapsa japonica* possess fucoxanthin (40–68% total carotenoids), fucoxanthiniol (30–42%), β-carotene (15–27%) and violaxanthin (1–2%) as the principal carotenoids, a pattern resembling the Chrysophyceae. Fiksdahl *et al.* (1984) conclude that there is a relationship, on the basis of pigments, between the Raphidophyceae, Xanthophyceae, and possibly the Chrysophyceae. The longer, anteriorly directed flagellum bears microtubular hairs, the shorter, posteriorly directed flagellum is smooth. Raphidophytes lack a cell wall, and oil is reportedly the principal storage product. An eyespot is also lacking. Specialized features include muciferous

Fig. 2.43 *Scytosiphon lomentaria* (Phaeophyceae, Scytosiphonales). The tubular, unbranched thalli are pseudoparenchymatously constructed. The hollow centre is constricted at intervals. Scale = 5 cm. (Photograph: R.C. Ficken.)

Fig. 2.44 *Sporochnus* sp. (Phaeophyceae, Sporochnales). Thalli are richly branched, with a conspicuous axis. Reproductive structures are borne in clusters on the branch tips. Scale = 5 cm. (Photograph: R.C. Ficken.)

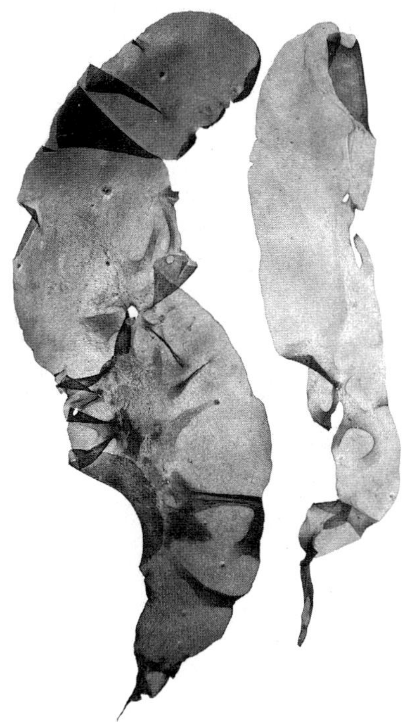

Fig. 2.45 *Punctaria plantaginea* (Phaeophyceae, Scytosiphonales). The parenchymatous thalli are foliose, and two-layered. Scale = 2 cm. (Photograph: R.C. Ficken.)

bodies (mucocysts) similar to those of the Prymnesiophyceae and Chrysophyceae; in *Gonyostomum*, trichocysts homologous with those in the Dinophyceae occur (Mignot, 1967). Sexual reproduction is unknown.

Cryptophyceae: cryptomonads

A class largely of flagellates of marine and freshwater habitats (Oakley & Santore, 1982). In photosynthetic forms the major carotenes are α-carotent and diatoxanthin. Also present are phycobiliproteins (phycoerythrin and phycocyanin) which differ from those of the Cyanophycota and Rhodophycota (Stainer, 1974), and which are located in the intrathylakoidal spaces and not in phycobilisomes, as in the Cyanophycota and Rhodophycota (Gantt *et al.*, 1971). Non-photosynthetic, colourless forms possess leucoplasts, suggestive of a photosynthetic ancestry. The storage product is starch-like, and found between the chloroplast envelope and the chloroplast ER. An eyespot may be present.

Cryptomonad cells are ovoid and dorsiventrally flattened. Ultrastructurally they are complex, and have been regarded as possibly ancestral to the

Fig. 2.46 *Dictyota* sp. (Phaeophyceae, Dictyotales). Thalli are parenchymatous and possess apical growth. The branches are flattened, dichotomously branched and terminate in individual apical cells. Scale = 5 cm. (Photograph: R.C. Ficken.)

Fig. 2.47 *Spatoglossum* sp. (Phaeophyceae, Dictyotales). Thalli are foliose, parenchymatous, and have apical growth. Scale = 5 cm. (Photograph: R.C. Ficken.)

Chromophycota. One unique structure is the nucleomorph, contained within a periplastidial compartment and surrounded by a double membrane; there is evidence that it is a subsidiary genome (Gillott & Gibbs, 1980). Two apically or laterally inserted, somewhat unequal flagella normally arise from the base of a depression, or gullet. They are tapered terminally, and bear microtubular hairs. The outer cell covering is a periplast, with proteinaceous plates arranged in a helical pattern (Hibberd *et al.*, 1971). Ejectosomes (trichocysts) occur, different in structure from those of the Dinophyceae, and probably most closely related to the R-bodies of the kappa particles of ciliates (Hovasse *et al.*, 1967). Growth is by binary fission: sexual reproduction is not known.

ORDER
Cryptomonadales.

EUGLENOPHYCOTA

Euglenophyceae: euglenoids
(Figs 2.62 & 2.63)

A class of *c.* 1000 species in 40–50 genera and six orders (Leedale, 1982a; Leedale *et al.*, 1982), in which the greatest amount of information has been derived from studies with *Euglena,* and specifically with *E. gracilis* (Buetow, 1968, 1982).

Fig. 2.48 *Tilopteris mertensii* (Phaeophyceae, Tilopteridales). Portion of a plant showing the main axis and the opposite, trichothallic distichous branching. Scale = 500 μm. From South (1975) with permission.

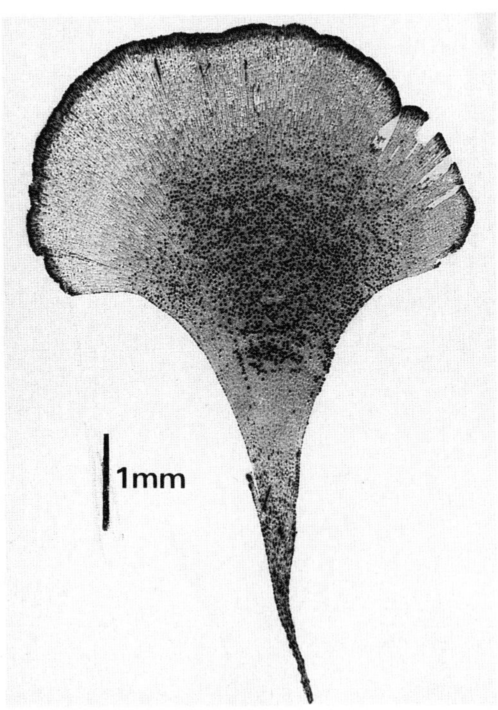

Fig. 2.49 *Syringoderma phinneyi* (Phaeophyceae, Syringodermatales). Whole plant, showing fan-like thallus with margin of apical cells. From Henry & Müller (1983) with permission.

Fig. 2.50 *Halopteris* sp. (Phaeophyceae, Sphacelariales). Habit of whole plant. Growth is apical and the main axes are heavily corticated. Scale = 2 cm. (Photograph: R.C. Ficken.)

The Euglenophyceae are characterized, in common with the Chlorophycota, by the possession of chlorophylls *a* and *b*. Chloroplasts also contain β-carotene, astaxanthin, antheraxanthin, diadinoxanthin and neoxanthin. They are disc- or plate-like, surrounded by a double envelope and one membrane of chloroplast ER, the latter not continuous with the nuclear membrane. Thylakoids are in stacks of three, and two thylakoid bands traverse the stroma of the pyrenoid. Paramylon (β-1, 3 glucan) is the storage product, forming a shield around the pyrenoid, but outside the chloroplast; it does not stain with iodine. Chrysolaminarin may also be stored. All green euglenoids investigated to date have been shown to be photoauxotrophic; they are able to grow photosynthetically in the light and heterotrophically in the dark. A number of species are heterotrophic or parasitic. All cells have two basal bodies, with one or two emergent flagella. Flagella are similar to those of the Dinophyceae, with a flagellar rod running the length of the flagellum

Fig. 2.51 *Desmarestia firma* (Phaeophyceae, Desmarestiales). Habit, showing plant in which laterals have taken over the main growth following damage to the main axis. Growth is trichothallic and the leafy axes and branches have a midrib. Scale = 5 cm. (Photograph: R.C. Ficken.)

and fibrillar hairs arranged in a single helical row along the length of the flagellum. The flagella emerge from an anterior chamber (reservoir). An eyespot is typically present, independent of the chloroplast but associated with a paraflagellar swelling. The cell covering is a proteinaceous pellicle, composed of a number of helically arranged interlocking strips. The pellicle may be flexible, resulting in the characteristic euglenoid

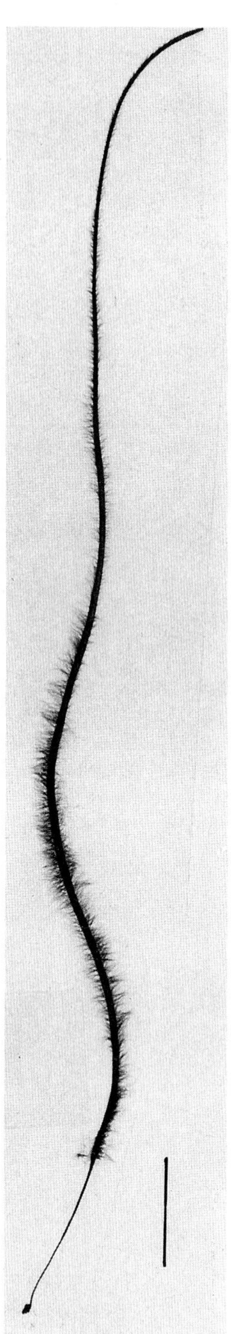

Fig. 2.52 *Chorda tomentosa* (Phaeophyceae, Laminariales). The simple unbranched thallus is parenchymatously constructed and clothed with assimilatory hairs. Attachment is by a small, discoid, rhizoidal holdfast. Scale = 5 cm. (Photograph: R.C. Ficken.)

Fig. 2.53 *Laminaria ochroleuca* (Phaeophyceae, Laminariales). The thallus comprises a holdfast with haptera, a terete stipe followed by the meristem (or transition zone), and a blade which is divided terminally.

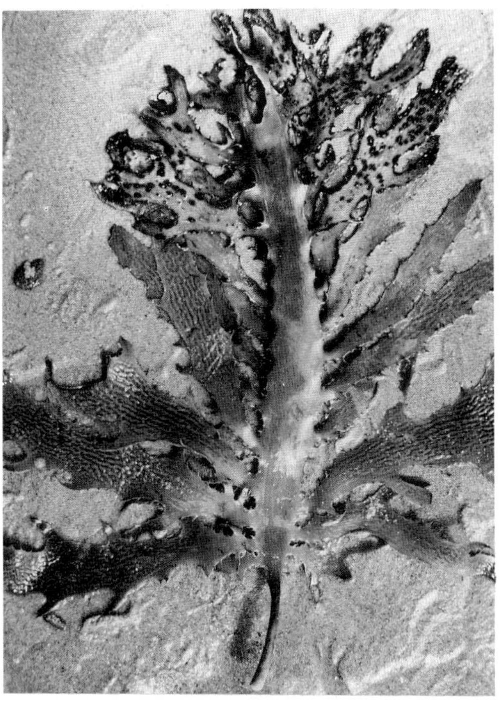

Fig. 2.55 *Ecklonia* sp. (Phaeophyceae, Laminariales). The blade is flattened in the central mid-rib region, and gives rise to numerous proliferous laterals.

Fig. 2.54 *Postelsia palmaeformis* (Phaeophyceae, Laminariales). The short, sturdy stipe supports numerous terminal blades which are ridged and toothed. Plants are restricted to highly exposed sites on the Pacific coast of North America.

Fig. 2.56 *Durvillaea antarctica* (Phaeophyceae, Durvilleales). Unlike members of the Laminariales, which it resembles, *Durvillaea* possesses blades with apical growth. Blades may be greatly subdivided into 'thongs', and their tissues are air-filled, giving the plant considerable buoyancy.

Fig. 2.57 *Ascoseira mirabilis* (Phaeophyceae, Ascoseirales). Plants superficially resemble *Durvillaea*, but the modes of growth and reproduction are different. Scale = 5 cm. (Photograph: R.C. Ficken.)

Fig. 2.58 *Fucus serratus* (Phaeophyceae, Fucales). Habit showing the perennial, parenchymatous, dichotomously branched thallus. Axes have a mid-rib and lateral, flattened 'wings' with serrate margins.

movement. The nucleus is mesokaryotic, with permanently condensed chromosomes; it undergoes a unique form of mitosis. Reproduction is solely by cell division; sexual reproduction is unknown.

ORDERS

Eutreptiales; Euglenales; Rhabdomonadales; Sphenomonaldales; Heteronematales; Euglenamorphales.

CHLOROPHYCOTA: GREEN ALGAE

This division incudes all algae possessing chlorophylls *a* and *b*, except the Euglenophycota (Silva, 1982). The features which characterize the Chlorophycota (as the Chlorophyta) are summarized by Hoek and Jahns (1978), Moestrup (1978), Hoek (1981) and Melkonian and Ichimura (1982). The chloroplasts are enclosed by a double membrane and lack an additional membrane of ER;

Fig. 2.59 *Himanthalia elongata* (Phaeophyceae, Fucales). The vegetative part of the plant comprises a small 'button' attached to the substrate by a short stalk. Visible are the thong-like reproductive receptacles, which arise from the 'button'.

Fig. 2.60 *Cystophora torulosa* (Phaeophyceae, Fucales). The richly branched thallus has air-bladders which assist in flotation. (Photograph: R.C. Ficken.)

Fig. 2.61 *Hormosira banksii* (Phaeophyceae, Fucales). This southern hemisphere species occupies a habitat comparable to that of *Fucus* in the northern hemisphere. The alternately inflated and constricted structure gives a bead-like appearance to the thallus.

thylakoids are in stacks of 2–6 or more. Pyrenoids when present are contained within the chloroplast, are often penetrated by thylakoids, and are surrounded by starch deposits, starch being the most important storage product. The eyespot, when present, is normally positioned within the chloroplast. Flagella are equal or unequal in length, usually apically inserted, and typically two in number. They lack hairs, but may be

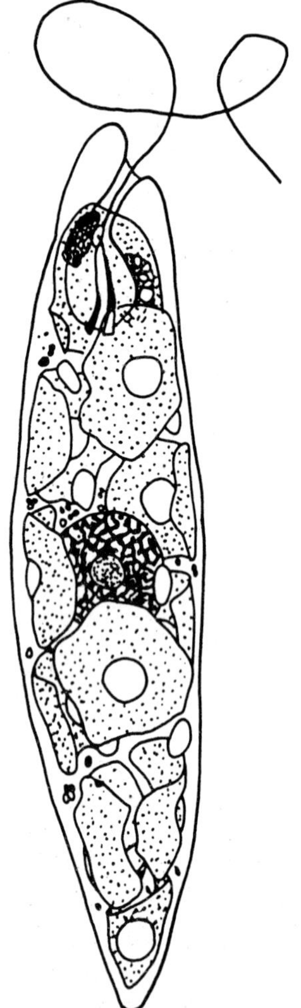

Fig. 2.62 *Euglena gracilis* (Euglenophyceae), Lateral view of a living cell. The delicate pellicular striations are not shown. Not to scale. Adapted from Leedale (1967).

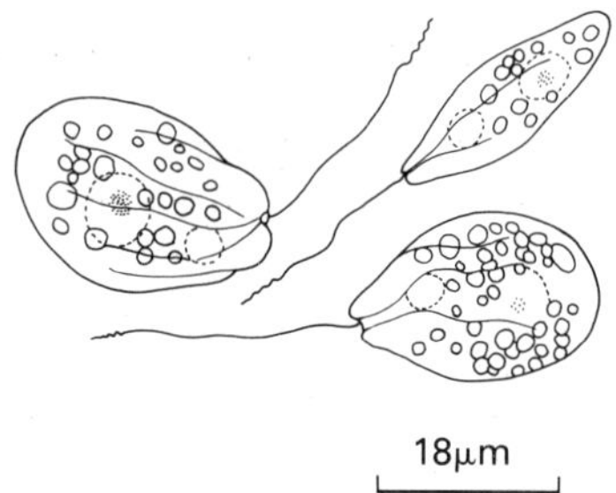

Fig. 2.63 A colourless biflagellate euglenoid (Euglenophyceae) recovered from a cryophilic habitat. Note the gullet, the heterodynamic flagella and the storage products. From Hoham & Blinn (1979) with permission.

covered with a delicate 'fur' or scales. The flagellar transition zone contains a nine-pointed stellate body (Moestrup, 1982). Cell walls are composed of cellulose, except in siphonous members, which have mannan or xylan, and members of the Prasinophyceae, which have galactan and uronic acid residues. Many Chlorophycota are calcified, especially tropical marine siphonous forms and members of the Charophyceae.

Asexual reproduction is by fragmentation, or by formation of zoospores, aplanospores, or autospores. Sexual reproduction may be isogamous, anisogamous or oogamous. A wide range of reproductive patterns and life histories occurs. With the exception of rhizopodial unicells and large, complex multicellular thalli, the Chlorophycota exhibit all morphological types found within the algae.

Modern studies in mitosis, cytokinesis and motile cell structure have led to proposals for the reclassification of the green algae (cf. Pickett-Heaps, 1975; Stewart & Mattox, 1975), these based on the recognition of several important phyletic lines. One of these (the 'chlorophytan' or 'chlamydomonad' line) is distinguished by the presence of a phycoplast during cytokinesis; another (the 'bryophytan' or 'charophytan' line) is distinguished by the presence of a phragmoplast. The second of these lines is believed to be ancestral to the higher plants (Steward & Mattox, 1975). While these lines are widely accepted in recent classifications, there is wide discrepancy in detail, and substantial 'fine-tuning' continues as new discoveries are published. For some recent reviews, proposals, and critiques consult Round (1971, 1981), Bold and Wynne (1978), Lee (1980), Christensen (1980), Deason (1984), Ettl (1981), Ettl and Komarek (1982), Hoek (1981), Moestrup (1982), Melkonian (1982), Melkonian and Ichimura (1982) and Irvine and John (1984).

Three classes are included here (Table 2.1). The Prasinophyceae are included with the Chlorophyceae by Hoek (1981), yet accorded a divisional status in Round (1981); most frequently, however, they are assigned to a separate class. Round (1981) assigned them to a division because of their very distinctive scaly wall ornamentation. Hoek (1981) used this same feature to show how the 'prasinophyte' type may represent an evolutionary link between a scale-covered archetypal green algal flagellate and chlorophytes with cellulose walls; he thus included them as members of the order Volvocales. The Charophyceae occupy an isolated position between the green algae and the mosses and liverworts (bryophytes) and are probably an evolutionary dead end. They are commonly assigned to a separate division, in recognition of their ancient divergence from the main lines of evolution in the Chlorophycota.

The majority of Chlorophycota are freshwater forms (90%), the remainder marine. The marine flora shows considerable uniformity in tropical seas, but rather less in cooler regions. The freshwater species are largely cosmopolitan. Fossil green algae are known from the Palaeozoic era, in the case of the Charophyta from the Silurian period (435–460 million years BP), and in the case of marine, calcified genera, from the Ordovician period (500–530 million years BP).

A summary of the three classes of Chlorophycota follows.

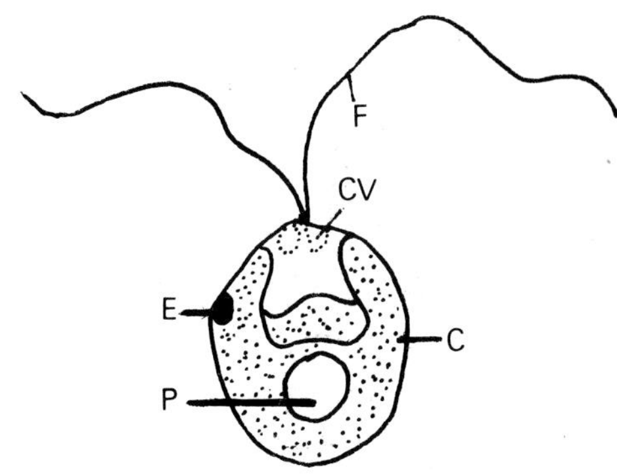

Fig. 2.64 *Chlamydomonas* sp. (Chlorophyceae). General view of the vegetative cell. F, flagella; E, eyespot; C, chloroplast; P, pyrenoid; CV, contractile vacuole. Not to scale.

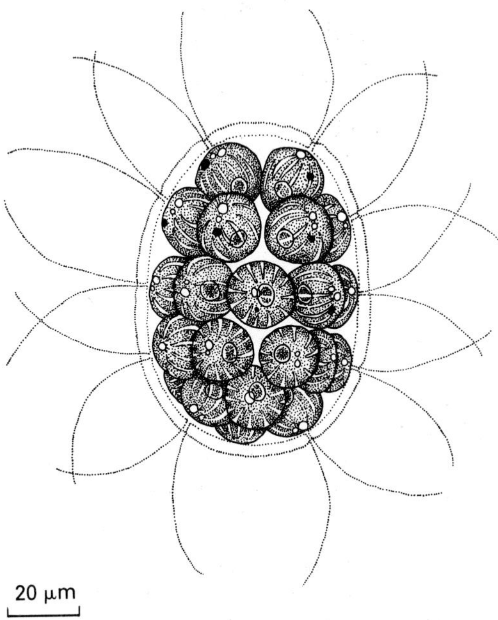

Fig. 2.65 *Pandorina unicocca* (Chlorophyceae, Volvocales). 32-celled colony. Note the lack of differentiation of the cells. From Nozaki (1981) with permission.

Chlorophyceae (Figs 2.64–2.75)

The Chlorophyceae *sensu lato* include about 560 genera and 8600 species according to Silva (1982) (although probably more than 20 000 species according to Melkonian and Ichimura [1982]). In addition to chlorophylls *a* and *b*, the chloroplasts contain various xanthophylls, including violaxanthin, zeaxanthin, antheraxanthin, and neoxzanthin; siphonoxanthin and siphonein are characteristic for the order Bryopsidales. Chloroplast features, pyrenoids, and starch storage are as described for the Chlorophycota. Motile cells normally possess two or four equal, smooth flagella, apically inserted and not arising from a pit. Flagellar scales or hairs are rare. The cell wall is composed, when present, principally of cellulose. It may be calcified, and may also contain hydroxyproline glycosides, xylan, and mannan as common constituents.

A complete range of morphological types occurs in the Chlorophyceae, with the exception of rhizopodial or large, complex multicellular forms. Silva (1982) describes three evolutionary lines

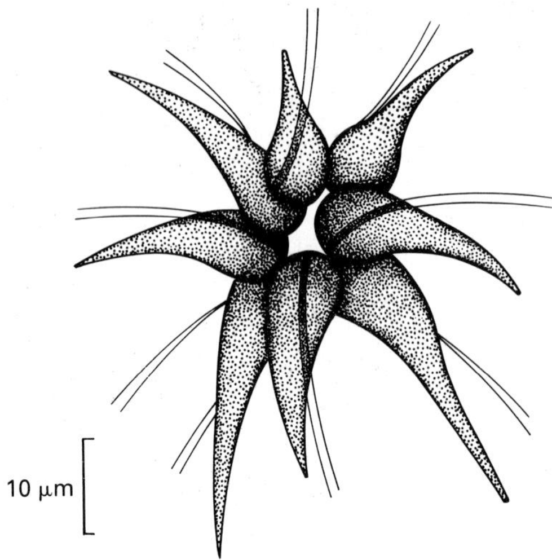

Fig. 2.66 *Pyrobotryis* (Chlorophyceae). Structure of an individual colony, showing differentiation (smaller anterior cells are uppermost); cells have two homodynamic flagella. From Sarma & Shyam (1974) with permission.

suggested by recent ultrastructural studies. Accordingly, the Chlorophyceae include those green algae 'characterized by a collapse of the interzonal spindle at telophase, presence of a phycoplast during cytokinesis, and motile cells with two or more apically inserted flagella, their basal bodies associated with four or more relatively narrow bands of microtubules.' This scheme would include the traditionally recognized volvocine, tetrasporine, and chlorococcine lines of evolution (Bold & Wynne, 1978) as well as members of the orders Chlorosarcinales, Microsporales, Oedogoniales and Chaetophorales *sensu stricto* (Silva, 1982). A second phyletic line includes the Ulvophyceae (= Ulvaphyceae of Sluiman *et al.* 1980; *see* O'Kelly & Floyd, 1983; Floyd & O'Kelly, 1984); Silva (1982) does not assign a formal rank to this group, in which the members lack a phycoplast. The independent status for the class is also brought into question by Lokhorst and Star (1983). Other possible diagnostic features assigned to the Ulvophyceae by O'Kelly and Floyd (1983) include: a 180° rotational symmetry of the flagellar apparatus and arrangement of its components in a counter-clockwise absolute orientation when viewed from above; and terminal caps with a bilobed construction enclosing

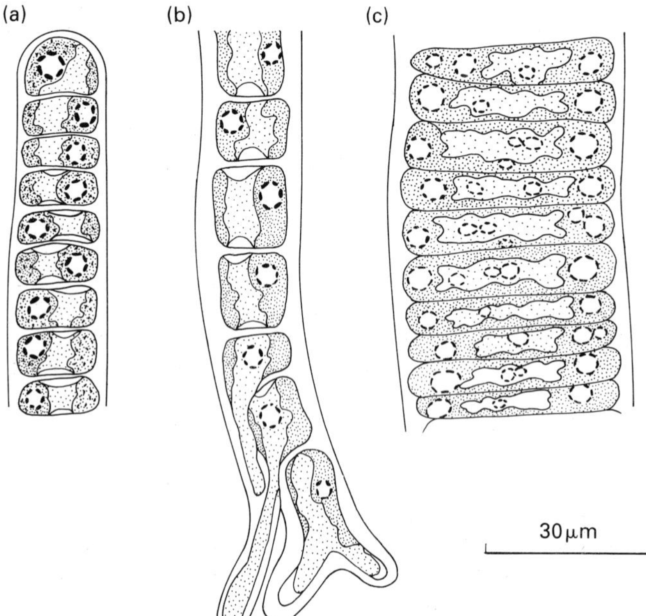

Fig. 2.67 *Ulothrix speciosa* (Chlorophyceae, Ulotrichales). Diagrams showing the apex (a), basal rhizoids and attachment (b) and the mature region of the filament (c). Each cell contains a parietal chloroplast, with one or two pyrenoids. From Berger-Perrot (1981) with permission.

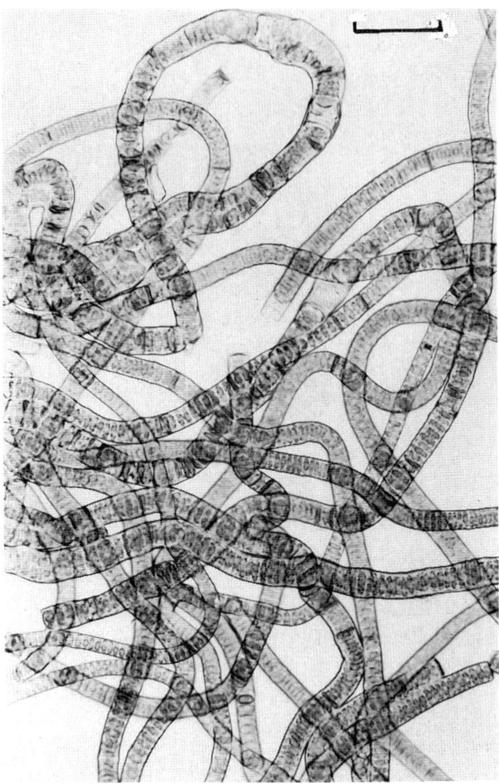

Fig. 2.68 *Rosenvingiella polyrhiza* (Chlorophyceae, Prasiolales). Vegetative filament showing uniseriate and polyseriate regions. Scale = 50 μm. From Hooper & South (1977) with permission.

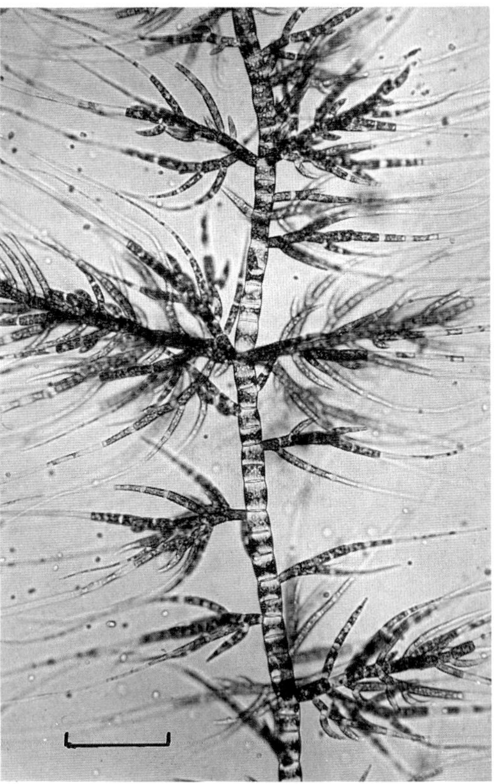

Fig. 2.69 *Draparnaldia* sp. (Chlorophyceae, Chaetophoraceae). Part of filament showing the main axial cells with girdle-shaped chloroplasts, and the lateral branches of limited growth, terminating in hairs. Scale = 50 μm.

the proximal ends of the basal bodies. The Ulvophyceae would include the orders Ulotrichales, Ctenocladales, Ulvales, Acrosiphoniales, Cladophorales, Siphonocladales, Bryopsidales and Sphaeropleales (Silva, 1982). The Charophyceae *sensu* Stewart and Mattox (1975) could be recognized as the third series, and in addition to the charophytes would include the Chlorokybales, Klebsormidiales, Zygnematales, Trentepohliales, and Coleochaetales. The siphonous green algae could be recognized as a fourth line which diverged early on from the ulvophycean line. For the purpose of this account, we have adopted the conservative approach taken by Silva (1982), recognizing that substantial changes in higher levels of classification will ultimately be required to accommodate the growing body of ultrastructural information.

ORDERS

Volvocales; Tetrasporales; Chlorococcales; Chlorosarcinales; Microsporales; Oedogoniales Chaetophorales; Ulotrichales; Ctenocladales; Ulvales; Prasiolales; Sphaeropleales; Acrosiphoniales; Cladophorales; Siphonocladales; Bryopsidales; Dasycladales; Chlorokybales; Klebsormidiales; Zygnematales; Trentepohliales; Coleochaetales.

Charophyceae: stoneworts (Fig. 2.76)

The class contains a single order and seven living genera (Daily, 1982; Grant & Sawa, 1982). Cells are uninucleate and typically contain numerous ellipsoidal chloroplasts arranged in twisted,

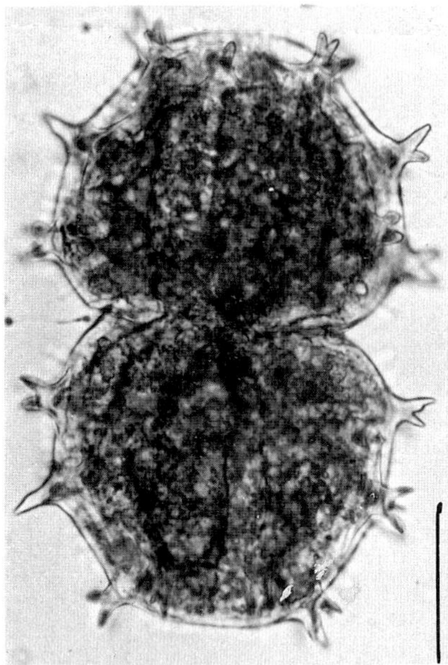

Fig. 2.70 *Xanthidium* sp. (Chlorophyceae, Conjugales). This desmid cell comprises two equal halves, or semi-cells, separated by the narrow isthmus which contains the single nucleus. The thickened walls are ornamented with spines. Scale = 25 μm.

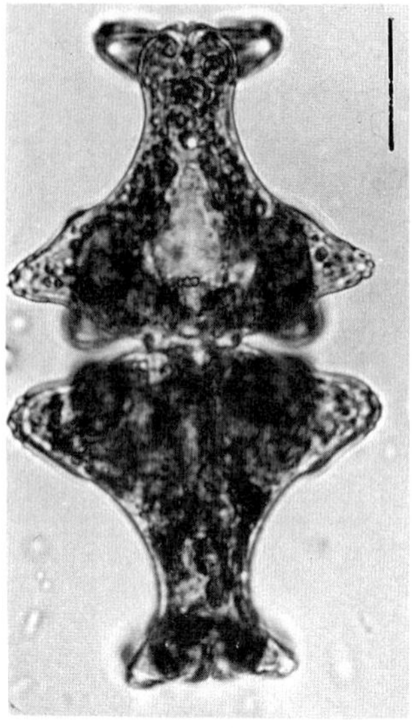

Fig. 2.71 *Euastrum insigne* (Chlorophyceae, Conjugales). While lacking spines, the walls of this desmid species are characteristically inflated, and extended at the polar regions. Scale = 25 μm.

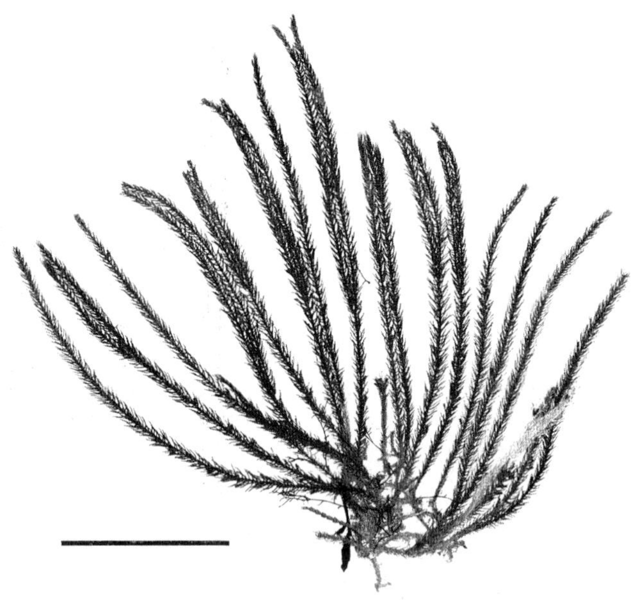

Fig. 2.72 *Caulerpa brownii* (Chlorophyceae, Caulerpales). The thallus is coenocytic and consists of a creeping rhizome (entangled here) which gives rise above to uprights that are densely branched, and below to attaching rhizoids. Scale = 5 cm. (Photograph: R.C. Ficken.)

longitudinal, parallel rows. The pigments and storage products are as described for the Chlorophycota. Cell walls are composed of cellulose, and are frequently heavily calcified; this feature leads to the deposition of marl ($CaCO_3$ and $MgCO_3$) in freshwater habitats where charophytes abound. Calcified remains also fossilize readily, especially the female organ, which is termed

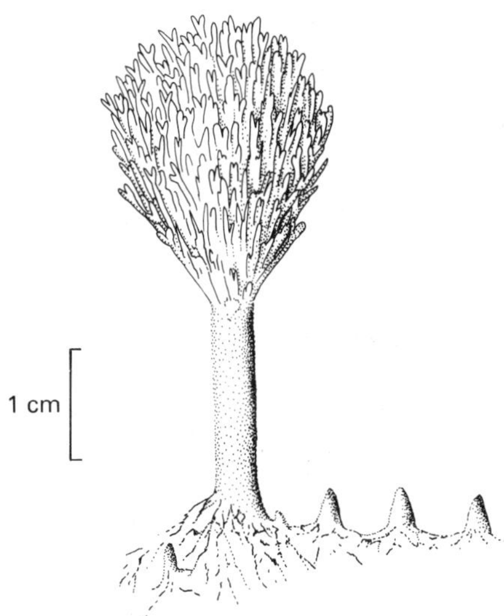

Fig. 2.73 *Penicillus capitatus* (Chlorophyceae, Caulerpales). The calcified thallus consists of a stalk which bears branched, terminal, siphonous branches. Attachment is by a dense mass of rhizoids buried in the soft substrate favoured by this tropical species. From Meinesz (1980) with permission.

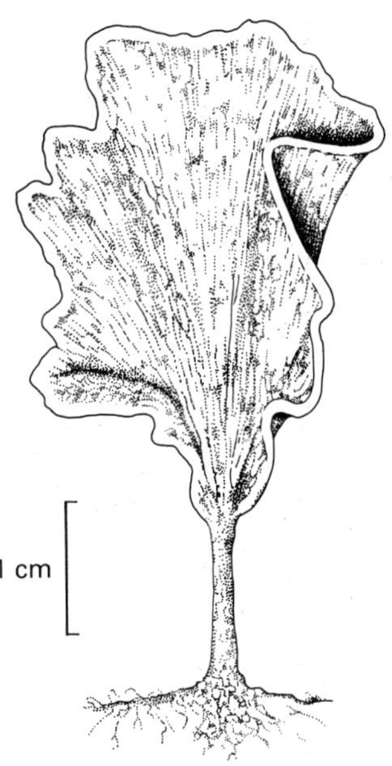

Fig. 2.74 *Udotea petiolata* (Chlorophyceae, Caulerpales). The calcified, siphonous thallus forms a fan-like upper portion borne on a stalk subtended from a rhizoidal base buried in the soft substrate. From Meinesz (1980) with permission.

Fig. 2.75 *Cladophora rupestris* (Chlorophyceae, Cladophorales). View of a whole plant, showing the dense branching. Structure is semi-coenocytic, the cells containing numerous nuclei. Scale 2 cm. (Photograph: R.C. Ficken.)

a gyrogonite in the fossilized state. Cell division occurs by a phragmoplast. Of the bryophyte-like features in this group, notable are the reproductive bodies which are enclosed within a sterile envelope (otherwise unique in the algae) and the sperm, which bears two somewhat unequal flagella, subapically inserted and covered by scales. Only one flagellar root occurs, which is generally associated with a multilayered structure (MLS) similar to that found in sperm cells of archegoniate plants (Moestrup & Ichimura, 1982; Moestrup, 1982). The sperm body is divided into three regions and is elongate, unlike that of any Chlorophyceae. Asexual zoospores are not formed by charophytes, although vegetative reproduction is effected by starch-filled (amylum) stars, by bulbils on the rhizoids, and by protonema-like outgrowths from the nodes.

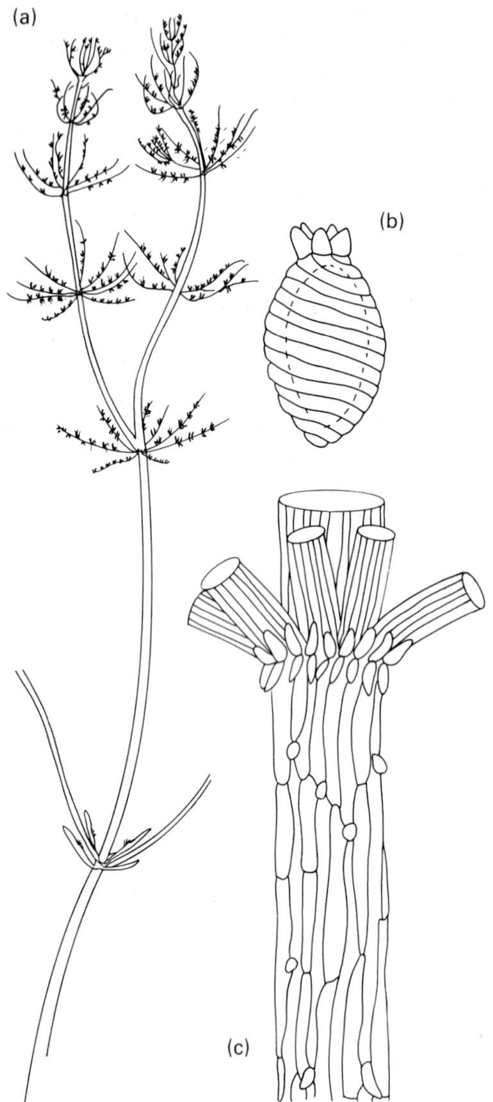

Fig. 2.76 *Chara vulgaris* (Charophyceae). (a) Whole plant showing main axis and whorls of simple branchlets at the nodes, separated by internodes. (b) Oogonium, or 'globule', the female reproductive organ. (c) Detail of corticated axis and nodal region. Not to scale. (After Wood 1967.)

Growth is apical, commencing from a protonemal stage (compare with bryophytes), and leading to a nodal–internodal structure. Whorls of branches are formed from the node, and the internodes are composed of a single enlarged cell that may be corticated (e.g. *Chara*).

The Charophyceae are primarily freshwater organisms, although a few species may occur in brackish water. That they diverged very early from the Chlorophyceae is evident from the fossil record; charophytes are known from the Palaeozoic era, as far back as the Silurian period (435–460 million years BP).

ORDER
Charales (usually divided into two tribes, the Nitelleae and Chareae).

Prasinophyceae

A class of coccoid or flagellate unicells; in flagellate species the flagella and cell body are covered with one or more layers of scales of several types. Cells are uninucleate and possess a large, lobed parietal chloroplast. Pigments may include siphonoxanthin (known elsewhere only in the Chlorophyceae). A pyrenoid is usually present, penetrated by thylakoids and, in some species, by lobes of cytoplasm (Norris, 1982a, b). Starch and mannitol are the principal storage products. Most cells possess an eyespot, and some possess trichocysts (Pearson & Norris, 1975; Norris, 1980, 1982b) comparable with the ejectosomes of the Cryptophyceae. There may be one, two, four, six, or eight flagella. In uni- and biflagellate species the flagella are laterally inserted, while in quadriflagellate species the flagella are of approximately equal length and arise from a pit-like depression in the apical region of the heart-shaped cell. Some species (e.g. *Halosphaera viridis*; Parke & Adams, 1961) have alternating motile and non-motile stages. A theca may enclose the cell (e.g. *Platymonas* and *Prasinocladus*). Asexual reproduction may involve zoospore-forming stages or simple division; sexual reproduction is unknown.

Most Prasinophyceae are free-living, marine, brackish water or freshwater organisms. Some are symbionts, such as *Tetraselmis*, which forms a well-known symbiosis with the marine flatworm *Convoluta*.

Much remains to be determined regarding the fine structure of members of the class; at present the taxonomy is tentative, as is the status of the Prasinophyceae (Parke & Green, 1976). Moestrup (1982) segregates the genera *Monomastix, Pedinomonas,* and *Scourfieldia* (and *Micromonas*) in the class Loxophyceae, not recognized by Norris (1980, 1982b, c).

ORDERS

Pedinomonadales; Monomastigales; Pyramimonadales; Pterospormatales; Chlorodendrales.

The 'Glaucophyta' or 'Glaucophyceae'

Parker (1982) does not include a division or class of glaucophytes, although recent work suggests that such recognition may be required. Skuja (1954) assigned a number of flagellates containing endosymbiotic blue-green algae to the division Glaucophyta (excluding the more recently recognized dinoflagellates containing endosymbiotic blue-green 'phaeosomes' [Taylor, 1973]). These composite organisms (known as syncyanoses) may represent, through a few remaining living representatives, examples of important steps in the evolution of chloroplast-containing eukaryotic cells. Extant species include those with naked cyanomes (host cells) and their derivatives, those surrounded by a wall (Geitler, 1959). The blue-green component (cyanelle) contains chlorophyll *a* and phycobiliproteins, as in the Cyanophycota.

The status of this group remains questionable, although Kies (1980) has suggested it forms a natural assemblage with some affinity with the green algae. This conclusion is substantiated in the findings of Moestrup (1982), who provides a summary based on flagellar structure. The Glaucophyceae (*sensu* Moestrup, 1982) could be grouped together with the euglenophytes and chlorophytes. Earlier suggestions that *Glaucocystis* could be a red alga (Robinson & Preston, 1971) have been refuted by Kremer *et al.* (1979).

CHAPTER 3

Cellular and subcellular organization

General features

Algal cells, like those of any other living organism, are composed of a protoplast, which is either naked or enclosed within a cell wall or envelope. The external living membrane is the plasmalemma. In prokaroytic algae (Cyanophycota and Prochlorophycota; Fig. 3.1) the living contents of the cell are lacking in membrane-bounded organelles; rather, they are divided into zones, with the nuclear material concentrated centrally (the centroplasm) and the photosynthetic lamellae peripherally (the chromoplasm). In eukaryotic algae the living material is compartmentalized in a number of intracellular, membrane-bounded organelles. These include the nucleus, the chloroplasts, the mitochondria, and the Golgi bodies. The endoplasmic reticulum (ER) comprises a further intracellular membrane system, or cisterna. Other cell inclusions include carbohydrate reserves (often associated with pyrenoids) and a variety of crystalline, lipid, or other materials. In many algae motility is effected with the aid of flagella, whip-like appendages whose bounding membrane is continuous with the plasmalemma. Internally, complex root structures occur, and in many motile species an eyespot (stigma) may be present, this often associated with the flagellar basal system and the chloroplast.

While there are fundamental differences between prokaryotic and eukaryotic cellular and subcellular organization, there is a widely held view (the endosymbiotic theory) that eukaryotic plastids and mitochondria arose from prokaryotes through endosymbiosis by prokaryote, photosynthetic forms (e.g. cyanelles) and bacteria (*see* Chapter 8 and Margulis, 1981; Giddings *et al.*, 1980; Brawley & Wetherbee, 1981). The theory thus provides for an evolutionary continuum between the pro- and eukaryotes, and among the eukaryotes themselves. The endosymbiotic

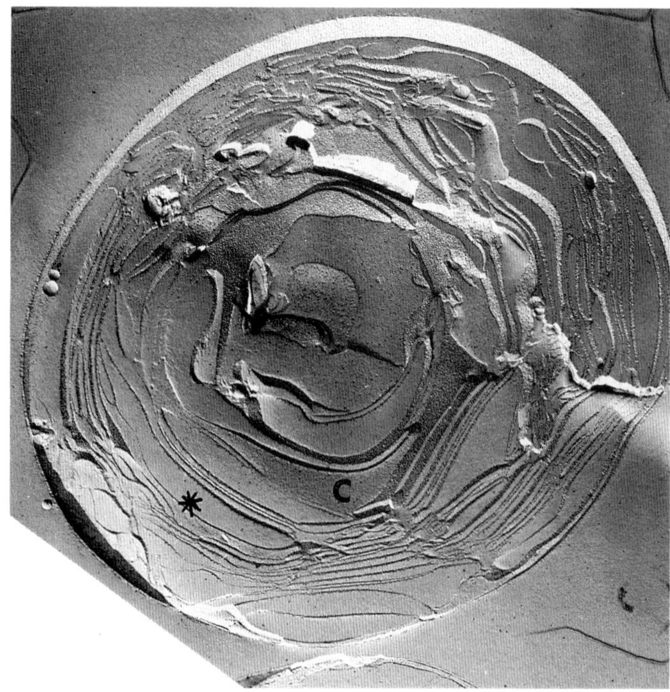

Fig. 3.1 *Prochloron* sp. (Prochlorophyta). Freeze–fracture replica of a cell showing general organization. Note the thylakoids whose lumen(*) can be distinguished from the cytoplasm(C) by the smoother texture (× 11 000). From Giddings *et al.* (1980) with permission.

Table 3.1 Summary of types of cell covering in the algae. (Modified from Dodge, 1973.)

Class	Naked		Modifications beneath the plasmalemma			Scales outside the plasmalemma			Silica frustule	Cell wall	
	Stage	Normal	Periplast	Pellicle	Theca	Organic	Calcite	Silica		Incomplete (lorica)	Complete
Cyanophyceae											+
Prochlorophyceae											+
Rhodophyceae		+									+
Chrysophyceae		+						+		+	
Prymnesiophyceae						+	+				
Xanthophyceae	+										+
Eustigmatophyceae	+										+
Bacillariophyceae	+								+		
Dinophyceae					+						+
Phaeophyceae	+										+
Raphidophyceae		+									
Cryptophyceae			+								
Euglenophyceae				+	+						
Chlorophyceae	+	+									+
Charophyceae	+										+
Prasinophyceae						+					

theory is not held by all (*see* discussion in Chapter 8 and Dillon, 1981), and its weakness is a dependence on circumstantial proof from extant species. Further study of very early eukaryote fossils may throw more light on this intriguing question in time (*see* Vidal [1984] for a popular discussion). While cellular and subcellular organization in eukaryotic algae are in many respects similar to that of other eukaryotic cells, there are also many distinct and unique features. It is the latter that will be highlighted here.

The cellular characteristics of the algal classes were summarized in Chapter 2. For a majority of the classes, the basic cell organization is thought to be derived from a motile unicellular 'ancestral' type (except in the prokaryotes and the Rhodophycota; *see* Chapters 4 and 8). Although cell organization and development are highly varied, tissue differentiation is poorly developed in algae. During the process of growth, maturation, and reproduction, marked changes in cellular and subcellular organization may occur (Pickett-Heaps, 1975). In considering cellular and subcellular organization, this dynamic state should be borne in mind.

Cell coverings: the cell wall

While many flagellates and algal spores and gametes are described as 'naked', the majority of algal cells are covered by one or more bounding layers of relatively inert material (principally carbohydrates), which may or may not be impregnated or layered with inorganic substances such as calcium carbonate and silica. Many of the walls are comparable with those found in higher plants, although specialized coverings (such as the pellicle and theca) may be distinctive for a particular group of algae. Detailed information on wall formation, chemistry, and structure is now available for a wide variety of algae; surprisingly, while wall features may be uniform within some well-circumscribed groups of algae, in others there may be marked differences in wall structure and chemistry at different stages of growth, or at different stages of the same life history.

The classification devised by Dodge (1973) remains a useful framework for discussion of algal cell coverings. In Table 3.1 a summary is given of the main types of cell covering found in the various algal classes.

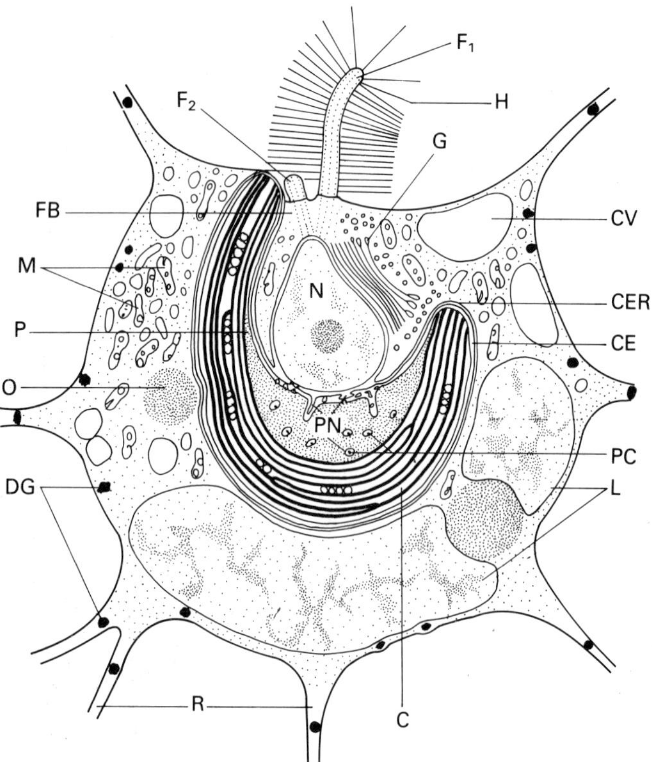

Fig. 3.2 *Chrysamoeba radicans* (Chrysophyceae). Schematic section of a naked, amoeboid stage. C, chloroplast; CE, chloroplast envelope; CER, chloroplast endoplasmic reticulum; CV, contractile vacuole; DG, dense globules; F_1, main flagellum; F_2, second flagellum; G, Golgi body; H, flagellar hair; L, leucosin vesicles; M, mitochondria; N, nucleus; O, oil globules; P, pyrenoid; PC, pyrenoid channels; PN, periplastidial network; R, rhizopodia. From Hibberd (1971) with permission.

Naked cells may occur as the principal vegetative state (some Bangiophycidae, Chrysophyceae [Fig. 3.2] and Chlorophyceae, and all Raphidophyceae) or as a transitory condition of asexual zoospores (Eustigmatophyceae), gametes (especially the male, e.g. Bacillariophyceae) or both zoospores and gametes (Chlorophyceae, Xanthophyceae and Phaeophyceae; Fig. 3.3). In naked cells the plasma membrane (plasmalemma) forms the outermost covering. In many naked forms the outer covering is flexible, allowing amoeboid or rhizopodial motion. Proteinaceous materials may be present outside the plasmalemma, and in *Porphyridium* (Bangiophycideae) it may be enclosed by a polysaccharide capsule (Ramus, 1972). In motile unicells of the Chrysophyceae and Raphidophyceae, a variety of projectile-releasing organelles are present. Two different types occur in chrysophycean cells; muciferous bodies and discobolocysts, the former being similar to muciferous bodies found in the Prymnesiophyceae, Raphidophyceae, and Dinophyceae (Lee, 1980). They both originate within the Golgi. Muciferous bodies are bounded by a single membrane, and on discharge form a fibrous network outside the cell. Discobolocysts have been described in detail from *Ochromonas tuberculatus* by Hibberd (1970). They are located in the outer layer of the cytoplasm and are a single membrane-bounded vesicle with a hollow disc in the outward-facing part. Discharge is explosive, releasing a thin thread with the disc at the tip. Muciferous bodies are abundant in the outer cytoplasm of the raphidophyte *Vacuolaria*, while in the raphidophyte *Gonyostomum* trichocysts similar to those in the Dinophyceae occur (Mignot, 1967).

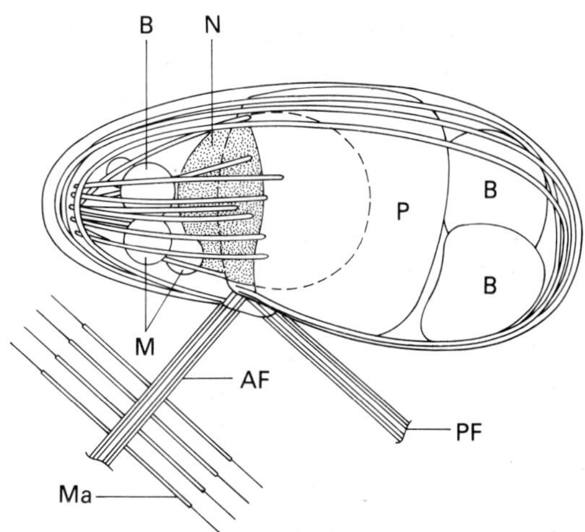

Fig. 3.3 Generalized diagrammatic impression of a laminarialean (Phaeophyceae, Laminariales) zoospore. The flagellar insertion, cytoskeletal microtubules and major organelles are shown in side view. AF, anterior flagellum; PF, posterior flagellum; N, nucleus; P, chloroplast; B, type 'B' vesicle; M, mitochondrion; Ma, hairs. From Henry & Cole (1982a) with permission.

In transitory naked phases, the naked condition is usually rapidly lost once zoospores or gametes have ceased swimming and have become attached to the substrate, since wall formation rapidly ensues.

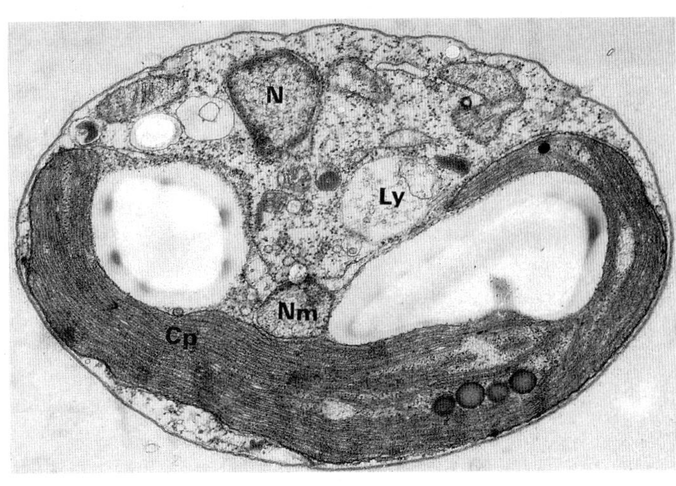

Fig. 3.4 Transverse EM section of *Hemiselmis brunnescens* (Cryptophyceae) at the anterior end of the nucleus. Note the proteinaceous plates of the periplast, as well as general cell details. Cp chloroplast; Ly, lysosomes; N, nucleus; Nm, nucleomorph (× 17800). From Santore (1982b) with permission.

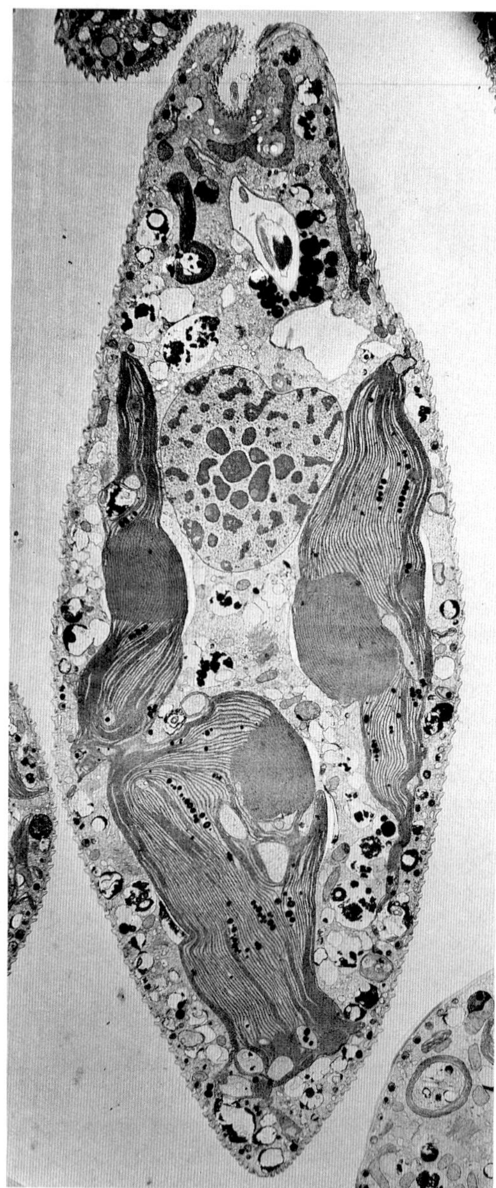

Fig. 3.5 Longitudindal EM section of *Euglena granulata* (Euglenophyceae) showing the anterior reservoir, the nucleus with chromosomes, chloroplasts, and a sectional view of the proteinaceous pellicle with numerous surface corrugations (× 5400). From Walne & Arnott (1967) with permission.

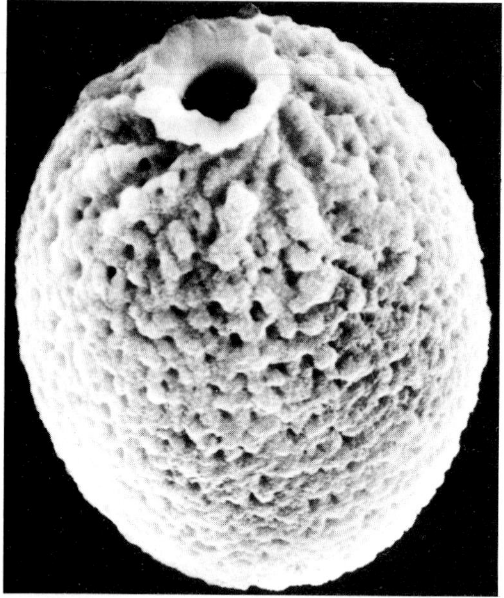

Fig. 3.6 Empty cell envelope of *Trachelomonas lefevrei* (Euglenophyceae) composed of anastomosed, mineralized mucilage strands. Note the rimmed flagellar opening (× 2800). Photograph courtesy of P.L. Walne.

Those cell coverings which feature various modifications, mostly beneath the plasmalemma, are the periplast, the pellicle, and the theca. They are characteristic for the classes Cryptophyceae, Euglenophyceae, and Dinophyceae respectively. In some accounts (e.g. Round, 1973; Bold & Wynne, 1978) algae with cell coverings of these types are described as 'naked'.

The periplast of the Cryptophyceae is composed of an outer fibrous layer with proteinaceous plates internal to the plasmalemma. The plates are arranged in a variety of patterns that are distinctive at the species level (Santore, 1982a, b; Fig. 3.4); they may include hexagonal (*Cryptomonas*) and rectangular (*Chroomonas*) types. Trichocysts are found in association with the plates.

The euglenoid pellicle (Fig. 3.5) (Leedale, 1967, 1982b) is comparable with the periplast, in that the material underlying the plasmalemma is proteinaceous (80% protein, the remainder lipids and carbohydrates). Walne (1980) included a review of current information on the pellicle; West *et al.* (1980a) postulate that the pellicle has

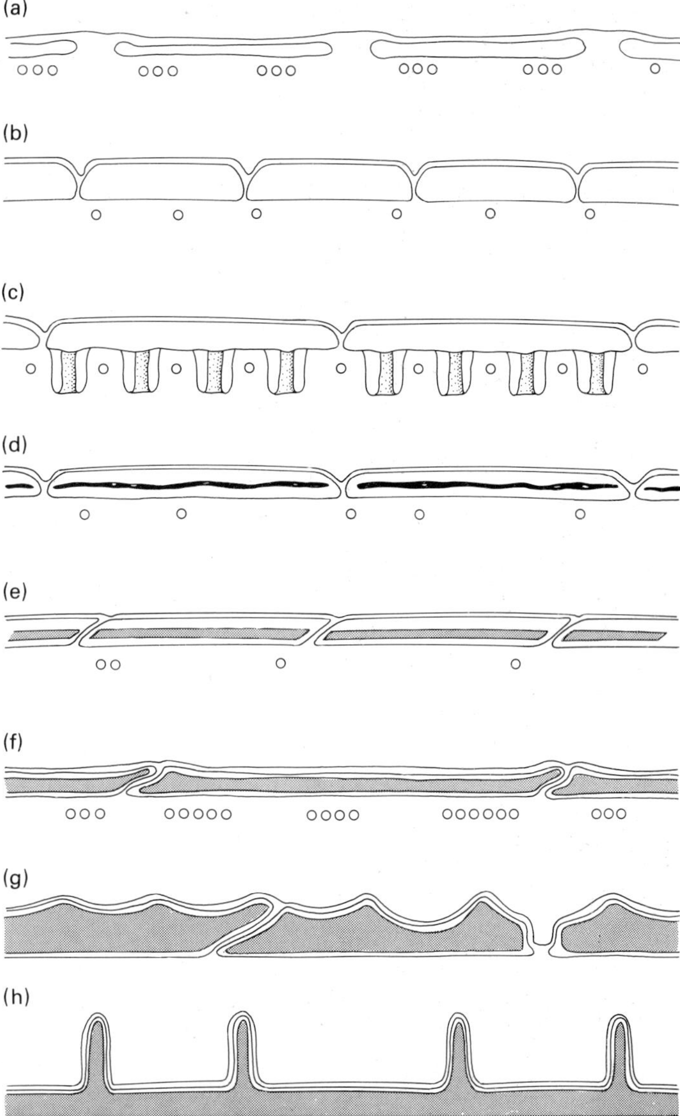

Fig. 3.7 Vertical sections of various types of dinoflagellate theca. (a) *Oxyrrhis*. (b) *Amphidinium*. (c) Some species of *Gymnodinium*. (d) *Katodinium*. (e) *Woloszynskia*. (f) *Glenodinium* and *Heterocapsa*. (g) *Ceratium* and some *Peridinium* species. (h) *Prorocentrum*. From Dodge & Crawford (1970) with permission.

evolved along two different lines: (i) towards increased complexity and (ii) towards simplification and reduction. In *Euglena* the surface appears corrugated in transverse section, this because of the presence of proteinaceous pellicular strips that are arranged in a helical or longitudinal pattern, and that articulate and interlock with one another. In *Trachelomonas* the ridges are absent, and the cell is enclosed in a mineralized lorica composed of mucistrands (Fig. 3.6; West & Walne, 1980) that contain extensive deposits (20–60% wt) of manganese (West *et al.*, 1980b).

In many euglenoids the pellicle may be pliable and capable of contracting and expanding movements (metaboly) when in a non-swimming condition. Beneath the proteinaceous layer is a network of microtubules and, often adjacent to the grooves of the pellicle, muciferous bodies that connect via narrow pores to the exterior.

The dinophyte theca (Fig. 3.7) (Dodge & Crawford, 1970) or amphiesma (Loeblich, 1970)

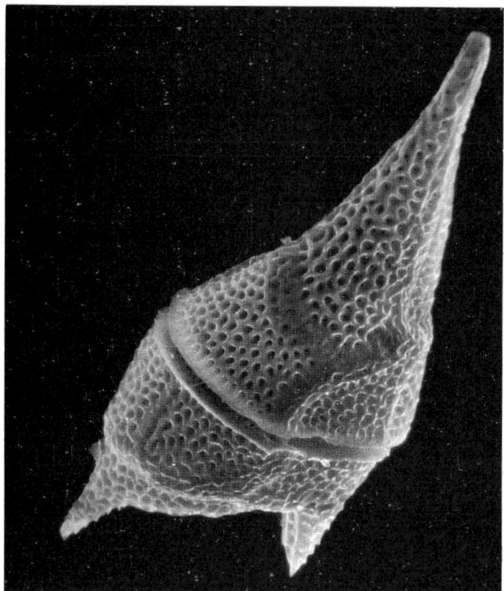

Fig. 3.8 *Ceratium brachyceros* (Dinophyceae). SEM of the cell showing the general arrangement and ornamentation of the thecal plates, and the girdle groove (× 1040). From Bourrelly & Couté (1976) with permission.

consists of several layers. There is considerable controversy as to whether the plasma membrane is the innermost or outermost (Bold & Wynne, 1978), although since the outermost membrane is continuous with that of the flagella it is likely that it is the plasmalemma. Dinoflagellates are described as 'naked' or 'armoured', the latter forms characterized by the presence of microfibrillar thecal plates that are laid down in association with a layer of flattened vesicles below the outer membrane. The vesicles are present in both types of theca, and Loeblich (1970) suggests the structure is essentially the same for both; thin plates may occur in some species described as 'naked'. The number and arrangement of the plates is of considerable taxonomic importance, and taxonomists have devised elaborate formulae for numbering the plates in a unique way for each species (Figs 3.8 & 3.9) The shape and morphology of the thecal plates are also taxonomically important. Plates may be ornamented with ridges or reticulations, or spines and other projections, some of which may be silicified. In many species, the plates are perforated to allow the discharge of trichocysts.

Members of the Chrysophyceae (Fig. 3.10),

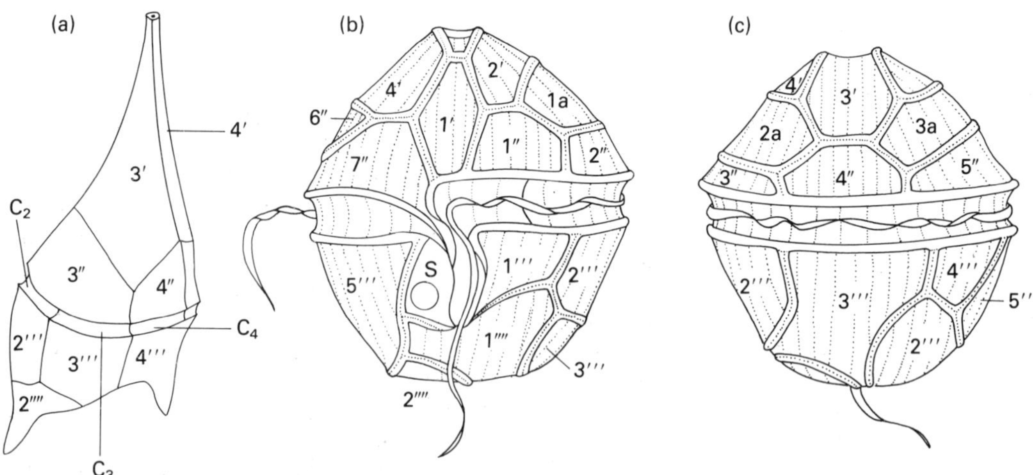

Fig. 3.9 Examples of the arrangement and numbering of thecal plates in dinoflagellates. (a) *Ceratium brachyceros* (Dinophyceae) — diagrammatic ventral view of the cell, showing the thecal plate numbering sequence. 1′, apical series; 1″, pericingular series; 1‴, postcingular series; 1⁗, antapical series; C_2, C_3, C_4, cingular. From Bourrelly & Couté (1976) with permission. (b, c) *Peridinium gregarium* (Dinophyceae) from ventral and dorsal views, showing the plate numbering system (series as in (a) above). S, sulcus. From Lombard & Capon (1971) with permission.

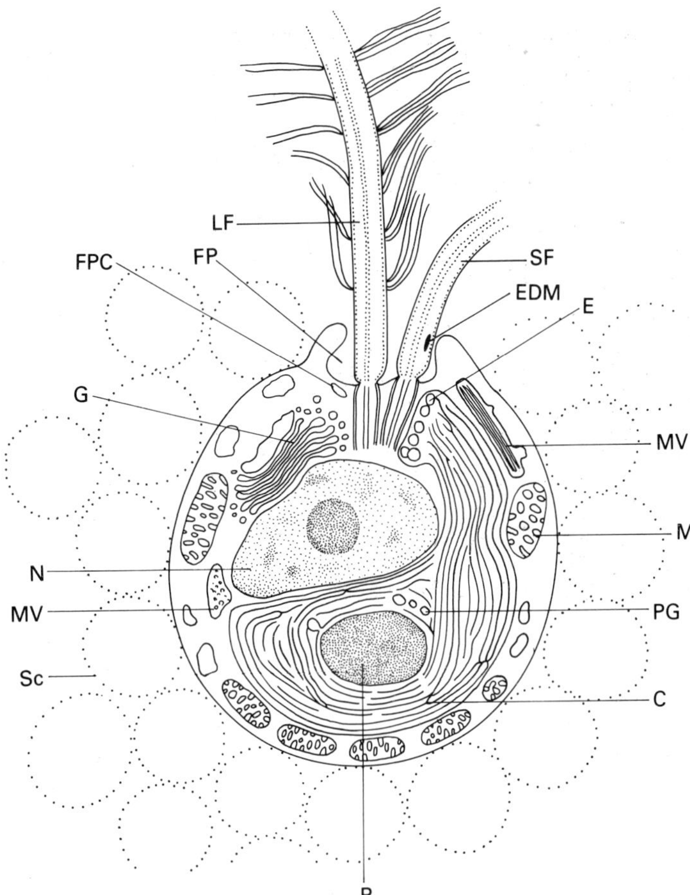

Fig. 3.10 *Syncrypta glomifera* (Chrysophyceae). Schematic section of the cell, showing the investment of scales. C, chloroplast; E, eyespot; EDM, electron-dense material; FP, flagellar pit; FPC, flagellar pit cisternum; G, Golgi body; LF, long flagellum; M, mitochondrion; MV, mastigoneme (flagellar hair) vesicle; N, nucleus; P, pyrenoid; PG, pigment granules; Sc, scales; SF, short flagellum. From Clarke & Pennick (1975) with permission.

Prymnesiophyceae, Prasinophyceae, and Dinophyceae (rarely) possess scaly coverings (body scales) deposited on the surface of the plasmalemma. More than one type of scale may be present. Their formation is a remarkable process, associated in most instances with the Golgi vesicles (first demonstrated by Manton & Parke, 1960). In some (e.g. the prasinophyte *Pyramimonas*) a special 'scale reservoir' has been described (Fig. 3.11). Scale morphology is highly varied, and has been revealed through some exquisite studies with the transmission and scanning electron microscopes. It is generally recognized that in algae possessing both body and flagellar scales (e.g. Prasinophyceae) the scales are of different types. They are, however, formed from the same Golgi vesicles and it is a puzzle as to how one type of scale is deposited on the flagellar plasmalemma and another on the body. In *Pyramimonas orientalis* and *P. parkeae*, for example, six different types of scales pass through the scale reservoir, each of which has a specific position on the flagella or body (Moestrup, 1982). Scale release in other algae is also commonly from the region of the flagellar base. In the dinophyte *Heterocapsa*, Morrill and Loeblich (1983) have shown how the membranes of the Golgi scale vesicles act as moulds for the developing scales; they suggest that scale production is a continuous process in this genus. The purpose of scales of such varied chemistry and morphology is uncertain, although protection, increase in size, and buoyancy may be involved.

Three principal types of scales have been described on the basis of chemical composition — organic, calcite, and silica (Table 3.1) — and they are discussed in turn. Other Golgi-derived flagellar or wall adornments include the 'knob scales' of the Pavlovales (Prymnesiophyceae; Green, 1980), which are electron-dense bodies,

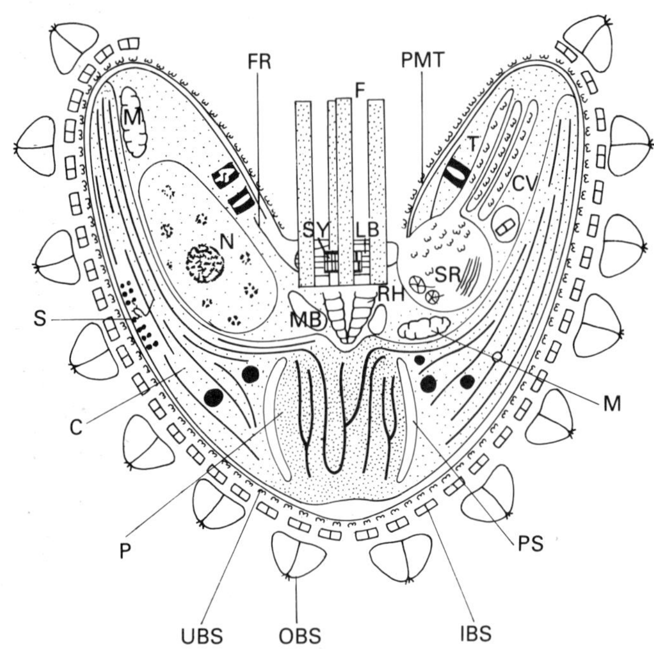

Fig. 3.11 Diagrammatic longitudinal section of *Pyramimonas lunata* (Prasinophyceae) showing ultrastructural features (except flagellar scales). C, chloroplast; CV, cylindrical vesicles; F, flagella; FR, flagellar microtubular roots; IBS, intermediate body scales; LB, lateral fibrous band; M, mitochondria; MB, microbody; N, nucleus; OBS, outermost body scales; P, pyrenoid; PMT, pit microtubules; PS, pyrenoid starch sheaths; RH, rhizoplasts; S, stigma; SR, scale reservoir; SY, synistome; UBS, undermost body scales (not to scale). From Inouye *et al.* (1983) with permission.

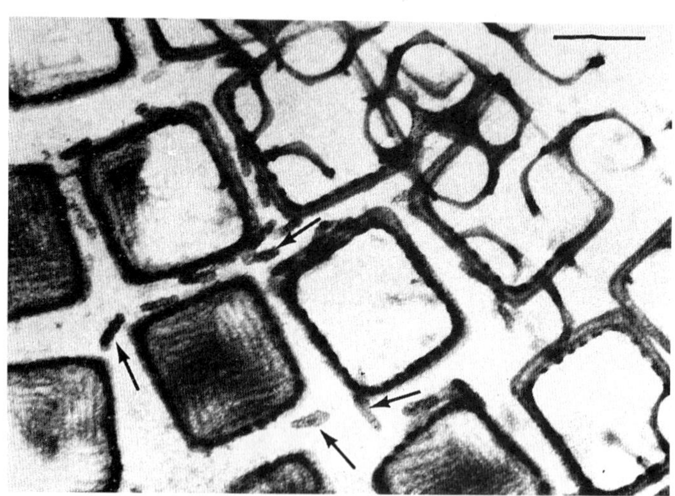

Fig. 3.12 Body scales of *Pyramimonas gelidicola* (Prasinophyceae). Included are box scales with footprint scales between them and crown scales (arrows) interlocking the box scale corners. Scale = 0.2 μm. From McFadden *et al.* (1982) with permission.

and the cylinder scales of *Chrysochromulina* (Prymnesiophyceae; Leadbeater, 1972; Manton *et al.*, 1981).

Organic scales are found in the Prasinophyceae, as body and flagellar scales (Moestrup, 1982), usually of different types, and as underlayer scales in coccolith-forming Prymnesiophyceae. Up to five different kinds of body scales may be present in prasinophytes (Fig. 3.12), each of a different morphology; Norris (1980) has suggested that the evolutionary complexities of the cell increase as the number of layers of scales increases. In the highly developed *Tetraselmis* (Fig. 3.13) the scale layers are fixed together to form a solid envelope (Norris, 1982). Organic scales have a fibrillar structure, but their chemical composition is poorly known. It is pectin-like, and for the organic underscales of *Apistonema* (Prymnesio-

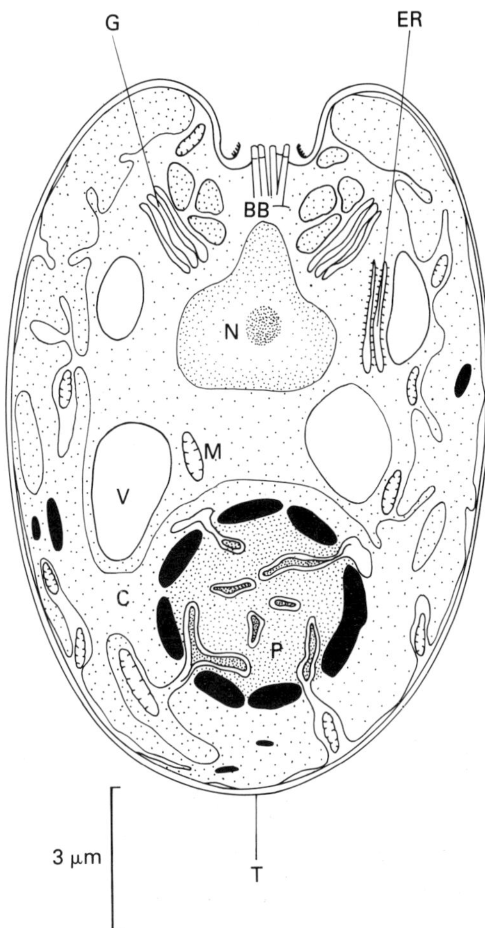

Fig. 3.13 *Tetraselmis astigmatica* (Prasinophyceae) showing the solid envelope (or theca, T) and general features in median section. BB, basal body; C, chloroplast; ER, endoplasmic reticulum; G, Golgi dictyosome; M, mitochondrion; N, nucleus; P, pyrenoid matrix; V, vacuole. From Hori *et al.* (1982) with permission.

phyceae) it has been described as cellulosic glycoprotein.

Calcite scales, or coccoliths, are characteristic for the coccolithophorids (order Coccosphaerales) of the Prymnesiophyceae (Norris, 1982). They were originally described from the Cretaceous period, when their algal nature was not realized. They consist of an organic scale with calcium carbonate (more rarely aragonite) deposited on one surface in a pattern characteristic for each species. Scale morphology is highly elaborate

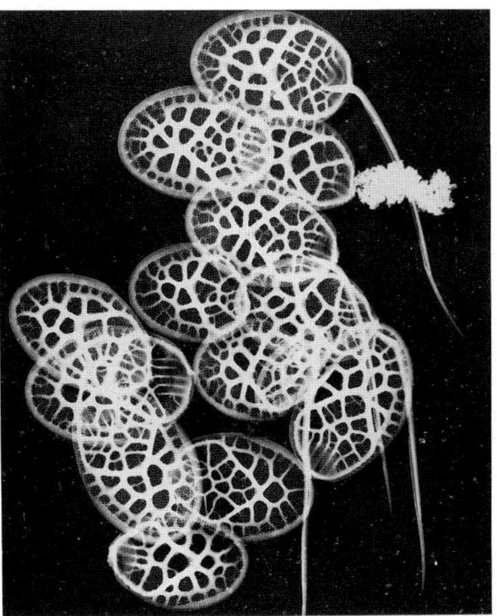

Fig. 3.14 Body scales and apical scales with bristles of *Mallomonas harrisea* (Chrysophyceae) (× 5500). From Takahashi (1975) with permission.

and is not necessarily the same for each stage of the same life history; spines are often present. The number of scales varies with the age of the cell. Calcite scales, like organic scales, are formed in the Golgi apparatus.

In the Chrysophyceae, silica scales may be deposited in a thin pectic envelope external to the plasmalemma. These silica body scales may be highly elaborate (as in *Mallomonas*; Fig. 3.14), and there may be several kinds on the same cell. Silica scales differ from organic and calcite scales in that usually they are manufactured not in the Golgi apparatus, but rather in vesicles associated with the endoplasmic reticulum, on the surface of the chloroplast. Moestrup (1982), however, described a second type of chrysophyte scale (as in, for example, *Sphaleromantis*), found on both the flagellar and body surfaces, and manufactured in the Golgi apparatus.

The diatom (Bacillariophyceae) cell wall or frustule is composed of silica, with an organic coating. It has been extensively studied with the aid of light microscopy and the electron microscope (especially through carbon replication and the SEM). Frustule structure and ornamentation form the basis of diatom classification, and are best observed in prepared and cleaned cells;

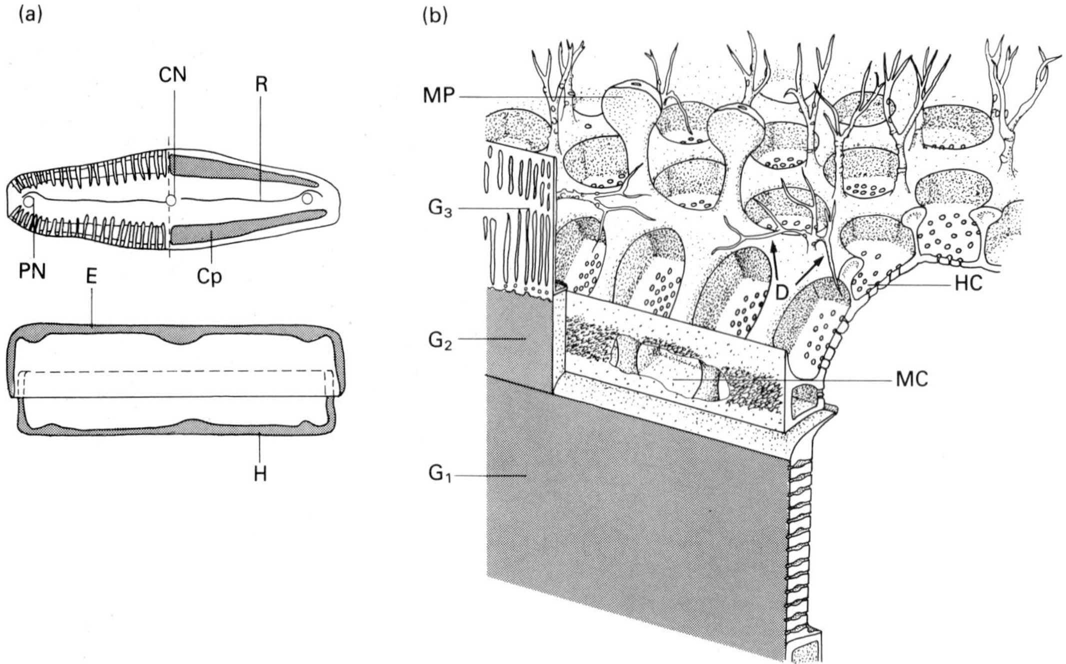

Fig. 3.15 (a) Generalized view of the diatom frustule, from the valve (upper) and girdle (lower) views. CN, central nodule; Cp, chloroplast; E, epitheca; H, hypotheca; PN, polar nodule; R, raphe. (b) Details of frustule structure of the valve/girdle band in *Triceratium favus* (Bacillariophyceae). D, dendritic structure; HC, hexagonal chamber; MC, marginal canal; MP, marginal process; G_{1-3}, girdle bands. Scale = 1 μm. From Miller & Collier (1978) with permission.

diatoms are thus routinely identified from non-living material.

The frustule comprises two almost-equal halves, fitted together like a Petri dish (Fig. 3.15a); the outer, more or less flattened, opposing surfaces are the valves, the larger epivalve and the smaller hypovalve. Joining them together is the girdle (Fig. 3.15b), composed of two overlapping parts, the epicingulum and the hypocingulum. The epivalve and epicingulum comprise the epitheca, the hypovalve and hypocingulum the hypotheca. In more complex diatoms the girdle may be extended by means of additional, intercalary girdle bands. Cells generally have radial (centric; Biddulphiales) or bilateral (pennate; Bacillariales) symmetry in valve view. Hendey (1964) recognized two additional types (Fig. 3.16): trellisoid, with the structure arranged margin to margin (e.g. *Eunotia*), and gonoid, with the structure supported by angles (e.g. *Triceratium*). Many pennate diatoms possess a raphe, a central, S-shaped slit along the length of one or both valves, divided by a central nodule and terminated by polar nodules. These raphid forms are capable of gliding movement, which is effected through the secretion of mucilaginous material derived from cystalloid bodies (Drum & Hopkins, 1966). Pennate diatoms lacking a raphe possess a central, unornamented area known as a pseudoraphe.

In addition to the raphe, a variety of complex ornamentations and perforations of the frustules give the diatom cell its intrinsic beauty and form. Perforations are simple pores, or elaborate loculi, often arranged in lines, or striae. The pores may be strengthened by silicified thickenings called ribs or costae. The loculi are chamber-like (Figs 3.17 & 3.18), and separate the wall into two layers in a honeycomb-like fashion. On one side is a perforated layer, or velum, and on the other an opening, or foramen. The valve surfaces are frequently extended into processes, and the margins into arms or setae. Of great interest are the labiate processes found in centric and pennate

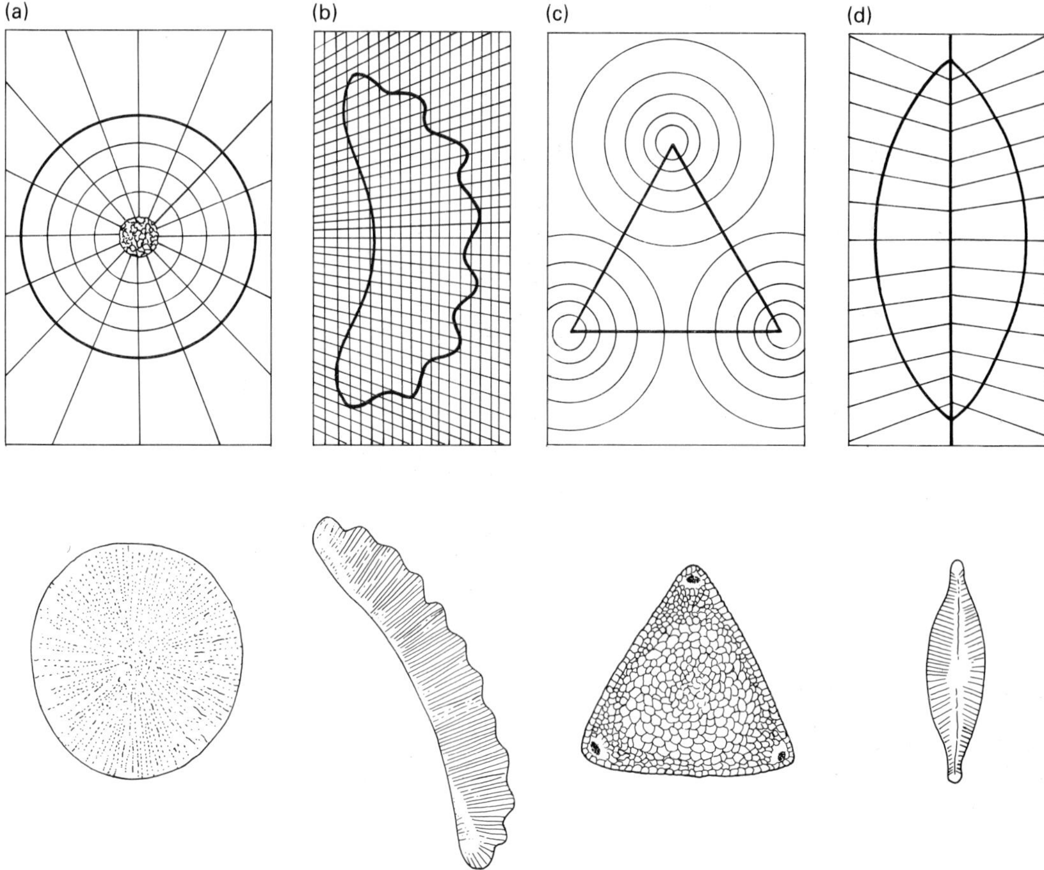

Fig. 3.16 Cell types in the Bacillariophyceae. From Hendey (1964) with permission. For explanation, *see* text.

forms. Hasle (1973) suggested that they may be predecessors of the raphe, since they are absent from pennate species with a well-developed raphe. The various processes greatly increase the surface area/volume ratio of the cell, and may assist in flotation as well as in the attachment of adjacent cells in colonial forms.

The degree of silicification of the frustule is variable (up to 95.6%); only in some rare instances is silica absent (e.g. *Phaeodactylum tricornutum*; Lewin *et al.*, 1958). Chemically, the silica is quartzite ($SiO_2 \cdot nH_2O$), and includes trace elements such as aluminium, magnesium, iron and titanium (Mehta *et al.*, 1961; Lewin, 1962). The 'skin' covering the frustule is of amino acids and sugars (Coombs & Volcani, 1968; Hecky *et al.*, 1973); a variety of extracellular, organic secretions have been described (Werner, 1977). The siliceous material is laid down in regular patterns; vesicles originating from the Golgi apparatus collect beneath the plasmalemma and fuse to form silica deposition vesicles, the silicalemma (Reimann *et al.*, 1966), one for each valve. Rapid silica deposition ensues and the material assumes the morphology characteristic of the particular species. In *Hantzshia amphioxys* the microtubule centre (MC) plays a major organizational role in the development of new valves following mitosis (Pickett-Heaps & Kowalski, 1981). Surprisingly, the new valve arises in the silicalemma along the centre of the cell; it is later moved to its proper position during a major reorganization within the cell, controlled by the MC. Subsequent to valve formation, new Golgi vesicles are formed and fuse to form the girdle band (Dawson, 1973). During silica deposition, the silica is taken up by the cell as silicic acid, through a process of active transport (Lewin

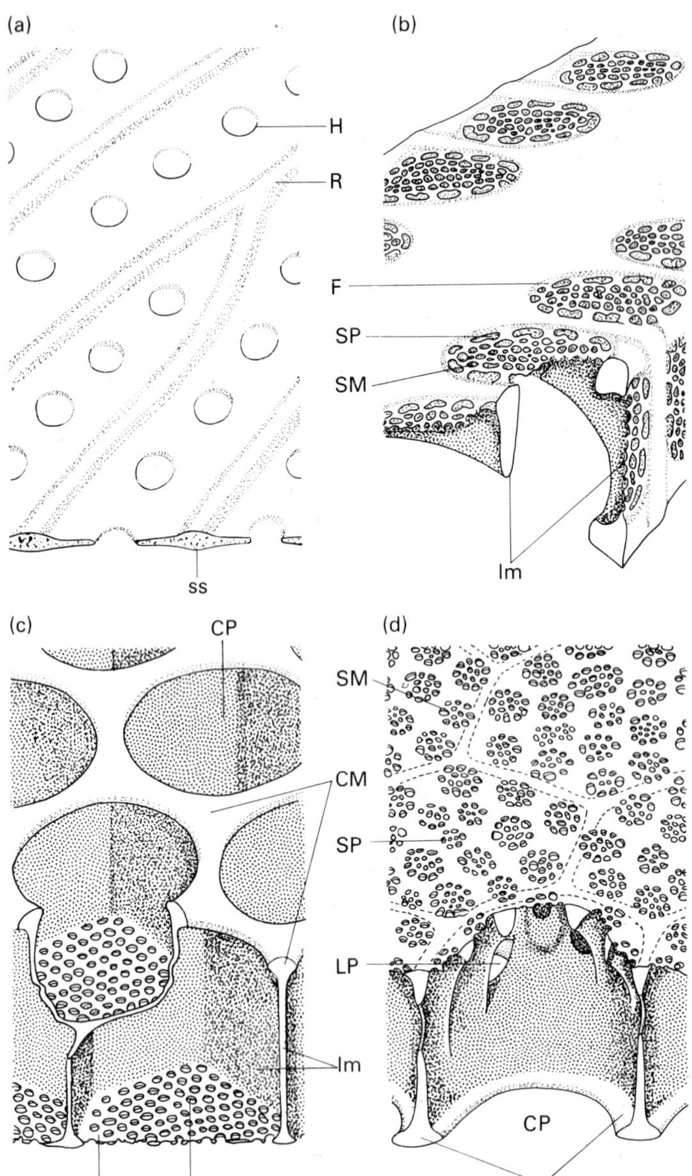

Fig. 3.17 Types of loculi in diatom frustule walls. (a) Hole or pore (*Chaetoceros didymus* var. *anglica*). (b) Incomplete loculus (*Synedra tabulata*). (c) Loculus opening outwards (*Coscinodiscus lineatus*). (d) Loculus opening inwards (*Coscinodiscus wailesii*). CM, cover membrane; CP, cover pore; F, fulcrum; H, hole; LP, lateral pore or pass pore; LM, Lateral membrane; R, rib; SM, sieve membrane; SP, sieve pore; ss, spongy porous structure. From Hendey (1971) with permission.

& Chen, 1968). The process is inhibited in the presence of germanium dioxide (Lewin, 1966).

The cell wall is always formed outside the plasmalemma (as in higher plants). Its structure is highly variable, and can be subdivided into two types: incomplete and complete. The former is characteristic for a variety of flagellates, the latter for all prokaryotic algae and most members of the Chlorophyceae, Xanthophyceae, Phaeophyceae, and Rhodophyceae (Table 3.1).

The incomplete cell wall is usually termed a lorica, a flask- or cup-shaped structure in which sits a flagellate cell. Examples are found in the Prasinophyceae, Chrysophyceae (Fig. 3.19) and Chlorophyceae. In the prasinophyte *Platymonas* the lorica (sometimes incorrectly termed a theca) is composed of polysaccharides; it is formed from stellate particles produced in the Golgi vesicles.

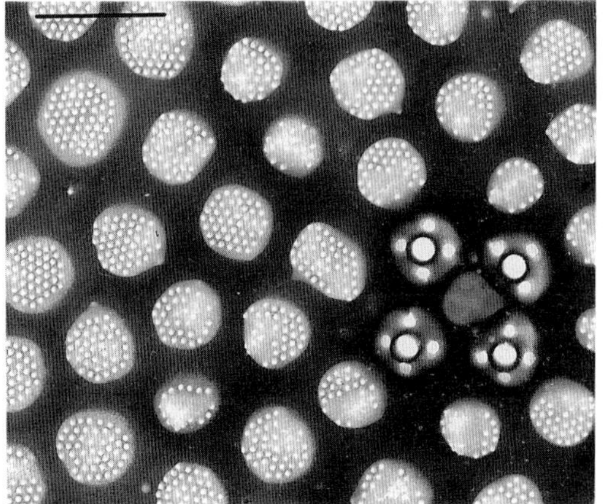

Fig. 3.18 View of the central strutted processes in *Thalassiosira incerta* (Bacillariophyceae). Scale = 1 μm. From Hasle (1978) with permission.

Microfibrillar, cellulosic–pectic loricas are common in the chrysophytes, and their formation is probably *in situ*, outside the cell (Dodge, 1973). Silica impregnataion may occur, and in some genera calcification. The chrysophyte lorica is highly variable in form, and of taxonomic importance. In the Chlorophyceae the lorica is characteristic for members of the Phacotaceae (Volvocales). Cells with a structure comparable with that of other Volvocales are contained in an iron-impregnated lorica. In *Dysmorphococcus globosus* the immature lorica includes fibrillar (outer) and granular (inner) components. During early development sequential biomineralization with iron occurs, and pores develop. Ultimately, secretion of mucilaginous materials occurs through the pores, which become obscured (Porcella & Walne, 1980).

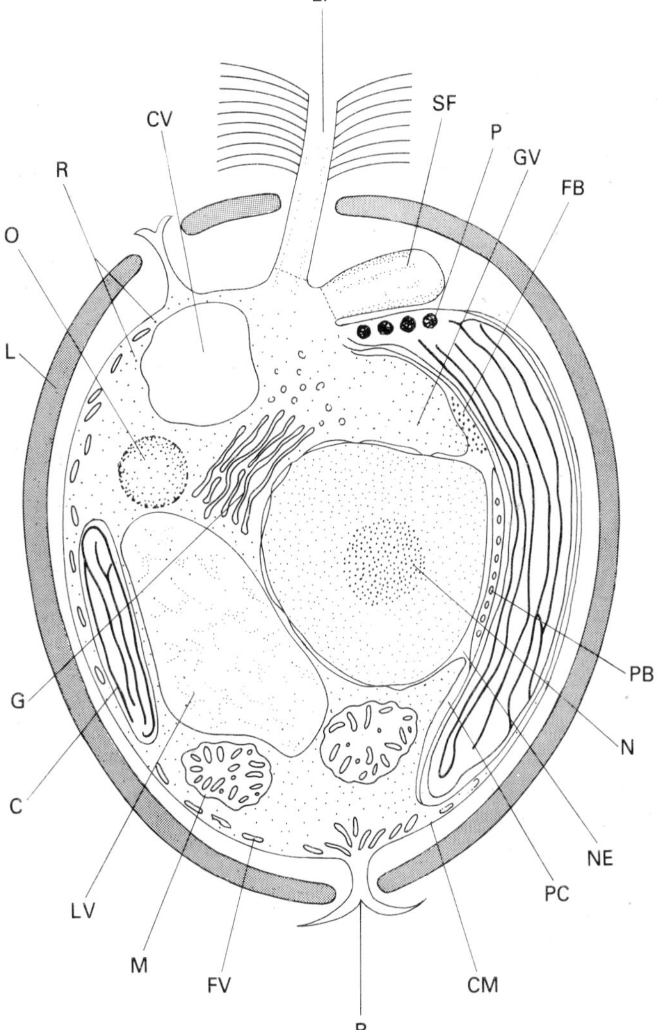

Fig. 3.19 *Chrysococcus rufescens*, a loricate member of the Chrysophyceae shown in schematic section. B, branched cytoplasmic process; C, chloroplast; CM, cell membrane; CV, contractile vacuole; FB, fibrillar bundles; FV, flattened vesicles; G, Golgi body; Golgi vesicles; LF, long flagellum; LV, leucosin vesicles; L, lorica; M, mitochondrion; N, nucleus; NE, nuclear envelope; O, oil globule; P, pigment chambers; PC, periplastidial cisternae; PB, periplastidial reticulum (ER); R, ribosomes; SF, short flagellum. From Belcher (1969) with permission.

The cell walls of the Cyanophycota and Prochlorophycota possess prokaryotic features. In the Cyanophycota the wall, or envelope, possesses many similarities with that of Gram-negative bacteria (Fig. 3.20; Allen, 1968). It lies between the plasmalemma and a variety of investments, referred to as a mucilaginous sheath, capsule, or 'shroud'. In earlier accounts the envelope was described as four-layered, the layers designated as LI, LII, LIII, and LIV (inner to outer). LI and LIII are electron-transparent, and LII and LIV electron-dense. The electron-transparent layers are now designated as spaces between the 'LII' and 'LIV'; any structure previously ascribed to them is attributed to fixation artifacts (Drews & Weckesser, 1982). The innermost layer (LII) contains the major constituent of peptidoglycan (up to 50% dry wt), comprising murein, glycopeptide, and mucopeptide. It is separated from the plasmalemma by an electron-transparent space (LI). The outer cell membrane (LIV) appears as a double-track structure in ultrathin EM preparations (Butler & Allsopp, 1972), and is separated from the peptidoglycan layer by an electron-transparent space (LIII).

Delicate plasmodesmata may cross the transverse wall in filamentous forms. The sheath serves to prevent desiccation and is important in gliding in many filamentous examples. While its chemistry is inadequately known, it is probably composed of pectic acids and mucopolysaccharides. Martin and Wyatt (1974) have suggested three different types: a sheath, visible without staining; a mucilaginous 'shroud' with indefinite boundaries; and a similar 'shroud' with well-defined limits. The sheath is often pigmented, and its thickness varies according to environmental and growth conditions.

The cell wall of the Prochlorophycota is comparable with that of the Cyanophycota (Whatley, 1977), and muramic acid is present (Moriarty, 1979).

True cell walls, in many respects comparable to those of higher plants, are found in the Rhodophyceae, Phaeophyceae, Xanthophyceae, Chlorophyceae, and Charophyceae; in addition, a true wall surrounds the cyst of some Dinophyceae and the vegetative cells of members of the Eustigmatophyceae. Generally, true cell walls are made up of two components, a microfibrillar framework (Fig. 3.21) embedded in an amorphous, mucilaginous material. Incrusting sub-

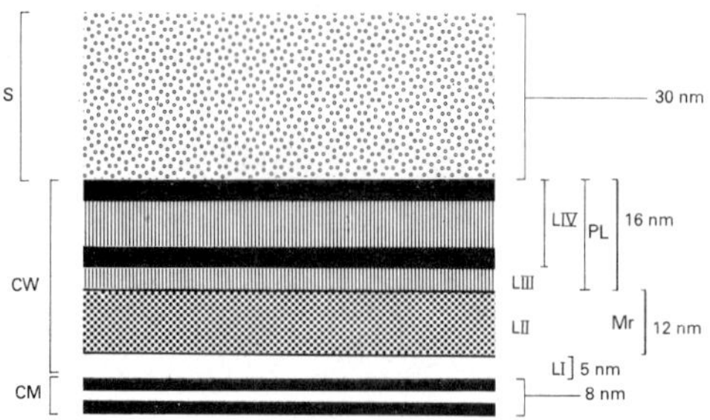

Fig. 3.20 Organization of the cell coverings of blue-green algae (Cyanophyceae), based on electron microscopy. CM, plasmalemma; CW, cell wall; S, sheath or slime layer; Mr, murein; PL, plastic layers; cell wall layers LI–IV. (Based on Carr & Whitton, after Jost.)

Fig. 3.21 The wall of *Valonia* (Chlorophyceae) showing two successive layers of fibrils oriented at almost 90° to each other (× 14000). From Dodge (1973) with permission.

stances such as silica, calcium carbonate, or sporopollenin may also be present. The chemistry and formation of cell walls have been reviewed by Siegel and Siegel (1973) and MacKie and Preston (1974). With the exception of selected Rhodophyceae and Phaeophyceae, especially those whose walls yield commercially important substances, and a number of Chlorophyceae (*see* Pickett-Heaps, 1975), the large majority of algal cell walls have not been critically studied. It is generally accepted, however, that as in the higher plants the Golgi bodies are involved in the production of microfibrillar wall material. Materials are assembled in Golgi vesicles beneath the plasmalemma and then passed to the exterior where they are synthesized into the microfibrillar components. Electron microscopy (especially the use of freeze–fracture methods) has added a great deal to the understanding of the microarchitecture of the wall; the precise mechanism by which the orientation of the microfibrils is determined is still, however, poorly understood. The assembly and synthesis of the amorphous wall materials may also involve the Golgi bodies, as their formation is often associated with increased Golgi activity. In the case of sporopollenin, however, the nuclear envelope has been suggested as the site of synthesis (Atkinson *et al.*, 1972). The accumulation of incrusting substances has been widely studied, particularly the process of calcification in the Rhodophyceae and Chlorophyceae (*see* Chapter 6). The process is highly significant ecologically, especially in tropical environments. Also, the preservation of calcified and silicified forms has proved important as a means of preserving good fossil records in those algae possessing this feature.

In view of the variation among true cell walls, the principal classes are examined separately.

RHODOPHYCEAE

In the Bangiophycidae the wall of the macroscopic phase (e.g. *Bangia; Porphyra*) is composed of a microfibrillar framework of β-1, 3-linked xylan; recent reports (Gretz *et al.*, 1980; Mukai *et al.*, 1981) have shown that the conchocelis phase of these algae possesses cellulose as a wall component. An outer cuticle of β-1, 4-linked mannan (Frei & Preston, 1964) or protein (Hanic & Craigie, 1969) has also been reported in some bangiophytes. In the Florideophycidae the wall framework is composed of cellulose, with the fibrils randomly arranged. The amorphous mucilages which make up the rest of the wall (up to 70%; Dixon, 1973) are usually composed of water-soluble sulphated galactans (Percival & McDowell, 1967). Among the mucilages are the commercially important agars and carrageenans, which are capable of forming gels under certain conditions. All members of the Corallinaceae (Cryptonemiales) and some Nemaliales (e.g. *Liagora; Galaxaura*) deposit calcium carbonate (as calcite in the Corallinaceae, and as aragonite in the Nemaliales). Magnesium and strontium carbonates are commonly deposited in smaller quantities along with the calcium carbonate. The calcification process has been linked to photosynthesis (Pearse, 1972), although it may occur in the dark (Okazaki *et al.*, 1970). There are several theories as to the biochemistry of the process (*see* Digby 1977a, b; Lee, 1980).

The pit plug (= pit connection, Fig. 3.22; Pueschel, 1980) is a wall feature of the Florideophycidae and the conchocelis phase of *Bangia* and *Porphyra* (Bangiophycidae). It has been reported from one marine chlorophyte (Brawley & Sears, 1980). In the rhodophytes primary pit plugs are formed between two cells during division, and secondary pit plugs develop between laterally adjacent cells. The plugs are usually biconcave, with a central constriction (Pueshel, 1977). Formation occurs through incomplete cytokinesis, with an accumulation of endoplasmic reticulum in the aperture between daughter cells, followed by accumulation of material to form the plug (Norris & Kugrens, 1982). The plug matrix contains protein (Ramus, 1971; Pueschel, 1980), and possibly lipids and polysaccharides. Pit plugs may provide structural strength (Kugrens & West, 1973; Pueschel, 1980), intercellular transport (Wetherbee, 1979) or parasite–host connections (Goff, 1979). In post-fertilization development of the gonimoblast, intercellular transport of metabolites is facilitated by dislodgement of the pit plugs to form fusion cells (Turner & Evans, 1978).

PHAEOPHYCEAE

Cell walls are at least two-layered, with an inner cellulosic, microfibrillar skeleton. The microfibrils are most commonly arranged randomly, although they are sometimes arranged longitudi-

(a)

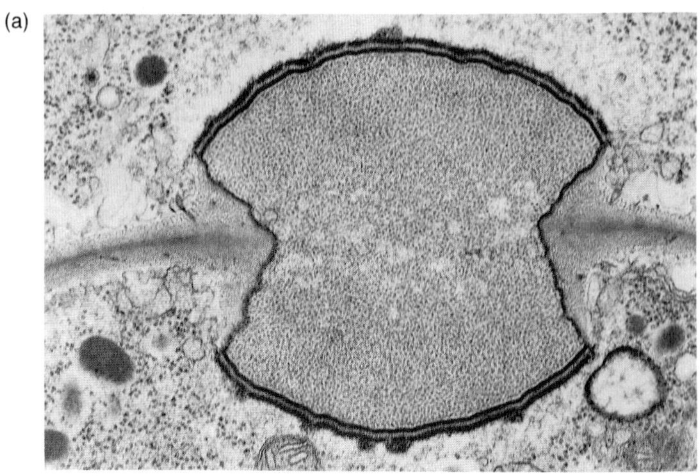

(b)

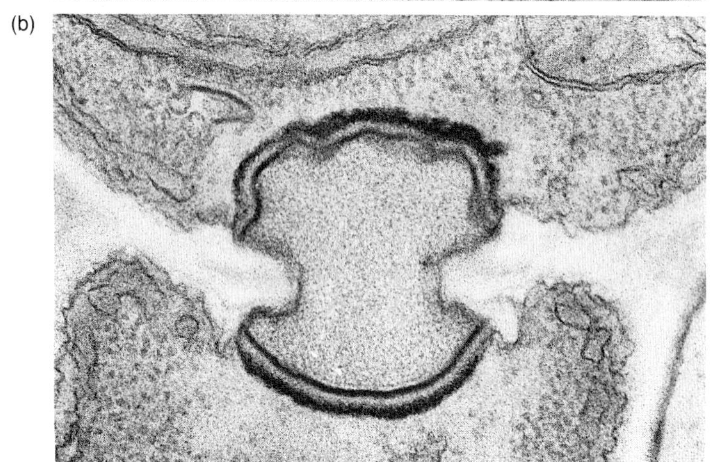

Fig. 3.22 Pit plug of (a) *Galaxaura* sp. (× 30400) and (b) *Pseudogloiophloea* sp. (Rhodophyceae) (× 64000). (EM micrographs courtesy of Dr C.M. Pueschel.)

nally or horizontally (Dodge, 1973). The amorphous components are principally alginic acid and fucoidan (sulphated polysaccharides; Mackie & Preston, 1974). The mucilage and cuticle are composed primarily of alginic acid, a polymer of five-carbon acids (D-mannuronic acid and L-guluronic acid) in variable proportions (Haug *et al.*, 1969). The alginic acid may constitute more than 30% of the dry weight of the alga. Alginates (salts of alginic acid) have valuable emulsifying and stabilizing properties, and are extensively used commercially (Chapter 9). The biosynthesis of alginic acid has been reported in Larsen (1981) and Percival (1979) and is reviewed in McCandless (1981).

Parenchymatous Phaeophyceae possess plasmodesmata or pores between most of their cells (Fig. 3.23), and in the Laminariales, Fucales and some Dictyotales the pores are grouped into primary pit areas. Sieve elements (Fig. 3.24) comparable to those of the higher plants occur in the Laminariales, where they facilitate rapid translocation of photoassimilates (Schmitz, 1981).

XANTHOPHYCEAE

Wall composition in selected Xanthophyceae (*Tribonema aequale;* Cleare & Percival, 1972: *Vaucheria* spp.; Parker *et al.*, 1963) is of randomly arranged microfibrillar cellulose (up to 90% in *Vaucheria*). In filamentous forms the wall comprises two overlapping halves, a feature not evident under the light microscope unless the cells are treated with a reagent such as concentrated potassium hydroxide. In *Ophiocytium majus* (Hibberd & Leedale, 1971) the halves consist of

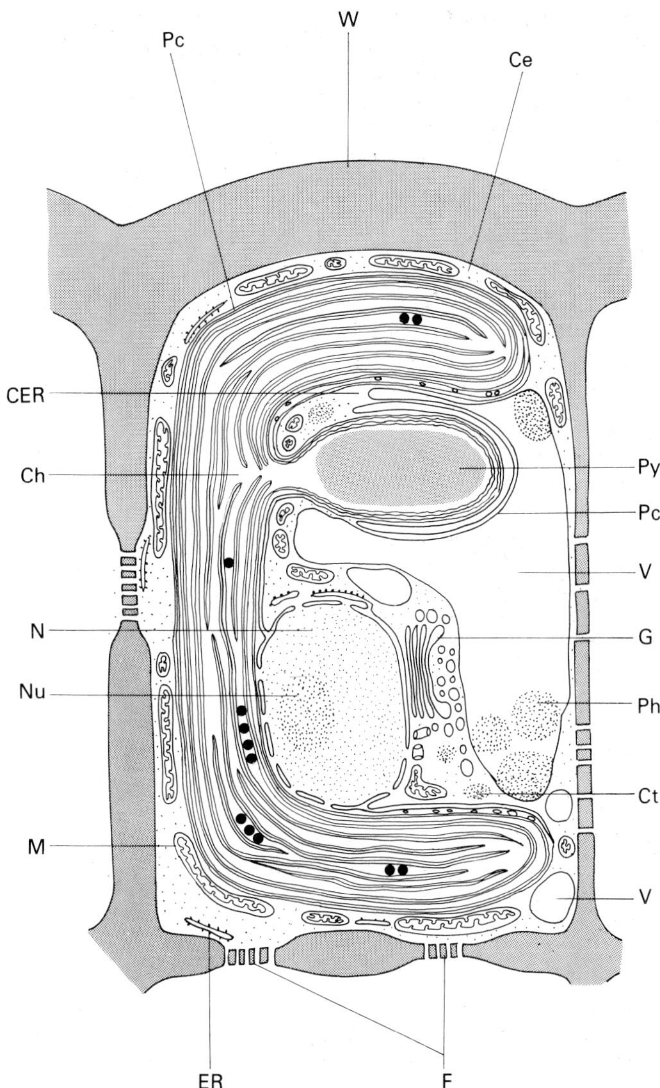

Fig. 3.23 Generalized diagram showing the ultrastructural features of the cell of *Scytosiphon* (Phaeophyceae). Ce, chloroplast envelope; CER, chloroplast endoplasmic reticulum; Ch, chloroplast; Ct, centriole; ER, rough endoplasmic reticulum; F, plasmodesma pit field; G, Golgi dictyosome; M, mitochondrion; N, nucleus; Nu, nucleolus; Pc, pyrenoid cap; Ph, physode; Py, pyrenoid; V, vacuole; W, wall. From Claytŏn & Beakes (1983) with permission.

a cap of fixed size fitted over a tubular basal portion which elongates during growth. In the filamentous *Tribonema* the walls are composed of H-shaped pieces of more or less equal size that alternately overlap one another.

CHLOROPHYCEAE

Although relatively few chlorophycean walls have been critically studied, there is great variation in structure and composition within the class. In many the wall is two-layered, with a firmer inner layer and an outer capsule or mucilaginous layer. Dodge (1973) listed three principal categories; those with a wall consisting mainly of cellulosic microfibrils (some Cladophorales, Siphonocladales); those with less distinct organization of cellulosic microfibrils and proportionately much more amorphous material (some Ulvales; Oedogoniales; many desmids and coccoid species); those with walls not based on cellulosic microfibrils (many siphonous species and some Volvocales). In the siphonous forms walls are composed of polymers of xylose (e.g. *Caulerpa*; Mackie & Percival, 1959) or mannose (*Acetabularia; Codium*; Frei & Preston, 1961) microfibrils embedded

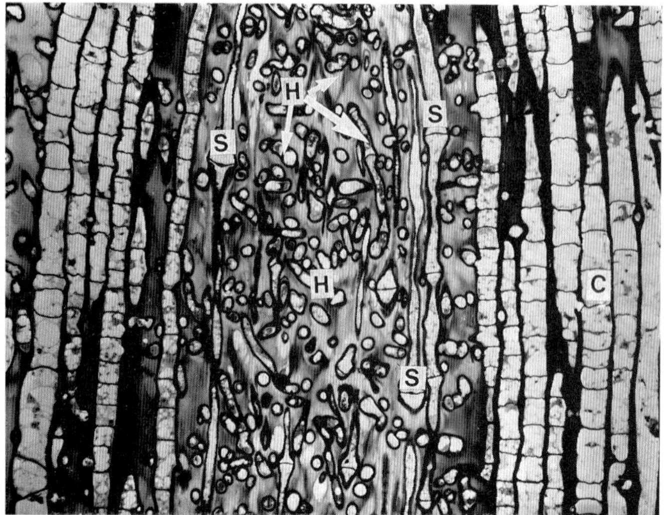

Fig. 3.24 Longitudinal section of *Alaria marginata* (Phaeophyceae). C, cortical cells; internal medulla of sieve (S) and hyphal (H) cells (× 170). From Schmitz & Srivastava (1975) with permission.

in a matrix which may consist principally of hemicelluloses. Many siphonous marine species have calcified walls; the process of calcification and deposition of calcium carbonate in rhizophytic algae such as *Halimeda* is a significant building process in shallow soft bottoms in tropical communities. The presence of chitin has been confirmed for the walls of *Pithophora oedogoniana* (Cladophorales) (Pearlmutter & Lembi, 1978). The walls of members of the Oedogoniales are unique in that they break in the upper region during cell division, leaving 'scars' or 'rings' (apical caps) (Hill & Machlis, 1968; Pickett-Heaps, 1975). The process is summarized in Fig. 3.25.

The wall may consist of up to six or seven layers (e.g. *Chlamydomonas reinhardtii*; Roberts *et al*, 1972). In *Valonia* it is rendered remarkably strong through alternate layers with the microfibrils arranged at right angles to one another (Fig. 3.21; cf. Roelofsen, 1966).

Remarkably, some green algae show a biochemical alternation of generations with respect to wall composition (cf. McCandless, 1981). The best reported example is *Bryopsis*, in which the walls of the gametophytes are composed of xylan and glucan, and those of the sporophyte are composed of mannans (Huizing & Rietema, 1975).

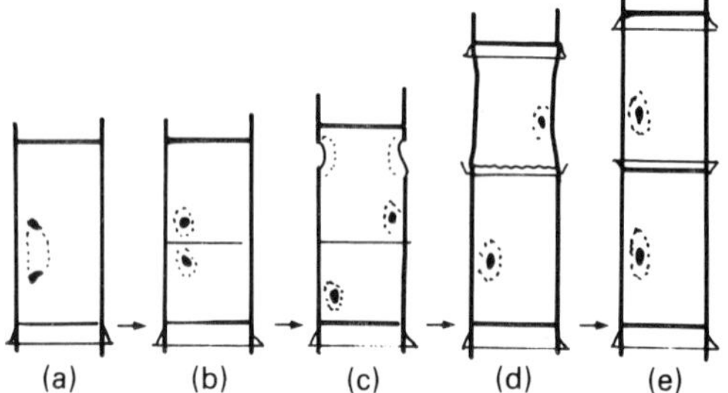

Fig. 3.25 Diagrammatic summary of cell division in *Oedogonium* (Chlorophyceae) demonstrating the splitting of the wall and formation of wall rings. (a) Telophase of mitosis and splitting of wall ring. (b) Cytokinesis. (c, d, and e) Rupture and splitting of wall ring as the protoplasts expand. (After Pickett-Heaps, 1975.)

CHAROPHYCEAE

Walls of the Charophyceae are composed of cellulosic microfibrils, and most are heavily calcified, hence the common name 'stoneworts'. In *Nitella* rhizoids (Chen, 1980) the microfibrils are aligned predominantly parallel to the cell's axis, whereas in internodal cells the alignment is transverse. Charasomes, complex membrane structures, have been described from the inner longitudinal walls of *Chara*; their function is uncertain, but is apparently not linked with bicarbonate (HCO_3^-) transport (Franceschi & Lucas, 1980).

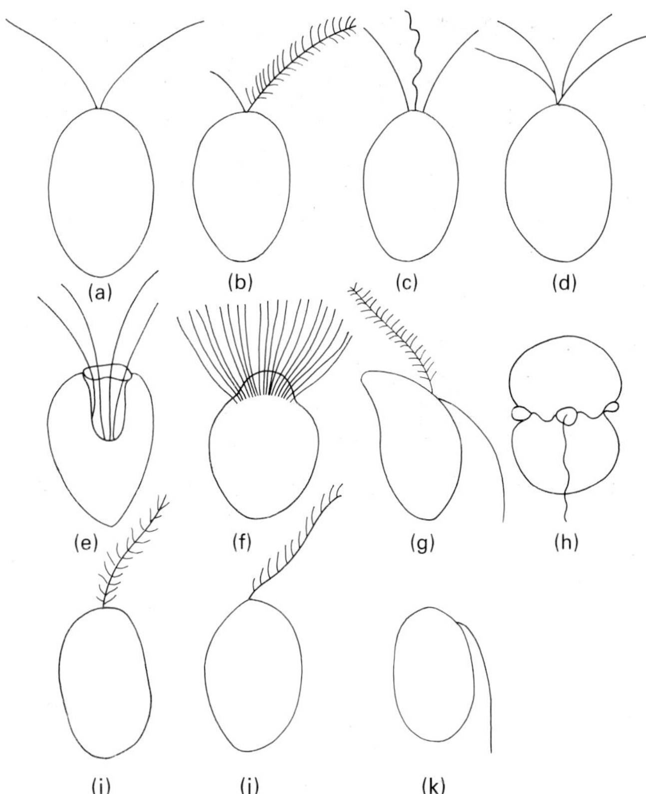

Fig. 3.26 Selected flagellar types and arrangements in the algae. (a) Isokont, homodynamic, smooth, apically inserted (e.g. *Chlamydomonas*). (b) Heterokont, unequal, apically inserted (e.g. *Tribonema*. Xanthophyceae). (c) Isokont, homodynamic, apically inserted, with a central haptonema (e.g. *Chrysochromulina*, Prymnesiophyceae). (d) Isokont, homodynamic, apically inserted. (e) Isokont (with scales, not shown), originating from a flagellar pit (e.g. *Pyramimonas*, Prasinophyceae). (f) Multiflagellate, or stephanokont (flagella inserted in a subapical ring) (e.g. *Oedogonium*, Chlorophyceae). (g) Heterokont, heterodynamic, laterally inserted (e.g. *Ectocarpus*, Phaeophyceae). (h) Transverse and trailing, heterokont flagella (hairs not shown), basally inserted (e.g. *Gymnodinium*, Dinopohyceae). (i) Single, apical, bilaterally hairy. (j) Single apical, unilaterally hairy. (k) Single, subapically inserted.

Flagella

Flagella occur in the majority of algal divisions. These locomotory structures are of fundamental phyletic importance; the electron microscope has revolutionized our knowledge of their structure and function. A comprehensive review of algal flagella is given by Moestrup (1982).

While they possess remarkably uniform internal microtubular structure, flagella are highly variable in external morphology, mode of insertion, and root systems. The evolutionary origins of flagella are still conjectural (*see* Margulis, 1981). Cavalier-Smith (1978) contends that the flagellar microtubules or their precursors arose with the nucleus, while Stewart and Mattox (1980) argue that the flagella came first, and that the nuclear spindle has evolved as an elaboration of the flagellar apparatus or its precursor. Flagella must be considered, nonetheless, as integral parts of the cell; their membrane is continuous with that of the cell plasmalemma, and their external structures are derived from the cell that bears them. Their role in cell motility is undisputed, yet only now are cell biologists coming to grips with the complex questions concerning the relationship between flagellar morphology and function.

Flagellar number, insertion, and organization may be fairly consistent within a division; a summary of common arrangements is given in Fig. 3.26. Motile algal cells are typically biflagellate, although in green algae quadriflagellate types are common; it is generally believed that the latter have been derived from the former (cf. Melkonian, 1982a). Uniflagellate forms are probably descended from biflagellate forms, while the unique triflagellate zoospores of the green alga *Entocladia wittrockii* may have been derived from a quadriflagellate ancestor. Multiflagellate types (stephanokont) exhibit a variety of complex modifications of the flagellar basal system or the internal structure of the flagellum (*see* Moestrup [1982] for discussion).

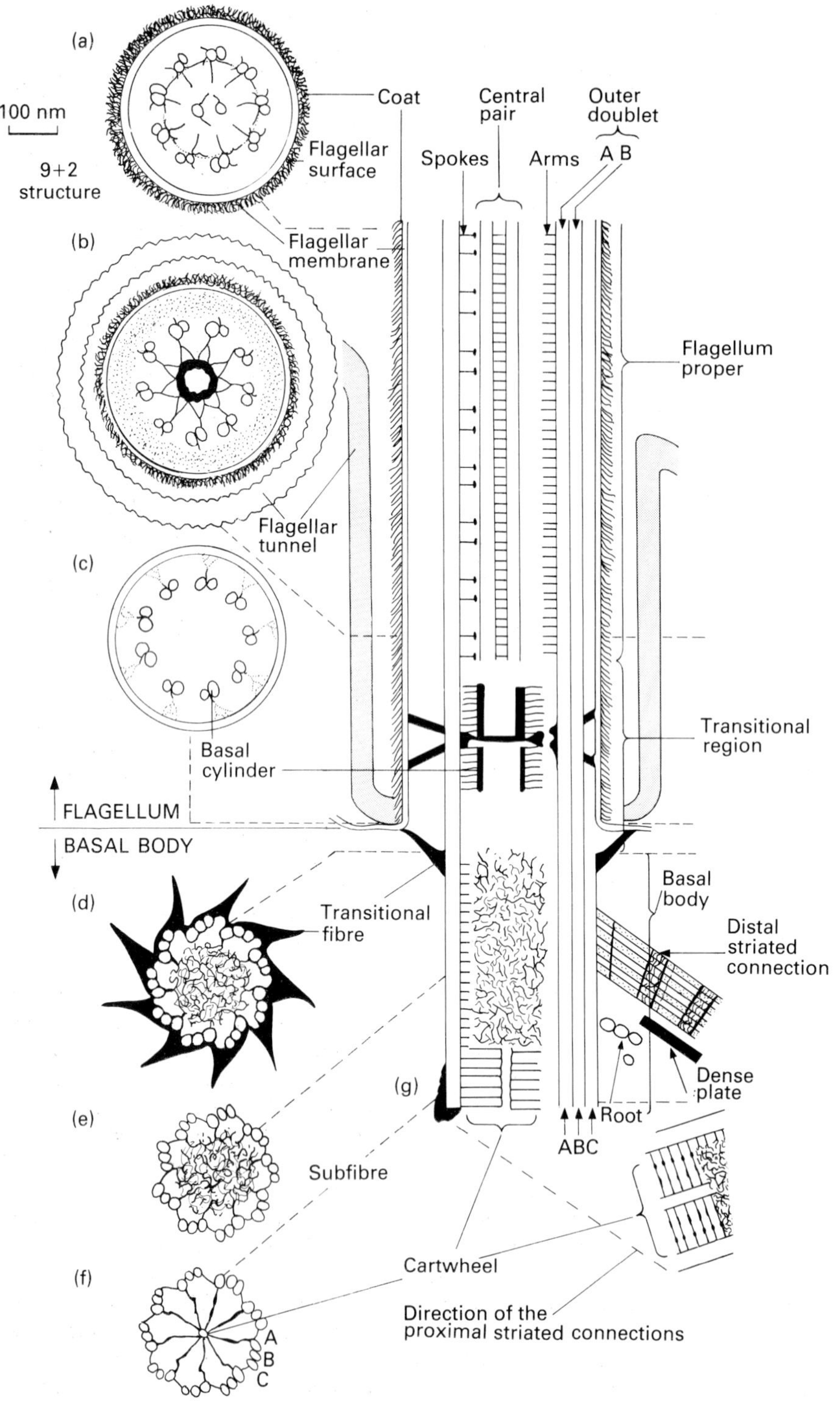

Fig. 3.27 Diagrammatic reconstruction of the flagellar axoneme, transition region, basal body, and elements of the root connecting fibres, etc. of *Chlamydomonas reinhardtii* (Chlorophyceae). From Cavalier-Smith T. (1974) with permission.

Where flagella are of equal length and appearance they are described as isokont; heterokont examples have dissimilar flagella (e.g. smooth and hairy) which may be of equal or unequal length (Moestrup, 1982). Where similar flagellar types of different lengths occur, the arrangement is designated anisokont. Flagellar movement may be co-ordinated (homodynamic) or independent (heterodynamic) on the same cell. Aberrant or unusual flagella occur in the classes Bacillariophyceae, Cryptophyceae, some Phaeophyceae, Dinophyceae, Euglenophyceae, and Prymnesiophyceae (Pavlovales).

True flagella consist of an external axoneme contained within the plasmalemma and comprising nine doublet microtubules surrounding two central microtubules (9 + 2) (Fig. 3.27). Modifications of the 9 + 2 microtubular architecture are rare, but include the spermatozoid of some diatoms (9 + 0) and the green alga *Golenkinia minutissima* (9 + 1; Moestrup, 1972), as well as the haptonema of the Prymnesiophyceae (Fig. 3.28) (seven microtubules surrounded by three concentric membranes; Manton, 1964a) and the 'pseudocilia' of tetrasporalean Chlorophyceae (9 singles + 0, e.g. *Tetraspora;* Lembi & Walne, 1971).

The morphology of the flagellar tip has become a matter of some phylogenetic importance. While it is likely that its appearance changes during motion, due to sliding of the internal microtubules (Melkonian, 1982b), it has been determined that in many examples the peripheral doublets terminate below the tip, with the central microtubules extending beyond (Fig. 3.29). In *Friedmannia israelensis* Melkonian (1982b) described the microtubule configuration towards the flagellar tip as 9 + 2 > 4 + 2 > 2 + 2 > 0 + 2. This attenuation can be referred to as a hairpoint, and consists of the extension of the central pair of microtubules (cf. Jonsson & Chesnoy, 1974) or of a single microtubule.

Hairpoints (Fig. 3.30) are common in Prymnesiophyceae, Chlorophyceae, and Prasinophyceae (Loxophyceae *sensu* Moestrup, 1982). Tips are normally blunt, of the 9 + 2 configuration, in Cryptophyceae, Dinophyceae, Euglenophyceae, Prasinophyceae (except the Loxophyceae *sensu* Moestrup, 1982), and Charophyceae. In heterokont groups the hairy flagellum is always blunt, except in the Phaeophyceae. In the latter the hairy flagellum frequently bears an extremely

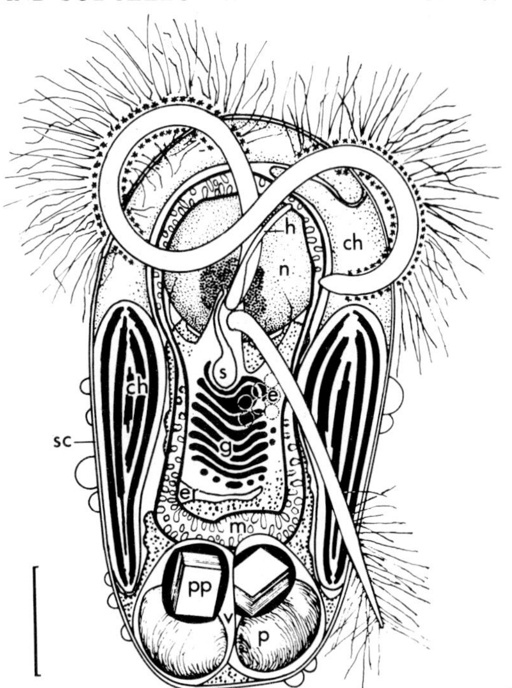

Fig. 3.28 *Pavlova calceolata* (Prymnesiophyceae). A semi-diagrammatic drawing with the flagellar appendages shown only where they do not interfere with other cell features. The ventral lobes of the chloroplast are cut away to show internal lamellae. h, haptonema; ch, chloroplast; e, eyespot; g, Golgi; s, pit; sc, superficial cisternae; m, mitochondrion; n, nucleus; v, vacuole containing polyphosphate (PP) and paramylon (p). Scale = 1 μm. From van der Veer (1976) with permission.

fine hairpoint, while the smooth flagellum possesses a two-stranded hairpoint. It has been speculated that the hairpoint of the male gamete (Fig. 3.31), serves as a mating structure (Müller & Falk, 1973), while in zoospores, where it may act as the first contact point with the substrate, it possesses adhesive qualities (cf. *Chorda tomentosa*; Toth, 1976).

The external axoneme grows through elongation; the process has been described in *Chlamydomonas* spp. and *Ochromonas danica* (Rosenbaum & Child, 1967; Rosenbaum, *et al*, 1969). New material is apparently added distally to the nine peripheral pairs of microtubules, while the central ones, which are attached to the terminal membrane of the flagellum, are extended by the proximal addition of new units.

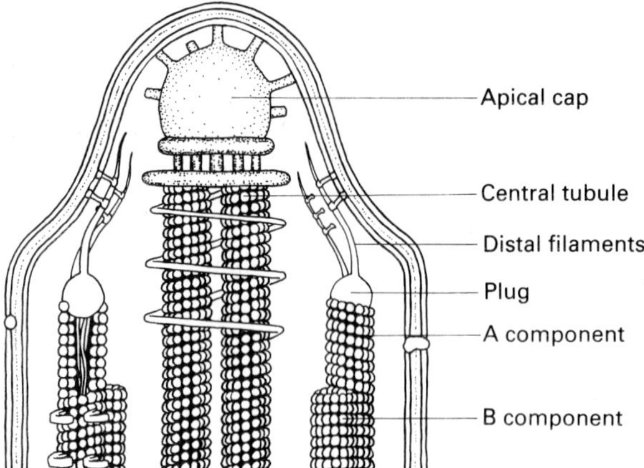

Fig. 3.29 Reconstruction of the flagellar tip, based on negative stain and thin section preparations of *Chlamydomonas* and *Tetrahymena* flagella. (After Dentler; *see* Dillon, 1981).

Fig. 3.30 Shadow-cast EM preparation of the spermatozoid of *Golenkinia minutissima*, showing two hairpoint flagella (× 5250). From Moestrup (1982) with permission.

Little is known about the formation and behaviour of the flagellar membrane. In many algae it is smooth, although a variety of ornamentations occur, including hairs and scales. The structure and formation of these ornamentations may or may not be parallel to that of similar ornamentations found on the cell body. They add to the diameter of the flagellum, but their precise function with respect to flagellar dynamics is uncertain.

Flagellar scales are confined to the algae and a few non-photosynthetic organisms that are perhaps related to algae (Moestrup, 1982). They occur in the Prasinophyceae, Charophyceae, and, rarely, in the Cryptophyceae and Dinophyceae. Scale formation is normally associated with the Golgi cisternae and Golgi vesicles (Fig. 3.32), although in *Rickertia sagittifera* (Moestrup, 1982) they are produced in vesicles located on the surface of mitochondria. Flagellar scales are deposited from the scale reservoir in Prasinophyceae. Several different types of scale may occur on the same flagellum, and they may be arranged in layers (Fig. 3.33a, b).

There is considerable confusion regarding flagellar hairs, and the terminology which predates the extensive use of EM in their study should be discarded (Manton, 1965). They are found in all chromophyte and chlorophyte classes, although in most Prymnesiophyceae and Chlorophyceae the flagella are generally naked. The function of flagellar hairs is not clearly understood, although it has been speculated that by increasing the diameter of the flagellum they amplify the stroke (Moestrup, 1982). Recent studies have sug-

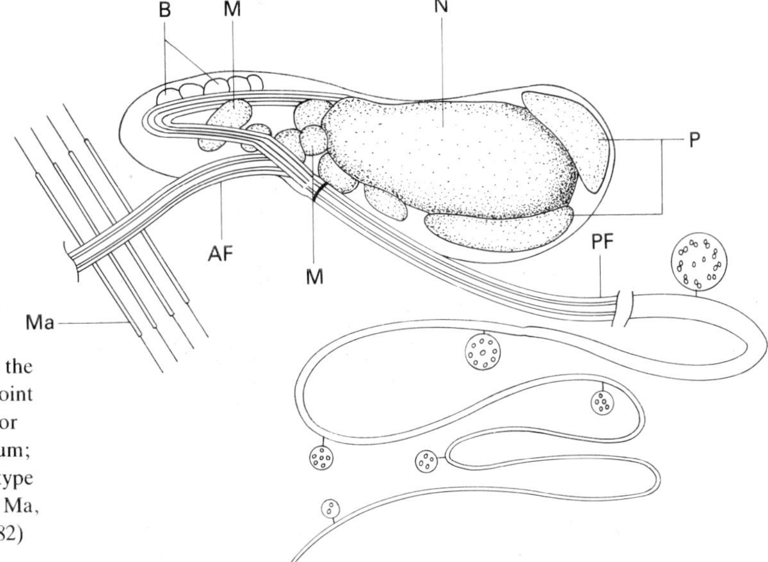

Fig. 3.31 Diagrammatic reconstruction of a generalized laminarialean (Phaeophyceae) sperm showing arrangement of the major organelles and the hairpoint posterior flagellum. AF, anterior flagellum; PF, posterior flagellum; N, nucleus; P, chloroplast; B, type 'B' vesicle; M, mitochondrion; Ma, hairs. From Henry & Cole (1982) with permission.

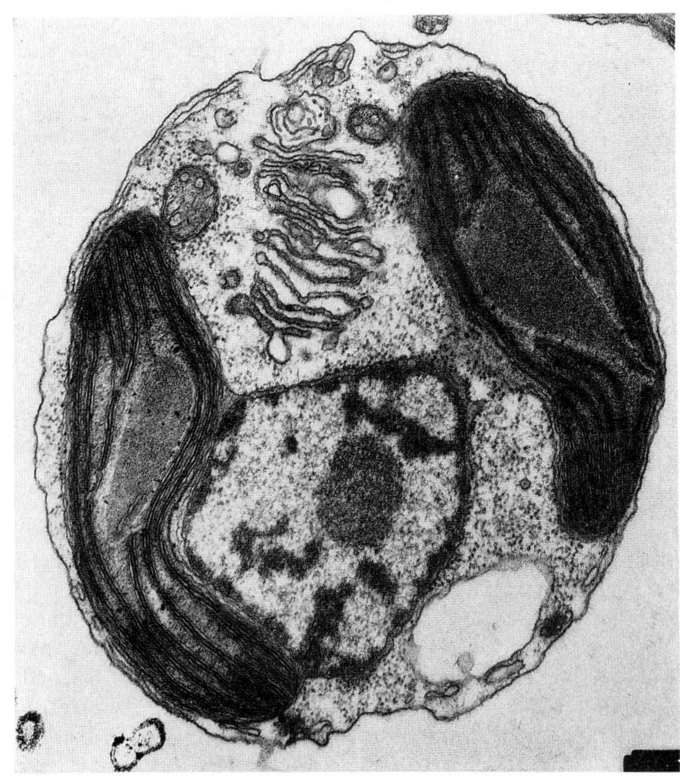

Fig. 3.32 Electron micrograph of *Chrysochromulina* (Prymnesiophyceae) showing the large scale-producing Golgi body positioned above the nucleus (× 37 100). (EM micrograph courtesy of Dr J.D. Dodge.)

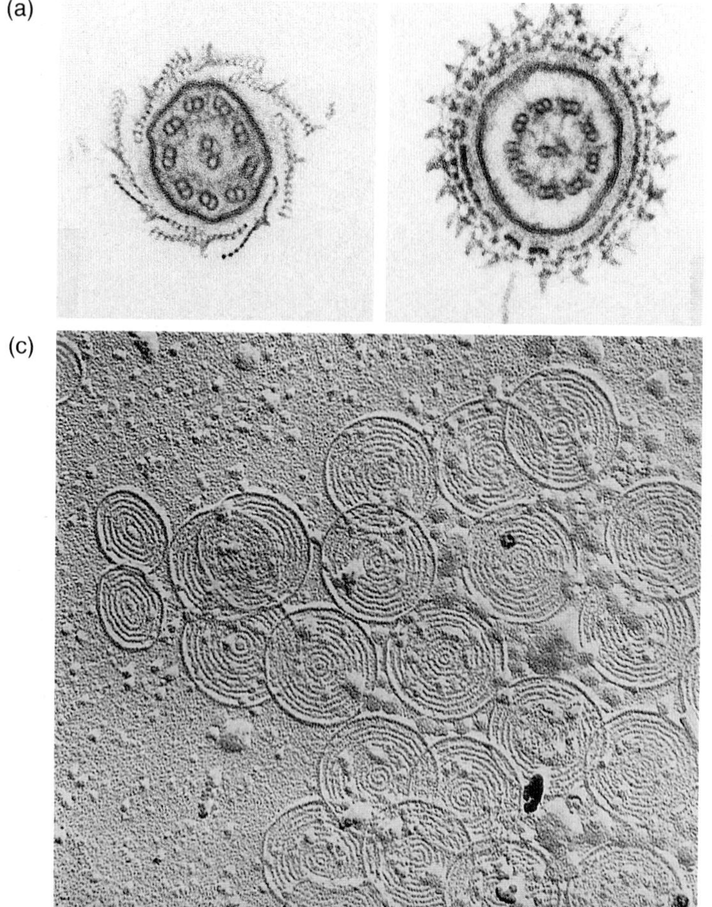

Fig. 3.33 (a) Scales of *Mamiella (Nephroselmis) gilva* (Prasinophyceae) arranged in eight rows in a single layer covering the flagellum (× 75000). (b) Scales of *Nephroselmis olivacea* arranged in three layers, in addition to the presence of thick hairs (× 60000) (c) Flagellar scales of *Dolichomastix nummulifera* (Prasinophyceae) (× 40000). From (a) Moestrup (1982), (b) Moestrup & Ettl (1979), and (c) Manton (1977) with permission.

gested two origins for flagellar hairs: the Golgi apparatus (e.g. Prasinophyceae; Moestrup & Thomsen, 1974; Melkonian, 1982c) and the endoplasmic reticulum (including the perinuclear space) (Moestrup, 1982). Moestrup (1982) grouped them into two types: tubular hairs and simple (non-tubular) hairs. The former he subdivided into cryptophycean hairs, tripartite hairs, and prasinophycean hairs.

The simple hairs are thin and very delicate and probably consist of a single row of subunits. Tubular hairs consist of two or more regions, with the proximal portion usually thick and tubular, and the distal portions resembling simple hairs.

Cryptophycean tubular hairs (Fig. 3.34) are arranged in two opposite rows on the longer flagellum, and in one row on the shorter. Unlike tripartite hairs, they lack a sharp distinction between a base and shaft. Only the chlorophyll *b*-containing Prasinophyceae and certain Charophyceae possess tubular hairs on more than one flagellum. Tripartite (heterokont) hairs are found in the Chrysophyceae, Bacillariophyceae, Phaeophyceae (Fig. 3.35), Eustigmatophyceae, Raphidophyceae, and Xanthophyceae; they are also known from some of the lower fungi (Oomycetes and Hyphochytridiomycetes; Moestrup, 1982). They are attached to the flagellum in two opposite ranks, and probably in the plane of the central tubules. Bouck (1969) illustrated three zones to tripartite hairs: the basal part; the hollow shaft; and the distal parts, which are very delicate and often lost in EM preparations. Prasinophyte hairs appear to have a diverse morphology, and more than one type may occur on the same

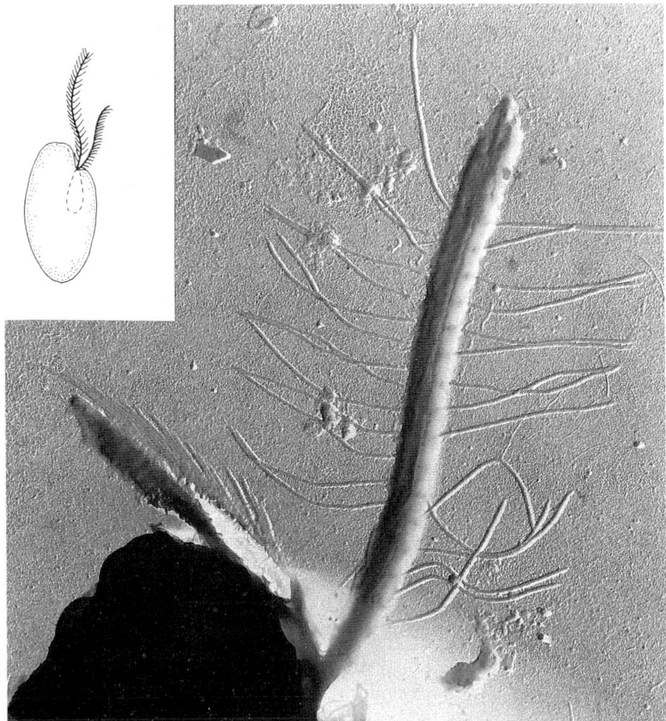

Fig. 3.34 Tubular flagellar hairs of a cryptophyte (insert × 750) arranged in two opposite rows on the longer flagellum and one row on the shorter flagellum (× 14625). From Moestrup (1982) with permission.

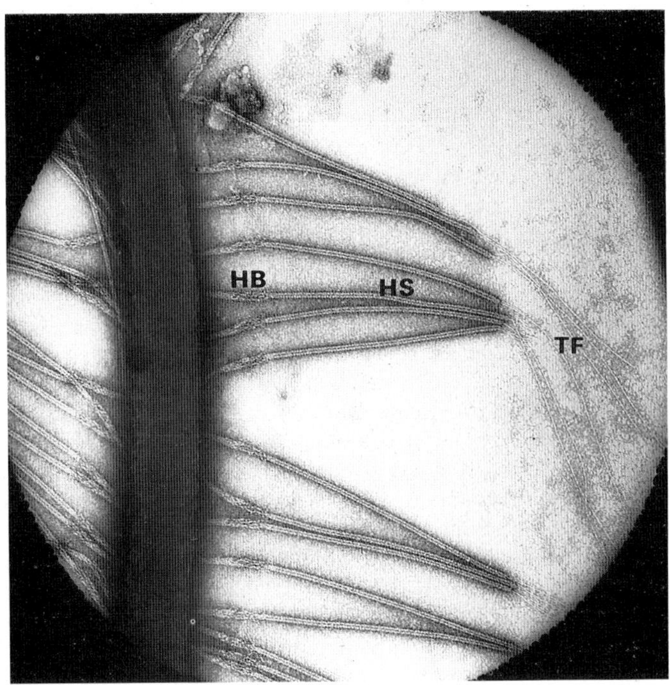

Fig. 3.35 Detail of the tripartite hairs on the anterior flagellum of the spermatozoid of *Laminaria digitata* (Phaeophyceae). HB, hair base; HS, hair shaft; TF, terminal filament (× 47600). From Maier & Müller (1982) with permission.

flagellum (e.g. *Mantoniella*; Manton & Parke, 1960). In the Charophyceae, flagellar hairs are found on zoospores of *Chaetosphaeridium* (Moestrup, 1974) and on zoospores and spermatozoids of *Coleochaete scutata* (Marchant in Pickett-Heaps, 1975; Marchant in Moestrup, 1978; Graham & McBride, 1979). They show affinities with hairs of the Prasinophyceae and other chlorophycean algae.

Simple or non-tubular hairs occur in a variety of groups. The hairs of Euglenophyceae and Dinophyceae are unique, although still poorly understood. Unlike tubular hairs, simple hairs are not differentiated into regions. In the Euglenophyceae they are arranged in a single row and in bundles (cf. Bouck *et al.*, 1978). In *Eutreptiella gymnastica* they are in fours (Throndsen in Moestrup, 1982) while they are in threes to sevens in *Euglena gracilis* (Moestrup, 1982). The hairs in *Euglena* are attached through the membrane to the paraxial rod (*see below*) and to one of the peripheral doublets of the axoneme. A dense felt of shorter hairs is also present on some euglenoid flagella; the component hairs may be arranged in parallel, transverse rows, or in spirals. The details of this felt have been revealed by Bouck *et al.* (1978) and show an extraordinary complexity (Fig. 3.36).

In the Dinophyceae, both flagella may bear hairs (Fig. 3.37), but of different types. In some species the transverse flagellum carries hairs unilaterally, apparently arranged in bundles (Clarke & Pennick, 1972; Leadbeater & Dodge, 1967) of different sized hairs. Hairs of the longitudinal flagellum are less well understood, and may be absent in some species.

Simple hairs described from groups other than euglenoids and dinoflagellates are not inserted in bundles. They are known from *Pedinomonas* (Loxophyceae, cf. Prasinophyceae; Manton, 1965), some species of *Chlamydomonas* (Chlorophyceae; Ringo, 1967; McLean *et al.*, 1974), *Pteromonas* (Chlorophyceae; Belcher & Swale, 1967) *Furcilla* (Chlorophyceae; Belcher, 1968), and *Gloeochaete* (Kies, 1976) and *Cyanophora* (Glaucophyta; Thompson, 1973). Very fine hairs form a dense covering on the anterior flagellum of the Pavlovales (Prymnesiophyceae) and in some Chrysophyceae (cf. Moestrup, 1982).

A covering other than scales or hairs, described as a 'tomentum' or 'wool', has been revealed in a number of groups as a result of EM studies. In some Chlorophyceae and Prymnesiophyceae, flagella previously regarded as smooth may be seen to be covered by a coating. In the Characeae the coating is beneath the outer layer of scales. Functions attributed to the coating, which is probably composed of glycoproteins, include agglutination in mating (Cavalier-Smith, 1975), protection against phagocytosis, and adhesion (Moestrup, 1982).

Fig. 3.36 Reconstruction of the complex felt of short hairs on the flagellum of *Euglena gracilis* (Euglenophyceae). (Modified from Moestrup, 1982; after Bouck, *et al.*, 1978.)

Additional features of the external flagellum are paraxial inclusions. These structures are contained within the flagellar membrane and can be grouped into flagellar swellings, paraxial rods and flagellar spines (Moestrup, 1982).

The majority of flagellar swellings are found at the flagellar base and are associated with eyespots (*see below*). Their lamellate or crystalline structure suggests that they function as part

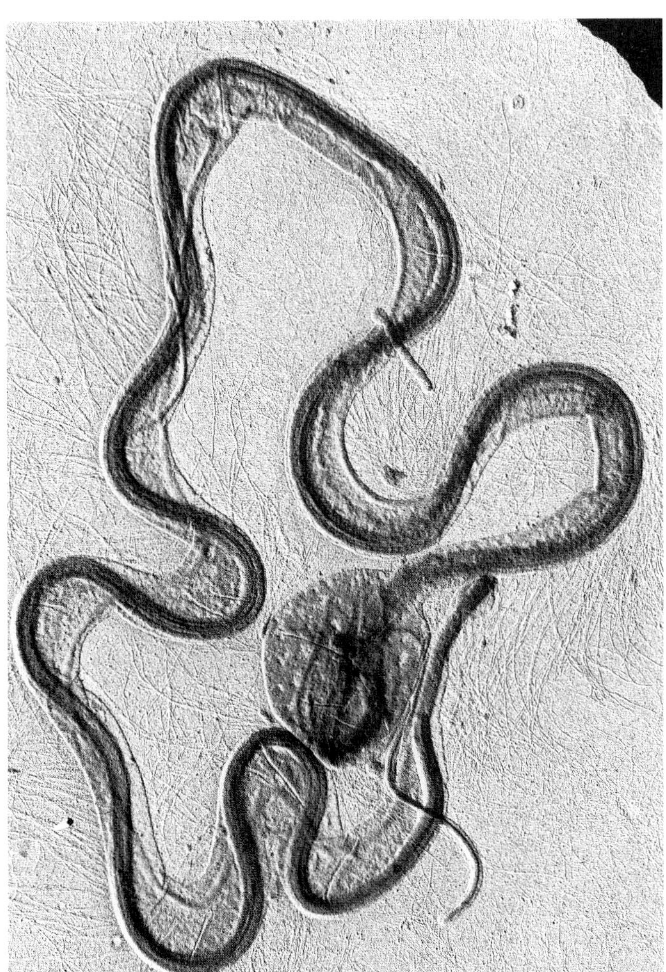

Fig. 3.37 Transmission EM of the transverse flagellum of *Prorocentrum minimum* (Dinophyceae) showing the axoneme (broad strand), the striated strand (paraflagellar body) and the flagellar hairs. (× 5600). From Dodge (1983) with permission.

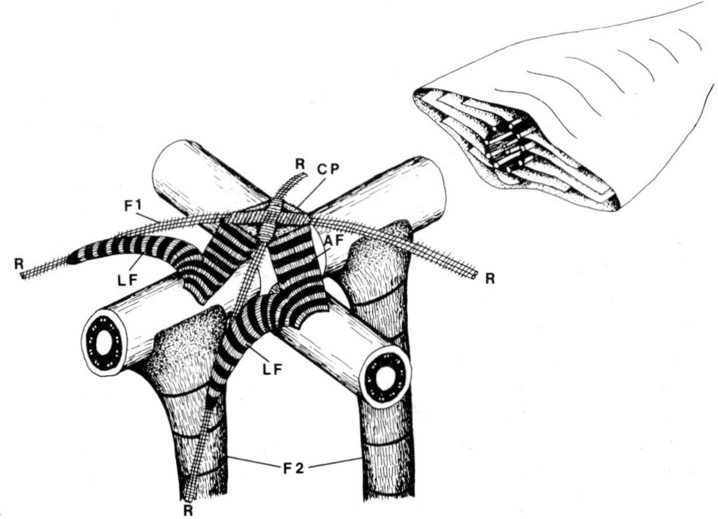

Fig. 3.38 Diagrammatic model of the flagellar apparatus of the zoospore of *Urospora penicilliformis* (Chlorophyceae). AF, ascending fibre; LF, lateral fibre; CP, capping plate; R, root with system I (F1) and system II (F2) striated root fibres. The internal structure of the wing-bearing portion of the flagellar axoneme is also shown. From Sluiman *et al.* (1982) with permission.

of the photoreception apparatus (*see* Robenek & Melkonian, 1983). In the heterokont groups, except the Eustigmatophyceae (Chrysophyceae, Xanthophyceae, Phaeophyceae), the swelling occurs at the base of the smooth flagellum. The Eustigmatophyceae are unique in that it is present at the base of the hairy flagellum. One prymnesiophyte, *Diacronema*, a non-photoreceptive species, possesses a flagellar swelling (Green & Hibberd, 1977). In the euglenoids the flagellar swelling (or paraflagellar body, PFB; cf. Robenek & Melkonian, 1983) is different from that in heterokonts. In the majority it occurs in association with the longer flagellum, between the transition region and the reservoir canal, some distance from the flagellar base. Very unusual flagellar swellings occur in *Urospora* (Chlorophyceae), a few Chrysophytes, and members of the Cryptophyceae. In *Urospora* (Fig. 3.38) each of the proximal parts of the four flagella of the zoospore supports a swelling or 'wings', these being attached internally by parallel flat plates to peripheral doublets (Kristiansen, 1974; Sluiman *et al.*, 1981).

Paraxial rods apparently give structural support to flagella in the Dinophyceae (Fig. 3.37), Euglenophyceae and some Chrysophyceae (Pedinellales *sensu* Moestrup, 1982). Elsewhere, they are known from flagella of bodonoids and trypanosomes (Protista). They normally extend for all or most of the length of the external flagellum, and may be cross-banded (e.g. the transverse flagellum of dinophytes), or hollow. The presence of ATPase along the paraxial rod in *Euglena* (Piccinni *et al.*, 1975) suggests that the rod is involved in an energy-consuming process.

Flagellar spines are uncommon, and are restricted to the flagella of male gametes of oogamous Phaeophyceae. In *Dictyota* they comprise a row of 12 short spines located on the hairy flagellum; they are firmly attached to one of the peripheral axoneme doublets (Manton, 1959). In some Fucales (*Fucus; Himanthalia; Xiphophora;* and *Hormosira*; cf. Moestrup, 1982) a single spine is present; it may be up to 1.0 μm long (e.g. *Himanthalia;* Fig. 3.39) and readily visible with the light microscope.

On entering the cell the axoneme terminates in a basal body. The transitional region (Figs 3.27, 3.40 & 3.41) between the axoneme and the basal body, however, is highly complex and has become one of the most useful indicators of phylogenetic relationships (Hibberd, 1979; Moestrup, 1982). Moestrup (1982) recognized six types of transitional region: the cryptophycean, dinophycean, heterokont (or *Ochromonas*), haptophycean (or *Prymnesium*), euglenoid (or *Entosiphon*), and green algal types (summarized in Fig. 3.41). Significant distinguishing features include the presence of one or more discs at the termination of the central tubules, the existence of a transitional helix (TH) found in some of the heterokont types (Chrysophyceae, Xanthophyceae, and Eustigmatophyceae, but absent from the Phaeophyceae, Raphidophyceae and Bacil-

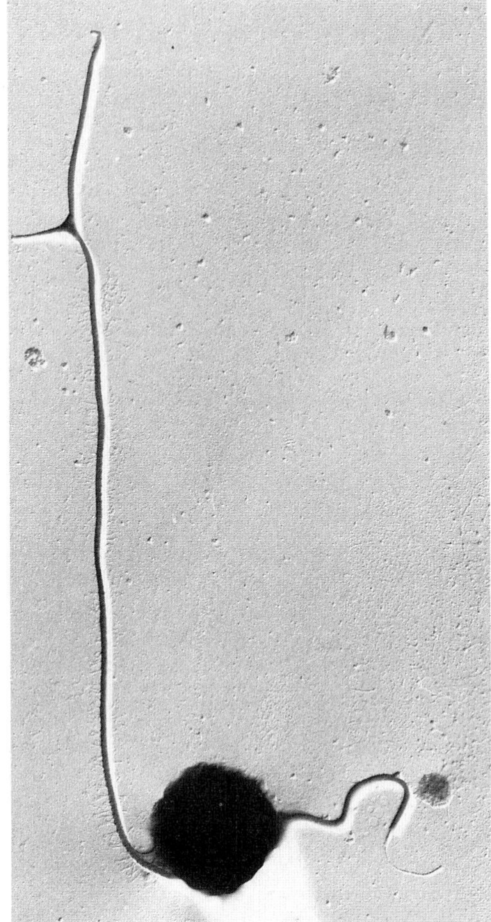

Fig. 3.39 Sperm of *Himanthalia* (Phaeophyceae) showing the prominent flagellar spine, which may exceed 1.0 μm in length (× 5400). From Moestrup (1982) with permission.

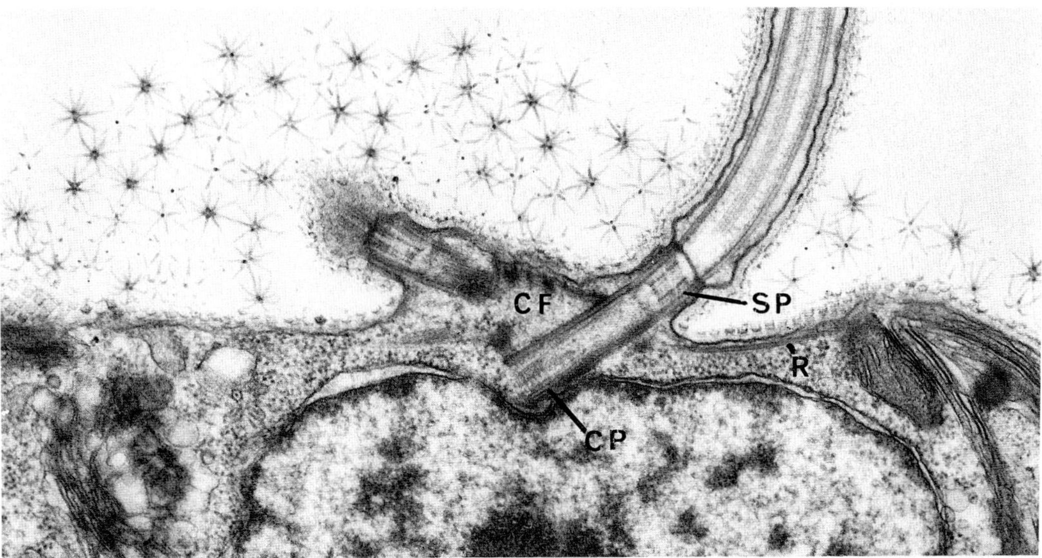

Fig. 3.40 Internal structure of the basal body and flagellar transition region of *Nephroselmis olivacea* (Prasinophyceae). CF, connecting fibre; CP, cartwheel pattern; R, microtubular flagellar roots; SP, stellate pattern (× 48000). From Moestrup & Ettl (1979) with permission.

lariophyceae), and the presence of stellate structures (Prymnesiophyceae, except Pavlovales, and all chlorophyll *b*-containing plants). The stellate region appears H-shaped in longitudinal section; in transverse section, the arms connect with the A-tubules of the peripheral axonemal doublets. *Pyramimonas* (Prasinophyceae) is unique in that it possesses a 'coiled fibre' or transitional helix (resembling the heterokont condition) and a stellate structure (characteristic for chlorophyll *b*-containing algae; Moestrup & Thomsen, 1974). In euglenoids, except for *Entosiphon*, the transitional region is apparently empty.

An understanding of the absolute orientation of the flagellar apparatus has been shown in some recent studies (Figs 3.42 & 3.43; Floyd *et al.*, 1980; Melkonian, 1981a, b; Melkonian & Berns, 1983; O'Kelly & Floyd, 1983; Floyd & O'Kelly, 1984) to be of fundamental importance in distinguishing between Ulvophycean and *Chlamydomonas* types among the green algae. Studies of brown algal flagellar base orientation (O'Kelly & Floyd, 1984a) have emphasized their differences from other classes of chlorophyll *c*-containing algae. Further investigations can be expected to clarify these relationships.

A basal body forms the termination of each flagellum within the cell. Distally, it forms the ends of the peripheral microtubules. Its basic structure shows little variation within the algal groups. Below the transitional region a third C-tubule is added to the peripheral doublets. These peripheral triplets are tilted to the radii at an angle of *c.* 130°. Distally they are normally attached by fine threads to a single central thread, a configuration which is described as the 'cartwheel' (Fig. 3.27). Proximally the lumen of the basal body may contain amorphous material or several discrete electron-dense bodies, it may be packed with ribosomes, or it may be empty (most euglenoids) (Moestrup, 1982). Connecting fibres may interconnect the basal bodies, but they are apparently absent from basal bodies which do not subtend a functional flagellum.

All algal flagella appear to possess flagellar roots, attached to the basal bodies and extending either superficially or internally into the cytoplasm. Modern EM studies have shown that the flagellar root systems of algal cells are highly complex. Internal roots may make contact with the nucleus, mitochondria, the Golgi apparatus, or the chloroplast. The connection between flagella and the nucleus so commonly illustrated in earlier texts, however, is often incorrect. They

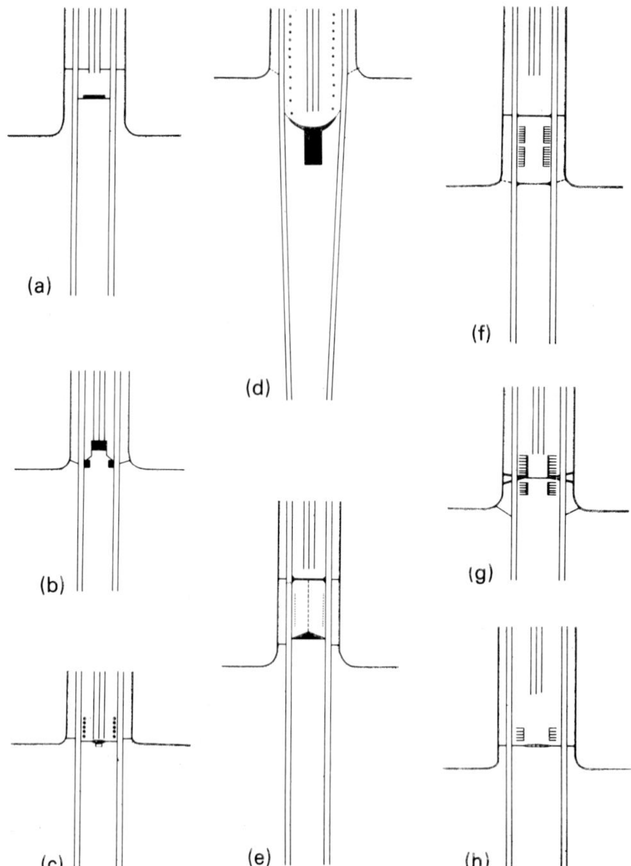

Fig. 3.41 Diagrammatic reconstructions of the flagella transition region in members of the Cryptophyceae (a, *Cryptomona cryophila*; b, *Amphidinium* sp.); Chrysophyceae (c, *Uroglena*), Euglenophyceae (d, *Entosiphon sulcatum*), Prymnesiophyceae (e, *Prymnesium parvum*), Prasinophyceae (f, *Nephroselmis olivacea*), Chlorophyceae (g, *Chlamydomonas reinhardtii*), and *Monomastix minuta* (h) (Loxophyceae *sensu* Moestrup, 1982). Scale = 0.2 μm. From Moestrup (1982) with permission. After Taylor & Lee (a); Dodge (b); Hibberd (c); Mignot (d); Manton (e, h); Moestrup & Ettl (f); and Cavalier-Smith (g).

are microtubular or fibrillar, often cross-banded structures (Fig. 3.44), the latter frequently termed rhizoplasts (Moestrup [1982] cautions against the use of this term). Flagellar roots have been ascribed a variety of functions, including anchorage, sensory transduction, stress absorption, and as skeletal or organizational structures. Enough is now known about them that generalizations of a phyletic nature are now possible.

Microtubular roots have been found in all algae examined to date; they may be rare, as in the Bacillariophyceae, or the only kind, as in the Phaeophyceae, Prymnesiophyceae, 'Glaucophyceae' (*sensu* Moestrup, 1982), Charophyceae, bryophytes, and vascular plants. Perhaps they achieve their greatest complexity in the Dinophyceae, where they are associated with a variety of ancillary structures (Dodge & Crawford, 1971), and the Raphidophyceae (Mignot, 1967, 1976), where they may be associated with an electron-dense plate and a multilayered structure. A special arrangement of the microtubular roots of the sperm in Fucaceae (Phaeophyceae) supports the proboscis, a unique anterior protrusion of the cell. A microfibrillar, cruciate flagellar root system is found in most green algae (with the exception of the Oedogoniales; *see* Moestrup, 1978; Melkonian, 1980), some loxophytes (*sensu* Moestrup, 1982), prasinophytes, and 'glaucophytes'. It consists of four roots spreading from the basal bodies, with the opposite roots normally possessing identical numbers of microtubules. Cruciate roots are among the most extensively studied, and have proved a useful feature for separation of the various 'chlorophyte' classes. Unlike the other chlorophyll *b*-containing algae,

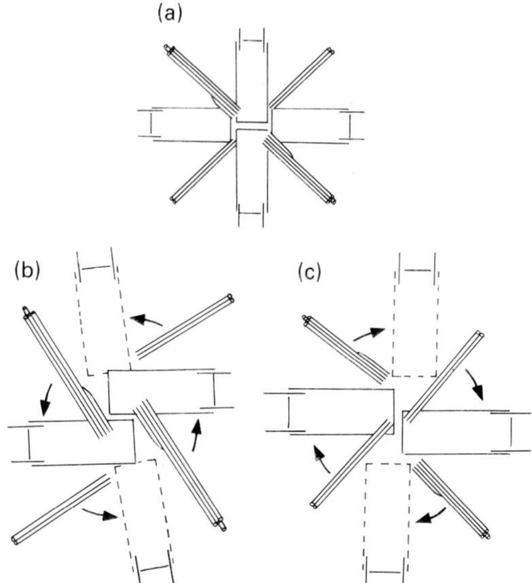

Fig. 3.42 Diagrammatic representation of absolute orientation patterns in flagellar structures with 180° rotational symmetry. (a) Strictly cruciate arrangement of basal bodies and microtubular rootlets. (b) Counterclockwise rotation, as in *Entocladia viridis* and Ulvophyceae *sensu* O'Kelly & Floyd. c) Clockwise rotation typical of Chaetophoraceae *sensu stricto* and Chlorophyceae. From O'Kelly & Floyd (1983) with permission.

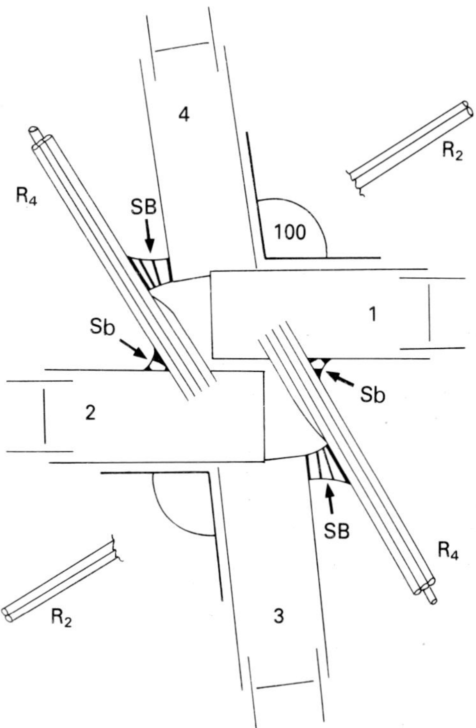

Fig. 3.43 Generalized representation of flagellar apparatus of a quadriflagellate zoospore of a member of the Ulotrichales or Ulvales (Chlorophyceae) as viewed from above. $R_{2,3,4}$, microtubular rootlets (with number of microtubules in the rootlet); SB, striated band connecting rootlets to the lower basal body pair; Sb, striated band connecting rootlets to the upper basal body pair. Basal bodies numbered 1–4. From Floyd & O'Kelly (1984) with permission.

the euglenoids apparently possess one microtubular root opposite each basal body and a single root in between (Moestrup, 1982).

Cross-banded fibrous roots are found in the Cryptophyceae, Dinophyceae, Raphidophyceae, Chrysophyceae, Xanthophyceae, Eustigmatophyceae, Euglenophyceae, 'Loxophyceae' (*sensu* Moestrup, 1982), Prasinophyceae (Fig. 3.44) and Chlorophyceae (Moestrup, 1982). They are variable in size and position, and in some respects resemble muscle in appearance and possibly function. In the Chrysophyceae they are associated with the nucleus, Golgi apparatus, or chloroplast; these types of root have been described as a rhizostyle. Non cross-banded fibrous roots have been described from the Prymnesiophyceae. Cross-banded roots normally occur in association with microtubular roots; in the Xanthophyceae the root arrangement is unique, and includes a combination of a descending root, cross-banded fibrillar root, and several microtubular roots (Hibberd & Leedale, 1971; Hibberd, 1980a).

Chloroplasts

In photosynthetic algae the photoreceptor pigments are located on or in membranes which form flattened vesicles known as thylakoids. In prokaryotes (Cyanophycota and Prochlorophycota) these are free within the cytoplasm, while in eukaryotes they are enclosed within bounding membranes to form a chloroplast. The matrix of the chloroplast is known as the stroma. Chloroplasts contain chloroplast DNA and chloroplast

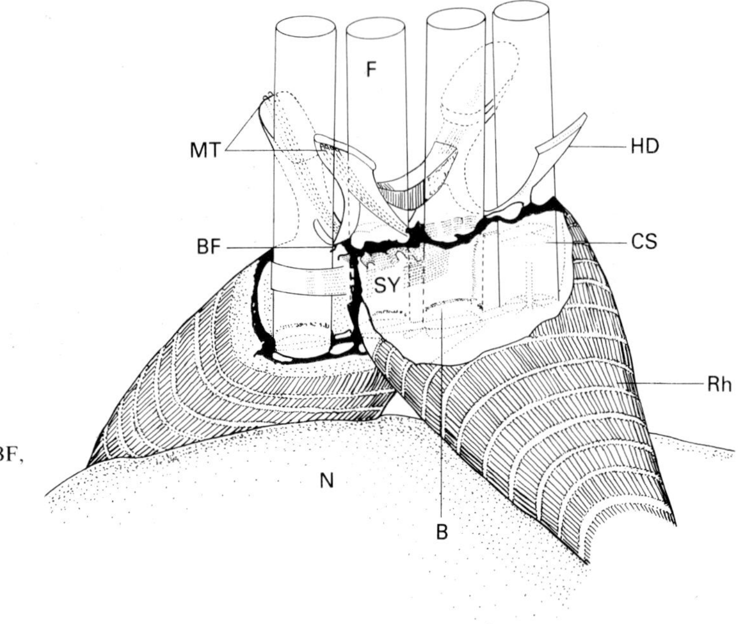

Fig. 3.44 Simplified diagrammatic interpretation of the basal body complex in *Tetraselmis* (Prasinophyceae). B, basal body; BF, binding fibrils; CS, cartwheel structure; F, flagellum; HD, half desmosome; MT, microtubules connected to half desmosomes; N, nucleus; Rh, rhizoplast; SY, synistosome. From Norris *et al.* (1980) with permission.

Table 3.2 Selected chloroplast features of eukaryote algal classes.

Class	No. of thylakoids per lamella	Girdle lamella present	Grana formed	No. of membranes around chloroplast	Starch in chloroplast
Rhodophyceae	1	–	–	2	–
Chrysophyceae	3	–	–	4	–
Prymnesiophyceae	3	–	–	4	–
Xanthophyceae	3	+	–	4	–
Eustigmatophyceae	3	–	–	4	–
Bacillariophyceae	3	+ (?)	–	4	–
Dinophyceae	(1) – 3	–	–	3	–
Phaeophyceae	3	+	–	4	–
Raphidophyceae	3	+	–	4	–
Cryptophyceae	(1) – 2	–	–	4	–
Euglenophyceae	3	–	–	3	–
Chlorophyceae	2 – many	–	+	2	+
Charophyceae	2 – many	–	+	2	+
Prasinophyceae	2 – 4	–	+ (?)	2	+

Table 3.3 General classification of chloroplasts, photosynthetic pigments, and storage products of algae. (Modified from Coombs & Greenwood, 1976.)

	PROKARYOTA	EUKARYOTA			
Compartments	Thylakoids in cytoplasm	Thylakoids in cytoplasm			
Membranes	Absent	Chloroplasts in cytoplasm Envelope of two membranes		Chloroplasts in subcompartment of cytoplasm Envelope + ER cisterna = four membranes (three in some — *see* *)	
Secondary pigments	Phycobilisomes *on* membrane		Chlorophyll *b* or *c*, and carotenoids *in* membrane		
Thylakoid arrangement	No stacks Thylakoids form simple lamellae		Stacks in cytoplasm Compound lamellae and grana	Compound lamellae or bands of 2–3 thylakoids Often very regular three-thylakoid bands ( ) = girdle lamellae	
Classification			CHLOROPHYCOTA	CHROMOPHYCOTA (and EUGLENOPHYCOTA)	
Secondary pigment groups (a) Phycobilins (b) Chlorophyll *b* (c) Chlorophyll *c*	a) PROCHLOROPHYCOTA CYANOPHYCOTA Cyanophyceae	RHODOPHYCOTA Rhodophyceae	Chlorophyceae Prasinophyceae Charophyceae	Dinophyceae * Cryptophyceae	Euglenophyceae * (b) Prymnesiophyceae Eustigmatophyceae (Raphidophyceae Chrysophyceae; phaeophyceae, bacillariophyceae, xanthophyceae)
Typical carbohydrate reserves	'Glycogen'	Amylopectin	Amylopectin and amyloses		Laminarin, paramylon, etc.
	α−(1, 4) glucans 'Starches'				β−(1, 3) glucans No starch

ribosomes; they are not, however, strictly autonomous from the nuclear genome. Protein and nucleic acid synthesis in chloroplasts is reviewed by Ellis (1976). Chloroplast development and division may be co-ordinated with that of the cell (especially in cells containing only a single chloroplast) or may proceed independently (especially in cells containing many chloroplasts). The complex questions of chloroplast inheritance and renewal are reviewed in Whatley (1982), and are further considered in Chapter 5.

General reviews of algal chloroplasts include those of Dodge (1973) and Coombs and Greenwood (1976). Any consideration of chloroplasts must include pigments and storage products (often associated with pyrenoids), as well as eyespots. The latter are dealt with separately. It is noteworthy that, while chloroplasts are structurally remarkably uniform among the vascular plants, in the algae their features are diagnostically important at the class level. A summary of algal chloroplast features is given in Tables 3.2 and 3.3.

PROKARYOTES

In the prokaryotes, thylakoids are arranged peripherally and are surrounded by a matrix of cytoplasm (Fig. 3.1). Following EM fixation with glutaraldehyde, the phycobilin pigments appear aggregated in phycobilisomes, comparable with those found in rhodophycean thylakoids, but different from those of the phycobilin-containing Cryptophyceae. In the heterocysts of nitrogen-fixing blue-green algae the thylakoids are reorganized into a sinuous reticular system; the phycobilisomes are lost and the chlorophyll *a* is retained. 'Stacking' of thylakoids does not occur in blue-green algae, but in the Prochlorophycota the chlorophyll *b*-containing thylakoids show some stacking, reminiscent of the stacking seen in all other algae except the Rhodophycota.

RHODOPHYCOTA

The structure and organization of rhodophyte thylakoids (Fig. 3.45) show a primitive homology

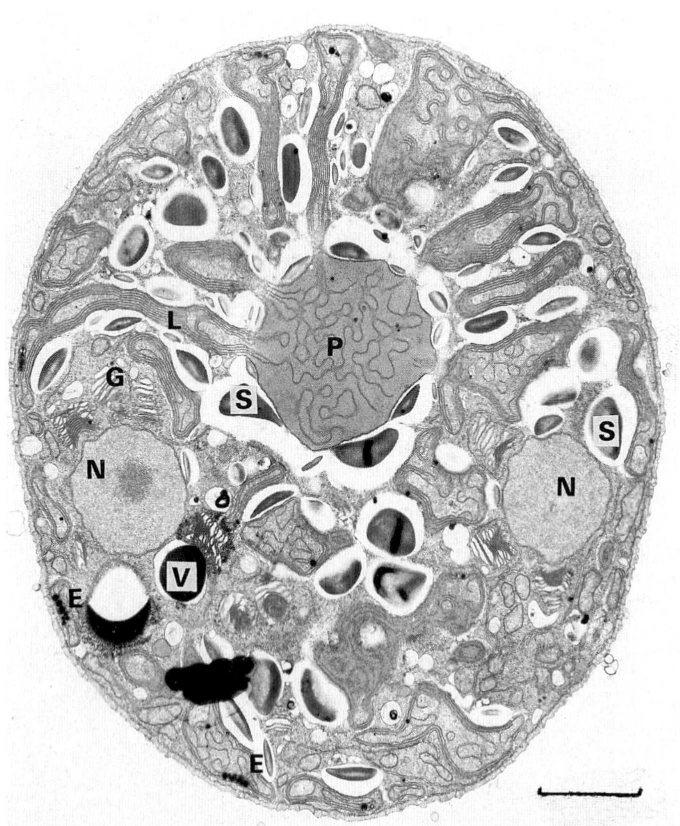

Fig. 3.45 EM section of the coccoid rhodophyte *Rhodella reticulata* showing arrangement of organelles including the chloroplast and pyrenoid. N, nuclei; P, pyrenoid, traversed by thylakoids; S, starch grains; E, osmophilic globules; G, Golgi apparatus; V, vacuole; L, radiating plastid lobes. Scale = 2.0 μm. From Deason *et al.* (1983) with permission.

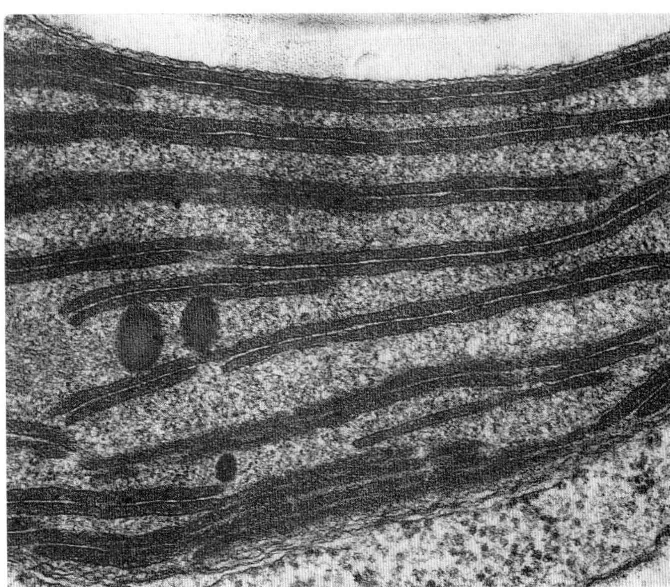

Fig. 3.46 Chloroplast of *Chroomonas mesostigmatica* (Cryptophyceae) showing thylakoids arranged in pairs. The thylakoid lumen is dense and filled with phycobilin pigment (× 81 000). From Dodge (1969b) with permission.

with those of blue-green algae, although there are substantial differences in function resulting from the former's compartmentalization within bounding membranes. The chloroplast envelope comprises two parallel membranes. The lamellae do not come into contact with each other (are not stacked) or with the bounding membrane. Phycobilisomes are present on the stromal surfaces of the lamellae. In simpler Rhodophyceae (Bangiophycidae) a pyrenoid is often present (e.g. *Porphyridium;* Coombs & Greenwood, 1976); the lamellae form a reticulum within the pyrenoid. In the Florideophycidae, pyrenoids occur only in the simple examples, and there the lamellae run parallel from margin to margin of the pyrenoid (Gibbs, 1962a, b). Starch, when stored, is held in the cytoplasm, outside the chloroplast. In these more advanced rhodophytes a peripheral lamella or thylakoid may lie inside the bounding membranes of the chloroplast; caution should be exercised, however, in assuming homology between this 'girdle lamella', as it has sometimes been called, and the similarly named structure in the Chromophycota (Coombs & Greenwood, 1976).

CHROMOPHYCOTA

There is considerable variation in chloroplast structure in the Chromophycota (*see* Tables 3.2 & 3.3). Chromophyte chloroplasts are segregated from the cell cytoplasm by two additional membranes (one in the Dinophyceae, also in the Euglenophycota). The outer membrane may be an extension of the nuclear membrane. A perichloroplastic space is present in all the classes except the Cryptophyceae and Dinophyceae (Coombs & Greenwood, 1976). In the Dinophyceae, a nuclear connection has not been demonstrated. The chloroplast subcompartmentalization of chromophyte cells is unique in eukaryotic organisms. The thylakoids are complexly arranged in the different chromophyte classes, and various patterns of 'stacking' have been described, with bands of two (Fig. 3.46) or three thylakoids most common. Girdle lamellae are found in the five classes indicated in Table 3.3. While it is likely that the lamellae within a single chromophyte chloroplast probably constitute a three-dimensional network, as in higher plant chloroplasts, the membranes are not invaginated in a granal pattern (Coombs & Greenwood, 1976).

The Cryptophyceae are isolated by their pigmentation and the structure and organization of their thylakoids (Fig. 3.46). Their phycobilin pigments, which are chemically different from those of the Cyanophycota, Prochlorophycota, and Rhodophycota, do not form phycobilisomes, but rather are located on the interior of the thylakoids. Their thylakoids are arranged in pairs. In

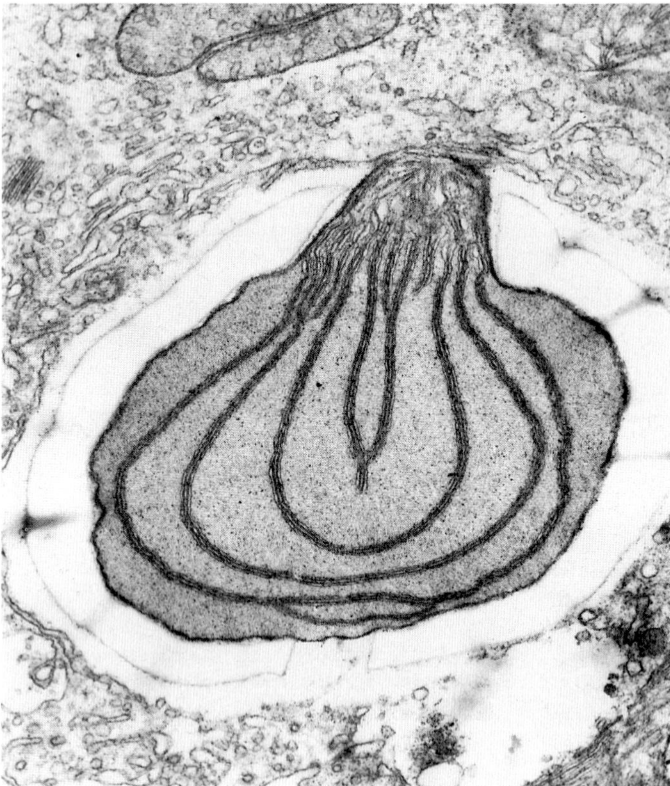

Fig. 3.47 Stalked pyrenoid of *Amphidinium carterae* (Dinophyceae) (× 22500). (Courtesy of J.D. Dodge.)

the Dinophyceae, thylakoid arrangement is the most varied in the algae, and pairs or even single lamellae are common.

The majority of chromophycotan classes possess intraplastidial pyrenoids. They are often stalked (a uniform characteristic in the Phaeophyceae; *see* Fig. 3.23 and Dodge, 1973). Thylakoid lamellae may traverse the pyrenoid matrix in some Prymnesiophyceae, Dinophyceae (Fig. 3.47), and Cryptophyceae. The Dinophyceae possess the greatest variation in pyrenoids in the Chromophycota (*see* Dodge [1973] for review). Starch is not formed in the chloroplast in the Chromophycota, but in the cytoplasm or, in some, in between the chloroplast membrane and the chloroplast ER.

CHLOROPHYCOTA

Chloroplasts in the Chlorophycota lack associated ER or nuclear membranes, thus resembling the Rhodophycota in the cell compartmentalization. Unlike the Chromophycota, they store starch within the chloroplast stroma and, often, as a sheath around the pyrenoid. Chlorophyte thylakoids are arranged in stacks of 2–6 or more; their multilayered arrangement may take on the appearance of grana (as in higher plants), with lamellar interconnections, or frets. Their arrangement in the Charophyceae is especially reminiscent of that in higher plants. Pyrenoids occur in most chlorophytes, and they are usually internal to the chloroplast. Large starch grains form on the pyrenoid surfaces. Pyrenoids, which are never stalked as they are in some Chromophycota, may be penetrated by a few to many thylakoids.

Eyespots (stigmata)

An eyespot (or stigma) is found in many motile algae, or the gametes and zoospores of non-motile algae. It consists essentially of osmiophilic, carotenoid lipid globules (plastoglobules), often associated with a photoreceptive, rod-shaped, or crystalline area near the flagellar axoneme. The photoreceptor is responsible for phototaxy.

Motile cells of the Raphidophyceae and Prymnesiophyceae normally lack eyespots; they are relatively uncommon in the Dinophyceae (which possess specialized types) and Cryptophyceae. Most flagellated cells in the Chlorophyceae, Prasinophyceae, Euglenophyceae, Xanthophyceae, Chrysophyceae, and Phaeophyceae (except zoospores of most Laminariales and the sperm of many oogamous species; Henry & Cole 1982a, b) possess them.

Dodge (1969a, 1973) classified eyespots into three principal types, and reserved a special category for the several kinds found in the Dinophyceae.

1 Chlorophyceae, Prasinophyceae, and some Chrysophyceae: the eyespot is part of a chloroplast but not obviously associated with flagella.
2 Cryptophyceae (Fig. 3.48), Xanthophyceae, and male gametes and zoospores of most Phaeophyceae (except for the zoospores of most Laminariales and the sperm of many oogamous species; Henry & Cole 1982b): the eyespot is part of a chloroplast and obviously associated with flagella.
3 Euglenophyceae, some Xanthophyceae, and Eustigmatophyceae: the eyespot is independent of the chloroplast but adjacent to the flagella.

In the Dinophyceae, eyespots range from collections of lipid globules containing carotenoids and lacking membranes to more complex arrays of lipid-containing globules surrounded by membrane envelopes (cf. Bold & Wynne, 1978). In *Glenodinium foliaceum* the eyespot is associated with an organelle comprising a stack of up to 50 flattened vesicles (Dodge & Crawford, 1969). Dodge (1984) described three types of dinoflagellate eyespot: independent and not membrane-bound; independent and surrounded by three membranes; and situated at the periphery of a chloroplast. All are positioned behind the sulcus and there is a strand of microtubules between the eyespot and theca; the strand is presumed to play a part in the transmission of directional stimuli from the eyespot to the flagellum. Dodge (1984) suggested that structurally the eyespots link the dinoflagellates with the euglenoids and chrysophytes. The most complex photosensitive organelle in the algae is the ocellus, found in a few marine dinophytes (e.g. *Warnowia*). It comprises a chamber lined by an invagination of the plasmalemma, superimposed by a refractive lens made up of closely packed, swollen vesicles. There is experimental evidence (Francis, 1967) that light from outside the cell is brought into focus in a layer lining the chamber known as the retinoid. In these respects the ocellus is the closest algal structure to light perception organs found in animals.

A great deal of attention has been paid, especially in the euglenophytes, to the question of the relation between eyespots, the chloroplast, flagella, and phototaxis. There are two views prevalent with respect to euglenophytes: one that the eyespot is the light receptor, the other that

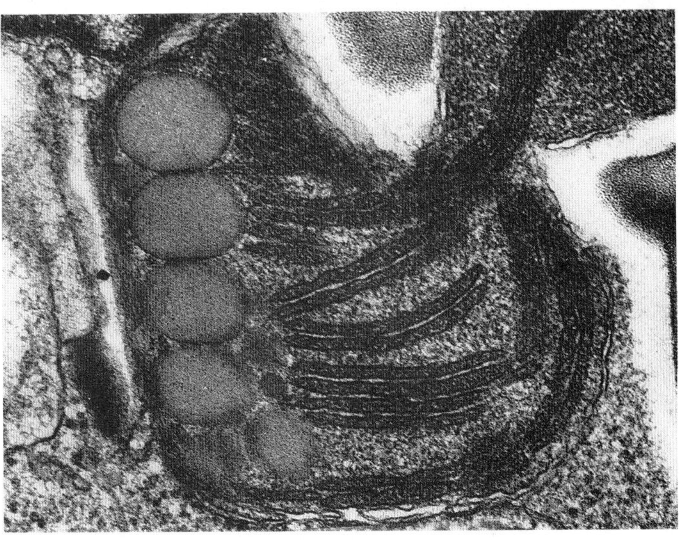

Fig. 3.48 Eyespot of *Chroomonas mesostigmatica* (Cryptophyceae) consisting of a single layer of plastoglobules situated at the end of a short spur of the chloroplast, projecting beyond the pyrenoid (at right) (× 44 100). From Dodge (1973) with permission.

the light reception takes place in the flagellar swelling or paraflagellar body (PFB). Current information suggests that the photoreceptor is a flavin-type pigment located in the PFB (Colombetti *et al.*, 1982). These authors suggest that the photosensory transduction system may involve two separate photoreceptor systems (step-up and step-down), coupled with a membrane-bound sodium/potassium transport system which, according to Robenek and Melkonian (1983), is located on the PFB membrane.

In the Chlorophyceae, Hartshorne (1953) was among the first to show that *Chlamydomonas* mutants lacking eyespots, while capable of showing positive phototaxis, respond much more slowly than wild-type cells possessing a normal eyespot. In the sperm of *Fucus,* Bouck (1970) attempted to portray the role of the eyespot in light reception. The eyespot occurs within the single chloroplast; Bouck (1970) suggested that the parabasal body and eyespot mediate light impinging on a photoreceptor bound to the plasmalemma. In the Chrysophyte *Dinobryon*, Kristiansen and Walne (1976, 1977) have described the possible photoreceptor–photoresponse mechanism in more detail. The two basal bodies are connected by a bridge, and the basal body of the smooth flagellum is connected to the eyespot region of the chloroplast by a structure arising from two of the triplets. This connection may conduct impulses from the eyespot to the flagellum (or flagella).

The eyespot is one of the most distinctive features of the Eustigmatophyceae. In the zoospores it is extrachloroplastidic (as in euglenoids) and comprises a group of globules at the anterior end of the cell, adjacent to the flagellar base. It is not surrounded by a membrane (cf. Lee & Bold, 1973).

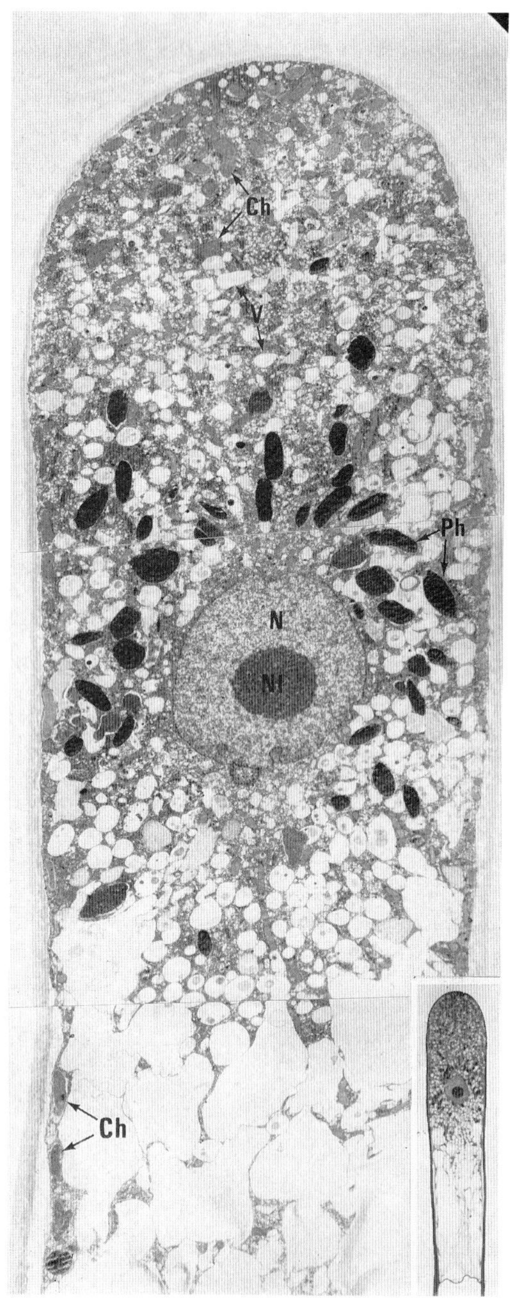

Fig. 3.49 Interphase apical cell of *Sphacelaria tribuloides* (Phaeophyceae) (whole cell inset). V, vacuoles; Ph, physodes; Ch, chloroplast; Nl, nucleolus; N, nucleus (× 2125; inset × 265). From Katsaros *et al.* (1983) with permission.

The nucleus, nuclear division, and cytokinesis

THE INTERPHASE NUCLEUS

The nucleus (Fig. 3.49) is the principal store of genetic information in the cell (the genome), and the controlling centre for selective expression of the stored information (additional genetic material is contained in the chloroplast as mentioned above (p.77) and in the mitochondria).

An exception, however, occurs in the Cryptophyceae, which possess the nucleomorph. This structure has been suggested as containing a subsidiary genome, and recent studies (Gillott & Gibbs, 1980) tend to confirm this. In both pro- and eukaryote algae, as in all other pro- and eukaryotes, the genetic information is stored in the triplet sequences of nucleotide bases on the double helix of polynucleotide chains of deoxyribonucleic acid (DNA) molecules. DNA replication and the processes of transcription and translation in the expression of genetic information are comparable in algae to other pro- and eukaryotes, and are not detailed here.

In the prokaryote algae the DNA is found in a ring-like configuration, centrally in the cell, and not contained within a membrane system. The process of replication occurs during cell division, as in eukaryotes, and the processes of transcription and translation occur in a tightly coupled manner. In eukaryote algae the DNA is found in the chromosomes, which in addition to DNA contain nuclear proteins and small amounts of RNA in a complex known as chromatin. The chromosomes are confined within the nucleus by a double membrane envelope (the nuclear envelope) which is continuous with the cell ER. The process of transcription occurs within the nucleus, whereas translation occurs outside the nucleus.

The algal interphase nucleus is generally comparable to the normal eukaryotic type found in most animals and plants. A general review was given by Dodge (1973), while specific aspects are discussed in general works on cell biology such as that of Dillon (1981). Exceptions are the nuclei of the Dinophyceae, Euglenophyceae, and Cryptophyceae, which possess unique features of the nucleus, nuclear division and cytokinesis (*see* reviews in Leedale, 1967, 1982b; Dodge, 1973). These types of nuclei have been termed mesokaryotic (Fig. 3.50) since, among other features, they possess chromosomes which remain condensed during all cell phases. Other features include the retention of the nuclear membrane during nuclear division (Dinophyceae and Euglenophyceae, but not Cryptophyceae) and persistence of the nucleolus (or endosome) during division (Euglenophyceae).

The membranes of the nuclear envelope are 7–8 nm thick, and are separated by a perinuclear cavity which may, in some instances, have specific functions such as flagellar hair construction. The nuclear envelope is perforated by numerous pores (Dillon, 1981), which may or may not be arranged in specific patterns (Dodge, 1973). In addition to the chromosomes, the nucleus contains the ground substance, or nucleoplasm, and one or more nucleoli, which are densely staining

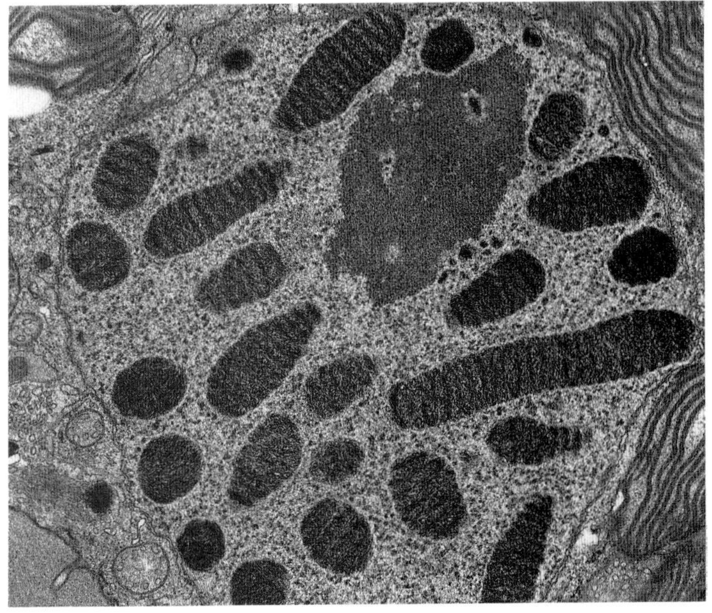

Fig. 3.50 Interphase mesokaryotic nucleus of *Heterocapsa* (Dinophyceae) showing the condensed, banded chromosomes (× 14 000). From Dodge (1971) with permission.

concentrations of basophilic material rich in ribonucleoprotein. They are intimately associated with the specific region of the chromosomal DNA that codes for ribosomal RNA. The transcription of ribosomal RNA is possible through a specific kind of RNA polymerase found only in the nucleolus (cf. Hopkins, 1978). Other possible roles of nucleoli are not as yet fully understood. The size and number of nucleoli are of value in identifying differences among algal nuclei (Dodge, 1973).

The majority of eukaryote algal cells are uninucleate. Multinucleate cells are prevalent, however, in the vegetative cells of most Florideophycidae (Rhodophycota), while in the siphonous Xanthophyceae (*Vaucheria*) and Chlorophyceae the multinucleate condition is normal. With some exceptions (e.g. Dinophyceae; Euglenophyceae) algal nuclei and chromosomes are small and, as a consequence, have been difficult to study using light microscopy. Much of the earlier information on algal chromosomes was reviewed by Godward (1966).

NUCLEAR DIVISION

A most important activity of cells is their ability to reproduce themselves. During vegetative growth in eukaryotic algae this is brought about by two processes: nuclear division, or mitosis, and cytokinesis, or cytoplasmic division. Mitotic events in algae have been extensively studied using light microscopy, although relatively few detailed ultrastructural accounts of mitosis have been reported. Considerable attention has been focused, however, on the features of nuclear division in the Dinophyceae, Euglenophyceae, Cryptophyceae, and Chlorophyceae. In the last group significant new light has been cast on relationships among algae and land plants (Pickett-Heaps & Marchant, 1972; Pickett-Heaps, 1975, 1976, 1979; Stewart & Mattox, 1975, 1980).

With the exception of the mesokaryotic forms, nuclear division in algae shows considerable similarity to that in animals, although the kinetochore, a universal feature of animal chromatids, and centrioles are not found in all algae.

Reviews of nuclear division in algae (e.g. Dodge, 1973; Brawley & Wetherbee, 1981) have highlighted those features which serve to distinguish the various algal classes from one another and from other eukaryotes. Important differences are evident in a number of mitotic features, and these are summarized in Table 3.4. Included are the behaviour of the nuclear envelope during mitosis (whether it remains intact, is provided with gaps or fenestrations, or breaks down), the polar structure as it relates to microtubule organization, and the nature of the spindle. Related to these are the presence or absence of kinetochores, centrioles, and specialized structures.

Fine structural information on mitosis in Rhodophyceae is principally based on studies of members of the Ceramiales (McDonald, 1972; Scott *et al.*, 1981; Phillips & Scott, 1981). Nuclei and chromosomes are very small, and chromosome numbers from n = 2 to n = 90 have been recorded. The most distinctive features are the polar rings, which arise during early prophase and remain throughout mitosis; it has been suggested that they may act as microtubule-organizing centres; they cannot be regarded as true centrioles, since all traces of flagella are absent in the red algae.

In the Chrysophyceae the rhizoplast plays an organizational role in mitosis (Slankis & Gibbs, 1972). During pre-prophase the basal bodies of the two flagella and the Golgi body replicate. A rhizoplast becomes associated with each pole, and the spindle microtubules appear to be attached to the rhizoplasts. The nuclear envelope may break down at prophase (*Ochromonas danica*; Slankis & Gibbs, 1972) or may remain relatively intact (*Hydrurus foetidus*; Vesk *et al.*, 1984).

In the Xanthophyceae, *Vaucheria* possesses a very long interzonal spindle and an intact nuclear membrane during mitosis (Ott & Brown, 1972); its mitotic features may be more typical of various phycomycete fungi than of other algae (Dodge, 1973).

The spindle is distinctive in the Bacillariophyceae; Manton *et al.* (1969a) described a spindle precursor consisting of a rectangular body made up of series of parallel plates in *Lithodesmium* (Biddulphiales); the precursor plays a role in the organization of the spindle prior to the breakdown of the nuclear envelope. A centrosome-like structure has been described for *Surirella* (Drum & Pankratz, 1963) although any role it might play in spindle organization is not clear.

In the Dinophyceae, the nucleolus and nuclear membrane persist during mitosis. The chromosomes, which appear to be attached to the nu-

Table 3.4 Distinguishing features of mitosis and cytokinesis in eukaryotic algae.

	Chromosomes*	Kinetochore	Nucleolus†	Nuclear envelope			Spindle type	Polar structure	Cytokinesis
				Intact	Polar gaps	Breaks down			
Rhodophyceae	–	+	–		+		Closed nuclear and cytoplasmic	Polar rings; no true centrioles	Annular wall plus septal plug
Chrysophyceae	–	–	–			+	Nuclear and cytoplasmic	Rhizoplast	Cleavage
Prymnesiophyceae	–	–	–			+	Nuclear and cytoplasmic	No centrioles	Cleavage
Xanthophyceae	–	–	–			+	Extended intranuclear	Centrioles	Cleavage
Eustigmatophyceae	–	–	–	+			Nuclear and cytoplasmic	Flagellar bases act as centrioles	Cleavage
Bacillariophyceae	–	–	–			+	Cytoplasmic?	Spindle 'precursor'	Furrowing
Dinophyceae	+ attached to nuclear envelope	–	–	+			Cytoplasmic tunnels		Cleavage
Phaeophyceae	–	+	–	(+)	+		Intranuclear	Extranuclear centrioles	Cell plate
Raphidophyceae	–	+	–	+			Intranuclear	Centrioles	Cleavage
Cryptophyceae	+ plus nucleomorphy	+ (simple)	+ large			+	Nuclear and cytoplasmic	Flagellar bases; no true centrioles	Cytokinetic ring
Euglenophyceae	+	+	+ (endosome)	+			Intranuclear and subspindles	?	Cleavage
Chlorophyceae	–	– or +	–	(+)	+	(+)	Nuclear and cytoplasmic or intranuclear	Centrioles	Phycoplast and phragmo-plast
Charophyceae	–	– or +	–		+		Nuclear and cytoplasmic	Centrioles	Phragmoplast
Prasinophyceae	–	– or +	–		+		Nuclear and cytoplasmic	Centrioles	Phycoplast or phragmo-plast and furrow

* + = persistently condensed; – = condensed at mitosis.
† + = persistent at mitosis; – = not persistent at mitosis

clear membrane, lack a kinetochore. Cytoplasmic microtubules, which penetrate the nucleus, give the appearance of a spindle, but they are not involved in the separation of the chromosomes and they do not appear to connect to any organizing structure. At the end of mitosis the nuclear envelope constricts following separation of the chromatids.

The spindle of the Phaeophyceae is reportedly intranuclear, with the centrioles extranuclear. The nuclear envelope remains intact during mitosis, except for polar gaps through which the spindle microtubules penetrate (Katsaros *et al.*, 1983).

Mitosis has been studied in only two cryptophytes (*see* Oakley & Santore, 1982). There appears to be no centriole; rather, spindle microtubules originate before prophase at the flagellar bases, from which they dissociate as the nuclear envelope breaks down. The kinetochore is a simple type, and at anaphase it is difficult to distinguish the individual chromosomes from the chromatin mass. The formation of spindle microtubules prior to prophase and the breakdown of the nuclear envelope is a feature that the cryptophytes share with the Prymnesiophyceae (Manton, 1964b).

Mitosis in the euglenophytes exhibits a number of unusual features, recently reviewed by Leedale (1982). The nuclear membrane remains intact and the endosome (nucleolus) divides into two. The chromosomes then come to lie along the division axis parallel to the dividing endosome. Segregation into chromatids may occur before, during, or after this process. The separation of chromatids at metaphase is staggered, and some

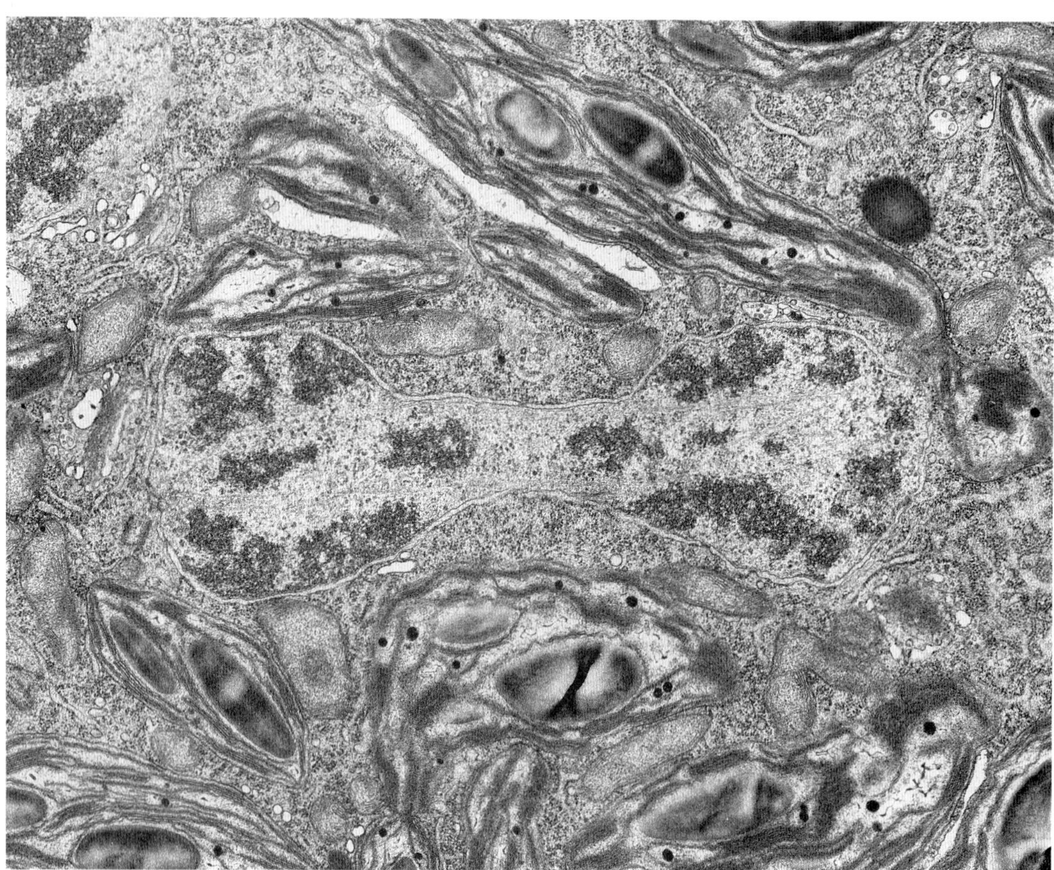

Fig. 3.51 Mitosis in *Cladophora* (Chlorophyceae) showing the completely closed spindle at mid-anaphase. (× 15720). From Scott & Bullock (1976) with permission.

may reach the poles while others remain at the equator. Nucleoplasmic microtubules form during prophase and are associated with the chromosomes; they terminate at the polar regions within the nuclear envelope and there are no associated cytoplasmic microtubules. Earlier reports of the absence of kinetochores (Leedale 1968; Dodge, 1973) are incorrect. Pickett-Heaps and Weik (1977) have recorded kinetochores in the mitotic chromosomes of *Euglena gracilis*, and Leedale (1982), on reinvestigation of his own earlier work, also found them. Remarkably, each chromatid appears to have its own miniature subspindle, composed of four centromeric microtubules associated with 8–12 larger microtubules.

The extensive work on mitosis in green algae (especially Chlorophyceae) has been summarized by Pickett-Heaps (1975, 1979) and Brawley and Wetherbee (1981). During mitosis the nuclear envelope may remain entire (Figs 3.51 & 3.52), intact with polar fenestrations, or may break down at prophase (e.g. desmids). The spindle is both nuclear and cytoplasmic, although in *Oedogonium* it is entirely intranuclear. *Oedogonium* is also distinctive in its possession of complex kinetochores.

The behaviour of the green algal spindle following nuclear division is of fundamental phyletic interest. It is discussed below.

CYTOKINESIS

Following nuclear division, cytokinesis normally occurs by one of two means: by growth of a new wall, or by constriction of the cell by means of a

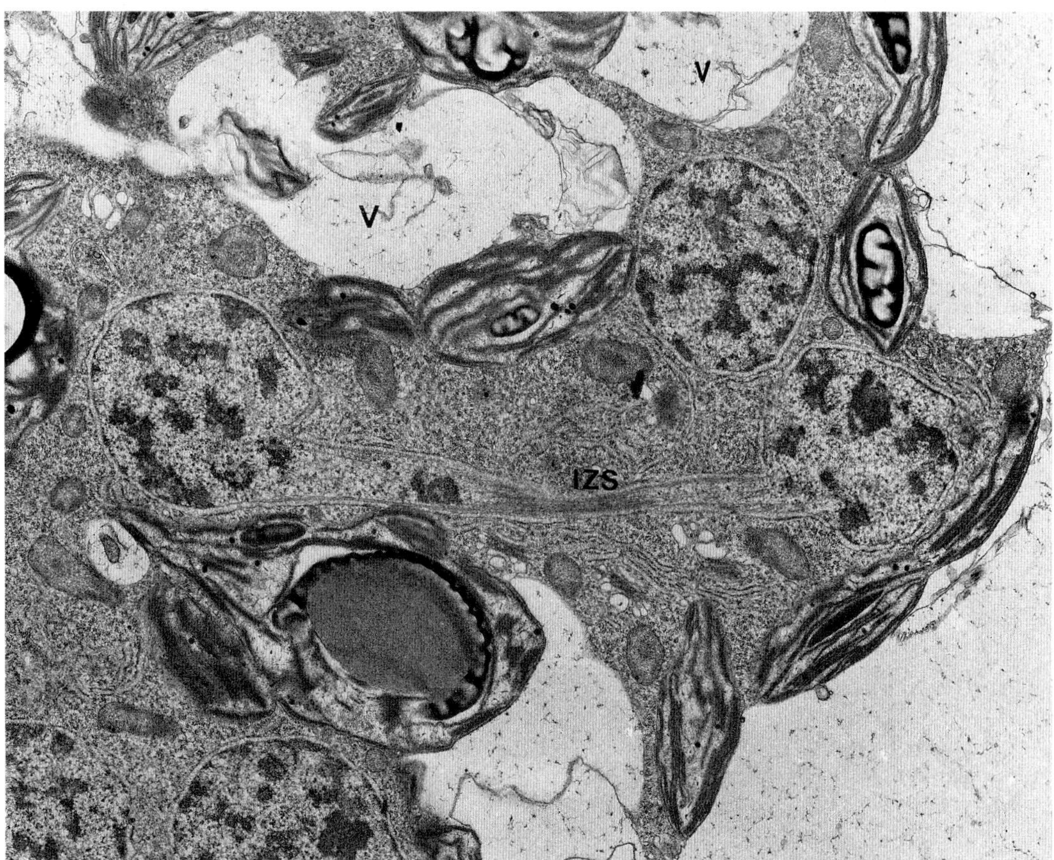

Fig. 3.52 Mitosis in *Cladophora* (Chlorophyceae) at late telophase (× 9000). V, vacuoles; IZS, residual interzonal spindle. From Scott & Bullock (1976) with permission.

furrow. Microtubules may play a role in the direction of this process, and the cell plate, a feature typical of higher plant cells, involves the participation of the Golgi apparatus in the deposition of new material.

Cytokinesis, or binary fission, is a relatively simple process in the prokaryotes (e.g. Cyanophycota) and is normally a consequence of cell expansion. Three types of cytokinesis have been described in the cyanophytes (Drews & Weckesser, 1982): a constrictive type; a septum-forming type; and a budding type. In the constrictive type, simultaneous invagination of walls occurs (Fogg *et al.*, 1973). In the septum type a septum is formed by invagination of the cytoplasmic membrane and the peptidoglycan layer, before invagination of the outer membrane commences. The thylakoids are pushed towards the centre and are cut across as the septum in completed. An unusual feature of cyanophyte cytokinesis is that septum formation is a continuous process, and three or more different septa may be initiated in sequence. Septum formation thus occurs independently of a specific replication cycle, a phenomenon quite different from typical eukaryote cytokinesis which always occurs at the end of a mitotic cycle (Dillon, 1981). The budding type of division has been demonstrated in *Chamaesiphon* (Rippka *et al.*, 1979).

The most primitive form of eukaryote algal cytokinesis occurs in flagellates, by longitudinal cleavage; this commences at the anterior (most) or posterior (e.g. some Dinophyceae) end of the cell. An important modification occurs in *Chlamydomonas*, where cleavage occurs only after the entire protoplast has made a quarter rotation within the cell wall; this step is regarded as a clue to the origins of transverse fission typical of filamentous forms of green algae (Dillon, 1981).

The behaviour of the mitotic spindle in relation to cytokinesis has been extensively studied in the green algae (Pickett-Heaps, 1975, 1979). If the spindle disperses, the new nuclei remain close together and a new set of microtubules perpendicular to the former spindle act as organizers for the new wall; the new microtubules (and the mode of cytokinesis) are referred to as a phycoplast. Alternatively, if the spindle persists, the daughter nuclei are held far apart and the new wall is formed by a phragmoplast (a coalescence of dictyosome vesicles containing wall components, and between the spindle microtubules),

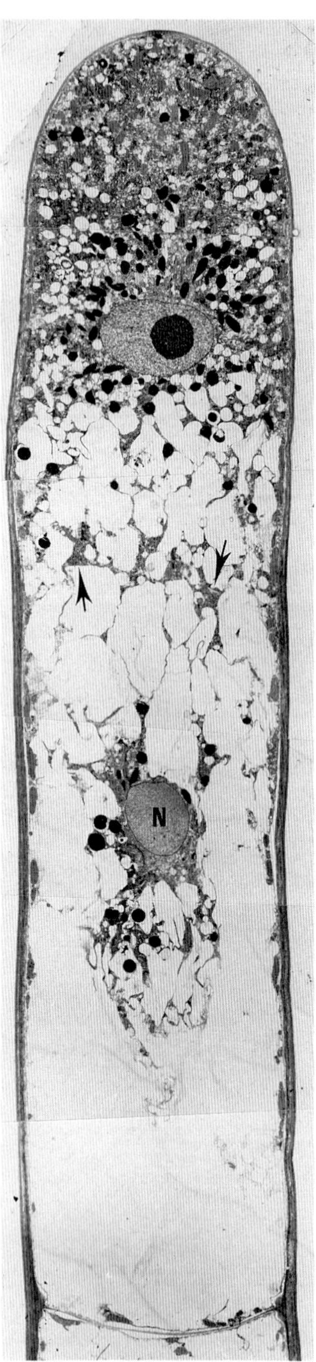

Fig. 3.53 Apical cell of *Sphacelaria tribuloides* (Phaeophyceae) in early stage of cytokinesis, showing the developing furrow (arrows) and the daughter nucleus (N) (× 900). From Katsaros *et al.* (1983) with permission.

or by a combination of a phragmoplast and furrowing of the cell plasmalemma. The phragmoplast type of cytokinesis is characteristic for the higher plants, and its occurrence in green algae (some Chlorophyceae, Prasinophyceae and all Charophyceae) has led to new proposals for their classification (Stewart & Mattox, 1975; Pickett-Heaps, 1975).

In the Rhodophyceae the early stages of cytokinesis result in incomplete septation between daughter cells and cytoplasmic continuity. Following this is the development of the pit plug (*see* p.62).

In the Phaeophyceae (Fig. 3.53) the cell plate develops centrifugally and becomes thickened with fibrillar material through the action of Golgi vesicles; the process closely resembles cell plate formation in higher plants (Markey & Wilce, 1975).

CHAPTER 4

Levels of organization

The range of morphology

An understanding of the organization of algae is fundamental to an appreciation of their evolution and classification. Modern microscopy has greatly expanded the potential for descriptive morphological accounts of algae, and corroborative physiological and ultrastructural details now greatly refine taxonomy.

Many algae exhibit great morphogenetic plasticity in response to environmental factors (Chapter 6), and many possess two or more entirely different morphologies at different stages of their life history (Chapter 5). The level of organization of the algae remains, however, comparatively simple and, with only a few exceptions among the larger Phaeophycota, differentiation into complex tissues is rare.

Morphological evolution has tended to trace parallel pathways in the various algal divisions. This parallelism was most eloquently discussed by Fritsch (1935). It assumes that from unicellular ancestors a comparable series of steps leading to colonial, filamentous, parenchymatous, and, in some, siphonous types was possible in each of the lines (now largely represented by divisions). In some divisions the sequence did not go far (or the more 'advanced' forms have been lost), in others the whole spectrum of morphologies can be seen in living species (e.g. Chlorophycota), and in some (e.g. Phaeophycota) the simplest forms may no longer exist, or are presently not recognized as brown algae. Where there is a fundamental degree of similarity among the species in a division, a single evolutionary line is suggested (monophyletic), while in others it is clear that more than one evolutionary line is present (polyphyletic, e.g. Chlorophycota). The evolutionary aspects will be further discussed in Chapter 8, but serve, however, as a useful framework for the following discussion of morphological variation.

Algae range from unicellular through to colonial, filamentous, pseudoparenchymatous, parenchymatous, siphonocladous (semi-coenocytic) and siphonous (coenocytic) forms (Table 4.1). The unicells may be non-motile, or coccoid (protococcoid in Round, 1973), amoeboid, or rhizopodial, and motile, or flagellate. Colonial forms range from palmelloid (an often temporary state resulting from loss of motility) to tetrasporal, a more permanent state when the products of cell division remain aggregated in a mucilaginous mass. In coenobial species the number of cells in the non-motile or motile colonies does not alter once they are formed; cells are organized in a definite arrangement.

Simple filamentous types become possible when cells remain joined together following cell division. Further planes of cell division allow the development of branched filaments and, when divisions are proportionately more abundant in the lower parts of the plant, distinct upright and prostrate portions may develop to give rise to heterotrichous types. Pseudoparenchymatous thalli develop through the branching of a single (uniaxial) or a number (multiaxial) of central filaments held together in a common matrix. Alternatively, the division of a filament in two or more planes results in a truly parenchymatous mode of growth. In the latter, cell divisions may occur anywhere in the thallus, or are localized in an apical or intercalary meristem. Coenocytic thalli result from repeated nuclear divisions with relatively few (siphonocladous; Hoek, 1981) or no (siphonous) cross-walls. As in filamentous forms, complex siphonous thalli can develop through the amalgamation of several siphons to form multiaxial types.

Unicellular organization — rhizopodial forms (Fig. 4.2)

Unicellular (or acellular) species are found in all classes of algae except the Phaeophyceae and Charophyceae. While it may be disputed that living, unicellular representatives of the algae are not necessarily the simplest forms, they characteristically contain the cytological and physiological features of their respective group.

Table 4.1 Summary of levels of organization in the algae.

	Unicellular			Colonial		Filamentous			Pseudoparenchymatous				
	Coccoid	Rhizopodial	Flagellate	Palmelloid or tetrasporal	Coenobial	Simple	Branched	Heterotrichous	Uniaxial	Multiaxial	Parenchymatous	Siphonocladous	Siphonous
Prokaryote													
Cyanophyceae	+			+		+	+				+		
Prochlorophycota	+												
Eukaryote													
Rhodophyceae	+						+	+	+	+	+		
Chrysophyceae	+	+	+	+	+		+				+		
Prymnesiophyceae			+	+									
Xanthophyceae	+	+	+	+		+	+						+
Eustigmatophyceae	+												
Bacillariophyceae	+			+									
Dinophyceae	+	+	+	+			+						
Phaeophyceae							+	+	+	+	+		
Raphidophyceae			+	+									
Cryptophyceae	+		+	+									
Euglenophyceae			+	+									
Chlorophyceae	+	+	+	+	+	+	+	+	+	+	+	+	+
Charophyceae									+				
Prasinophyceae	+		+	+									

Unicellular forms which lack (or have temporarily discarded) their flagella, have a naked protoplast, and move in an amoeboid manner, may be described as rhizopodial. They also possess cytoplasmic projections known as pseudopodia (which may assist in entrapment of food particles) and rhizopodia. They are most abundant in the Chrysophyceae and Xanthophyceae, but also occur in the Dinophyceae and several other classes (Table 4.1). The rhizopodial type may also be extended to include those forms in the Chrysophyceae and Xanthophyceae that consist of a naked cell resting in a theca, from which cytoplasmic filaments project.

The chrysophytes with a dominant amoeboid phase are placed in the Chrysamoebidales. The family Chrysamoebidaceae includes *Chrysamoeba,* which possesses a flagellate stage in which the secondary flagellum is reduced to a basal body. Other amoeboid forms in the Chrysamoebidales lack a flagellate stage (placed in the family Rhizochrysidaceae in Bourrelly, 1981). Networks of cells interconnected by rhizopodia develop in *Chrysostephanosphaera.* The loricate species are found in the family Stylococcaceae; they possess mostly amoeboid rather than flagellate reproductive stages. *Myxochrysis* is a remarkable member of the Chrysamoebidales, as it exists in the vegetative state as a multinucleate plasmodium. Its life history, which includes a flagellate swarmer stage, shows a close resemblance to that of slime moulds (Myxomycota). It has been suggested that these types of chrysophytes gave rise to the slime moulds (Hollande & Enguinet, 1955).

Rhizopodial xanthophytes show a comparable range of forms to those in the Chrysophyceae, from free-living amoeboid types possessing pseudopodia, or filopodia (e.g. *Heterochloris*), to loricate (e.g. *Rhizochloris*) and plasmodial forms (e.g. *Myxochloris*).

In the Dinophyceae the rhizopodial habit is exemplified by *Dinamoeba varians* (order Dinamoebales). A predominant amoeboid stage occurs, and the gymnodinioid motile stage is short-lived.

Unicellular organization — coccoid forms
(Figs 4.1 & 4.2)

The coccoid habit refers to those unicells where a non-motile state predominates and where motility is either entirely absent (e.g. the azoosporic chlorophytes such as *Chlorella*) or restricted to reproductive stages (*Chlorococcum*, Chlorophyceae). Coccoid forms occur in the majority of algal classes, and in some are the predominant, (70% of Xanthophyceae; Ott, 1982) and in others the only (all Eustigmatophyceae; Hibberd, 1982c), type of organization.

In the Prochlorophyta the single genus *Prochloron* is coccoid. Many cyanophytes are coccoid (e.g. *Aphanocapsa*; *Synechococcus*), especially members of the order Chroococcales (Chroococcaceae; Bold & Wynne, 1978). In many examples (e.g. *Chroococcus; Gloeocapsa*) there is a tendency for the products of binary fission to remain partially attached, or embedded in a common gelatinous matrix. There is considerable uncertainty as to the number of species of coccoid cyanophytes. Drouet & Daily (1956) reduced the number of genera to six, although their approach, based largely on morphological characters, is disputed by Stanier *et al.* (1971) and Rippka *et al.* (1979) who argue that on biochemical grounds, the coccoid Cyanophyceae comprise several distinct and widely separated groups.

In the Rhodophyceae the coccoid habit is rare and limited to members of the Porphyridiales (Bangiophycidae *sensu* Bold & Wynne, 1978), comprising some eight genera. The best known example is *Porphyridium* (Gantt & Conti, 1965; Gantt *et al.*, 1968) in which the spherical individuals are solitary or in aggregations held in mucilage. Amoeboid gliding has been described in *Petrovanella* (Pringsheim, 1968) and *Porphyridium* (Lin *et al.*, 1975; Sommerfeld & Nichols, 1970). Chloroplast shape is used as a generic distinction among coccoid rhodophytes (single and stellate, *Porphyridium;* numerous, discoid, *Rhodospora;* cup-shaped with 3–7 lobes, *Rhodosorus*). In other rhodophytes, a coccoid habit has been described only as a culture artifact: in the marine filamentous *Asterocytis* the thallus may become disorganized into a unicellular state in low-salinity culture (Lewin & Robertson, 1971), when the cells resemble *Chroothece* (Porphyridiales).

In the Chrysophyceae, coccoid forms are found in some Chrysocapsales and most Chrysosphaerales (Kristiansen, 1982). Cells of the Chrysocapsales (e.g. *Chrysocapsa*; *Phaesaster*) are embedded in mucilage, and while some are coccoid there is a greater tendency to a palmelloid con-

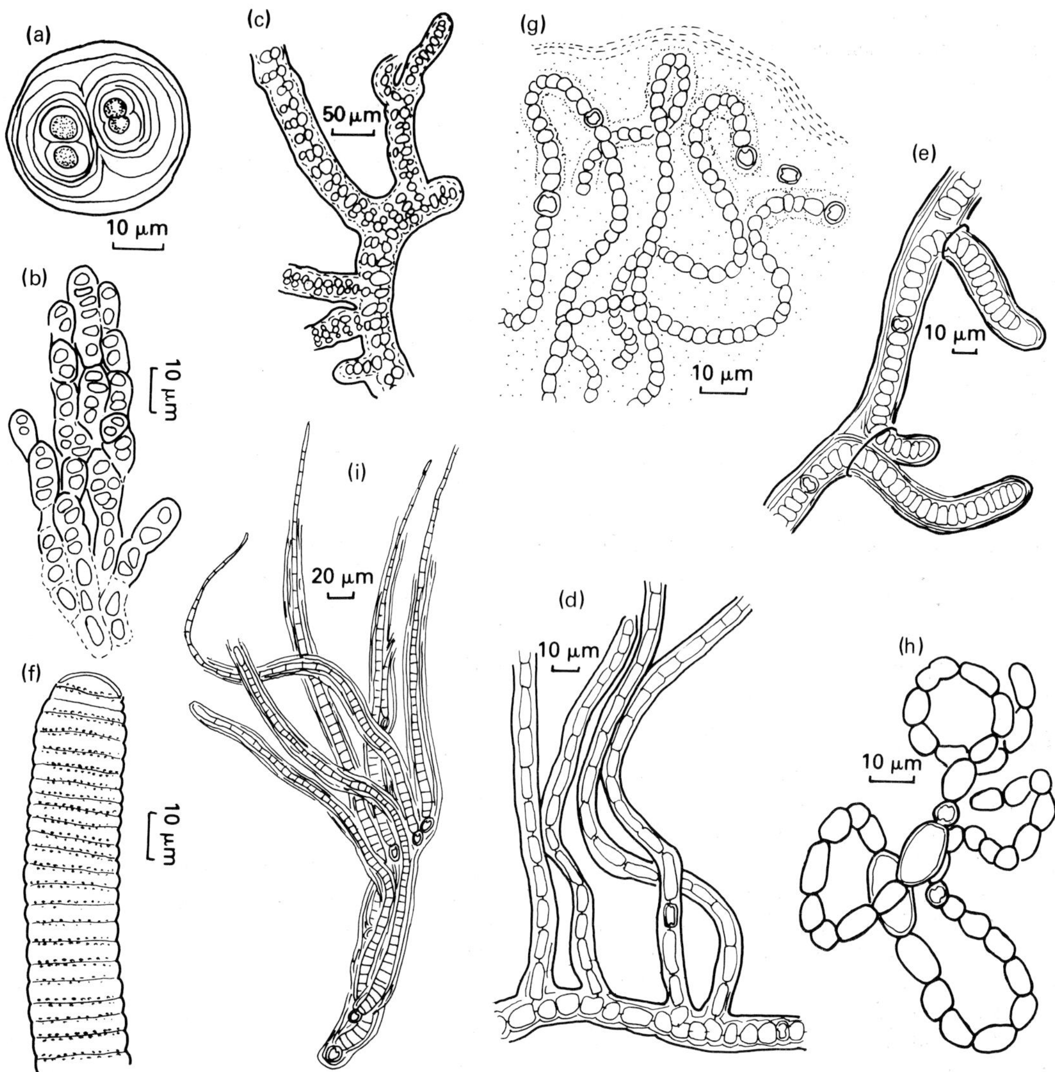

Fig. 4.1 Levels of organization in the Cyanophyceae. (a) *Gloeocapsa ralfsiana*, coccoid. (b) *Pleurocapsa cornuana*, palmelloid, showing a tendency to form simple filaments. (c) *Stigonema mamillosum* and (d) *Fischerella maior*, filamentous, with true branching. (e) *Scytonema millei*, filamentous, with false branching. (f) *Oscillatoria margaritifera*, simple unbranched filament. (g) *Nostoc piscinale*, colonial, with trichomes and numerous heterocysts embedded in a gelatinous mass. (h) *Anabaena flos-aquae*, trichomes with large intercalary heterocysts, planktonic. (i) *Calothrix orsiniana*, colonial, the trichomes with basal heterocysts and a conspicuous sheath. (After Bourrelly, 1970.)

dition; motile zoospores resembling *Chromulina* (Ochromonadales) are formed. In the Chrysosphaerales the range of coccoid forms parallels that of the Chlorococcales (Chlorophyceae) (Fritsch, 1935). The cells of *Chrysosphaera* are spherical, have a definite wall, and contain two parietal chloroplasts. Following division the daughter cells may remain aggregated in fours. Uniflagellate zoospores are formed (Starmach, 1972). *Epichrysis*, an epiphyte on filamentous freshwater algae, forms dense clusters of individual cells.

Most coccoid species in the Xanthophyceae are placed in the order Mischococcales (Hibberd,

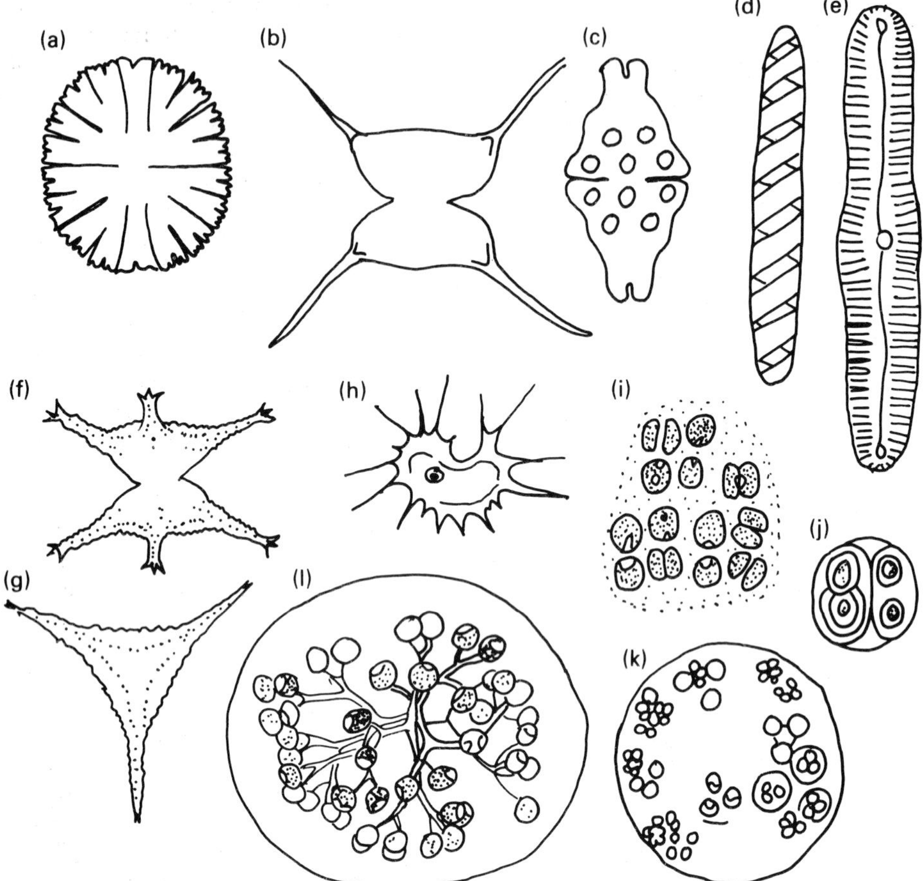

Fig. 4.2 Coccoid (a–g), rhizopodial (h) and palmelloid (i–k) organization. (a) *Micrasterias denticulatus* (Chlorophyceae). (b) *Staurodesmus validus* (Chlorophyceae). (c) *Euastrum didelta* (Chlorophyceae). (d) *Spirotaenia condensata* (Chlorophyceae). (e) *Pinnularia* sp. (Bacillariophyceae). (f) *Staurastrum anatinum* (Chlorophyceae) (face view). (g) *S. anatinum*, polar view. (h) *Rhizochrysis* sp. (Chrysophyceae). (i) *Tetraspora* sp. (Chlorophyceae). (j) *Gloeocystis naegeliana* (Chlorophyceae). (k) *Sphaerocystis schroeteri* (Chlorophyceae). (l) *Dictyosphaerium pulchellum* (Chlorophyceae). Not to scale. (After West, Fritsch, Doflein, Reinke, Artari, Chodat, and G.M. Smith.)

1982a). The vegetative cells may contain a single (e.g. *Characiopsis*) to many (e.g. *Ophiocytium*) nuclei, and lack contractile vacuoles and eyespots. In the Heterogloeales the cells, which are solitary, or more commonly in palmelloid, mucilaginous colonies, retain contractile vacuoles and eyespots (Hibberd, 1982a). The cell wall may be entire (*Botrydiopsis*) or in two pieces (e.g. *Ophiocytium*), and the surface is smooth or sculptured. Cell shape ranges from spherical through to ellipsoidal. Motile zoospores, when produced, are biflagellate with heterokontic, heterodynamic flagella.

The vegetative stage of all Eustigmatophyceae is composed of coccoid unicells (Hibberd, 1982c). Cell shape ranges from spherical to polyhedral, stellate or ovoid, and size varies from two to 18 μm (Hibberd, 1982b). A single deeply lobed chloroplast is present, and the cells have a typically yellow-green colouration. When formed, zoospores possess the highly characteristic structure for the class (Chapter 2).

Coccoid forms in the Dinophyceae occur either as free-living species (Phytodiniales, 13 genera) or as symbionts (Zooxanthellales, three marine genera) (Loeblich, 1982). Other non-motile unicells in the class are highly specialized and many are parasitic. In the Phytodiniales the vegetative cells are uninucleate and contain numerous small chloroplasts. They may be epiphytic, with a variety of modes of attachment, or free-living. Cells vary from spherical to crescent-shaped. Vegetative cells cannot divide, but reproduce by means of non-motile autospores, gymnodinioid zoospores, or thecate zoospores (e.g. *Stylodinium*). The genera and species are distinguished from one another principally on the basis of cell morphology and mode of reproduction. Members of the Zooxanthellales occur intracellularly as symbionts in a variety of marine invertebrates, and most lack a free-living form. Cell shape is spherical to elongate, and vegetative reproduction is by binary fission. When formed, zoospores are biflagellate and gymnodinioid (Loeblich, 1982).

Strictly coccoid forms are absent from the Raphidophyceae, although a palmelloid condition may occur (Heywood, 1982). The majority of cryptophytes are flagellate, but some may be coccoid or palmelloid; cell shape changes from the normally ovoid, dorsiventral form to spherical in the non-motile condition (Santore, 1977). The Euglenophyceae, primarily flagellate, may encyst or become palmelloid (Bold & Wynne, 1978); coccoid forms do not occur.

A wide variety of coccoid species is found in the Chlorophyceae (principally the orders Chlorococcales and Chlorosarcinales, whose members are mostly soil-dwelling). Coccoid Chlorococcales or Chlorosarcinales are either zoosporic (capable of producing motile zoospores) or azoosporic (not capable of producing motile zoospores). Some authors (e.g. Bold & Wynne, 1978) place the azoosporic types in the order Chlorellales. The Chlorococcales and Chlorosarcinales are distinguished from one another by their mode of multiplication. In the Chlorococcales the parental cells do not directly produce their progeny by cell division, whereas they do in the Chlorosarcinales. In the latter the mode of cytokinesis can be described as vegetative cell division, a term first introduced by Fritsch (1935). The process was later termed desmoschisis (Groover & Bold, 1969). Essentially, in desmoschisis the products of division incorporate components of the parent cell wall. As stated by Fritsch (1935) this mode of division is an essential starting point towards the development of a true filamentous habit; it is characteristic of all the advanced algae and higher plants. In the alternative mode of multiplication, as demonstrated by the Chlorococcales, the products of cell division are either naked, or surrounded by new walls. Groover & Bold (1969) referred to this mode of multiplication as eleutheroschisis.

Chlorococcum (*c.* 38 species; Archibald & Bold, 1970) and *Trebouxia* are well-known zoosporic, coccoid members of the Chlorococcales. In *Chlorococcum* the cells are spherical or ellipsoidal, uninucleate, and contain a single parietal, sometimes cup-shaped, chloroplast with one or more pyrenoids. The cells are not polarized and lack stigmata and contractile vacuoles, except when motile stages are formed. The species may be distinguished on the basis of a variety of biochemical features (Archibald & Bold, 1970). *Trebouxia* is a phycobiont (algal symbiont) in lichens (Ahmadjian & Hale, 1973). Cells contain a large chloroplast and a single pyrenoid. The ubiquitous *Chlorella* is the best-known azoosporic chlorococcalean alga. Its small cells (2–12 μm) are spherical or ellipsoidal and distinguished from *Chlorococcum* by their azoosporic nature and the presence of sporopollenin in the wall (Atkinson *et al.*, 1972). The taxonomy of *Chlorella* is complex, and there is no real agreement on the number of species. Physiological, biochemical (Shihira & Krauss, 1964; Vinayakumar & Kessler, 1975) and morphological (Fott & Novakova, 1969) features have been employed to separate species, with very different results. A colourless relative of *Chlorella, Prototheca*, causes the disease protothecosis in animals and man (*see* Klintworth *et al.* [1968] and Chapter 9).

Highly specialized coccoid unicells are found in the desmids (Chlorophyceae, Zygnematales) and diatoms (Bacillariophyceae) (Fig. 4.2). The desmids are abundant freshwater species, commonly placed into two families (orders, according to Mix, 1975). In one, the saccoderm-type (family Mesotaeniaceae) the unicells are round in cross-section, rod-shaped to oblong, and lack a median constriction. The single nucleus is centrally positioned and there may be two spiral chloroplasts (e.g. *Spirotaenia*), a plate-like chloroplast (e.g. *Mesotaenium*), or two stellate chloroplasts (e.g. *Cylindrocystis*). The wall is composed of a single piece and is smooth and lacking in pores. In the other, the placoderm-type (family Desmidiaceae), the cells are mostly flattened in cross-section and possess three planes of symmetry (front, side, and end) at right angles to one another. The cell wall is composed of two halves closely fitted together (it may be extremely difficult to detect the two halves using the light microscope). The walls may lack a constriction (e.g. *Closterium*) or (in the vast majority of placoderm species) there may be a marked constriction (the sinus) between the two halves (or semi-cells), which are joined at the isthmus (e.g. *Cosmarium; Euastrum; Micrasterias*). The single nucleus is usually positioned in the region of the isthmus. The walls are ornamented with ridges, spines, warts, and projections, and numerous pores are present. The axile chloroplasts, usually one in each semi-cell, are often intricately lobed or dissected, and contain pyrenoids. No flagellate stages occur in desmids, and multiplication is by vegetative cell division, a process that is complicated by the bipartite nature of the cell; it has been described in detail by Pickett-Heaps (1975). Following mitosis and cytokinesis the two semi-cells separate and each then buds a new semi-cell, which enlarges to the mature size as a mirror-image of the parent semi-cell.

The silica walls of the diatoms are also composed of two halves; their structure and morphological variation are described in Chapter 3. Examples of centric (radially symmetrical, order Biddulphiales) diatoms include *Biddulphia*, *Cyclotella*, and *Melosira*, while examples of bilaterally symmetrical types (order Bacillariales) are *Nitschia*, *Navicula*, and *Pinnularia*.

The principal mode of multiplication in diatoms is through cell division which, as in the desmids, results in the retention of one half of the parent cell wall in each of the progeny. The consequence of this in most diatoms is, unlike desmids, a gradual reduction in the cell size. (Chapter 5).

Both desmids and diatoms are capable of gliding movements, the former through the secretion of copious mucilage from the pores, and the latter through the raphe system in pennate species.

Unicellular organization — flagellates (Fig. 4.3)

The vegetative phase of many algae is a motile, flagellate unicell. Flagellate, vegetative unicells may be referred to as phytoflagellates, although the difficulty of this term from the phycological standpoint has been discussed by Dodge (1979). Phytoflagellates are found in all eukaryotic algal classes except the Rhodophyceae, Bacillariophyceae, Eustigmatophyceae, Phaeophyceae, and Charophyceae (the last three, however, possess flagellate gametes and/or zoospores). Flagellate unicells are the principal vegetative phase in the Dinophyceae, Raphidophyceae, Cryptophyceae, Euglenophyceae, and Prasinophyceae. Within many classes there is considerable uniformity in the features of the flagellate cells, whether vegetative or reproductive; the features common within each class were summarized in Chapter 2, and cytological details were given in Chapter 3.

Motile unicells are commonly spherical, elongate, or ovoid, and round in cross-section. Flattening may occur, resulting in a dorsiventral construction. Cell contents are usually polarized, with the forward (anterior) end of the cell commonly the site of flagellar insertion and the location of photoreceptive organelles and contractile vacuoles. The majority contain a single nucleus and one to several chloroplasts positioned laterally or posteriorly. A variety of storage products may be found. Flagellate unicells may be naked, may possess a variety of coverings beneath the plasmalemma, or may possess a true wall (Chapter 3). Departures from the above construction are common; the flagellate unicellular types possess, however, the unique combinations of morphological and biochemical features characteristic of their class.

A number of distinctive unicellular types are found in the Chrysophyceae (Bourrelly, 1981; Kristiansen, 1982). They include those possessing two heterokontic, heterodynamic flagella (e.g. *Ochromonas*), or a single emergent flagellum with the second reduced to a short stub or peduncle (e.g. *Chromulina*), or a single emergent flagellum with a strengthening band and wave-like conformation (cf. the transverse flagellum of the Dinophyceae) (Pedinellales), or a single emergent flagellum combined with a cytoplasmic, silicified skeleton (Dictyochales = silicoflagellates). Biflagellate forms possess one hairy flagellum and one smooth flagellum, the hairy member normally bearing tripartite hairs (p.70); the hairy flagellum is the emergent member when single. Cells may be naked (e.g. *Ochromonas*), covered with silica (e.g. *Mallomonas*) or organic scales (e.g. *Apedinella*), loricate (e.g. *Chrysococcus*), or enclosed in a tight-fitting wall. The unique,

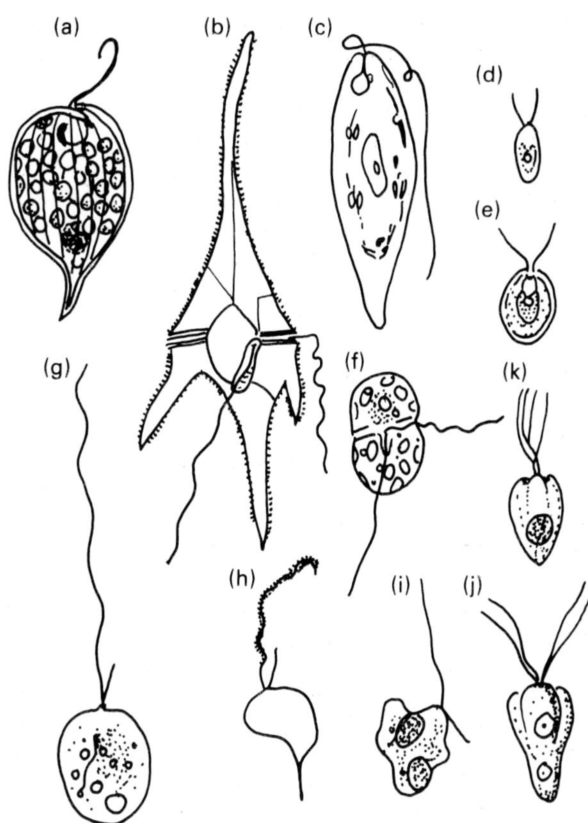

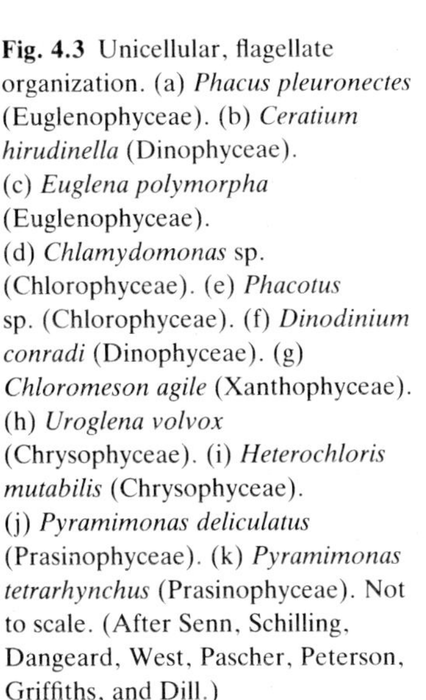
Fig. 4.3 Unicellular, flagellate organization. (a) *Phacus pleuronectes* (Euglenophyceae). (b) *Ceratium hirudinella* (Dinophyceae). (c) *Euglena polymorpha* (Euglenophyceae). (d) *Chlamydomonas* sp. (Chlorophyceae). (e) *Phacotus* sp. (Chlorophyceae). (f) *Dinodinium conradi* (Dinophyceae). (g) *Chloromeson agile* (Xanthophyceae). (h) *Uroglena volvox* (Chrysophyceae). (i) *Heterochloris mutabilis* (Chrysophyceae). (j) *Pyramimonas deliculatus* (Prasinophyceae). (k) *Pyramimonas tetrarhynchus* (Prasinophyceae). Not to scale. (After Senn, Schilling, Dangeard, West, Pascher, Peterson, Griffiths, and Dill.)

usually colourless choanoflagellates, which resemble the choanocytes of Porifera, and are included in the Chrysophyceae by Bourrelly (1981), should be referred to the animal kingdom (Leadbeater, 1972; Leadbeater & Manton, 1974; Kristiansen, 1982; Kristiansen & Takahashi, 1982).

The majority of chrysophytes (but not the Pedinellales) produce cysts (statocysts or stomatocysts). They comprise two unequal segments, a bottle-shaped, larger piece fitted with a smaller plug. The cyst wall is always siliceous regardless of the nature of the swimming stage (Bourrelly, 1957).

Characteristics of the flagellate unicells, which comprise the large majority of the Prymnesiophyceae, are as described for the class (p.20). Species lacking a haptonema or possessing one that is reduced are placed in the order Isochrysidales (Green & Pienaar, 1977), whose members also possess equal or subequal flagella and organic scale-covered cells. In some Isochrysidales (family Prinsiaceae) the organic scales act as templates for calcification and the formation of coccoliths (e.g. *Gephryocapsa*). The majority of coccolith-bearing species, however (or coccolithophorids), are placed in the order Coccosphaerales (Norris, 1982a). More than 150 genera of coccolithophorids have been described (Loeblich & Tappan, 1966; Tappan, 1980), many of which are fossil forms. Two basic kinds of coccoliths have been described: holococcoliths, in which rhombohedral calcite crystals form over the organic template, and heterococcoliths, which possess morphologically diverse crystals, usually of calcite but occasionally of aragonite. The assembled coccoliths may fit together to form a coccosphere (e.g. *Calcidiscus*). A complex classification of coccoliths, based on their shape, has been proposed (Braarud *et al.*, 1955; Tappan, 1980), although that such a classification is artificial is evident from a growing body of studies demonstrating the existence of coccolith- and non coccolith-bearing stages in the same life history (e.g. Parke & Adams, 1960; von Stosch, 1967; Leadbeater, 1970; Klaveness, 1973). Very little is known about the living cells of most coccolithophorids, but there is a large amount of literature detailing the fine structure of the scales (*see* Boney & Green, 1982).

The haptonema (p.67) is well developed in the family Prymnesiaceae (Prymnesiales, Prymnesiophyceae); it may coil in response to mechanical stimuli. The cells are covered with one to several layers of unmineralized scales (e.g. *Prymnesium*). The Pavlovales, distinctive in their possession of two unequal flagella, one of which is scaly (p.67), are also unique among the Prymnesiophyceae in their possession of a stigma. A great deal of attention has been focused on their fine structure (Green, 1980). Additional special features of this natural assemblage include knob-scales on the longer flagellum and a canal entering the cell close to the flagellar insertion (Boney & Green, 1982).

Vegetative, flagellate cells included in the Xanthophyceae are placed in the order Chloramoebidales (Ettl, 1978; Hibberd, 1982a). According to Hibberd (1982a), they likely belong in other classes, including the Chrysophyceae, Prymnesiophyceae, and Prasinophyceae.

The majority of the Dinophyceae are biflagellate unicells; they are distinctive in the structure of their flagella (pp.73 & 77), theca (p.26), and mesokaryotic nucleus (p.85), and in their possession of trichocysts, the pusule, and the ocellus. Trichocysts consist of a neck attached to the theca and a body containing paracrystalline proteinaceous material (Dodge, 1979). They differ from the ejectosomes of the Cryptophyceae in that, when discharged, they shoot out a long, cross-banded thread which is square or rhomboidal in section (Bouck & Sweeney, 1966; Dodge, 1973). In *Nematodinium* and *Polykrikos* nematocysts resembling those of the Coelenterata are found (Mornin & Francis, 1967; Greuet, 1972). The ocellus (p.83) is found in some larger species such as *Warnowia* (Greuet, 1977). The pusule is a permanent osmoregulatory organelle, and various types have been described from the dinophytes (Dodge, 1972). It is an invagination of the plasmalemma from the base of the flagellar canal, open to the exterior and characterized by the presence of a thick wall composed of two membranes.

While most dinoflagellates possess the characteristic thecal structure, with ventrally inserted flagella (e.g. *Gymnodinium; Peridinium*), some (e.g. *Prorocentrum*) possess anteriorly inserted flagella and a theca divided into equal right and left halves. Among the former examples, some possess an asymmetric theca, in which the transverse groove (cingulum) may be anteriorly positioned (e.g. *Amphidinium pacificum*), or in which the hypocone is reduced in size (e.g. *Katodinium rotundatum*).

Thecal plates are characteristic for the armoured dinoflagellates; up to 100 plates may be present. The number and arrangement of the plates have been employed as a taxonomic criterion, and plate formulae can be used to characterize genera and species. Plates of the epitheca and hypotheca (comparable to the epicone and hypocone of unarmoured forms in some classifications) are tabulated, resulting in a unique formula for a given taxon (Fig. 3.9). Three orders of armoured forms may be distinguished using such a system: Peridiniales (variable plate number; *see* Loeblich, 1970), Dinophysiales (18 plates, with a larger hypotheca and a smaller epitheca; Kofoid & Skogsberg, 1928; Dodge, 1979), and the Prorocentrales (two lateral plates, with anteriorly inserted flagella; Loeblich, 1982).

The majority of the swimming cells of dinoflagellates are haploid. Their common means of vegetative multiplication is by cell division, which may occur by longitudinal, transverse, or oblique partitioning of the cell following cytokinesis (Bold & Wynne, 1978). In naked forms division occurs by constriction; a new amphiesma may be formed around each half, or the old one may be shed (by ecdysis) prior to division. In most other forms the two halves of the parental theca separate following cytokinesis, and each of the progeny forms the missing half.

Resistant cysts of one of two kinds are commonly formed by dinoflagellates: those which resemble the vegetative cells (e.g. *Peridinium*), and those that are very different, usually spherical and covered with hollow projections (e.g. *Gonyaulax*). The cysts rupture to form an opening (the archaeopyle), which is taxonomically important. Many cysts are known only from fossil records (as hystrichospores; Sarjeant, 1974;

Tappan, 1980); they are discussed elsewhere (p.258).

A number of the Raphidophyceae (or Chloromonadophyceae; *see* Bold & Wynne, 1978; Dodge, 1979) are flagellate unicells, of very distinctive structure (p.29). Cells of *Gonyostomum* are large, covered by a thin periplast, dorsiventrally compressed, and possess two heterodynamic, unequal flagella. They contain numerous discoid chloroplasts, many trichocysts, and contractile vacuoles (Heywood, 1982). Reproduction is by longitudinal division while the cell is still in motion.

Most of the Cryptophyceae are flagellates and are unique in the ultrastructure of their chloroplasts, flagella, and ejectosomes (Dodge, 1979). Well-known genera include *Cryptomonas, Chroomonas, Chilomonas,* and *Hemiselmis*. Cells bear two anterior flagella which arise from a gullet, which is divided into a vestibular region and an ejectosome-lined channel extending posteriorly into the cell (Oakley & Santore, 1982). Flagella are more or less the same length; in *Cryptomonas* one bears two rows of tubular hairs and has a swelling at its base, whereas the other has only one row of hairs (Hibberd *et al.*, 1971). Taxa have traditionally been separated on the basis of the number of rows of ejectosomes in the gullet, and on pigmentation, although both characteristics are unreliable (*see* Bold & Wynne, 1978; Oakley & Santore, 1982). Cells are covered by a delicate periplast; plate-like structures lying beneath the plasmalemma are of a different shape in the various genera (e.g. hexagonal in *Cryptomonas*; Hibberd *et al.*, 1971; rectangular in *Chroomonas*; Gantt, 1971). Specific arrangements of ejectosomes may accompany the plates.

Ejectosomes (trichocysts, or taeniobolocysts in some works; Dodge, 1979) are found in all cryptophytes except *Hillea*. While analagous to those of the dinophytes and ciliates, they differ in structure. They are tapered threads when discharged, but internally they consist of a reel of membranous material.

The normal mode of multiplication of cryptophytes is by longitudinal division, during which cells may remain motile or non-motile. When in a palmelloid phase, the cells retain their flagella tightly coiled around the cells in the mucilaginous colonies.

Most euglenophytes are flagellated unicells, united as a distinctive class on the basis of a number of unique cytological and biochemical features (p.33). All possess two basal bodies and one (e.g. *Euglena*) or two (e.g. *Eutreptia*) emergent flagella of a highly specialized structure (Leedale, 1967). Colourless (e.g. *Peranema*) and pigmented species occur; Leedale (1982b) suggested that pigmented forms are most likely derived from colourless forms, through an endosymbiosis with chlorophycean chloroplasts, a sequence in reverse from that proposed in many previous accounts. Some colourless species (e.g. *Peranema trichophorum*) are phagocytic and ingest other cells and detritus. The 'mouth', or cytostome is equipped with parallel, hooked rods attached to a stiffened rim. The best-studied genus is *Euglena* (Leedale, 1967; Buetow, 1968, 1982), in which the species are separated on the basis of cell size and shape, length and motion of flagella, numbers and sizes of chloroplasts, pyrenoids, pellicular rigidity, muciferous bodies, and the form of paramylon granules (Leedale, 1982b).

Bovee (1982) categorized four kinds of euglenoid movements: swimming, contracting, crawling, and gliding. The last three involve various forms of metaboly.

The mucocysts, or muciferous bodies, are arranged in rows beneath the pellicle, and their secretion is a water-soluble polysaccharide. Cells are permanently coated in a thin slime layer (Rosowski, 1977). The mucilage secretions may develop as stalks (e.g. *Colacium*), complex cell envelopes (e.g. *Trachelomonas*), and cysts. The cysts in *Euglena* and *Distigma* are in the form of a thickened mucilaginous sheath, although in some genera the regular slime sheath may serve as temporary resistance against desiccation. Division may proceed within a slime layer (e.g. *Eutreptia*) and lead to the formation of extensive palmelloid colonies (Leedale, 1967). In *Euglena* the cells may become non-motile and greatly enlarged with age, and accumulate orange and black pigment particles.

Multiplication in euglenoids is entirely asexual, by cell division. After loss of the flagella, mitosis and replication of the basal bodies occurs; two daughter canals are formed, and the anterior end of the cell pushes inwards between them. The pellicular strips duplicate and the cell cleavage progresses backwards, following the helical arrangement of the pellicle. In forms with an elastic pellicle there is intense cell metaboly during the dividing process.

Chlorophycean flagellate unicells and their colonial derivatives are placed in the order Volvocales; the unicellular species have two, four or eight isokont, smooth, homodynamic flagella. The cells are naked (family Polyblepharidaceae, e.g. *Dunaliella*) or, as in the majority of species, surrounded by a cell wall that may be complex in structure (e.g. *Chlamydomonas*). The flagella normally emerge anteriorly through canals in the cell wall and, in the biflagellate species, at 180° to one another on either side of a papilla. Cells may be enclosed in a lorica, which is bipartite in *Phacotus*, the two halves separating during reproduction. In *Haematococcus* the wall is thickly gelatinous and penetrated by radiating strands of cytoplasm; in *Brachiomonas* it has projecting basal lobes. The remarkable *Medusochloris* (Pascher, 1917) resembles a medusa, the cell being a convex hemisphere with four processes each bearing a flagellum at the lower edge.

Volvocalean cells are normally uninucleate and lack a large central vacuole; instead they possess one or more osmoregulatory contractile vacuoles (absent in marine species) positioned directly beneath the papilla (e.g. *Chlamydomonas*), by the flagellar bases, or distributed throughout the cell (e.g. *Chlorogonium*). The conspicuous chloroplast is variable in form, and includes cup-shaped, parietal, asteroidal, reticulate, lamellate, H-shaped, and other types. An eyespot and one or more pyrenoids are commonly present, embedded in the chloroplast. Not all Volvocales are pigmented; both biflagellate (*Polytoma*) and quadriflagellate (*Polytomella*) colourless species occur.

Asexual reproduction is a result of mitosis and cell division. In naked species (e.g. *Dunaliella*) division may occur while cells remain motile, although most walled species (e.g. *Chlamydomonas*) temporarily lose their motility during mitosis and cytokinesis. During division duplication of organelles such as chloroplasts and pyrenoids occurs, although some organelles (e.g. contractile vacuoles, eyespots) apparently arise *de novo* in the progeny (Bold & Wynne, 1978). Divisions often occur in darkness, and may be sufficiently rapid and repeated that two, four, eight or more progeny form within the parent cell wall. The progeny are generally smaller, but after release gradually increase to the cell size typical for the species before undergoing further vegetative divisions.

The predominantly flagellate, unicellular Prasinophyceae are principally distinguished from the Chlorophyceae by their scale-covered flagella and cell body. Uniflagellate (e.g. *Pedinomonas*) and biflagellate species possess anterolaterally inserted flagella and a depression or groove. Quadriflagellate (or eight-flagellate) species (e.g. *Platymonas*) bear the flagella anteriorly, where they arise from a pit-like depression. The pigments and internal structure of prasinophytes are similar to *Chlamydomonas* (Chlorophyceae), with a cup-shaped chloroplast, an eyespot, and a large pyrenoid. A unique feature of some (e.g. *Pyramimonas; Halosphaera; Monomastix*) is the presence of trichocysts (Norris, 1982b), which closely parallel those of the cryptophytes (Dodge, 1979).

Prasinophyte reproduction is exclusively by vegetative cell division. According to Norris (1982b) the most primitive types remain motile during division, whereas the more advanced species lose their flagella. Sequential (more primitive) or synchronized (more advanced) duplication of organelles occurs at the onset of mitosis (Norris, 1982b). In some species (e.g. *Pterosperma; Pachysphaera*) a resistant, sporopollenin-containing phycoma stage may occur, this of importance in the interpretation of microfossils assigned to the Prasinophyceae (Norris 1980, 1982b).

Colonial organization — palmelloid, tetrasporal, and dendroid types (Figs 4.1, 4.2 & 4.4)

Most normally flagellate or coccoid unicellular algae may enter an (often temporary) palmella stage, a condition where the flagella are lost (if characteristically present) and the individuals undergo successive vegetative divisions while embedded in a common gelatinous matrix. Named after the volvocalean (Chlorophyceae) genus *Palmella*, the term may be strictly applied to those algae where the cells will readily revert to a motile condition (Bold & Wynne, 1978) or may be expanded to include all algae where the palmelloid habit is more permanent (Fritsch, 1935). Round (1973) and others, however, described the more permanently palmelloid species as tetrasporal (from *Tetraspora* of the Chloro-

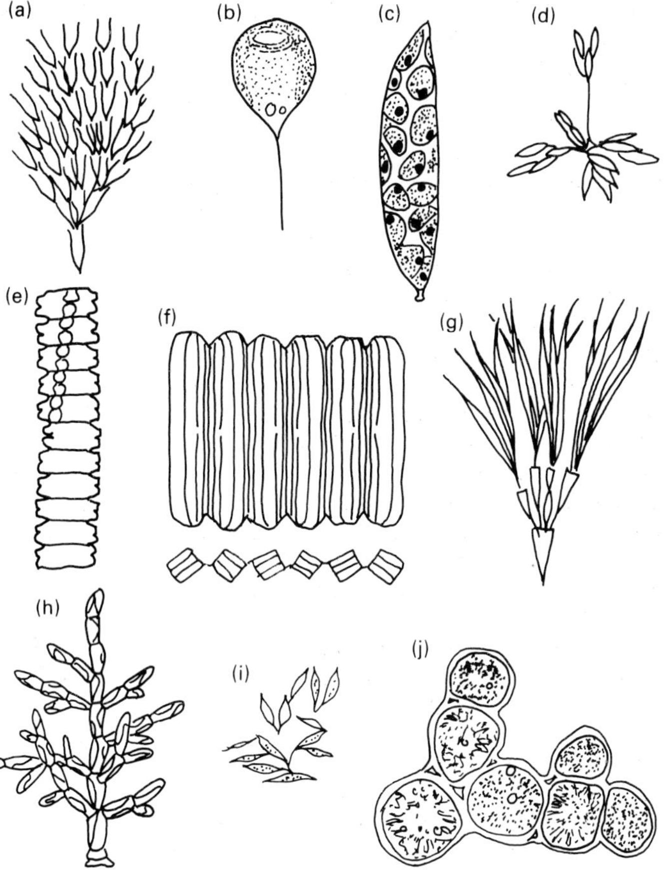

Fig. 4.4 Colonial organization — dendroid forms and simple filaments (a) *Dinobryon sertularia* (Chrysophyceae). (b) *Stylosphaeridium stipitatum* (Chlorophyceae). (c) *Characiopsis saccata* (Xanthophyceae). (d) *Cholorangium stentorium* (Chlorophyceae). (e) *Desmidium schwartzii* (Chlorophyceae). (f) *Pinnularia* sp. (Bacillariophyceae). (g) *Ankistrodesmus falcatus* (Chlorophyceae). (h) *Phaeothamnion* sp. (Chrysophyceae). (i) *Dactylococcus bicaudatus* (Chlorophyceae). (j) *Dinothrix paradoxa* (Dinophyceae). Not to scale. (After Lemmermann, Geitler, Carter, Cienkowski, West, Hustedt, and Pascher.)

phyceae). In tetrasporal types the motile stages (if present at all) are restricted to the reproductive cells. The palmelloid and tetrasporal forms are characterized by a lack of definite form, differentiation, or specific size limits to the colony.

Some of the more regular colonies of the Cyanophyceae (e.g. *Merismopedia* and *Halopedia*, as flat plates of cells; *Eucapsis*, with colonies in cubical masses) are included in the tetrasporal types. In some Cyanophyceae colonies comprise a few cells in a common matrix (e.g. *Chroococcus; Gloeocapsa*), little different from coccoid species, but may include species with extensive sheet-like gelatinous masses.

In some specialized examples, where the mucilage production is restricted to the base of the cell, dendroid colonies which resemble simple filaments may result (e.g. *Prasinocladus*, Prasinophyceae; *Dinobryon*, Chrysophyceae; *Colacium*, Euglenophyceae); these colonies readily revert to the motile condition.

In addition to some Cyanophyceae, tetrasporal forms (*sensu* Round, 1973) are found in the Chrysophyceae (e.g. *Stichogloea*), Prymnesiophyceae (e.g. *Phaeocystis; Sarcinochrysis*), Xanthophycae (e.g. *Gleochloris*), Dinophyceae (e.g. *Gloeodinium montanum*) and the order Tetrasporales of the Chlorophyceae (e.g. *Gloeocystis; Palmodictyon; Tetraspora*). A feature of *Tetraspora* is the presence of mucilaginous pseudocilia.

The development of extensive colonies occurs in some Bacillariophyceae; these may be capsular in nature (e.g. some *Navicula* species), or in the form of a mucilaginous tube (e.g. *Amphipleura*). The individual cells continue to move and divide within the mucilage, and the tube is generally attached to a firm substrate; it may reach several centimetres in length. The Bacillariophyceae (especially the Biddulphiales) feature a variety of additional specialized colonies, these dependent on mucilaginous attachments between cells following division. They may be filamentous (e.g. *Melosira, Fragilaria, Eunotia*). If loosely attached the frustules may glide over one another (e.g. *Bacillaria*), or they may be offset in a zigzag filament (e.g. *Grammatophora; Tabellaria*). Stellate colonies are characteristic of some species of *Tabellaria* and *Diatoma*. In a number of epiphytic diatoms (e.g. *Licmophora*) the cells may be interconnected with each other and attached to the substrate by mucilaginous stalks in a branching, dendroid colony.

Many eukaryote, tetrasporalian species may reproduce by simple fragmentation of the colony, but under the right conditions (e.g. a sudden change in temperature) will form motile zoospores.

Colonial organization — coenobia (Fig. 4.5)

Organized colonial forms, or coenobia, comprise aggregations of flagellate or non-motile cells. The former are found in a number of eukaryote classes (Table 4.1), while the latter are restricted to the Chlorophyceae. Motile coenobia generally differ from palmelloid, tetrasporal, or dendroid forms in that the number of individuals is fairly constant for a given species, there is a definite form to the colony, and often there is co-ordination and some degree of differentiation among the individuals.

The most complete range of motile colonial species is found in the Volvocales (Chlorophyceae). The coenobia are composed of cells, which often resemble a *Chlamydomonas*, in multiples of two and contained within a common gelatinous envelope. Colonies range from those with relatively few cells (e.g. *Gonium*, 4–32) to those

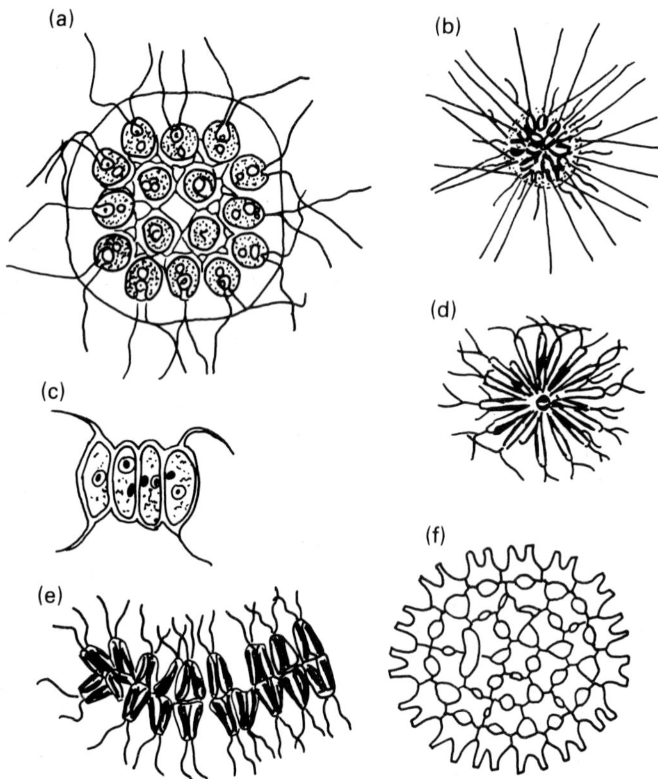

Fig. 4.5 Colonial organization — coenobia. (a) *Gonium pectorale* (Chlorophyceae). (b) *Chrysosphaerella longispina* (Chrysophyceae). (c) *Scenedesmus quadricauda* (Chlorophyceae). (d) *Synura adansonii* (Chrysophyceae). (e) *Chlorodesmus hispidus* (Chlorophyceae). (f) *Pediastrum duplex* (Chlorophyceae). Not to scale. (After Hartmann, Lauterborn, G.M. Smith, Conrad, and Philipps.)

with more (e.g. *Eudorina,* 16–128), culminating in *Volvox* (500–40 000). Within the colonies the cells are specifically oriented with respect to one another; colonies may be planar (*Gonium*; *Platydorina*), ball-like (*Pandorina*) or spheroid (*Eudorina*; *Volvox*), with the cells arranged in a single layer. As colony size increases there is an increasing tendency for the development of polarity. This is expressed in an anterior–posterior orientation and an anteroposterior gradation in the size of the cells and eyespots. Polarity also results in directional swimming, and in the formation of specialized cells for the formation of asexual and sexual reproductive bodies. The motile coenobia reproduce asexually by the formation of autocolonies, which contain the parental cell number. The parental colony size is achieved by subsequent cell enlargement. In the simpler species (e.g. *Pandorina*) all the cells are capable of autocolony formation, whereas in the advanced species (e.g. *Volvox*) only specialized, aflagellate, enlarged cells (gonidia) can undergo this process. Autocolony formation in *Volvox* is summarized in Fig. 5.3.

The non-motile chlorophycean coenobia are found in the order Chlorococcales, families Hydrodictyaceae (e.g. *Pediastrum; Sorastrum; Hydrodictyon*) and Scenedesmaceae (e.g. *Scenedesmus; Coelastrum; Dictyosphaerium*). The coenobia of *Pediastrum* and *Sorastrum* are flattened and one cell thick, and the uninucleate cells are arranged in concentric circles, with the marginal (and sometimes inner) cells provided with horn-like projections. The coenobia multiply via asexual reproduction. Biflagellate zoospores are formed in each of the cells, following synchronous mitoses. The parental cell wall then tears open and the then active zoospores are released, still contained within their common vesicle. After a brief period of motility the zoospores withdraw their flagella and adhere to one another in the characteristic colony arrangement for the species. The cells then undergo wall formation and the vesicle ruptures. In *Hydrodictyon* a cylindrical, net-like structure (up to 1 m long!) forms the coenobium. It is composed of coenocytic, tubular cells containing a large central vacuole and many minute nuclei. The cells are arranged in a polygonal configuration, the cells being joined end to end (usually six cells in a group). The coenobium reproduces asexually by the formation of numerous uninucleate, biflagellate zoospores in each of the cells. The zoospores gather in a vacuolar envelope, between the vacuole and the cell wall, and become arranged in a thin layer. After a period of active motility they withdraw their flagella and become permanently joined in groups characteristic for the species, the groups in a tubular configuration and lining the parental cell wall. The new coenobia are released following breakdown of the cell wall (Pickett-Heaps, 1975).

The coenobia of *Scenedesmus* are made up of cylindrical cells joined laterally in groups of four, eight or 16. The outer (and sometimes other) cells possess one or more spines. Vegetative multiplication involves a modified form of autospore formation (Pickett-Heaps & Staehelin, 1975). Motile zoospores are rare, and not associated with autocolony formation. Coenobia in *Coelastrum* are hollow spheres of 4–128 cells (Bold & Wynne, 1978). In *Dictyosphaerium* there is an entirely different organization of the spherical, gelatinous coenobia. Here the *Chlorella*-like cells are in groups of four and borne on stalks that may form an extensive dendroid system within the spherical colony.

Filamentous organization — simple filaments
(Figs 4.1 & 4.4)

A simple filament consists of a single row of cells firmly attached to one another, and with no branching. It represents the most basic form of multicellular algal thallus (Fritsch, 1935). Simple filaments are found in only a few classes (Table 4.1); they can be derived from motile or non-motile unicellular forms in which (in eukaryotes) modifications in spindle orientation, cytokinesis (e.g. desmoschisis) and wall formation have occurred. The normally uninucleate cells (e.g. *Ulothrix*; Chlorophyceae) form filaments that may range from only a few cells in length (e.g. *Stichococcus*; Chlorophyceae) to many (e.g. *Oscillatoria,* Cyanophyceae). Both free-living (e.g. *Ulothrix*; *Oscillatoria*) and attached types occur.

In the Cyanophyceae the simple filament is the trichome. The individual cells may be extremely small (< 1.0 µm diameter in in *Lyngbya*, Cyanophyceae). They may be of equal width throughout its length (e.g. *Oscillatoria*), with only slight modification of the terminal cell, or markedly tapered from base to apex, with the terminal portion hair-like (e.g. *Scytonema, Rivularia*) and the basal cell a heterocyst. Tapered trichomes may be grouped in a manner suggestive of branching (e.g. *Rivularia*), may radiate from a central point (e.g. *Gloeotrichia*), or may be embedded in a sheet-like mass. Colonies of short trichomes in a common gelatinous matrix are characteristic for *Aphanizomenon* and *Nostoc*. In some smaller, planktonic species (e.g. *Spirulina*) the trichome is helically twisted. Gliding movement occurs in most simple filamentous cyanophytes, and may be smooth, jerky, or swaying in nature (Fogg *et al.*, 1973); it is associated with pores and the secretion of mucilage.

Simple filamentous cyanophytes readily fragment into hormogonia. These result either from natural breaks, or from special biconcave cells known as separation discs, or necridia. The hormogonia are normally liberated from the parent sheath, but may remain within it (e.g. *Microcoleus*).

Tribonema is the best-known simple filamentous Xanthophyte. The unattached filaments are made up of normally uninucleate cells that are uniformly cylindrical or barrel-shaped, their walls consisting of two equal and slightly overlapping halves. The filaments thus have open ends, and may dissociate into H-shaped pieces. Asexual reproduction is effected through the formation of one or two heterokont zoospores in unspecialized cells, or through the formation of aplanospores or thick-walled hypnospores.

A wide variety of simple filamentous species occurs in the Chlorophyceae, in the orders Microsporales (e.g. *Microspora*), Oedogoniales (e.g. *Oedogonium*), Ulotrichales (e.g. *Ulothrix*), Sphaeropleales (e.g. *Sphaeroplea*), Klebsormidiales (e.g. *Klebsormidium*), and Zygnematales (e.g. *Spirogyra*; *Zygnema*; *Desmidium*).

The filaments may be free-floating (e.g. *Microspora*; *Sphaeroplea*; *Klebsormidium*; most Zygnematales) or attached. Attachment is commonly effected through simple modification of the basal cell, which may form a holdfast through expansion of the end wall (e.g. *Oedogonium*). The uninucleate cells have continuous walls (except in *Microspora*, where they are in H-shaped pieces, and in the Oedogoniales, where the walls split during division to form cap cells; p.64). A mucilaginous sheath is often present (especially Zygnematales). Apart from the holdfast cell there is generally little differentiation of the filament. In *Desmidium* the filaments are twisted.

Identification of the many simple filamentous Chlorophyceae is often difficult (e.g. *Oedogonium*, several hundred species; *Spirogyra*, > 275 species; Bold & Wynne, 1978). Separation at the generic level, however, is facilitated by the often distinctive features of the chloroplast and pyrenoids.

In the majority of simple Chlorophyceae filaments, growth is diffuse, with divisions occurring anywhere. Fragmentation or dissociation of the filament is the commonest means of vegetative reproduction; in some (e.g. *Klebsormidium*) dissociation may result even in single cells. Formation of motile zoospores is also common (except in the Zygnematales), and may take place in any cell (except the holdfast). Aplanospores or thick-walled hypnospores are also frequently produced, the latter as a means of survival during unfavourable conditions.

Filamentous organization — branched filaments (Figs 4.1, 4.4 & 4.6–4.13)

There are three principal modifications of branched filamentous thalli. In the first, a branched upright filamentous thallus is attached by a simple disc derived from the basal cell. In the second, the heterotrichous type, the thallus consists of two parts: a prostrate creeping base, and an erect, branched upright system. In the third either a single (uniaxial) or numerous (multiaxial) branched filaments become aggregated in a pseudoparenchymatous thallus. Accompanying these modifications is an increasing tendency for localization of cell division, for more cell specialization (and hence differentiation), and for the restriction of reproductive functions to special structures (e.g. sporangia and gametangia).

Branched filamentous systems occur in eight of the classes considered in this text (Table 4.1). In the Cyanophyceae, two modes of branching occur, the more primitive being false branching, which results from interruptions in the filament (e.g. by an heterocyst or separation disc). One (e.g. *Tolypothrix*) or both (e.g. *Scytonema*) portions of the trichome push out from the sheath as branches. No new division plane is involved. In the second, more advanced type, true branching, branches result from a second plane of division (e.g. *Hapalosiphon*).

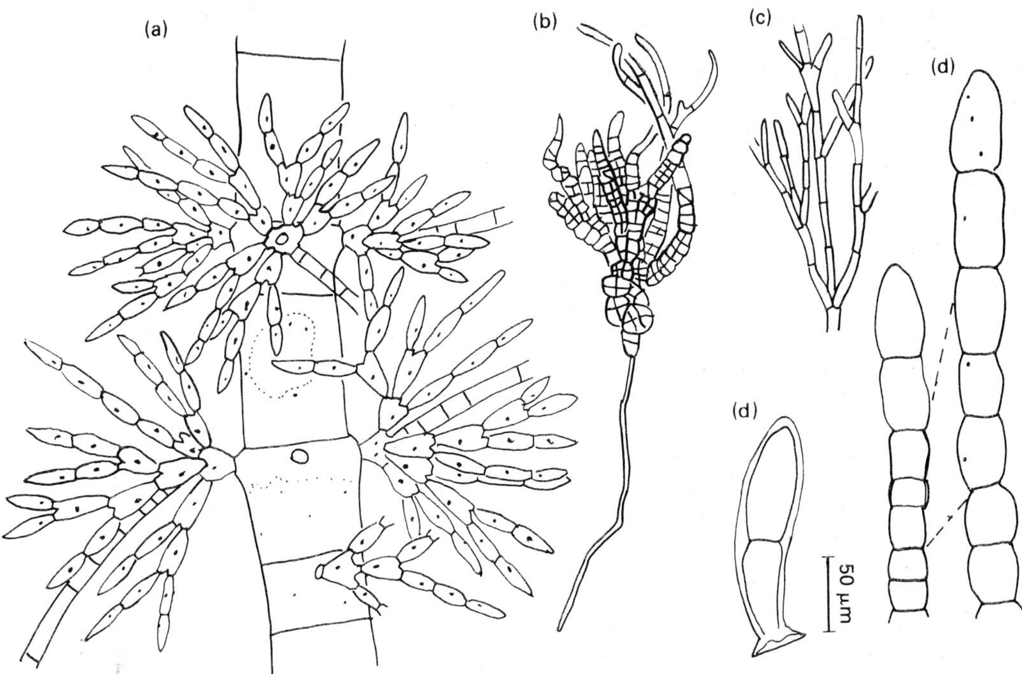

Fig. 4.6 Filamentous (a), heterotrichous (b), and siphonocladous (c, d) Chlorophyceae. (a) *Draparnaldia iyengarii*, portion of the thallus showing large cells of the main filament and whorled lateral branches with hairs. (b) *Fritschiella tuberosa*, a heterotrichous species with parenchymatous basal system producing rhizoids, and filamentous uprights. (c) *Cladophora* sp., a branched, siphonocladous species. (d) *Chaetomorpha darwinii*, an unbranched, siphonocladous form. Young plant (left) with basal holdfast cell and filaments (right) showing cell expansion following cell division. a, b, c not to scale. (After Tiwari, Pandey & Pandey, Iyengar, Hoek, and Migula.)

Branched filaments are rare in the Dinophyceae (e.g. *Dinothrix*), Xanthophyceae (e.g. *Heterodendron*), and Chrysophyceae (e.g. *Phaeothamnion*). Some heterotrichous examples also occur in each of these classes (e.g. *Dinocladium, Heterococcus,* and *Phaeodermatium*, respectively).

The heterotrichous filament is the simplest form of thallus in the Phaeophyceae. While growth is diffuse in the simpler forms (e.g. *Ectocarpus*), the more advanced examples show an increasing tendency towards intercalary cell division (e.g. *Pilayella*) or trichothallic growth (where divisions are concentrated in the subterminal portion of apical hairs, e.g. Tilopteridales). In many small species the prostrate system predominates (e.g. *Streblonema*). The heterotrichous species and their pseudoparenchymatous derivatives are generally grouped as the haplostichous Phaeophyceae. Among these are uniaxial examples such as *Acrothrix*, where growth commences from a solitary trichothallic filament and progresses to an apical mode, and *Desmarestia*, in which both terete and foliose thalli develop from a primary trichothallic axis (Chapman, 1972a, b).

Multiaxial types are more prevalent (e.g. members of the Chordariales, Cutleriales, and Sporochnales). They may be gelatinous and amorphous (e.g. *Leathesia*), branched (*Eudesme*), crustose (e.g. *Ralfsia*), or bladed (e.g. *Cutleria*). Considerable differentiation of the cells may result in distinct tissues (e.g. cortex, medulla, assimilatory layer), all derived from a strictly filamentous origin. Branching in multiaxial types is possible when a group of filaments and their associated branches diverge from the main axis. Branched filamentous stages may occur as part of the life history of otherwise parenchymatous

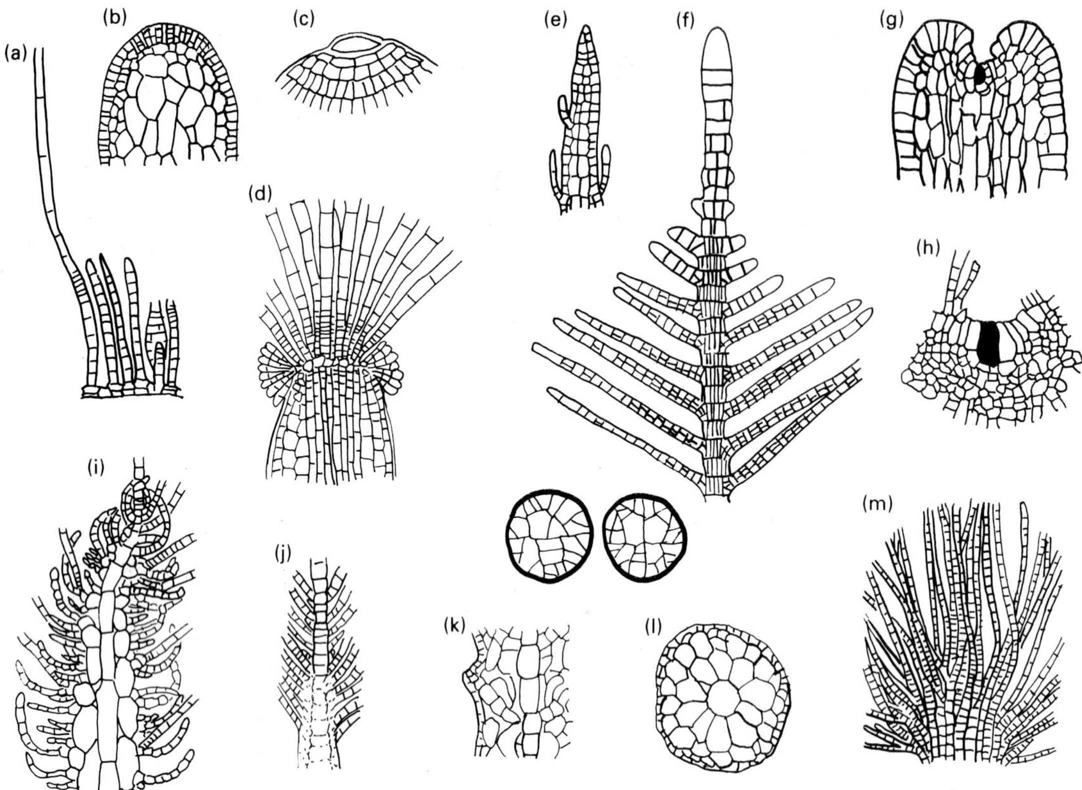

Fig. 4.7 Pseudoparenchymatous (haplostichous) and parenchymatous (polystichous) Phaeophyceae. (a) *Myrionema strangulans*, showing the two-layered basal system and the heterotrichous uprights with random division, and hairs. (b) *Chnoospora obtusifolia* — apical region of the parenchymatous thallus, and small-celled apical meristem region. (c) Surface view of apex of the dorsiventrally flattened thallus of *Dictyota dichotoma*, showing the single apical cell and pseudoparenchymatous development. (d) Section of the apex of *Carpomitra costata*, showing meristematic hairs from which are derived a multiaxial, pseudoparenchymatous thallus. (e) Longitudinal section through the apex of *Dictyosiphon foeniculaceus*, showing the single apical cell and its derivatives, leading to the parenchymatous thallus. (f) Apex of *Sphacelaria plumula*, with the main and lateral apical cells and parenchymatous axes (shown in cross-section). (g) Median longitudinal section of the apical region of a young parenchymatous sporeling of *Fucus* sp., showing the single apical cell (shaded) in the apical groove. (h) Median longitudinal section of an older sporeling of *Fucus* sp. (i) Apical region of *Acrothrix gracilis*, showing the haplostichous development. (j) Apical view of the thallus of *Desmarestia aculeata*; the haplostichous thallus is derived from trichothallic growth and has a single axial filament. Lateral branches and subsequent investing branchlets, as shown for *Desmarestia ligulata* (k) lead to either a bilaterally pseudoparenchymatous structure or, as in *Arthrocladia villosa* (l), a radial structure in which the axial filament remains readily visible in cross-section. (m) Longitudinal section of the apex of *Myriogloia sciurus* showing the haplostichous structure developed from a multiaxial trichothallic meristem. Not to scale. (After Sauvageau, Kuckuck, Cohn, Murbeck, Fritsch, Reinke, and Jonsson.)

Phaeophyceae (*see below*), such as the microscopic gametophytes of members of the Laminariales.

The branched filamentous habit is characteristic of most Rhodophyceae; only the Bangiophycidae include truly parenchymatous examples. Thallus construction in the Florideophycidae is based entirely on a filamentous organization (Dixon, 1973) derived from a primary heterotrichous condition (Fritsch, 1945). As in the Phaeophyceae, both uni- and multiaxial modes of growth occur, and have led to a remarkable

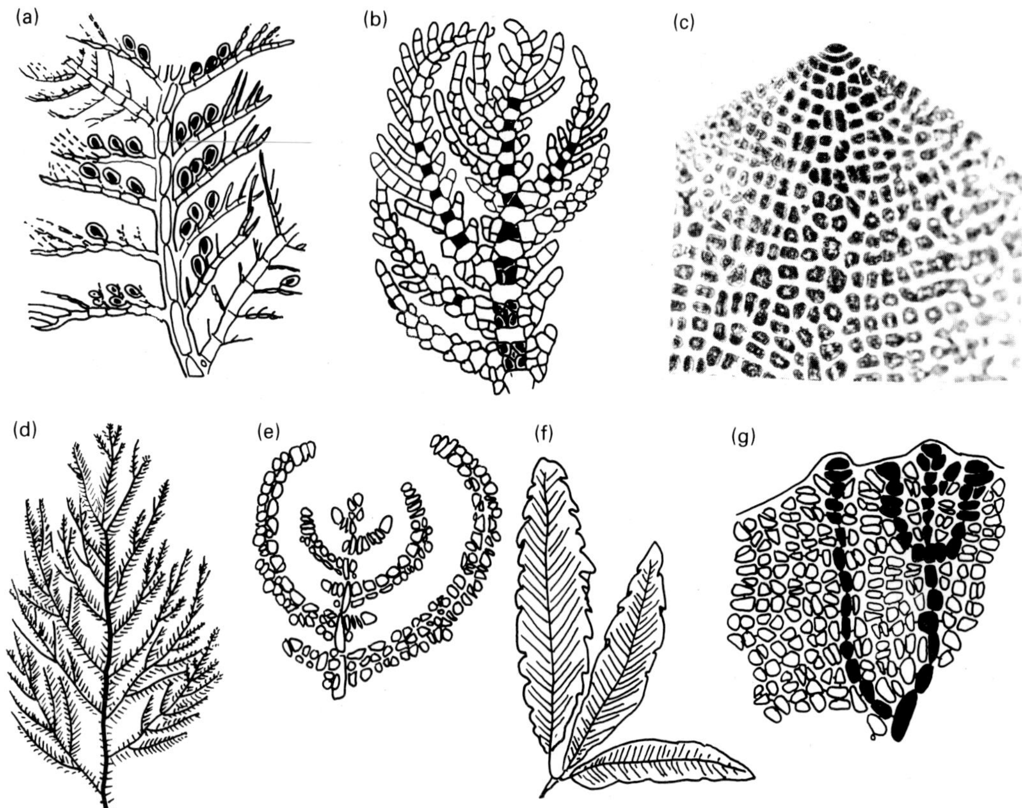

Fig. 4.8 Uniaxial filamentous Rhodophyceae. (a) *Antithamnion defectum* showing the main axis of unlimited growth and distichously arranged opposite branches of limited growth bearing young tetrasporangia. (b) Apical region of *Plumaria elegans*, showing the axial cells (shaded) and branches of limited growth. (c) *Deleseria decipiens*, photograph of the apical tip showing the single apical cell. The axial filament becomes obscured by corticating cells which form a thickened mid-rib (*see* f). (d) General view of *Bonnemaisonia asparagoides*, showing the branching pattern, with detail (e) of the apical region. (f) General view of the foliose thallus of *Delesseria sanguinea*, showing the mid-rib and veins. (g) Surface view of the apex of *Phycodrys rubens* with the main axial filament and derived laterals of limited growth (shaded). Not to scale. (After Kylin, Cramer, and Newton.)

range of external morphologies. Internal differentiation, however, is rather restricted and much less developed than in the Phaeophyceae. Red algal filaments are strictly apical in growth, being derived from a single apical cell (uniaxial types) or a group of apical cells (multiaxial types). The apical cell is typically dome-shaped and cuts off segments parallel to the base to give rise to axial cells. Intercalary cell division is rare in the Florideophycidae, and is restricted to some Cryptonemiales and Ceramiales. The axial cells increase in size by elongation and expansion; they cut off two or more pericentral cells which (except in the Rhodomelaceae of the Ceramiales) act as the potential apical cells of laterals. This pattern may be repeated in laterals of second, third, etc. orders. In the Rhodomelaceae (e.g. *Polysiphonia*) the pericentrals (known as siphons) are formed sequentially and lengthen with the axial cell from which they are derived, enclosing it. Extensive cortication of the siphons occurs in many Rhodomelaceae (e.g. *Rhodomela; Laurencia*), and foliose (e.g. *Symphyocladia*) and essentially prostrate thalli (e.g. *Placophora*) occur.

The filaments are normally richly branched in a monopodial manner; true dichotomous branching does not occur in uniaxial types, as the apical cells do not divide vertically (compare with api-

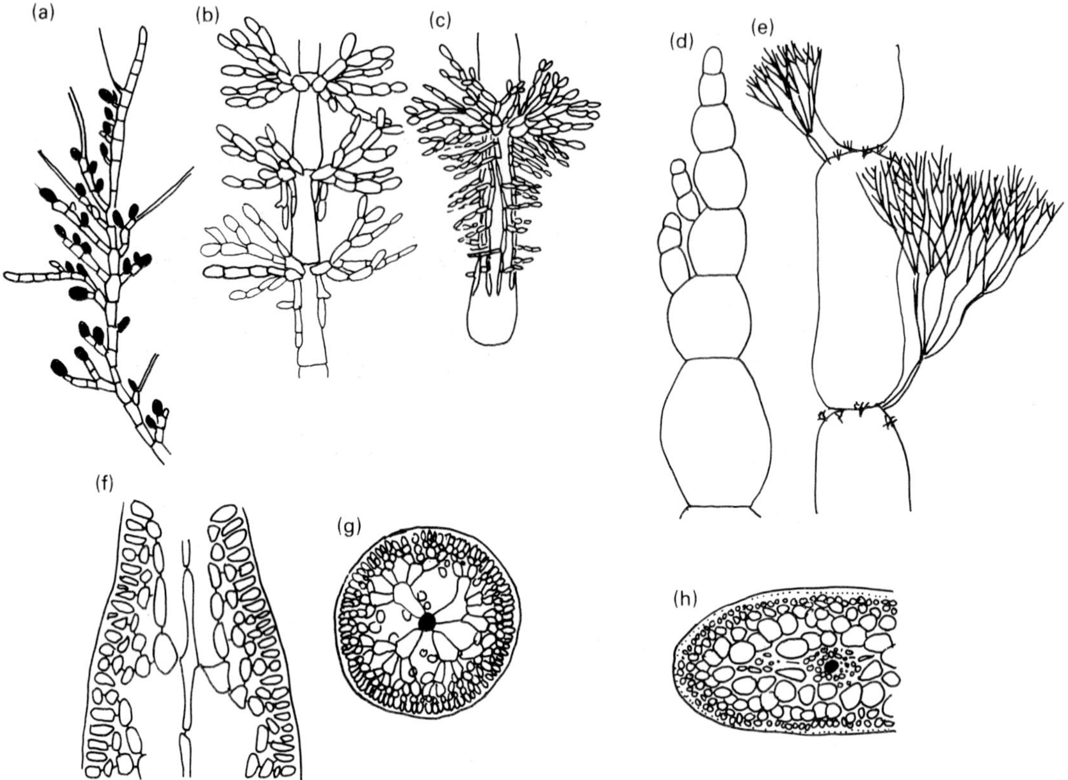

Fig. 4.9 Uniaxial Rhodophyceae *continued*. (a) *Audouinella secundata*, showing hairs, simple laterals and monospores. (b & c) Views of *Sirodotia suecica* with large axial cells and filaments of limited growth developing either as free branches, or as cortication. (d) *Griffithsia corallina* apical region, and (e) showing the large barrel-shaped cells bearing reduced or longer branches of limited growth. (f) *Dumontia incrassata*, longitudinal section showing the axial filament, with the medullary and cortical layers derived from branches of limited growth. (g) Cross-section of *Gloeosiphonia capillaris* showing the single axial row and four radially arranged laterals of limited growth, resulting in the terete form of the thallus. (h) Cross-section of *Sphaerococcus coronopifolia*, showing uniaxial construction (axial cell shaded) and laterally compressed thallus. Not to scale. (After Kylin, and Bornet & Thuret.)

cal, parenchymatous Phaeophyceae). Pseudodichotomy occurs when laterals close to the apical cell develop equally with the main axis, and true dichotomy can occur in multiaxial types where equal numbers of axial filaments diverge during branching. Generally, however, the main axis is described as possessing unlimited (or indeterminate) growth, and the laterals as possessing varying degrees of limited (or determinate) growth; these terms are indicative of the existence of apical dominance (*see* Chapter 7). The ultimate morphology of the thallus is determined by the relative growth of primary and lateral filaments. The axial cells and their derivatives are produced in a highly regular sequence, which has been analysed in detail for some Rhodophyceae (Dixon, 1973). The presence of primary and secondary pit plugs facilitates such growth analyses as it allows the following of discrete cell lineages, starting from the apical region.

The simplest heterotrichous Florideophycidae are typified by *Audouinella* (Woelkerling, 1983), where the basal attachment ranges from a single holdfast cell to a creeping or coalesced prostrate system. *Batrachospermum* exemplifies the uniaxial type of thallus construction in which the axial

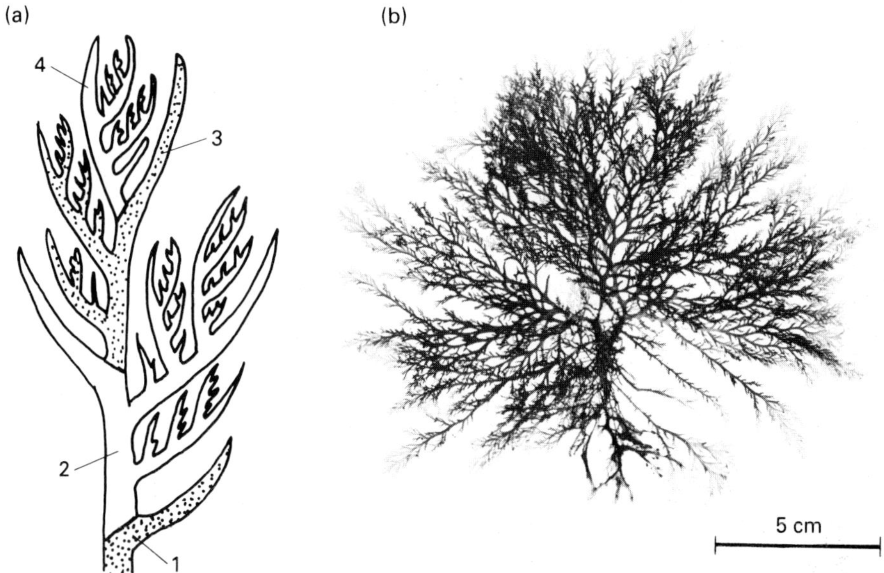

Fig. 4.10 Uniaxial Rhodophyceae *continued*. (a) Apex of *Plocamium cartilagineum* showing sequential branching pattern (1–4); each sequence comprises two branched and one simple (basal) lateral. A single apical cell terminates each branchlet. (b) Photograph of whole plant of *P. cartilagineum*. (After Naegeli.)

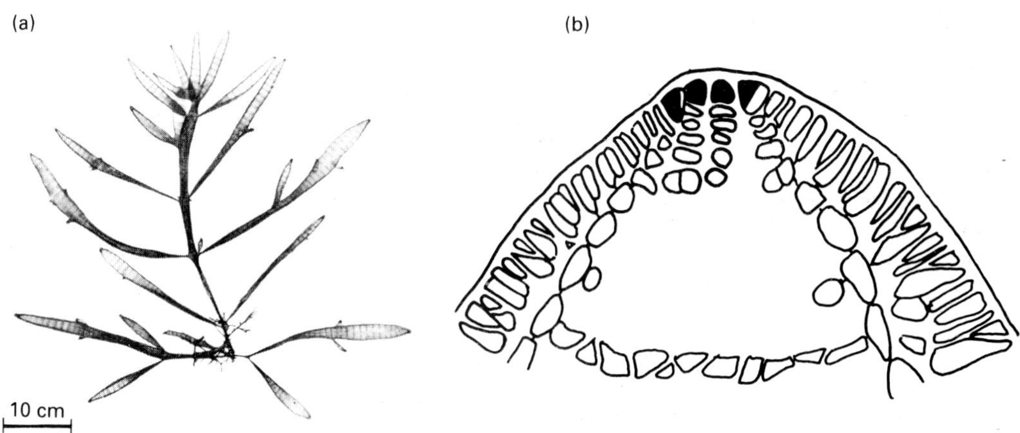

Fig. 4.11 Multiaxial Rhodophyceae. (a) *Chylocladia* sp., photograph of whole plant showing terete, hollow thallus with interior incompletely or completely partitioned by diaphragms. (b) Median longitudinal section of the apical region of *Chylocladia reniformis* showing the apical cells (shaded) and a diaphragm formed by transversely growing laterals. (b after Kylin.)

cells produce determinate branches in whorls. In addition, the pericentral cells may also produce down-growing rhizoidal filaments that form an investiture of corticating cells which, in turn, may produce secondary whorls of branches (Fritsch, 1945). In *Gloiosiphonia* the uniaxial thallus is much firmer in construction. The laterals produce secondary branches that give rise to progressively smaller cells, these ultimately remaining joined together to form a continuous

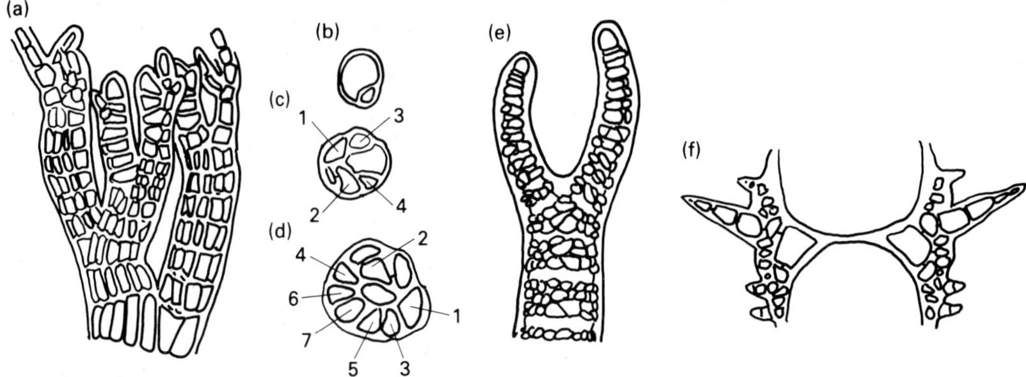

Fig. 4.12 Polysiphonous and corticated uniaxial Rhodophyceae. (a) *Polysiphonia nigrescens* — apical region showing apical cells and development of bands of 'siphons' from each axial cell. (b, c & d) Cross-sections of *Polysiphonia* sp. to show the sequence of pericentral cell ('siphon') development (numbered 1–7). (e) Apex of *Ceramium deslongchampii* showing apical cells and pseudodichotomy. (f) *Ceramium* sp., detail of cortication of branches of limited growth, resulting in partial bands at the juncture between axial cells. Not to scale (After Kylin, W. R. Taylor, and Cramer.)

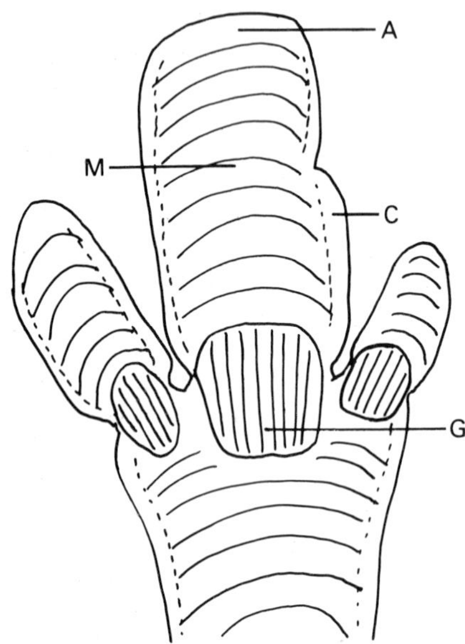

Fig. 4.13 Diagram of the apex (A) of an articulated coralline alga. The uncalcified genicula (G) are composed of uncalcified axial filaments, while the intervening calcified sections are composed of a medulla (M) and cortex (C) around the axial filaments.

surface layer. The internal cells differentiate to form a cortex. In *Dumontia* the internal cells are further differentiated into hyphae, narrow elongate cells which run vertically in the thallus and form secondary pit connections with the other cells. An incredible array of uniaxially constructed foliose thalli is found in the Delesseriaceae (Ceramiales); they possess mid-ribs and veins superficially like those in the leaves of higher plants (e.g. *Delesseria*; *Membranoptera*; *Phycodrys*; *Polyneura*). Most delesseriaceous species have conspicuous apical cells throughout their growth (e.g. *Delesseria*), although in some (e.g. *Nitophyllum*) the apical cell is obscured in older plants. Four pericentral cells are formed, and the lateral pair continue to divide, producing the membranous expansion of the leafy thallus. This is usually monostromatic, although up to five cell layers are present in some species (e.g. *Hemineura*; Kylin, 1956). The dorsal and ventral pericentrals possess more suppressed growth, and develop to form the mid-rib. Unlike the majority of Florideophycidae, a number of Delesseriaceae exhibit extensive intercalary divisions.

Suppression of the upright system has led to a wide range of crustose species, including those in which the reduced upright filaments, which may be simple or branched, are arranged in a gelatinous matrix (e.g. *Petrocelis*), are densely aggregated (e.g. *Hildenbrandia*), or are heavily calcified (e.g. *Clathromorphum*). The calcified species may be small and regular in outline (e.g. *Fosliella*) or spreading and irregular (e.g. *Lithothamnion*), with either a fairly smooth surface (e.g. *Phymatolithon*) or one bearing short branched or unbranched uprights (e.g. *Lithothamnion glaciale*). Crusts are attached by short rhizoids.

A large array of uniaxial species is found in the Ceramiaceae (Ceramiales), all richly branched and ranging in form from entirely uncorticated species (e.g. *Antithamnion*) to those with large-celled axial filaments that are partially or completely invested with bands of determinate corticating filaments (e.g. *Ceramium*) and those which are heavily corticated throughout (e.g. *Ptilota*).

The simpler multiaxial construction is typified by *Nemalion*, which has soft dichotomously branching gelatinous thalli. The central colourless axial filaments are readily separable under gentle pressure; their numerous determinate branches enclose the thallus in a layer of small, photosynthetic cells, most of which terminate in a hyaline hair. A similar structure occurs in *Galaxaura*, except that the apical cells of the axial filaments are grouped in an apical depression. The thallus is strongly calcified, giving it a rigid structure.

Greater multiaxial specialization is found in the heavily calcified articulated corallines (Cryptonemiales), such as *Corallina, Amphiroa, Jania* and *Calliarthron*. The determinate branches are formed in interrupted series, between which the lightly calcified axial filaments form the articulating points of the thallus.

Much more complex multiaxial types include terete, fleshy, or foliose species of dense construction; in these the multiaxial derivation of the thallus and the presence of the apical cells is difficult to distinguish except in the early developmental stages. Terete forms include *Ahnfeltia, Gymnogongrus*, and *Agardhiella*, while large foliose species include *Gigartina* (which may reach 1 m in length), *Iridaea*, and *Rhodoglossum*. In some Rhodymeniales a specialized mode of growth occurs in which the hollow, tubular thalli are chambered, the chambers separated by transverse medullary hyphae (diaphragms) (e.g. *Gastroclonium; Lomentaria; Champia*).

Differentiation in even the most advanced multiaxial Florideophycidae does not result in complex tissues. Terete and foliose thalli, whether gelatinous or calcified, are commonly differentiated into surface, cortical, and medullary cells; hyphae are frequently formed internally to the medulla, or among the cortical cells. Specialized structures include hairs, tendrils (e.g. *Hypnea*), and gland cells (e.g. *Antithamnion; Turnerella*).

The branched filamentous habit is well developed in the Chlorophyceae, and is typical for some Oedogoniales and all Chaetophorales and Ctenocladiales. Heterotrichy is a marked feature of the latter two orders, which contain a remarkable variety of forms.

Thalli of *Bulbochaete* (Oedogoniales) are richly branched, the successive branches arising unilaterally. Most cells possess a colourless bristle with a bulbous base. Cell division is basal and intercalary. Among the heterotrichous Chaetophorales and Ctenocladiales are those where both the prostrate and upright systems are developed (e.g. *Stigeoclonium*; *Trentepohlia*), those in which the postrate system is scarcely developed and the upright system, which predominates, is relatively elaborate (e.g. *Chaetophora*; *Draparnaldia*), and those where the prostrate system dominates, either as a discoid structure (e.g. *Chaetopeltis*; *Pringsheimiella*; *Pseudulvella*; *Ochlochaete*) or as a creeping prostrate filament (e.g. *Aphanochaete*; *Acrochaete; Bolbocoleon*). In prostrate species the uprights may be reduced to one or two cells (e.g. *Acrochaete*) or restricted to hairs or setae (e.g. *Bolbocoleon, Coleochaete*).

In the Charophyceae (e.g. *Chara, Nitella*) the precisely organized, predominantly upright thalli are terminated by a dome-shaped apical cell of indeterminate growth. Cells cut off from the apical cell are alternately nodal and internodal. The nodal cells give rise to 6–20 pericentral cells which produce whorls of branches of determinate growth, while the internodal cells do not divide further, but elongate considerably (up to several centimetres in some species), becoming multinucleate as a result of amitotic division (Shen, 1967). Corticating rhizoids derived from the basal cells of some of the determinate branches form an investiture over the internodal cells in some species. Thalli are anchored to the substrate by multicellular rhizoids. Development of new plants begins with a protonema, a prostrate filamentous phase comparable with the protonema of members of the Bryophyta. Charophytes readily undergo asexual reproduction, either through the production of uprights from the protonemal basal system, or through the production of special propagule-like structures formed on the basal or upright system. Asexual zoospores are not formed.

Parenchymatous organization (Figs 4.14 & 4.15)

Parenchymatous organization occurs when cells of the primary filament divide in all directions; any essentially filamentous structure is thus lost early on. The parenchymatous habit is characteristic of many Phaeophyceae, although there are some simpler representatives in the Rhodophyceae (Bangiophycidae, e.g. *Porphyra*; *Smithora*), Chrysophyceae (e.g. *Phaeosaccion*) and Chlorophyceae (e.g. *Ulva*; *Enteromorpha*). Cells comprising most parenchymatous thalli are uninucleate; in the Phaeophyceae a high degree of tissue differentiation can occur in the more advanced species (e.g. members of the Fucales, Laminariales, and Durvillaeales). Growth in the simpler examples is diffuse (e.g. *Desmotrichum*, Phaeophyceae; *Porphyra*; *Ulva*) and may result in leafy thalli of one cell layer (monostromatic, e.g. *Desmotrichum*; *Monostroma*, Chlorophyceae; some species of *Porphyra*) or two cell layers (distromatic, e.g. *Ulva*; *Punctaria*, Phaeophyceae; some species of *Porphyra*). In other simple types a tubular (e.g. *Enteromorpha*) or initially saccate (e.g. *Omphalophyllum*, Phaeophyceae; *Monostroma*, Chlorophyceae) thallus is found. In these leafy or tubular examples there is little cell differentiation, except for the production of rhizoids from the lower cells to form a simple holdfast. Cells may, however, be arranged in regular groups (e.g. *Prasiola*, Chlorophyceae) or rows (e.g. some *Blidingia* and *Enteromorpha* species, Chlorophyceae). Many of the simpler parenchymatous algae show a lack of differentiation of the reproductive cells; almost any cell may form asexual spores, although in the foliose species spore production may be restricted to the marginal areas of the thallus (e.g. *Ulva*).

Apical growth from single apical cells which terminate the apex and individual branches leads to a variety of terete or foliose types in the Phaeophyceae, including the dichotomously branched thalli of the Dictyotales (e.g. *Dictyota*) and the Fucales (e.g. *Fucus*). In the latter order, the apical cell (or cells; *see* McCully, 1966) is located in an apical groove or pit, and is three- or four-sided, and has three (more primitive; Jensen, 1974) or four lateral cutting faces and a basal cutting face. It is noteworthy that the apical cell of *Fucus* has its embryonic origin at the base of a trichothallically growing hair (*see* Fritsch, 1945). Median, perpendicular divisions of the apical cell result in the true dichotomous branching typical for many genera in the order. Thalli are either bilaterally or radially branched, and may be flattened or terete.

The Durvillaeales (e.g. *Durvillaea*) possess large *Laminaria*-like thalli derived from an apical margin of apical cells. In the most advanced parenchymatous phaeophycean development intercalary meristems, consisting of an actively dividing meristematic tissue, are responsible for the formation of the blade, stipe, and holdfast structure (e.g. *Laminaria*; *Alaria*). This meristem (or transition zone) may exhibit seasonal activity in perennial species such as *Laminaria hyperborea*. Classification at the family level in the Laminariales is based on the nature of the transition zone (e.g. split in the Lessoniaceae, not split in the Laminariaceae and Alariaceae). The surface layers of these advanced species continue division throughout the growth of the thallus, and are referred to as a meristoderm. In the highly differentiated Fucales and Laminariales well-developed cortical and medullary tissues are developed. In the Laminariales (e.g. *Macrocystis*, which may reach 100 m in length) transport of metabolites is effected through specialized hyphae (trumpet hyphae) which have sieve areas and are homologous with phloem in higher plants (Schmitz, 1981).

Siphonocladous organization (Figs 4.6 & 4.16)

Siphonocladous organization (Hoek, 1981) is restricted to members of the Chlorophyceae in which the unbranched (e.g. *Urospora*, Acrosiphoniales; *Chaetomorpha*, Cladophorales) or branched (e.g. *Acrosiphonia*, Acrosiphoniales; *Cladophora*, Cladophorales) filaments are composed of multinucleate (semi-coenocytic) cells. Also included are members of the Siphonocladales (grouped with the Cladophorales in some treatments; *see* Hoek, 1981) such as *Valonia*, *Siphonocladus*, *Anadyomene*, and *Microdictyon*. Some of the Siphonocladales are also distinctive in their possession of segregative cell division (e.g. *Dictyosphaeria*; *Siphonocladus*) in which the protoplast cleaves into walled, rounded portions which expand into new segments.

Valonia occurs as a single, spherical vesicle up to 10 cm in diameter — it has been described as the largest plant cell, although this is not strictly

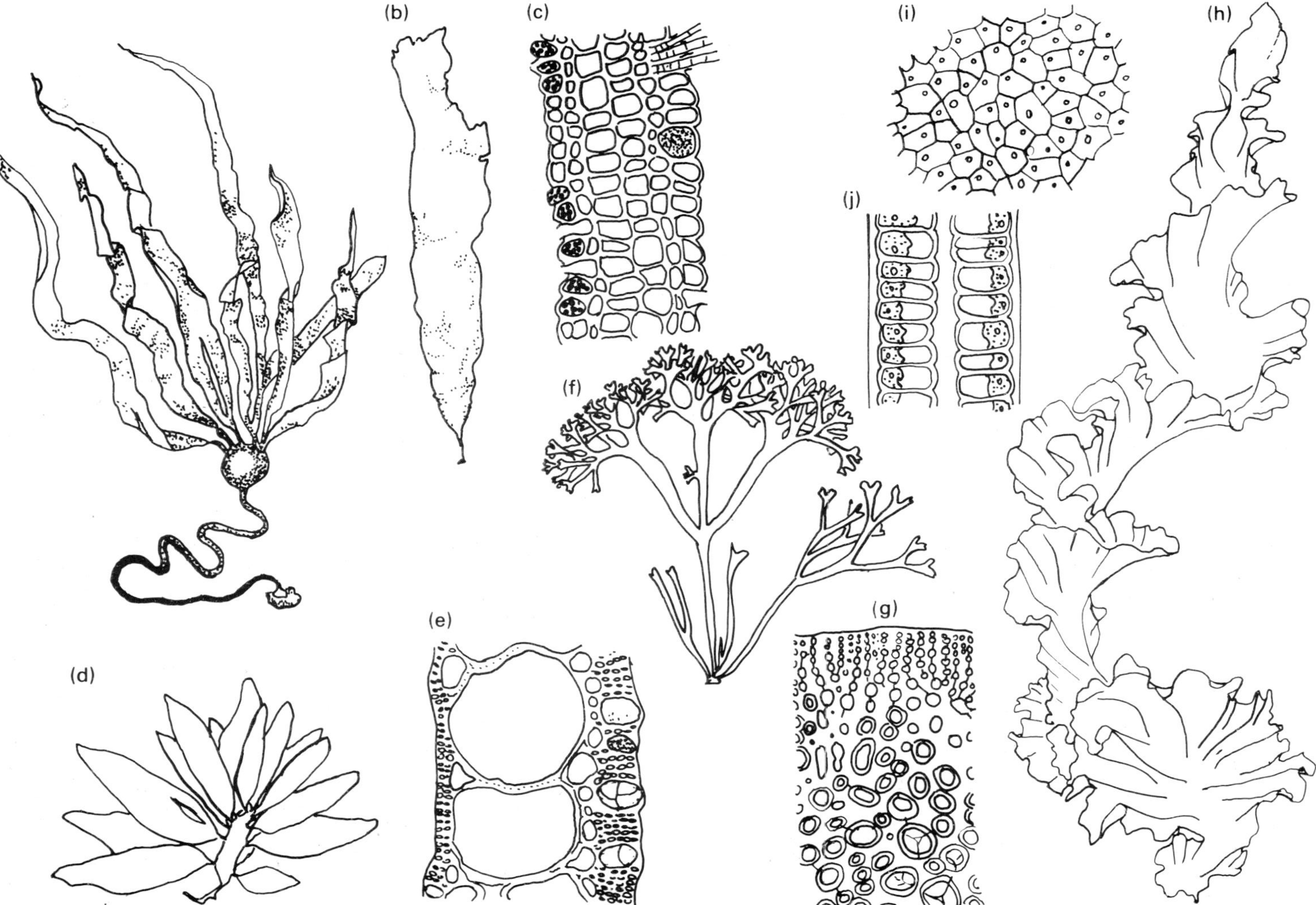

Fig. 4.14 Parenchymatous and pseudoparenchymatous organization. (a) *Nereocystis luetkeana* (Phaeophyceae). (b) *Punctaria plantaginea* (Phaeophyceae) — whole plant and in section (c). (d) *Palmaria palmata* (Rhodophyceae), whole plant and in section. (e), showing large medullary cells, and small-celled cortex with tetrasporangia. Apart from members of the Bangiophyceae, Rhodophyceae which appear to be parenchymatous are derived from a filamentous origin and are thus best referred to as pseudoparenchymatous. (f) *Chondrus crispus* (Rhodophyceae), whole plant, and in section (g), showing thick-walled medullary hyphal filaments and a densely compacted, small-celled cortex. (h) Whole plant of *Ulva lactuca* (Chlorophyceae) with detail of the surface view (i) and the two-layered cross-section (j). Not to scale. (After Postels & Ruprecht, Rosenvinge, and W.R. Taylor.)

Fig. 4.15 Parenchymatous organization. The thalli of *Fucus* species (*F. vesiculosus* above, and *F. distichus* spp. *edentatus*, below) exhibit some of the highest levels of differentiation and organization in the algae.

so as many small cells are produced by segregative division in the peripheral zone (Egerod, 1952); those formed in the lower region protrude as attaching rhizoids.

In the simpler, unbranched siphonocladous species (e.g. *Urospora*) there is little differentiation of the filament, although the basal cells produce rhizoids that grow downwards and coalesce to form a holdfast. There is a tendency for divisions to occur more frequently in the basal region, and the cells are larger in the more distal parts of the filament. Any of the cells, except those at the very base, may undergo zoosporogenesis. Growth in the branched *Cladophora* is apical, intercalary, or both (Hoek, 1982), and the thallus organization may be acropetal, irregular or, rarely, dorsiventral. In *Acrosiphonia (Spongomorpha)* a rope-like entwining of the lower axes is effected by down-growing rhizoids that arise from the lower cells, or by special hooked laterals.

The strikingly beautiful foliose thalli of *Anadyomene* and *Microdictyon* are constructed from branched, multinucleate, semi-coenocytic filaments which have become anastomosed to form a monostromatic thallus. Microscopically,

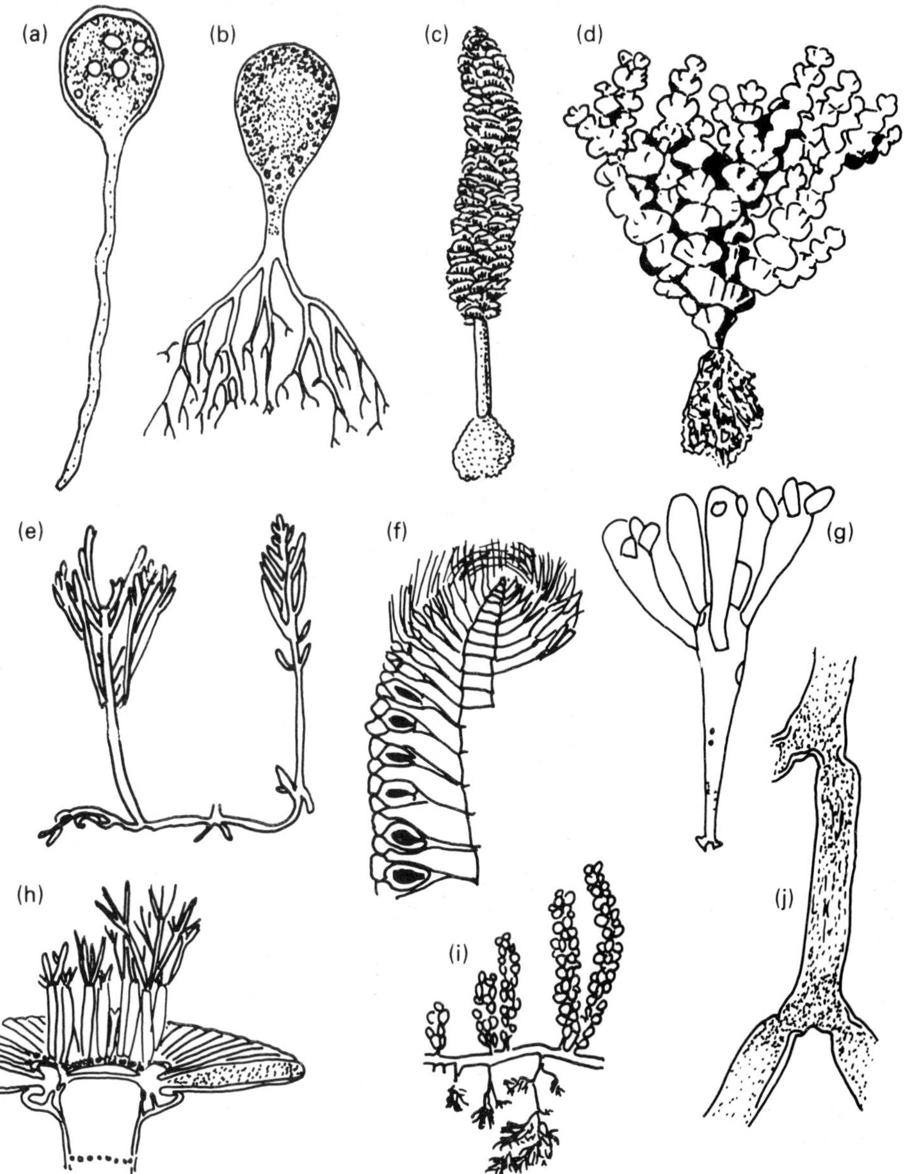

Fig. 4.16 Siphonous organization in the Chlorophyceae and Xanthophyceae. (a) *Protosiphon botryoides* (Chlorophyceae). (b) *Botrydium granulatum* (Xanthophyceae). (c) *Rhizocephalus phoenix* (Chlorophyceae). (d) *Halimeda incrassata* (Chlorophyceae), a calcified species. (e) *Bryopsis corymbosa* (Chlorophyceae). (f) Vertical section of the apex of the calcified species *Neomeris annulata* (Chlorophyceae). (g) *Valonia utricularis* (Chlorophyceae). (h) *Acetabularia mediterranea* (Chlorophyceae) — section of the cap of a fertile plant showing the terminal hairs and radially arranged gametangia. (i) *Caulerpa racemosa* (Chlorophyceae). (j) *Dichotomosiphon tuberosus* (Xanthophyceae). Not to scale. (After Klebs, Rostafinski & Woronin, Gepp, Boergesen, Cramer, Schmitz, Oltmanns, and Ernst.)

the cells are seen to be closely contiguous in *Anadyomene*, with the lower cells much larger than those at the margin, while in *Microdictyon* they are similarly sized and cylindrical throughout.

Siphonous organization (Fig. 4.16)

In a number of marine Chlorophyceae (orders Siphonales, Bryopsidales, Dasycladales) and some Xanthophyceae (e.g. *Botryidium; Vaucheria*), enlargement and elaboration of the thallus proceeds in the absence of septa. Nuclear divisions are not followed by cytokinesis (free nuclear division) and the result is a coenocytic, multinucleate thallus and a siphonous organization. The siphonous habit has not, however, restricted thallus elaboration, as a remarkable range of coenocytic forms is found. Fossil evidence shows that in some orders (e.g. Dasycladales) the living representatives are only remnants of a once much larger assemblage of species.

Siphonous organization ranges from saccate (e.g. *Botryidium*) to uniaxial (e.g. *Vaucheria; Bryopsis*) and multiaxial (e.g. *Codium*) forms. Complex radial branching patterns are exhibited by members of the Dasycladales (e.g. *Neomeris*), and foliose forms are found in some Caulerpas. In many, support of the aseptate thalli is assisted through calcification or by the presence of trabeculae (singular trabecula), which (e.g. *Caulerpa*) are extensive ingrowths of the cell wall. With the exception of some uniaxial species, and members of the Xanthophyceae, the majority of the siphonous algae do not undergo asexual zoosporogenesis. In the Chlorophyceae, the siphonous groups possess distinctive xanthophyll pigments. Siphonous organization in the Chlorophyceae may well have evolved from forms comparable to the saccate *Protosiphon* (order Chlorococcales), in which a coenocytic condition develops, as in the siphonous groups, through free nuclear division.

The siphonous habit in the Xanthophyceae is restricted to the saccate *Botryidium* and the filamentous *Vaucheria*, soil and mud-dwelling algae. In *Vaucheria*, septa are laid down only as a means of segregating gametangia or zoosporangia from the main thallus. The irregularly branched filaments elongate from the tips, where cytoplasmic zoning is evident (Ott & Brown, 1974).

In *Bryopsis* the delicate, feather-like uniaxial thalli possess free branches which arise acropetally, and are pinnately or radially arranged on the main axis. The uprights are borne on a creeping rhizomatous base, which is attached to the substrate by rhizoids. An interesting wound-healing response has been discovered, where protein bodies in the cytoplasm migrate to the site of an injury and become involved in the formation of a plug that prevents further loss of cytoplasm (Burr & West, 1971). In *Derbesia* the uniaxial thalli are more irregularly branched; here the siphonous, sporophytic plants alternate with a saccate (*Halicystis*) sexual phase. The simpler species of the widespread tropical genus *Caulerpa* show a basic structure not unlike that of *Bryopsis* (e.g. *C. fastigiata; C. verticillata*), although the majority exhibit considerably more differentiation of the thallus. The stoloniferous basal system gives rise to uprights which show a range of form that simulates the shoots of a range of higher plants (Fritsch, 1935). They may be verticillate (*C. verticillata*), foliose (e.g. *C. prolifera*), or with branches in two rows (e.g. *C. taxifolia*) or radially arranged (e.g. *C. racemosa*). Although *Caulerpa* is not calcified, a considerable degree of support to the aseptate thalli is provided by the extensive trabeculae.

In the Dasycladales the thalli are calcified, radially symmetrical, uniaxial, and possess whorled branches. Other distinctive features include reproduction by operculate cysts and the presence of diaphysis, in which new growth pushes through the old, leaving a scar (Bold & Wynne, 1978). The Dasycladales are essentially unicellular, as they remain uninucleate until the time of reproduction. The single, so-called 'giant', or primary, nucleus is located in a basal rhizoid when the plants are in the vegetative condition. The whorled branches may be loosely arranged (e.g. *Neomeris; Batophora; Cymopolia*), or closely adpressed, resulting in a compact surface layer of dilated branches. In *Acetabularia* the branch whorls of the mature plant are restricted to the upper end of a relatively long stalk, which is otherwise bare. The primary nucleus remains in the basal region, beneath the substrate. *Acetabularia* has been the subject of extensive morphogenetic research using experimental merotomy, in which the nucleus is removed and the enucleate plant studied under controlled conditions. The work has been useful in investigating the role of the nucleus in the control of morphogenesis (*see* Chapter 7).

Multiaxial siphonous forms are found in the families Codiaceae (e.g. *Codium*) and Udoteaceae (e.g. *Avrainvillea*; *Udotea; Penicillus; Rhipocephalus; Halimeda*). The spongy, non-calcified thalli of *Codium* are composed of a medulla of colourless, dichotomously branched siphons, and a cortex of inflated, club-shaped vesicles (utricles) containing the chloroplasts. The utricles terminate forks of the apically growing medullary threads. *Codium* includes species with a crustose, spherical, branched erect, or blade-like thallus; the most important diagnostic feature, however, is the shape of the utricles (Silva, 1951). Annular thickenings of the siphons, which may almost occlude the lumen, give a measure of strength to the thallus. Reproductive structures of the Codiaceae are borne laterally on the utricles.

Members of the Udoteaceae differ from the Codiaceae on two grounds: in their possession of specialized gametangia, and in their possession of both chloroplasts and amyloplasts (heteroplastidy). In *Avrainvillea* and *Udotea* the calcified, multiaxial thalli are a fan-shaped (flabellate) blade borne on a simple or branched stalk and anchored to the substrate by a rhizoidal mass, or a stoloniferous base. Internally they are differentiated into a medulla of intertwining siphons, with perpendicular, only slightly inflated laterals closely adpressed on either side to form the cortex. In the remarkable *Rhipocephalus*, the unbranched stalk bears a brush-like mass of numerous, crowded flabellae, whereas in *Penicillus* the similarly brush-like top is made up of free, branched, calcified filaments. In *Halimeda* the plants are composed of a branching series of usually flat, strongly calcified segments that are separated from one another by uncalcified joints.

CHAPTER 5

Reproduction and life cycles

Introduction

There are a number of functions associated with reproduction in the algae apart from an increase in the numbers of individuals. These include mechanisms to aid in dispersal and survival under adverse conditions, and mechanisms to bring about the enhancement of species fitness through genetic recombination. Although it is convenient to designate reproduction as asexual or sexual, many life cycles obligately combine both processes.

In many unicellular species, cell division, or binary fission, is the only mechanism for asexual reproduction. This can be extended to multicellular forms, which may reproduce either by fragmentation or by the production of specialized multicellular structures termed gemmae or propagules. It is seen at its most complex in those coenobial and colonial forms which reproduce asexually by autocolony formation. These processes are often termed vegetative reproduction, with asexual reproduction being reserved for events which include the formation of specialized unicellular spores.

Asexual spores are functionally distinguished from gametes in that they develop without syngamy, i.e. they do not fuse to produce a zygote. Motile spores which possess flagella are termed zoospores, reflecting the historical belief that motility is an animal-like characteristic. Their non-motile homologues are aplanospores, essentially zoospores which omit the motile period, and autospores, which are ontogenetically aflagellate spores, usually a miniature of the cell from which they are derived. Some workers do not make this distinction and regard all nonmotile spores as aplanospores. Exospores and endospores are found in the Cyanophyceae and are formed respectively by abstriction of the cell contents or internally within the parent cell. In a like manner the entire contents of a cell of some eukaryotic algae may be released as a single non-motile monospore. Many other types of spores in the algae are known to be associated with sexual processes, for example tetraspores in the Phaeophyceae and Rhodophyceae or auxospores in the diatoms. Unfused gametes may develop parthenogenetically and thus function as spores.

Spores or specialized vegetative cells may enable survival under conditions adverse to vegetative growth. These usually have thickened or otherwise modified cell walls and reduced metabolic activity. These include the akinetes, which are derived from vegetative cells, and resting spores, cysts, statospores, and hypnospores, which ontogenetically are resistant spores. Zygotes in some algal groups may also remain dormant and are termed hypnozygotes or zygospores. Many flagellates (Table 4.1) are capable of forming temporary non-motile 'palmelloid' stages with cells embedded in a gelatinous matrix. This may be a response to adverse conditions, but as both cell division and spore formation may occur in the palmellae they cannot be regarded as resting stages.

Sexual reproduction involves two functionally opposite processes.

1 Syngamy, a two-stage process involving the cytoplasmic fusion of gametes (plasmogamy) followed by the pairing of chromosomes (karyogamy).

2 Meiosis, the interchange of genetic information (in synapsis) and the independent assortment of chromosomes into the resultant cells.

Syngamy is termed isogamous when the gametes are morphologically identical and anisogamous when one gamete (the male) is smaller and more motile than the other (the female) gamete. In isogamous syngamy the gametes frequently show sexually differentiating behavioural or physiological attributes, as in *Ectocarpus* where one gamete, the female, settles first and produces a pheromone which attracts the male (Müller, 1981). In non-flagellated gametes, such as occur in the Zygnematales, one gamete may be motile while the other remains within the confines of the cell in which it was formed. There may also be ultrastructural differences between apparent isogametes such as in *Chlamydomonas reinhardtii* (Triemer & Brown, 1975). Oogamous syngamy refers to the fusion of a sperm and a non-flagellated female gamete, the egg. Sexually reproducing rhodophytes are oogamous and the aflagellate male gametes are termed spermatia. A fourth type of syngamy which is not so readily observed is autogamy, the fusion of two gametic nuclei within an individual.

One of the most important aspects of sexual reproduction is the site of meiosis in relation to syngamy. In the algae four such sites can be distinguished.

1 During the germination of the zygote, which is the only diploid phase; such meiosis is termed zygotic.

2 During spore formation, the spore then germinating to form a haploid plant which produces gametes by mitosis; such meiosis is termed sporic.

3 During the formation of gametes, which are the only haploid phase; such meiosis is termed gametic.

4 During vegetative cell division, which is not immediately followed by spore or gamete formation. One part of the thallus is thus haploid the other diploid; such meiosis is termed somatic.

The cytological and morphological sequence of events involved in sexual reproduction is usually termed the life cycle, which is frequently, but erroneously, regarded as synonymous with life history. Life histories are broader in scope, and may be defined as the sum of an organism's adaptations to survival and reproduction. These include physiological and ecological considerations, such as the partitioning of resources between growth and reproduction. The life cycles of many algae have been completed in culture, but for only a few of these have cytological data, determining the level of ploidy and the site of meiosis, been obtained. Algal chromosomes are generally small in comparison to those of flowering plants and accurate counts are difficult to obtain. Direct measurement of nuclear DNA by microspectrophotometry, and particularly the use of DNA-specific fluorochromes, holds much promise and has already been used in detecting changes in the level of ploidy (Goff & Coleman, 1984).

Algal life cycles are classified according to the site of meiosis, i.e. zygotic life cycle, sporic life cycle, gametic life cycle, and somatic life cycle (Fig. 5.1). The distribution of these among the algal classes is given in Table 5.1, together with the types of syngamy exhibited. As with morphology (Table 4.1), it is the Chlorophyceae which show the greatest variety of reproductive mechanisms.

Zygotic life cycles are generally regarded as primitive, and occur most frequently among the primarily unicellular taxa and their colonial or palmelloid counterparts.

Sporic life cycles have a spore-producing phase, the sporophyte, which is usually diploid, and a usually haploid, gamete-producing phase, the gametophyte. The recurring sequence of sporophyte and gametophyte is termed alternation of generations. The generations are termed isomorphic where the two phases have identical vegetative morphologies, and heteromorphic where they differ. Algae with differing morphological phases but the same chromosome complement are known (i.e. no syngamy or meiosis occurs); this is termed pleiomorphism. The evolution and success of such heteromorphic/pleiomorphic species has been attributed to the exploitation of different physiological capacities of the different forms (*see* Chapter 7), particularly relating to photosynthesis and growth, the so-called 'functional form hypothesis' (Littler & Littler, 1980). Morphologically reduced phases are also less susceptible to predation by herbivores than upright fleshy thalli (Lubchenco & Cubit, 1980).

The term alternation of generations is perhaps best avoided in discussing algal life cycles, as there are many exceptions to an obligate alternation. Many species are capable of reproducing one or both phases by means of asexual spores, while others undergo pleiomorphic sequences. It is also inappropriate to use this term for those members of the Rhodophyceae in which a secondary diploid carposporophyte phase is interpolated between the normal haploid–diploid sequence.

Gametic life cycles which characterize the animal kingdom and the Oomycota are not common in the algae, and appear to have arisen independently in several algal groups (Table 5.1). Somatic meiosis is even rarer, being known with certainty from the Prasiolales (Chlorophyceae) and in some freshwater members of the Batrachospermales (Rhodophyceae).

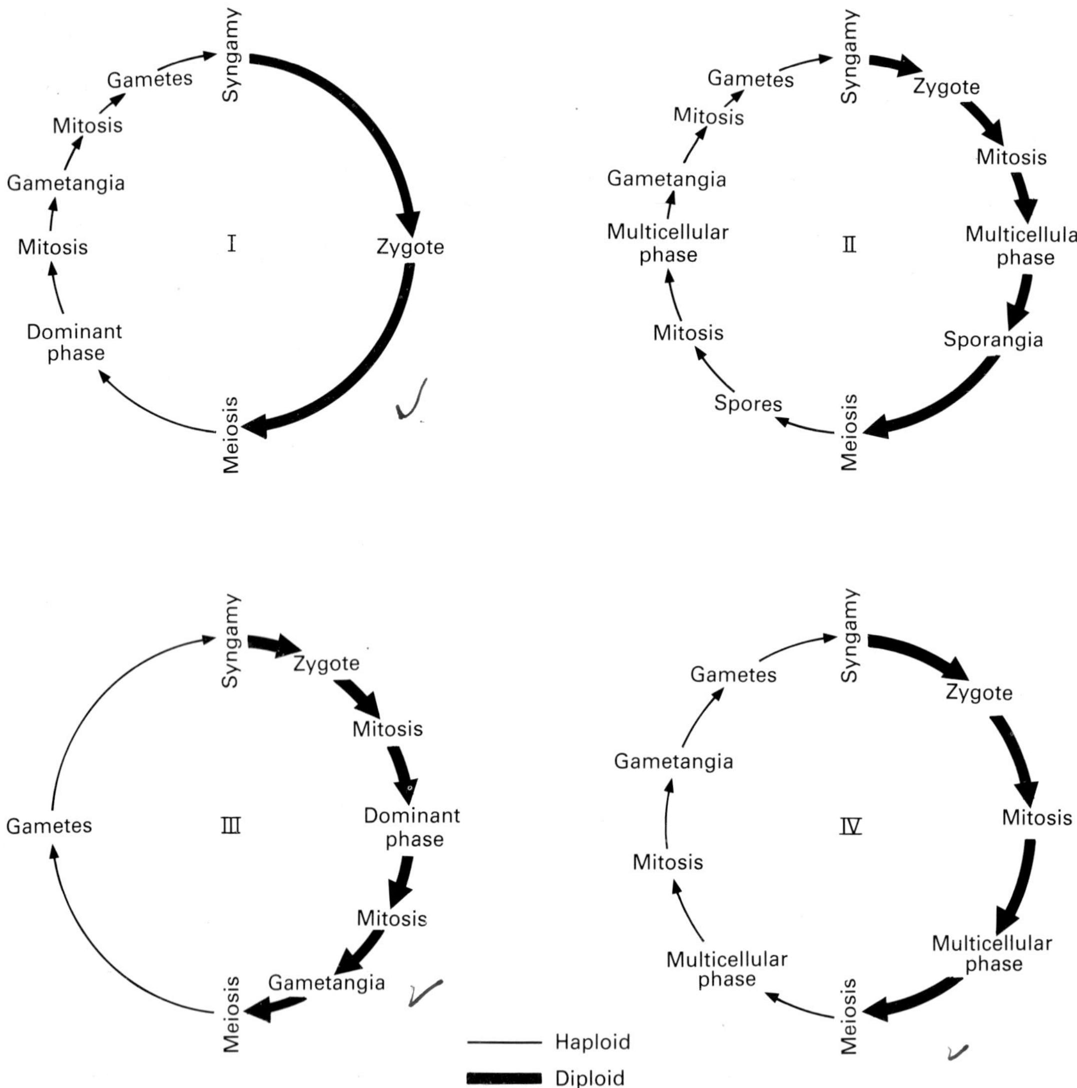

Fig. 5.1 Life cycles exhibited by algae. I, zygotic; II, sporic; III, gametic; IV, somatic. The dominant phase in I and III may be multicellular, or if unicellular the most persistent and capable of mitotic reproduction.

Vegetative reproduction

Reproduction via simple mitotic division has been reported in all the algal divisions with unicellular representatives. The simpler unicellular members of the Chlorophyceae, however, reproduce asexually solely by spores. Such division is termed eleutheroschisis (Groover & Bold, 1969) and differs from true cell division, or desmoschisis, in that walls are formed *de novo* and do not incorporate parts of the parent cell wall.

Some groups of algae have evolved specialized cytokinetic processes to maintain distinctive morphologies in their offspring cells.

In the diatoms the two valves move apart in cytokinesis and new valves form in the cleavage furrow. Both epitheca and hypotheca of the parent become the epivalves of the offspring; thus in actively growing diatom populations there

Table 5.1 Sexual reproduction and life cycles in the algae.

	Form of syngamy						Site of meiosis			
	Absent or unreported	Autogamy	Isogamy	Anisogamy[1]	Oogamy	Zygotic	Isomorphic sporic	Heteromorphic sporic	Gametic	Somatic
Cyanophyceae	+									
Prochlorophycota	+									
Rhodophyceae					+		+	+		+
Chrysophyceae		+	+			+				
Prymnesiophyceae			+					+[2]		
Xanthophyceae				+	+	+	+			
Eustigmatophyceae	+									
Bacillariophyceae		+	+		+				+	
Dinophyceae			+	+		+[3]			+[4]	
Phaeophyceae			+	+	+		+	+	+[5]	
Raphidophyceae	+									
Cryptophyceae	+									
Euglenophyceae	+									
Chlorophyceae			+	+	+	+	+	+	+	+
Charophyceae					+	+[6]				
Prasinophyceae	?[7]									

1 Includes behavioural/biochemical anisogamy.
2 Documented in some instances, e.g. *Hymenomonas carterae*, but others may be examples of pleiomorphism with no alternations of ploidy.
3 Planozygotes may be formed and meiosis is frequently post-zygotic, being delayed until after the germination of the cyst.
4 Reported in *Noctiluca miliaris* (Zingmark, 1970).
5 Occurs only in the Fucales and related orders, however some workers regard the oogonium as a reduced gametophyte and hence would classify meiosis as sporic rather than gametic.
6 Microdensiometric measurement of DNA shows the same levels in sperm and somatic cells, hence meiosis cannot be gametic.
7 Huber and Lewin (1986) report electrophoretic phenotypes of *Tetraselmis* characteristic of heterozygotes. They are presumably diploid and this could indicate the occurrence of sexual reproduction.

is a progressive diminution of cell size (Fig. 5.2). Usually a return to maximum size is brought about via sexual reproduction and auxospore formation, though instances of size regeneration via extrusion of cytoplasm and the regeneration of larger valves have been reported (Stosch, 1965a).

The armoured species of the Dinophyceae display a number of cytokinetic patterns (Walker, 1984). In some genera (e.g. *Peridinium* and *Glenodinium*), the theca may be shed prior to division, necessitating its total re-formation in the offspring. In other species, such as *Ceratium* spp. and *Pyrodinium bahamense*, the parent

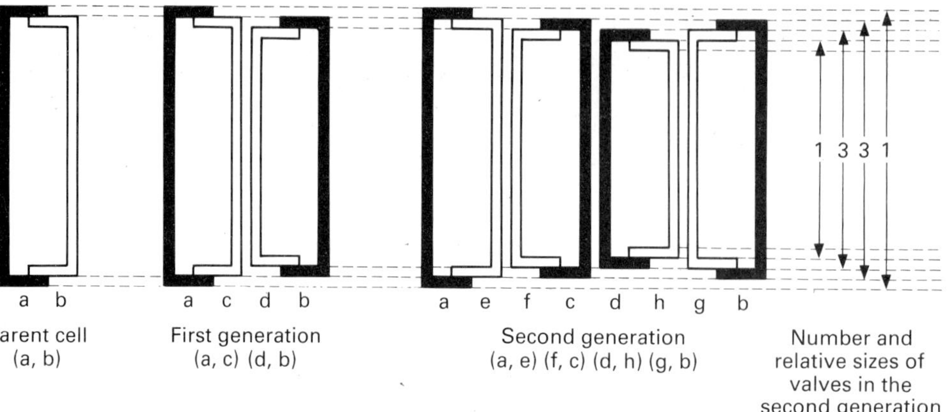

Fig. 5.2 Diagrammatic representation of a diatom in girdle view showing progressive decrease in size with cell division. (After Hendey, 1964.)

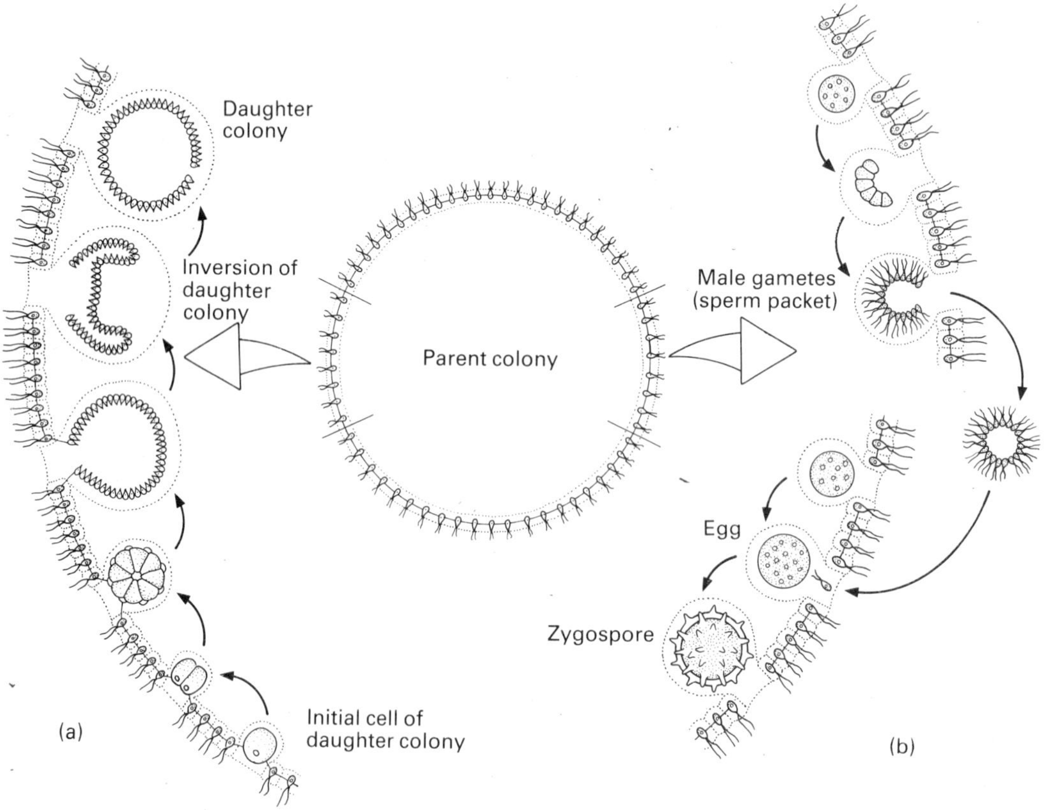

Fig. 5.3 Diagrammatic illustration of the life history of *Volvox*. (a) Asexual reproduction through the formation of daughter colonies. (b) Sexual reproduction. (After G.M. Smith, 1955.)

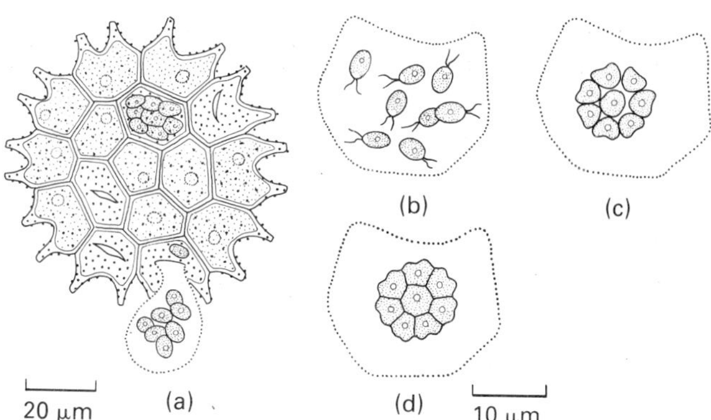

Fig. 5.4 Autocolony formation in *Pediastrum*. (a) Mature vegetative plant producing zoospores. (b) Motile zoospores. (c & d) Zoospores cease movement and aggregate to form a new colony. (After G.M. Smith, 1955.)

theca is divided between the daughter cells and only the missing plates are re-formed.

Placoderm or true desmids are characterized by the possession of semi-cells, and cytokinesis occurs in the region of the isthmus, after which each cell buds a new semi-cell.

Volvocalean colonies reproduce by autocolony formation brought about by division of all or some (in the more advanced genera) of the cells of the colony to form an embryonic colony that differs from the parent in cell size, but not cell number. On liberation the cells increase in size to that characteristic of the parent (Fig. 5.3). In *Volvox* (Starr, 1969; Karn *et al*., 1974), division is restricted to a series of cells which produce a hollow sphere within the parent colony, and with each mitosis its cells become smaller. The new colony everts, its cells form flagella at their apical poles, and it is released by rupture of the parent sphere. Reproduction by autocolony formation also occurs in the colonial members of the Tetrasporales such as *Hydrodictyon* and *Pediastrum* (Marchant, 1974). The protoplast of some or all of the cells of the colony undergoes divisions to form biflagellate zoospores. These are not liberated but aggregate to form a new colony within the parent cell wall (Fig. 5.4).

Vegetative reproduction through fragmentation is widely reported from all algal divisions in both freshwater and marine habitats, and is especially common in filamentous species. It appears to have an important role in maintaining populations in habitats such as estuaries (Nienhuis, 1974) or saltmarshes (Norton & Mathieson, 1983) or at the extremes of geographical ranges (*Asparagopsis*; Dixon, 1965: *Callithamnion*; Whittick, 1978: and *Codium*; Fralick & Mathieson, 1972). One planktonic population of *Pilayella littoralis* is maintained by fragmentation, aided by the filament-weakening effects of infection by a chytrid *Eurychasma dicksonii* (Wilce *et al*., 1982).

Some algae produce specialized vegetative propagules. Hormogonia (Fig. 5.5c) consisting of short filaments of undifferentiated cells are formed by the disintegration of some filamentous cyanophytes (Nichols & Adams, 1982). Multicellular asexual structures termed gemmae are described from the freshwater rhodophyte *Hildenbrandia rivularis* (Nichols, 1965), and propagules resembling spermatangial branches occur on male, female, and tetrasporangial plants of *Polysiphonia ferulacea* (Kapraun, 1977). In the Phaeophyceae, some members of the Sphacelariales form club-shaped, bi- or triradiate propagules from modified branches (Fig. 5.6). *Fucus* has the potential of reproduction via the formation of adventive embryos on the rhizoids (McLachlan & Chen, 1972). In the Charophyceae, bulbils occur on rhizoids, allowing perennation and initiating a protonema from which the upright phase arises (Grant & Sawa, 1982).

Asexual reproduction by spores

EXOSPORES AND ENDOSPORES

Exospores and endospores (Fig. 5.5d, e) are characteristic of the Chaemosiphonales and Pleurocapsales of the Cyanophyceae (Nichols & Adams, 1982). Exospores develop externally by cleavage from the ends of vegetative cells, while

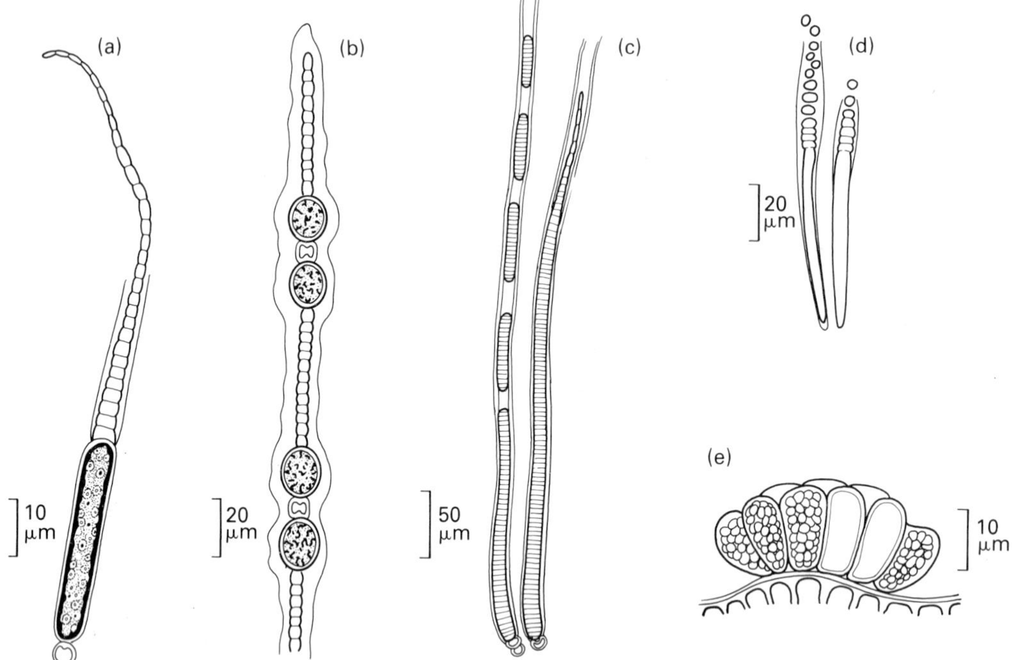

Fig. 5.5 Reproductive bodies of the Cyanophycota. (a) *Gloeotrichia*, akinete developing next to a basal heterocyst. (b) *Wollea*. (c) *Calothrix*, production of hormogonia. (d) *Chamaesiphon*, production of exospores by budding from the free ends of filaments. (e) *Dermocarpa*, endospores within a parent cell. (After Fogg *et al.*, 1973)

endospores are formed by the fission of enlarged vegetative cells.

MONOSPORES, BISPORES AND PARASPORES

Monospores (Fig. 5.7) are the commonest asexual spores of the Rhodophyceae and are distinguished from spermatia by their generally larger size and the possession of plastids (Dixon, 1973). They are especially important in the Bangiales, being the only reproductive cells known in the simpler filamentous members (Dixon, 1973). In *Bangia* and *Porphyra* monospores occur on both the conchocelis and macroscopic phases; in the latter they are essentially sloughed off vegetative cells. Similar cells are also reported in *Ulva* (Chlorophyceae; Bonneau, 1978).

In the Nemaliales (Rhodophyceae), some species of *Audouinella* reproduce exclusively by monospores, while in others with sexual reproduction, for example *Pseudogloiophloea confusa*, both the gametophyte and sporophyte generations may be recycled (Ramus, 1969). Monospores have been described from culture studies of a number of more advanced red algae (Dixon, 1973), but their role in nature is unclear. The

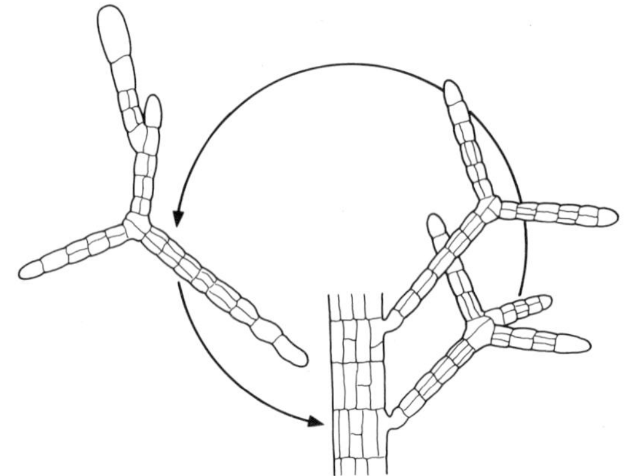

Fig. 5.6 Vegetative propagules in *Sphacelaria*. (After van den Hoek & Flinterman, 1968.)

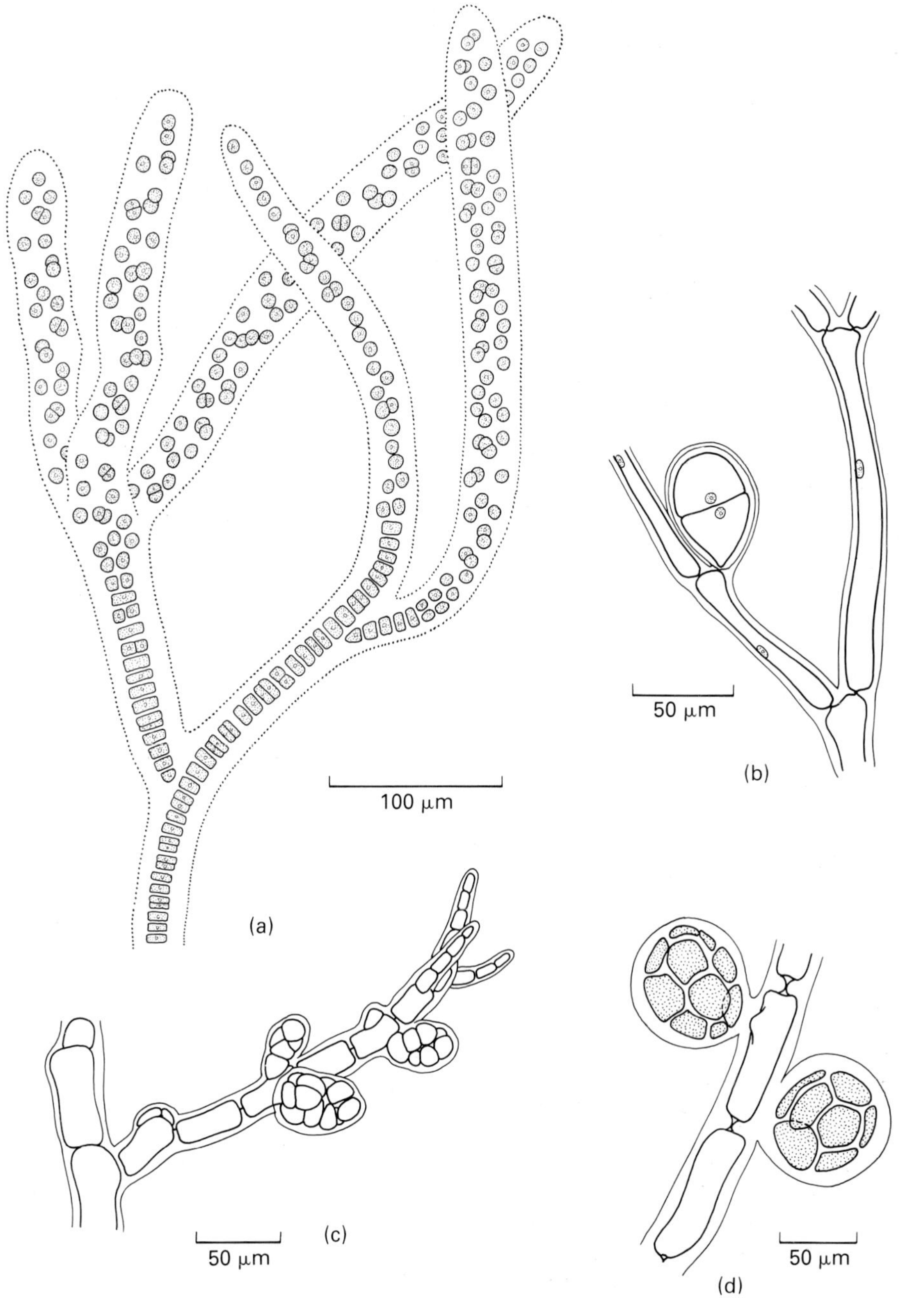

Fig. 5.7 (a) Monospore release in *Goniotrichum* (Rhodophyceae). (After Abbott & Hollenberg 1976). (b) Bisporangia in *Callithamnion*. (c) Parasporangia in *Callithamnion*. (After Rosenvinge 1924). (d) Polysporangia of *Ptilothamnion*. (After Wollaston, 1984.)

monospores reported from the Tilopteridales (Phaeophyceae) are now known to be parthenogenetically developing eggs or tetranucleate apomeiotic sporangia (Kuhlenkamp & Müller, 1985). Monospores may also be formed on diploid parts of the thalli of *Prasiola* spp. (Chlorophyceae) and recycle this phase (Friedmann, 1959).

In the Rhodophyceae, bisporangia (Fig. 5.7b) are commonly reported on tetrasporophytes, and in some instances may be stages in the development of tetrasporangia (Dixon, 1973). They are known from crustose members of the Corallinales and in *Lithophyllum litorale* and *L. corallinae*. Suneson (1950) showed that the binucleate sporangia were apomeiotic. Members of the Corallinales are regarded as having *Polysiphonia* types of life cycles (West & Hommersand, 1981), and the bisporangia are presumably a mechanism for the asexual reproduction of the tetrasporophyte phase.

Paraspores (Fig. 5.7c) found in some Rhodophyceae may be regarded as irregular masses of monospores. They are best known from *Plumaria elegans* (Drew 1939; Whittick, 1977) and *Ceramium strictum* (Rueness, 1973), where they occur on the asexual triploid generations found in the northern range of these species.

ZOOSPORES, APLANOSPORES AND AUTOSPORES

Zoospores or aplanospores are formed by the subdivision of the parent cell protoplast, and are known from the Chlorophyceae, Prasinophyceae, Xanthophyceae, Eustigmatophyceae, Chrysophyceae, Dinophyceae, Prymnesiophyceae, and Phaeophyceae.

Among Chlorophyceae, in *Chlamydomonas* two, four, or eight daughter protoplasts are produced and form cell walls while within the parent (Fig. 5.8a) and escape by rupture or lysis of the parental wall (Schlosser, 1976). Members of the vegetatively non-flagellate orders Chlorococcales and Chlorosarcinales are capable of reproducing by biflagellate or quadriflagellate zoospores which resemble *Chlamydomonas*. In contrast, members of the aflagellate Chlorellales do not produce zoospores and reproduce asexually by autospores (Fig. 5.8b).

Belcher (1966) described a coccoid prasinophyte *Prasinochloris* that produced quadriflagellate, scaled, motile cells resembling *Pyramimonas*. The free-living stages of the cyst-forming genera may be interpreted as zoospores (Norris, 1980).

In the unicellular Xanthophyceae the palmelloid forms may produce zoospores and aplanospores, but many of the coccoid species produce autospores (Hibberd, 1980a). A similar range of spores occurs in the Eustigmatophyceae (Hibberd, 1980b). The coenocytic xanthophyte *Vaucheria* produces sporangia which develop a single multinucleate zoospore bearing numerous pairs of flagella (Fig. 5.8c, d) (Ott & Brown, 1974b). On settling, the flagella are withdrawn, a cell wall is formed, and the spore develops directly into a coenocytic filament.

The multicellular genera of the Chrysophyceae, Dinophyceae, and Prymnesiophyceae produce zoospores which resemble the free-living members of the class. In *Phaeothamnion* (Chrysophyceae), individual cells produce 4–8 zoospores resembling the free-living *Ochromonas*. The filamentous dinoflagellate *Dinothrix* releases zoospores which resemble *Gymnodinium*, but on subsequent division the cells do not separate. The pseudofilamentous '*Apisthonema*' stage of the Prymnesiophyte *Hymenomonas carterae* produces non coccolith-bearing zoospores (Fig. 5.9) (Stosch, 1967).

Zoospores are frequently found in the Phaeophyceae and in the multicellular Chlorophyceae as part of a sexual life cycle. It is usually only in the more primitive taxa that reproduction via zoospores occurs independently of sexual reproduction. In the Phaeophyceae zoospores may be produced in either plurilocular or unilocular sporangia (Fig. 5.10). In the former a single cell enlarges and undergoes mitotic division to produce a number of smaller cells (locules), each of which produces a single zoospore. In the unilocular sporangium the first division may be either meiotic or mitotic, subsequent divisions are not associated with wall formation, and the zoospores mature within a single cell (locule). In the many members of the Ectocarpales zoospores from plurilocular sporangia regenerate the sporophyte generation (e.g. *Pilayella*; Knight, 1923: *Ectocarpus*; Müller, 1967: *Laminariocolax tomentosoides;* Russell, 1964:, *Giffordia mitchelliae*;

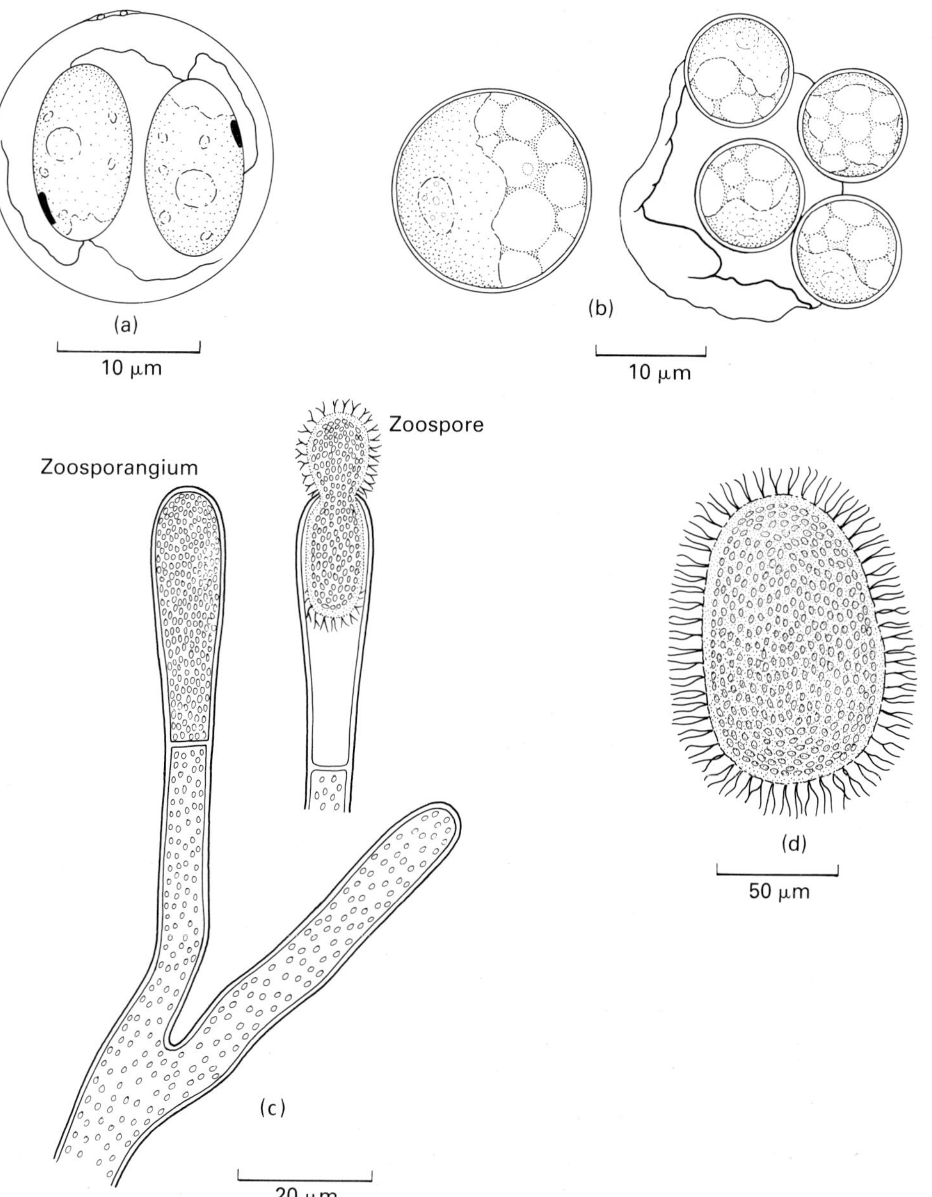

Fig. 5.8 (a) Zoospore formation in *Chlamydomonas eugametos*. (b) Vegetative cell and autospore formation in *Chlorella*. (After Bold, 1973). (c & d) Asexual reproduction in *Vaucheria*. (c) Development of a zoosporangium and release of zoospore. (d) Zoospore. (After G.M. Smith, 1955.)

Edwards, 1969). There are also many reports, particularly from members of the Ectocarpales, Chordariales, and Scytosiphonales, of the failure of meiosis in the unilocular sporangium and the consequent repetition of the parent phase (Wynne & Loiseaux, 1976; Pedersen, 1981).

Reproduction by asexual zoospores is common in the simple filamentous members of the Chlo-

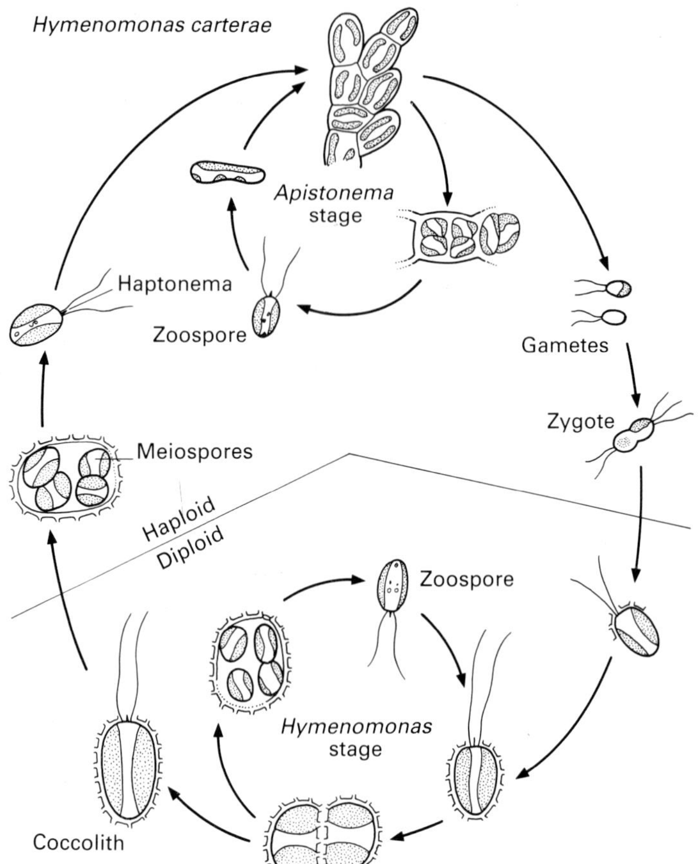

Fig. 5.9 Life cycle of *Hymenomonas carterae* (After Lee, 1980; from von Stosch, 1967; Leadbeater, 1970.)

rophyceae. In the Ulotrichales many genera produce either bi- or quadriflagellate zoospores. In both freshwater and marine species of *Ulothrix* quadriflagellate spores develop directly into new filaments identical to the parent (Lokhorst, 1978). Similar processes are seen in the Chaetophorales in *Stigeoclonium, Chaetophora* and *Coelochaete* (Bold & Wynne, 1985). Biflagellate spores which reproduce the parent phase are reported in the Cladophorales *Cladophora* and *Rhizoclonium*; in the latter it appears to be a common method of propagating the sporophytic phase under estuarine conditions (Nienhuis, 1974). Similar asexual reproduction via bi- or quadriflagellate zoospores occurs in the filamentous gametophyte of *Urospora*, particularly at low temperatures (Kornmann, 1961). Some species of *Enteromorpha* (Innes & Yarish, 1984) and *Blidingia* (Tatewaki & Iima, 1984) reproduce by zoospores, but in *Ulva* reproduction of the sporophyte by quadriflagellate zoospores is unknown (Tanner, 1981). Quadriflagellate diploid zoospores (Fig. 5.11) are reported from *Ulva mutabilis* (Fjeld & Lovlie, 1976), but these produce diploid gametophytes rather than sporophytes.

Stephanokontic zoospores, possessing a ring of flagella, are produced in *Oedogonium* (Hoffman & Manton, 1962) and other members of the Oedogoniales (Bold & Wynne, 1985), where they reform the parent generation (Fig. 5.12). Elsewhere in the green algae such zoospores occur only in *Derbesia* and closely related genera. In many of these they are formed by meiosis and occur only as part of a sexual life history (Tanner, 1981), but an asexual isolate of *Derbesia marina* is known to reproduce directly by apomeiotic stephanokontic zoospores (Fig. 5.13d) (Kornmann, 1970a).

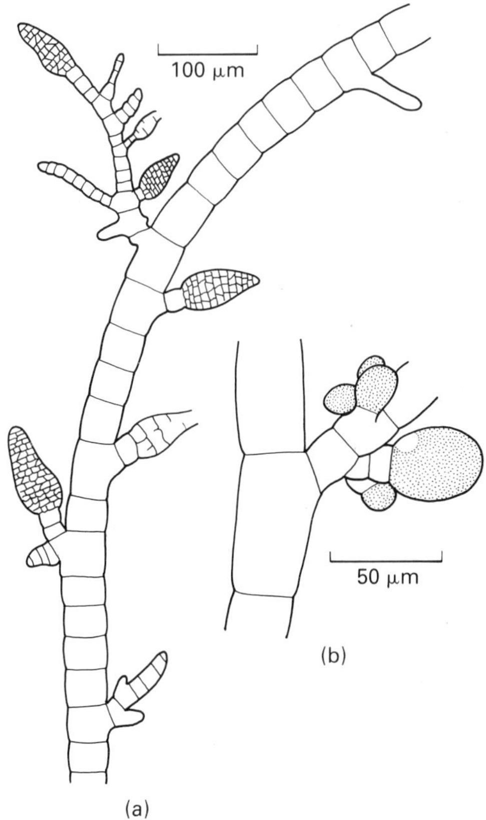

Fig. 5.10 *Ectocarpus*. (a) Plurilocular reproductive structure, (b) unilocular reproductive structure. (After Müller, 1972.)

Resting stages and perennation

Specialized resting stages are found in many algal groups (Fryxell, 1983). They are particularly common in fresh waters, where they may have to withstand complete desiccation or be exposed to wider variations in temperature and nutrients than are their marine counterparts. The most notable marine examples are the dinoflagellate hypnozygotes (Dale, 1983), which occur in the benthos of the inshore and may be brought into the surface waters by upwelling (*see* Chapter 9). Many marine centric diatoms also possess resting spores as part of their life histories (Fig. 5.14b) (Hargraves & French, 1983).

In the marine macrophytes spores usually germinate immediately, though thick-walled overwintering carpospores are reported in *Nemalion helminthoides* (Rhodophyceae; Martin, 1969). *Acetabularia* (Chlorophyceae) also forms thick-walled cysts (Tanner, 1981).

Akinetes (Fig. 5.5a, b) are single cells responsible for perennation in several orders of the Cyanophyceae, including some unicellular forms (Nichols & Adams, 1982). They have thickened walls, are usually larger than vegetative cells, and possess abundant reserves of stored nitrogen in cyanophycean bodies. In some genera (e.g. *Cylindrospermum*; Wolk, 1982), the proximity of heterocysts appears to stimulate the formation of akinetes, but this is not a general phenomenon. The environmental factors leading to akinete formation are not well understood, but in some instances are associated with nutrient deficiencies. In batch cultures they may appear at the end of the exponential growth phase (Sutherland *et al.*, 1979), and germination may be induced by the introduction of new culture medium (Miller & Lang, 1968). In nature increases in both irradiance and day length (Wildman *et al.*, 1975) are also implicated. Miller and Lang (1968) present a detailed account of the germination of akinetes of *Cylindrospermum*, where the cell may divide prior to emergence through a pore in the cell wall.

Under unfavourable conditions, particularly of dessication, many of the Chlorophyceae produce thick-walled resting cells (Fig. 5.14c) (Coleman, 1983). Akinetes are modified vegetative cells in which the wall becomes thickened. Hypnospores and hypnozygotes also have thickened walls, but these are produced *de novo* by protoplasts which previously separated from the walls of the parental cell. In the laboratory such cells may remain viable for periods in excess of 20 years (Coleman, 1983). Many of the unicellular volvocalean green algae produce thick-walled resting cells (Lembi, 1980). Akinetes occur in freshwater filamentous green algae, in *Stigeoclonium* (Chaetophorales) (Cox & Bold, 1966), and in the Zygnematales, in *Zygnema*, but are apparently absent in *Spirogyra*, which perennates exclusively by dormant zygotes (Bold & Wynne, 1985). Aplanospores of the freshwater *Ulothrix fimbriata* may also develop into hypnospores. In the Prasinophyceae asexual cysts of palmella stages probably occur in most freshwater and some marine species (Norris, 1980).

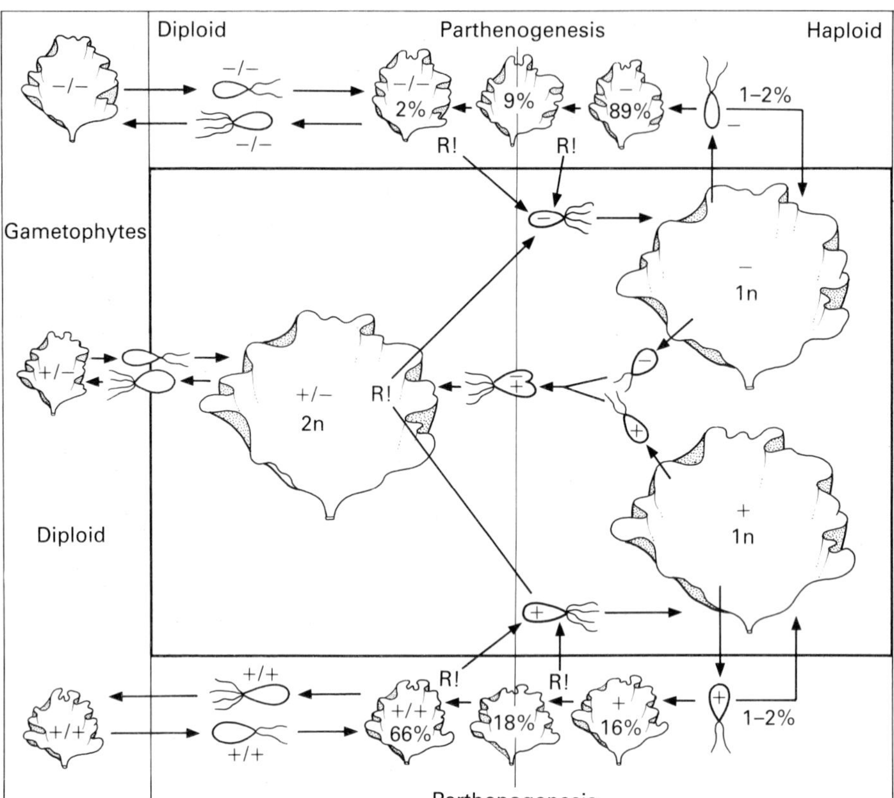

Fig. 5.11 Life cycle of *Ulva*. The inner box illustrates the basic alternation between isomorphic diploid sporophytic and haploid gametophytic generations. The marginal boxes show alternate reproductive events. Gametophytes may develop parthenogenetically into gametophytes or sporophytes. Parthenosporophytes are haploid, diploid or partially haploid and diploid. Percentages are for *Ulva mutabilis*. R! = meiosis. (After Tanner 1981, in part from Fjeld & Lovlie 1976.)

One of the characteristic features of the chrysophytes is their ability to form siliceous statospores or resting cysts (Pienaar, 1980; Sandgren, 1983). During formation the cytoplasm of the parent cell migrates into this cyst, which is then sealed by a plug (Fig. 5.15). On germination the plug dissolves and one or more motile cells emerge. In *Dinobryon cylindricum* nuclear division occurs prior to wall formation, resulting in a binucleate cell, and there is evidence that autogamic recombination may occur (Sandgren, 1980, 1981). Prymnesiophyte cysts are not well known and are probably not produced by those species which have an alternate benthic phase in their life histories (Hibberd, 1980c). In *Prymnesium* the cells are ovate in shape, with a wide pore which does not appear to be closed by a separate plug. Statospores are also produced by some motile xanthophytes and these have bipartite overlapping walls (Hibberd, 1980a).

Diatoms form heavy-walled resting spores and resting cells (Fig. 5.14b), which differ from vegetative cells primarily in their darker pigmentation. In *Stephonopyxis palmeriana* the formation of resting spores is triggered by phosphate depletion (Drebes, 1966). Resting spores may survive considerable periods of darkness, up to three years in *Melosira italica* subsp. *antarctica* (Lund, 1954).

Many free-living members of the Dinophyceae are capable of forming encysted stages in response to unfavourable conditions (Dale, 1983), and in

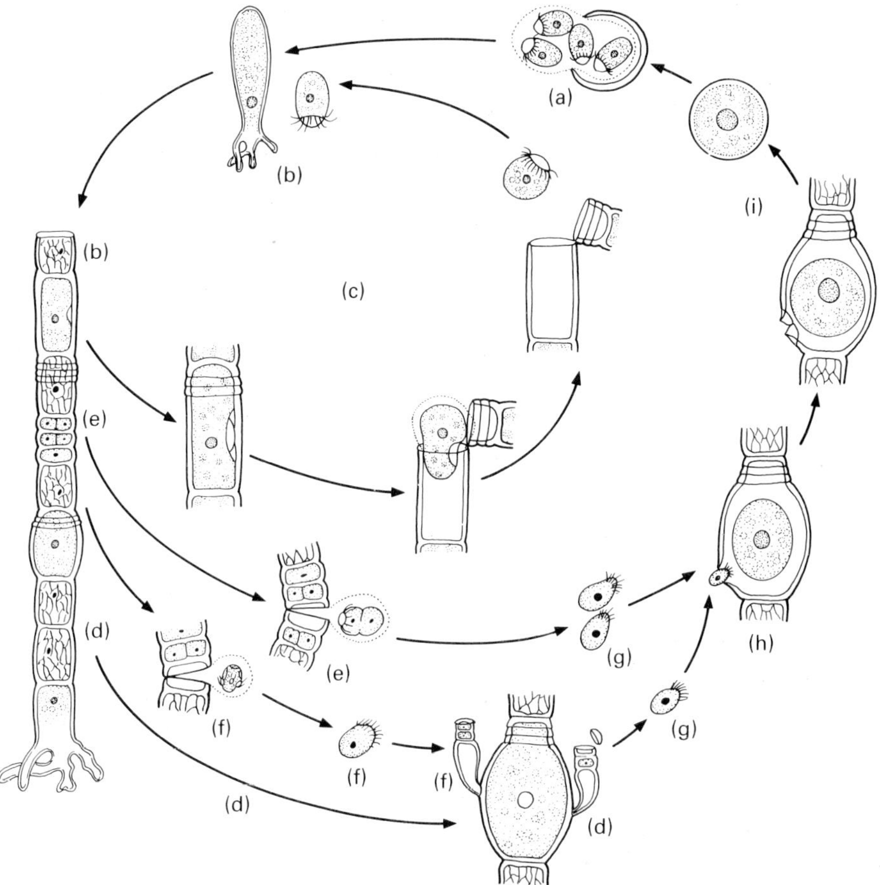

Fig. 5.12 Life cycle of *Oedogonium*. (a) Germination of zygote (meiotic) and release of stephanokontic zoospores. (b) Development of filament. (c) Asexual reproduction via zoospores. (d) Development of oogonium. (e) Spermatogenesis in a macrandrous species. (f) Nanandrous species; androspores settle on the oogonium and develop into dwarf males. (g) Sperm. (h) Egg with sperm penetrating the oogonial wall. (i) Development and liberation of the zygote. (After Esser, 1982.)

various species these may or may not resemble the parent cell (Fig. 5.14a). The latter type are now known (Evitt, 1963) to include the hystrichospheres which are abundant in the fossil record. Resting cysts are also known in the parasitic genus *Dissodinium* (Drebes, 1981).

Sexual reproduction

Sexual reproduction is absent or uncertain in the Prochlorophycota, Eustigmatophyceae, Rhaphidophyceae, Cryptophyceae, and Euglenophyceae (Table 5.1). It does not occur in the Cyanophyceae, although gene transfer between individuals is well documented. Bazin (1968) showed that strains of *Anacystis nidulans* resistant to either of the antibiotics polymyxin or streptomycin produced a double-resistant strain when grown together. The mechanism was transformation, whereby DNA freed into the medium was taken up by the recipient cell. Subsequent experiments involving the incubation of Cyanophyceae with extracted DNA (Devilly & Houghton, 1977; Stevens & Porter, 1980) have confirmed the

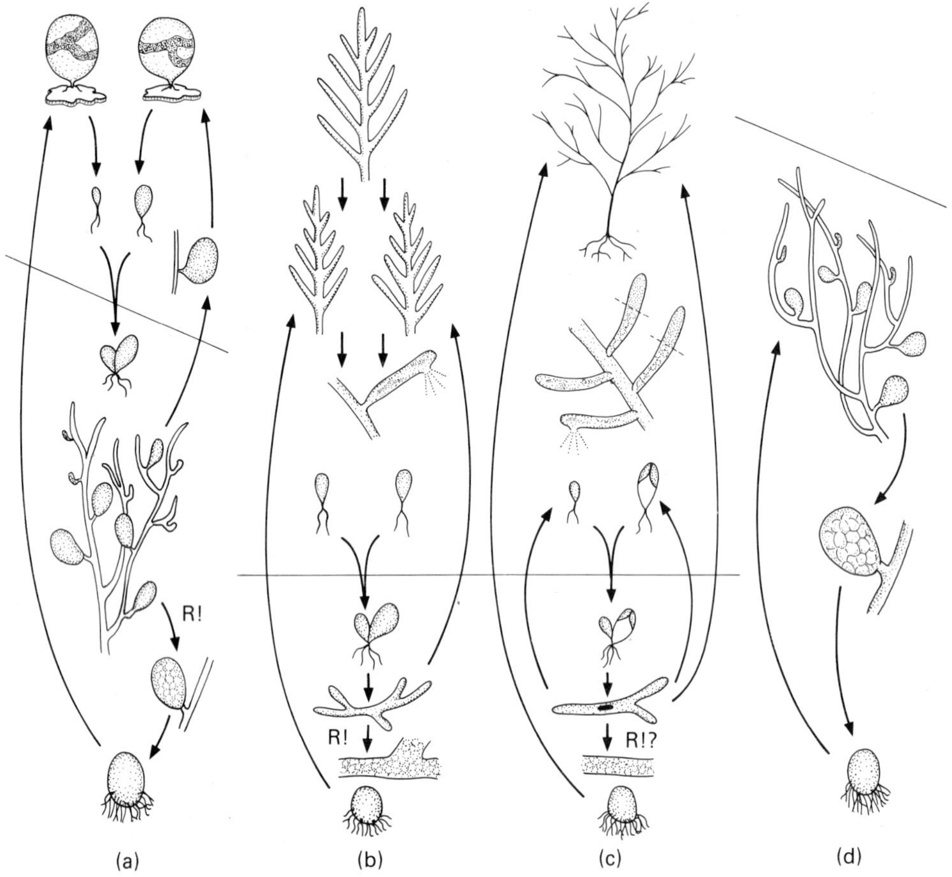

Fig. 5.13 Life cycles in the *Derbesia–Halicystis–Bryopsis* complex (Chlorophyceae). (a) Alternation between gametophytic *Halicystis* and sporophytic *Derbesia*. (b) Alternation of *Bryopsis* with a *Derbesia*-like sporophyte. (c) Alternation of gametophytic *Bryopsis hynoides* with a reduced protonemal sporophytic stage. (d) Asexual reproduction of *Derbesia* by stephanokontic zoospores. R! = meiosis. (After Tanner, 1981.)

donation of functional genes. Even though cyanophages are well known (Safferman, 1973), there is no conclusive evidence for transduction in the Cyanophyceae, and conjugation is also unknown, although heteroclone formation through filament anastomosis is reported in *Nostoc muscorum* (Delaney *et al.*, 1976).

ZYGOTIC LIFE CYCLES

Sexual reproduction occurs in the Chrysophyceae and is best known in the loricate genera because of the obvious zygote formation (Pienaar, 1980). Loricate cells attach at their openings and the protoplasts fuse to form a zygote (Fig. 5.15). Syngamy is anisogamous in *Dinobryon cylindricum*, though autogamy in binucleate resting cysts has also been reported (Sandgren, 1980, 1981).

Sexual reproduction in the Xanthophyceae has been rarely reported and appears largely confined to the phylogenetically advanced siphonous forms such as *Vaucheria* (Hibberd, 1981b). The multinucleate antheridium produces a large number of colourless sperm (Moestrup, 1970), which enter through a pore in the wall of the uninucleate oogonium (Fig. 5.16). A thick-walled resting zygote is formed, and meiosis occurs at germination to produce a siphonous filament.

Walker (1984) lists 22 species of Dinophyceae

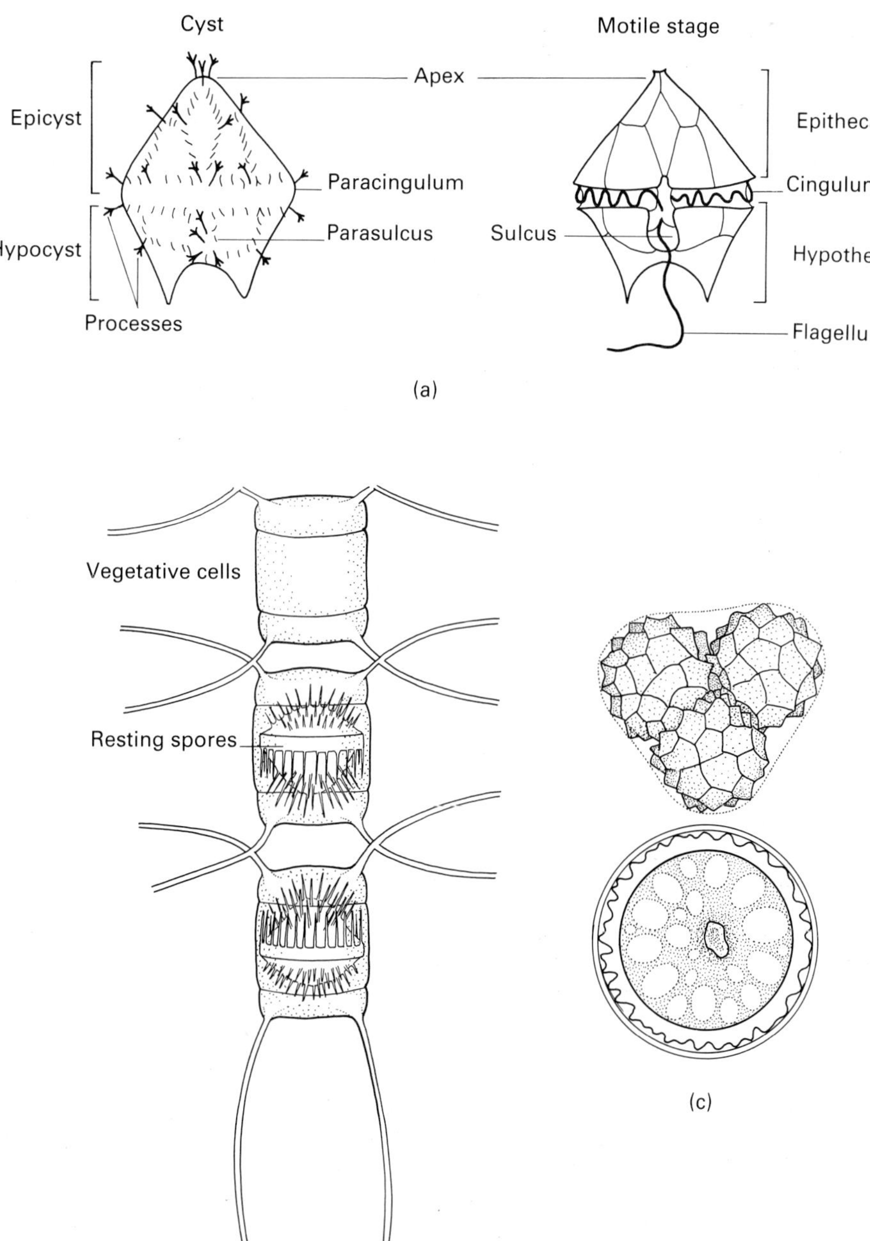

Fig. 5.14 (a) Comparison of the morphology of a resting cyst and motile cell of a thecate dinoflagellate (After Dale, 1983). (b) Resting spores in a filamentous colony of *Chaetoceros* (Bacillariophyceae), with vegetative cells (After Hendey, 1964). (c) Hypnospores of *Chlorococcum hypnosporum* (Chlorophyceae). (After Coleman, 1983.)

with sexual cycles, and all except *Noctiluca miliaris* appear to be zygotic. Most are isogamous with the gametes smaller than the vegetative cells, and those of armoured species may be naked (Fig. 5.17). Syngamy results in a motile planozygote which may undergo meiosis, but in

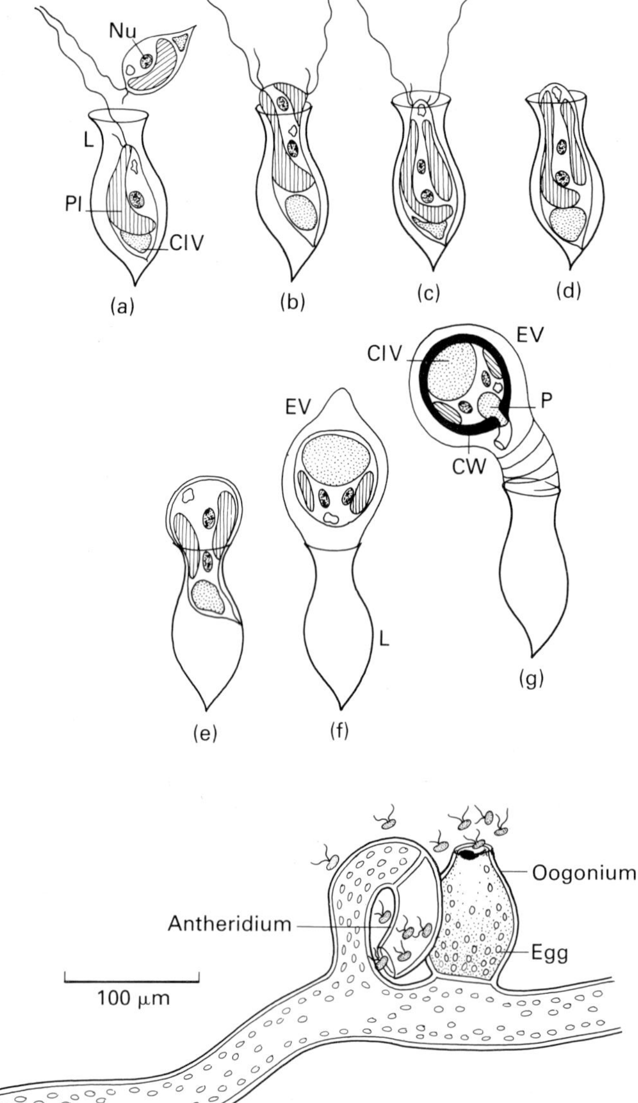

Fig. 5.15 Sexual reproduction in *Dinobryon cylindricum*. (a) Initial gamete contact. (b) Plasmogamy. (c) Binucleate zygote (Nu). (d) Commencement of cyst formation; flagella have been lost. (e) Secretion of the encystment vesicle (EV) and emergence from the lorica (L). (f) Immature statospore prior to silica deposition. (g) Mature hypnozygote with siliceous cyst wall (CW) and plugged pore (P). CIV, Chrysolaminarin storage vacuole; Pl, plastid. (After Sandgren, 1981.)

Fig. 5.16 Antheridium and oogonium of *Vaucheria*. (After G.M. Smith, 1955.)

most instances develops into a thick-walled dormant hypnozygote. Hypnozygotes require a period of dormancy ranging from a few hours in *Peridinium gatunense* (Pfiester, 1977) to several months in *Protogonyaulax tamarensis* (Anderson, 1980). Meiosis is detected by the observation of a cyclosis in which the nucleus appears to rotate. This is attributed to chromosome pairing (Stosch, 1973) and may occur within the zygote or on excystment. Generally two cells are produced, depending on the species, but in *Gymnodinium pseudopalustre* a single biflagellate cell is released, which subsequently divides meiotically to form four daughter cells (Stosch, 1973).

Nowhere among the algae has sexual reproduction been investigated more thoroughly than

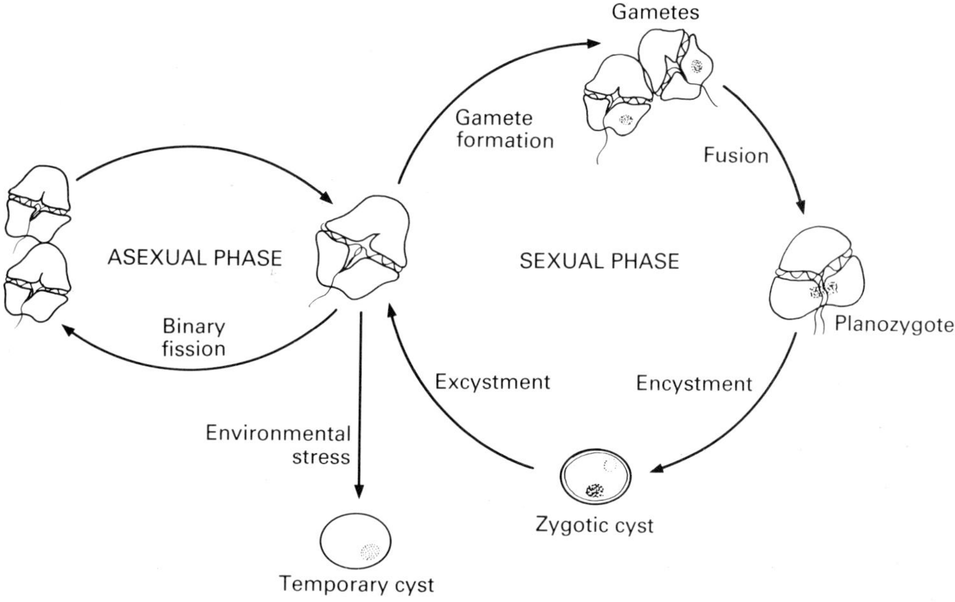

Fig. 5.17 Life cycle of a free-living dinoflagellate. (After Walker, 1984.)

in *Chlamydomonas* and related unicellular green flagellates (Fig. 5.18). Most species are isogamous, although anisogamy and oogamy are known (Tschermak-Woess, 1959, 1962). Despite appearing morphologically similar, isogamous species are biochemically anisogamous, and only complementary mating types fuse. Depending on the species, clonal populations derived from a single individual may be homothallic, producing compatible gametes, or heterothallic, requiring the presence of a second clone. Sexual reproduction in *Chlamydomonas reinhardtii* may be induced by nitrogen depletion, and two mitotic divisions lead to the formation of four gametes per cell. The clumping and eventual fusion of gametes are mediated by the presence of high molecular weight sulphated glycoproteins, gamones, on the flagellar tips (Wiese, 1974). After agglutination of the flagella tips lysis of the gamete wall occurs (Goodenough & Weiss, 1975) and the cells fuse. Within 24 hours the zygotes expand and form a thickened cell wall to produce a zygospore (Cavalier-Smith, 1976). After a period of dormancy the zygotes germinate via meiotic division to produce 4–8 haploid motile vegetative cells (Buffaloe, 1958). There has been considerable work on the genetics of *Chlamydomonas*, especially relating to nutrition, photosynthesis, and motility (Lewin, 1976a). Hybridization studies between *C. moewusii* and *C. eugametos* have shown that recombination occurs in the chloroplast DNA and that a strong selection occurs, favouring chloroplasts most compatible with the hybrid nucleus (Lemieux *et al.*, 1981). This may indicate that nucleoplastid incompatibilities are an important source of lethality in such crosses.

Coleman (1979) and Starr (1980) have reviewed the sexual life cycles of the colonial members of the Volvocales. *Gonium* and *Pandorina* are isogamous, *Eudorina* (Goldstein, 1964) is anisogamous, and *Volvox* oogamous (Fig. 5.3) (Starr, 1980). Thus, in this group, evolution in vegetative complexity is paralleled by evolution in reproductive processes. In *Volvox*, homothallic strains producing both sperm and eggs in the same colony and heterothallic species in which they are formed in different colonies are known. Minute colourless sperm are produced in a packet of 16–512 individuals. In heterothallic species the sperm packets are released and swim to a female colony, dissolve a hole in the matrix, and dissociate into individual sperm which fertilize the eggs. In homothallic species the sperm packets dissociate within the colony in which they were formed. The zygotes form thickened cell walls within the

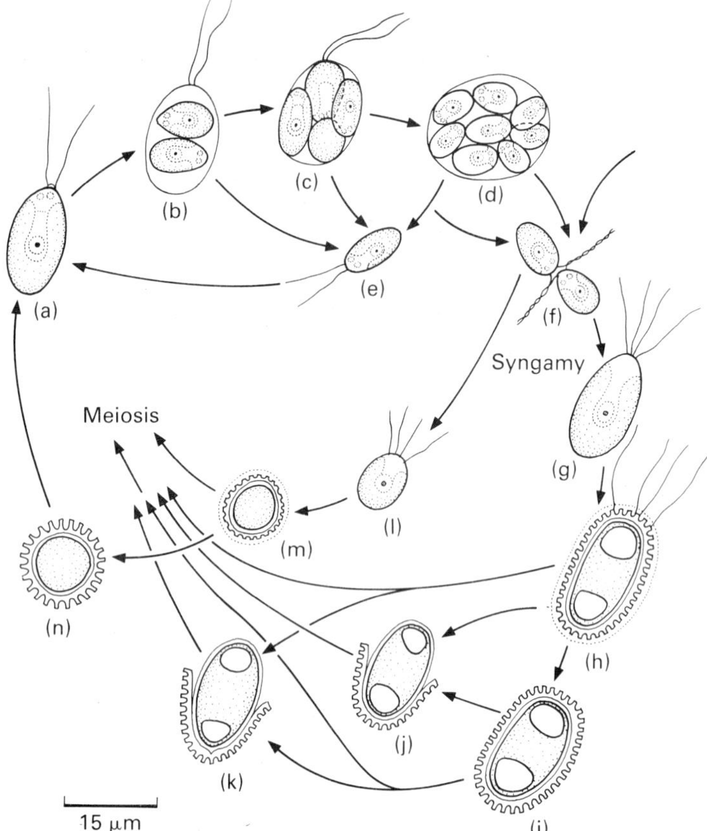

Fig. 5.18 Zygotic life cycle of a green snow alga *Chloromonas brevispina*. (a) vegetative cell. (b–d) Zoospore formation. (e) Zoospore. (f) Syngamy. (g–n) Various stages in the development of the resting zygospore. These stages have frequently been identified as separate species. (n) Meiosis occurs during the germination of the zygospore. (After Hoham *et al.*, 1979.)

parent colony, which eventually breaks down to release them. Starr (1975) has shown both cytologically and genetically that meiosis in *Volvox* occurs during germination of the zygotes. The resulting protoplast is biflagellate and initially remains within the inner wall of the spore, where it behaves much as an asexual gonidium and produces a new colony through cell division and subsequent enlargement. Zygotic life cycles are also the rule in the Tetrasporales, Chlorococcales, Chlorosarcinales, and Chlorellales (Bold & Wynne, 1985). In the coenobial forms such as *Pediastrum*, sexual reproduction is by isogamous biflagellate gametes. On germination a zoospore is formed, which divides within the zygospore wall to form a new colony (Davis, 1967).

Zygotic life cycles are presumed for many species of the filamentous Ulotrichales and Chaetophorales, especially those from fresh waters, though cytological evidence is usually lacking. In *Coleochaete*, which is variously placed within the Chaetophorales (Bold & Wynne, 1985) or associated with the Charophyceae (Stewart & Mattox, 1975), sexual reproduction is oogamous and the eggs resemble vegetative cells or may possess a small receptive trichogyne. The biflagellate sperm are formed within antheridia. The zygote is retained within the parent plant and becomes covered with an overgrowth of vegetative cells (Graham, 1985).

Two other groups of green algae, the Oedogoniales and the Zygnematales, show more specialized developments of the zygotic life cycle. In *Oedogonium* (Fig. 5.12), sexual reproduction is oogamous, with each oogonium producing a single cell. The antheridia are smaller, and each produces 2–4 multiflagellate stephanokontic sperm. Some species are bisexual, with antheridia and oogonia on the same thallus, while others (termed macrandrous) have vegetatively isomorphic filaments of separate sex. In nanandrous species, extreme sexual dimorphism occurs, with a dwarf male filament growing epiphytically on the female. The dwarf males arise from androspores, which are intermediate in size between sperm and normal zoospores. They are formed in androsporangia, which may be found on the same filament as the oogonia in gynandrosporous species or arise on separate filaments in idioandrosporous species. In idioandrosporous species the androspore is chemically attracted to the female filament (Rawitscher-Kunkel & Machlis, 1962). The released sperm of macrandrous species are chemotactically attracted to oogonia (Machlis *et al.*, 1974), whereas in nannandrous species the development of the dwarf male stimulates development of an oogonium in close proximity, producing a mucilaginous sheath enveloping the apex of the dwarf male. The sperm enter the oogonium via a papilla and the zygote thickens to produce a dormant zygospore, with meiosis occurring during germination (Hoffman, 1965).

Motile cells are lacking in the Zygnematales. In *Spirogyra* projections from adjacent filaments make contact and the contiguous walls dissolve to form a conjugation tube (Fig. 5.19). One nucleus migrates and undergoes syngamy with the other non-motile nucleus, and the resulting zygote develops a thickened wall. In *S. crassa*, meiosis occurs prior to germination and three of the resulting nuclei degenerate (Godward, 1961). The mechanism in *Mougeotia* is similar, except that isogamy occurs and the zygote is formed in the conjugation canal, whereas in *Sirogonium* there is direct fusion of filaments and no conjugation canals are formed (Hoshaw, 1965).

Starr (1955) described methods for isolating potentially sexual strains of desmids which although common in peatland pools are rarely observed to reproduce sexually. Coesel and Texeira (1974) suggested that this is not due to environmental constraints, but is an innate feature of many strains. The filamentous members of the saccoderm desmids, such as *Desmidium* and *Hyalotheca*, produce conjugation canals, and depending on the species the zygote may be formed within the canal or in one of the parent cells. Conjugation canals are formed in some placoderm genera. In *Micrasterias papillifera*, a conjugation canal is formed and the zygote is formed within it (Coessel & Texeira, 1974). Pairs of cells of compatible mating types of *Closterium* produce a mucilaginous matrix, each then opens at the isthmus to release the protoplast, and the two fuse to produce a zygospore (Fig. 5.20). At germination the zygospore protoplast divides meiotically to form two cells of opposite mating type. Sexual reproduction in saccoderm desmids is similar, with broad conjugation canals developing. In *Mesotaenium* (Starr & Rayburn, 1964) and *Cylindrocystis* (Biebel, 1973), zygote germination is again meiotic, but in contrast to the placoderms four offspring cells may be produced.

Karyological observations and DNA measurements (Sawa, 1965; Shen, 1967) suggest the occurrence of a zygotic life cycle in the Charophyceae. However, from a morphological perspective, sexual reproduction in the Charophyceae is perhaps the most advanced in the algae, with both the male and female structures surrounded by specialized vegetative cells and thus showing some of the characteristics of the archaeogoniate plants (Graham, 1985). *Chara* spp. may be monoecious or dioecious (Proctor, 1971), with

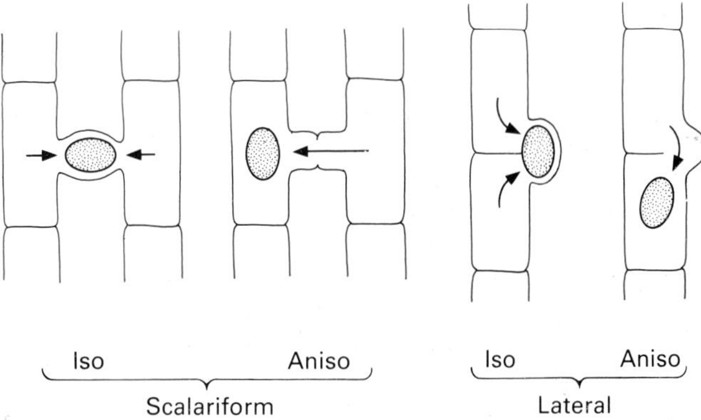

Fig. 5.19 Conjugation strategies in the Zygnemataceae. (After Esser 1982.)

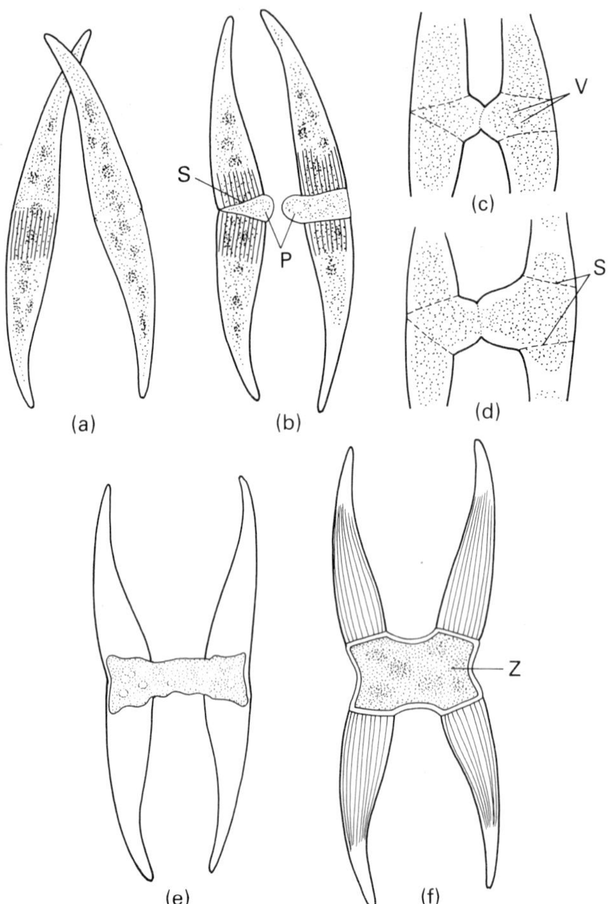

Fig. 5.20 Conjugation in *Closterium rostratum*. (a) Initial pairing of cells. (b) Protrusion of papillae (P) as semi-cells separate at suture lines (S). (c) Contact of papillae with vacuolated tips (V). (d) fusion of protoplasts in conjugation tube. (e & f) Development of the zygospore (Z). (After Brook, 1981.)

the antheridia and oogonia developing at the branch-bearing nodes (Fig. 5.21a). The antheridia are supported by a columnar stalk, the pedicel, and are enclosed by shield cells (Fig. 5.21b). At its apex the pedicel bears eight capitular cells; these produce secondary and tertiary capitula, which form spermatogenous filaments of thin-walled cells, each of which produces a single sperm (Fig. 5.21c). In the oogonium, a row of cells develops from a nodal initial and the distal cell develops into a single egg or oosphere plus sterile cells (Sawa & Frame, 1974). The basal cell forms the attachment pedicel, which bears a cell, ultimately producing the five spiral sheath cells surrounding the developing egg (Fig. 5.21d). Each sheath cell produces a small apical cell forming the corona of the oogonium. At maturity, splits appear in the sheath cells beneath the corona, allowing sperm to enter. After fertilization the walls of the sheath and zygote thicken to form a dark oospore (Fig. 5.21e, f). On germination a filamentous protonema develops, which differentiates to produce nodal cells at each end; those at the basal pole produce rhizoids, while an apical cell is formed at the other. Apart from some differences in the cellular arrangement in the reproductive organs the reproduction and life cycles are essentially identical in the other members of the class.

SPORIC LIFE CYCLES

All but a few species of the sexually reproducing Rhodophyceae exhibit sporic life cycles, and all are oogamous (West & Hommersand, 1981). In the Florideophycidae the majority of species have a triphasic *Polysiphonia*-like life cycle (Fig. 5.22).

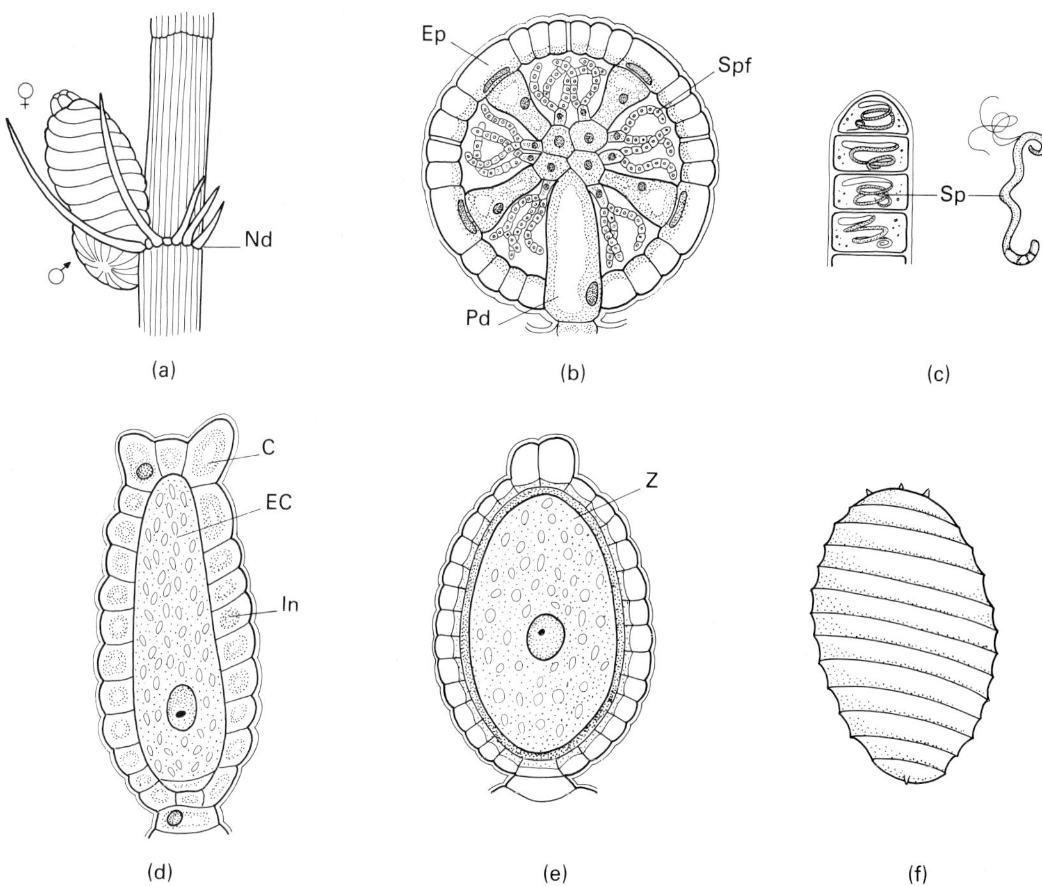

Fig. 5.21 Development of reproductive structures in *Chara*. (a) Node (Nd) bearing oogonium and antheridium. (b) Section through mature antheridium: Pd, pedicel; Ep, epidermal cell; Spf, spermatogenous filament. (c) Tip of spermatogenous filament and sperm (Sp). (d) Section through oogonium: C, corona; EC, egg cell; In, integument cell. (e) Section through oogonium showing zygote (Z) with thickening wall. (f) Mature zygote with thickened wall. (After G.M. Smith, 1955.)

The dioecious gametophytes and the tetrasporophyte are most often isomorphic, while the carposporophyte is reduced and remains attached to the female gametophyte. This cycle is found in all the orders of the Florideophycidae accepted by Kylin (1956), but it is lacking in the Palmariales (van der Meer & Todd, 1980) and Bonnemaisoniales (Chihara & Yoshizaki, 1972). The male gametes, which are aflagellate and lack cell walls, are termed spermatia and are formed singly in each spermatangium, although regeneration of the spermatangium may occur after gamete release (Dixon, 1973). The oogonium (carpogonium) has its apex extended in a hair-like process, the trichogyne, which receives the spermatium during fertilization.

The location of the carpogonium and the post-fertilizational changes in the development of the carposporophyte have been the accepted basis for the classification of the Florideophycidae at the ordinal level (Kylin 1956). In the Nemaliales the carposporophyte develops directly from the fertilized carpogonium and a mass of carpospores is formed. In other orders the zygotic nucleus is transferred to an auxiliary cell (Fig. 5.23); the transfer may involve the development of a long

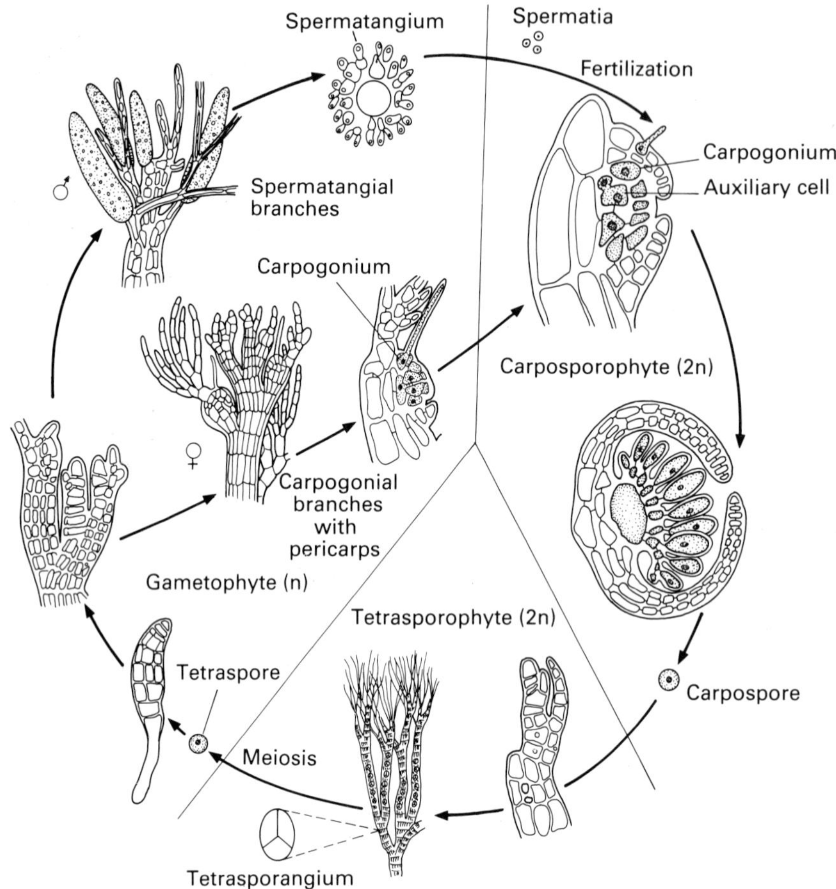

Fig. 5.22 Life cycle of *Polysiphonia*. Note tetrahedral division of the tetrasporangium. (After West & Hommersand 1981.)

connecting filament or ooblast. Secondary ooblast filaments may arise from the auxiliary cell and fuse with other auxiliary cells. The auxiliary cells, which may fuse with other vegetative cells, ultimately produce a series of gonimoblast filaments, which in turn produce the carpospores. The auxiliary cell and its derived cells make up the carposporophyte. The term cystocarp is often used synonymously with carposporophyte, but also has been used to include accessory associated female gametophyte tissue, such as the flask-shaped pericarps which surround the carposporophytes in some members of the Bonnemaisoniales and the Ceramiales.

The released carpospores germinate to produce the diploid tetrasporophyte phase, which at maturity produces tetraspores by meiotic division within a tetrasporangium. Synaptonemal complexes, regarded as evidence of meiosis, have been detected ultrastructurally (Kugrens & West, 1972). Three basic types of tetrasporangial cleavage occur: tetrahedral (Fig. 5.22), cruciate, and zonate (Fig. 5.24a, b). Usually a given species has only one form of division (Guiry, 1977). Tetrad analysis from a number of species shows that sex is genotypically determined during meiosis, and this has been confirmed in *Gracilaria* in an analysis of pigment mutants and sex. There is evidence that a pair of alleles showing incomplete dominance are responsible for the

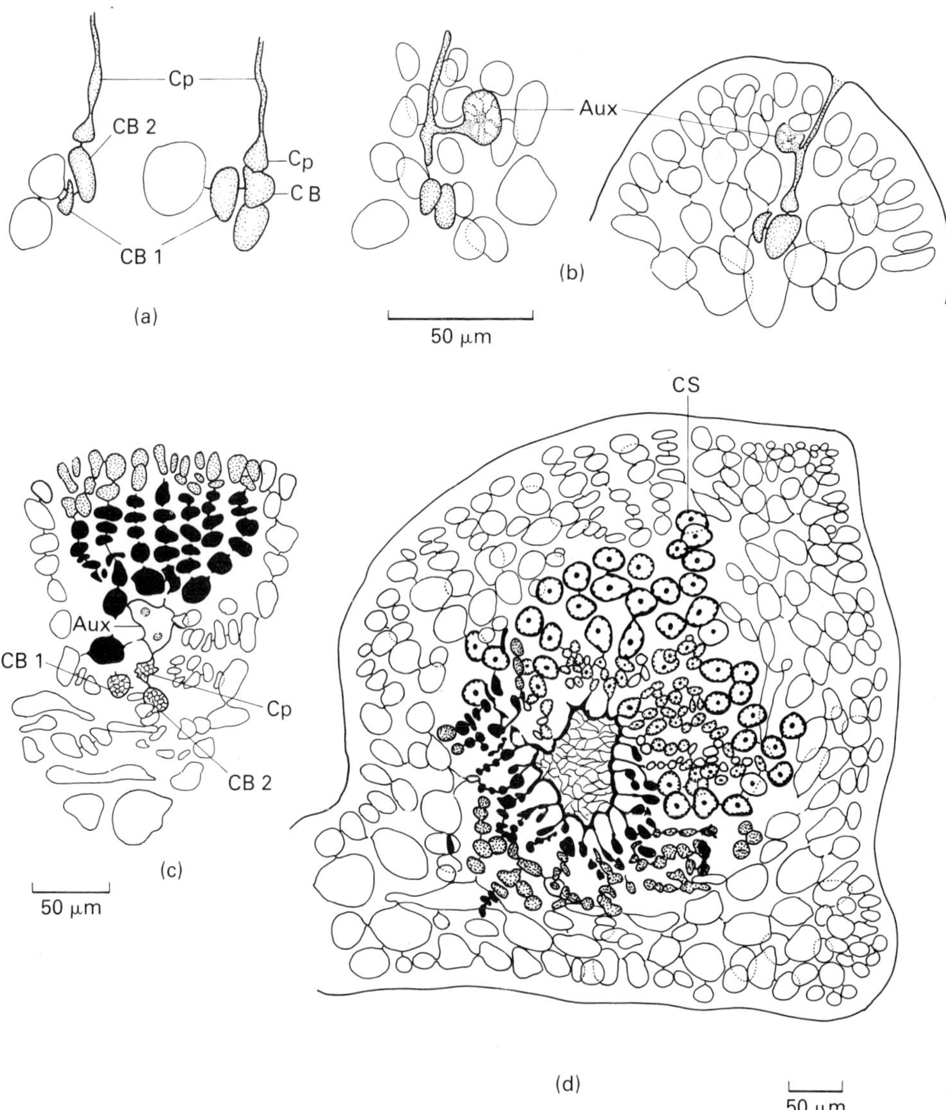

Fig. 5.23 Development of carposporophyte generation in *Fimbriophyllum*. (a) Carpogonial branch: CB, carpogonial branch cell, Cp, carpogonium with elongate trichogyne. (b) Development of auxiliary (Aux) cell and transfer of diploid nucleus from fertilized carpogonium. (c) Auxiliary cell producing gonimoblast filaments. (d) Maturing carposporophyte producing chains of carposporangia (CS). (After Hansen 1980.)

determination of sex in the Rhodophyceae; diploid tetrasporophyte cells would therefore be heterozygous and neuter (van der Meer & Todd, 1977). Mitotic recombination may occur to produce paired islands of homozygous tissue which, while diploid, are male and female and may in part explain the occasional occurrence of gametophytic reproductive organs on tetrasporophytes. *Griffithsia pacifica* has large multinucleate cells which can be induced to undergo somatic fusion. When cells from male and female plants are so fused the resulting plant bears tetrasporangia-

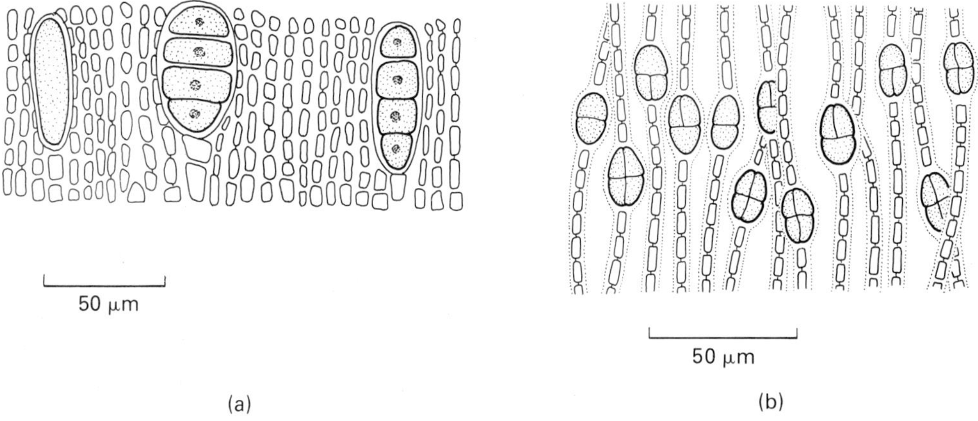

Fig. 5.24 (a) Zonate tetrasporangia in *Haematocelis*. (b) Cruciate tetrasporangia in *Petrocelis*. (After Abbott & Hollenberg 1976.)

like structures, even though there is no evidence for nuclear fusion. This work on the red algae indicates that the level of ploidy is not the primary mechanism determining whether a plant is gametophyte or a sporophyte: this also occurs in the Phaeophyceae (Nakahara & Nakamura, 1973; Henry & Müller, 1983) and the Chlorophyceae (Fjeld & Lovlie, 1976).

In addition to the *Polysiphonia* life cycle, the red algae show a number of examples of heteromorphic life cycles. These usually involve the alternation of a large, erect gametophyte with a reduced filamentous or crustose tetrasporophyte generation. In many instances these two phases had been previously recognized and described as separate species.

Audouinella pectinatum shows a heteromorphic sequence of tetrasporophytes 1–2 cm high and dwarf filamentous dioecious gametophytes < 0.5 mm high (West, 1968). Other species of Nemaliales, *Nemalion multifidum* (Fries, 1969), *Liagora farinosa* (Stosch, 1965b), *Pseudogloiophloea confusa* (Ramus, 1969), have larger gametophytes with *Audouinella*-like tetrasporophytes. Members of the Bonnemaisoniaceae (Chihara & Yoshizaki, 1972) show heteromorphy, with the tetrasporophyte generation being reduced to either a profusely branched prostrate filamentous stage (*Hymenoclonium serpens*), or a less prostrate, creeping filamentous *Trailliella intricata* or *Falkenbergia* phase.

The Cryptonemiales and Gigartinales contain a number of genera in which the tetrasporophyte is reduced, usually to a crustose phase, and these include *Farlowia* spp. (DeCew & West, 1982a), *Pikea* (Scott & Dixon, 1971), and *Gloiosiphonia* (Edelstein & McLachlan 1971). Carpospores from *Acrosymphyton purpuriferum* develop into a *Hymenoclonium* phase and from *Halymenia floresia* into an *Audouinella*-like phase (Cortel-Breeman & Hoek, 1970). Other members of the Gigartinales show a variety of heteromorphic life cycles, and these are best exemplified in *Mastocarpus* spp. 'formerly regarded as a section of the genus *Gigartina* (Guiry *et al.*, 1984), whereas *Gigartina* spp. *sensu stricto* have essentially *Polysiphonia* life cycles. Tetraspores from the crustose *Petrocelis* develop into *Gigartina papillata* and *G. jardinii* (West, 1972), and a similar relationship has been shown between *Mastocarpus stellata* and *Petrocelis cruenta* (Chen *et al.*, 1974). The *Mastocarpus* life cycle is complicated in some populations, in that carpospores may develop directly into carposporophyte-bearing female gametophytes (Polanshek & West, 1977; Rueness, 1978).

Members of the Phyllophoraceae show a wide variety of life cycles. Crustose phases are reported in the life cycle of *Gymnogrongus leptophyllus* (DeCew & West, 1981). *Phyllophora pseudoceranoides* (Newroth, 1972) has a *Polysiphonia*-like cycle and *P. truncata* is tetrasporoblastic (Newroth, 1971). In this latter species, and in *Gymnogrongus griffithsiae* (Cordeiro-Marino & Poza,

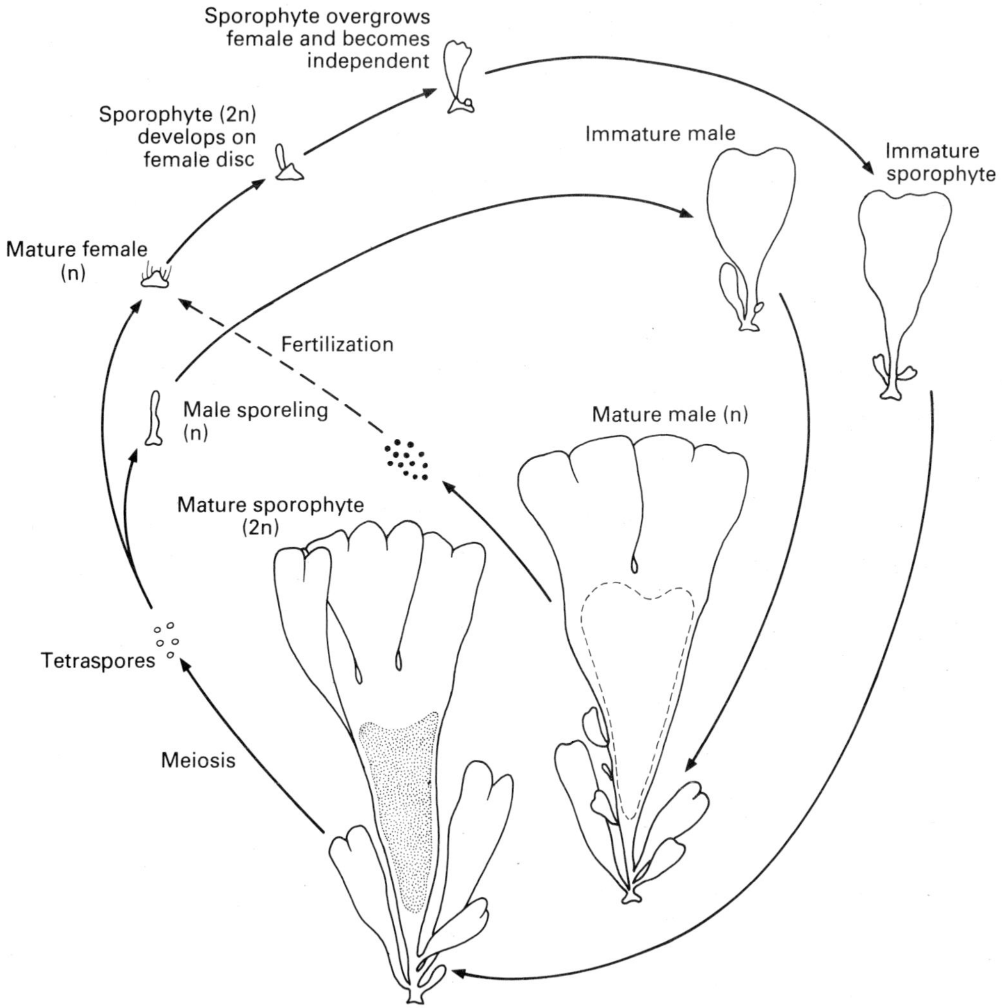

Fig. 5.25 Life history of *Palmaria palmata*. (After van der Meer & Todd 1980.)

1981), carposporangia formed within the female gametophyte produce filaments which bear tetrasporangia on the surface of the thallus. In the related *Gymnogrongus chiton* McCandless and Vollmer (1984) have presented immunocytological evidence (based on the distribution of carrageenan types) that meiosis occurs during tetrasporogenesis. Guiry (in McCandless & Vollmer, 1984) argues that these reduced phases should be regarded as tetrasporophytes rather than carpotetrasporophytes, which are found in some *Liagora* spp. and *Helminthocladia* spp. Carpotetrasporophytes as exemplified by *Liagora tetrasporifera* resemble the carposporophyte of carpospore-producing *L. farinosa* (Stosch, 1965b), but they produce meiotic tetrasporangia at the apex of the gonimoblast filaments.

In the Palmariales (Fig. 5.25) the female gametophyte is reduced and the carposporophyte is lacking. Tetrasporophytes and males were reported for *Palmaria palmata*, but female plants were unknown. Colour mutants, used to distinguish gametophyte tissue from sporophyte tissue, showed that spore tetrads from tetrasporophytes segregated 2:2 into large-bladed male gametophytes, isomorphic with the tetrasporophytes, and dwarf discoid females 0.1 mm diameter at maturity (van der Meer & Todd, 1980). The female gametophytes produce numerous carpogonia with elongate trichogynes, which after fertilization develop directly into the bladed tetrasporophyte.

The life cycle of the crustose *Rhodophysema elegans* shows a still further reduction with the absence of a tetrasporophyte generation. The crust is a monoecious gametophyte, and the fertilized carpogonia develop directly into meiotic tetrasporangia, the spores from which reproduce the crust (DeCew & West, 1982b).

Members of the Rhodymeniales and the Ceramiales basically exhibit a *Polysiphonia* type of life cycle with many minor deviations, usually as a result of failure of meiosis in the tetrasporangium. Examples include *Lomentaria orcadensis* (Svedelius, 1937) and *Antithamnion boreale* (Sundene, 1962), which directly recycle the tetrasporophyte generation. Reports of tetrasporangia and gametangia borne on the same thalli are also widespread in this group (Knaggs, 1969), although little is known of the cytological events underlying these occurrences. Several genera, e.g. *Pleonosporum* and *Tiffaniella*, produce polysporangia (Fig. 5.7d), which are homologous with tetrasporangia, in that meiosis occurs and the resulting spores undergo further mitotic divisions within the sporangium (Drew, 1937).

Members of the Bangiales may display a reproductive cycle involving a shell-boring alga, *Conchocelis rosea* (Fig. 5.26). This was first demonstrated in *Porphyra umbilicalis* by Drew (1949) and subsequently shown to occur in *Erythrotrichia* (Heerebout, 1968), *Bangia* (Richardson & Dixon, 1968), *Smithora* (Richardson & Dixon, 1969), and *Porphyropsis* (Murray *et al.*, 1972). It is now generally accepted that the life cycle of *Porphyra* and *Bangia* is essentially sexual, with the *Conchocelis*-phase presumably the diploid sporophyte generation (Cole & Conway, 1980). However, there are many species in which the chromosome number of the foliose and conchocelis phases are reportedly identical, and fertilization and meiosis are unrecorded. The conchocelis phase produces rows of terminally liberated conchospores, which germinate to produce the thallose phase. Depending on the species, 16–256 spermatia, also termed β-spores, are produced in cells on the surface of the thallus. The liberated spermatia fuse with small protrusions, prototrichogynes, found on the thallus surface. The cell which bears them, assumed to be the carpogonium, divides to produce 4–16 carpospores, also known as α-spores; these germinate to produce the *Conchocelis*. Chromosome studies suggest that meiosis occurs during the development of the conchospores. In an elegant ultrastructural study, Hawkes (1978) has unequivocably demonstrated syngamy. It should, however, be emphasized that the sexual reproduction has only been shown for a few of these species.

Life cycles in the Prymnesiophyceae are confusing. Most if not all of the coccolithophorids have more than one cell type in their life cycles (Hibberd, 1980c), and as such they could be regarded as possessing sporic life cycles, but unequivocal demonstrations of the site of meiosis are usually lacking and pleiomorphism is likely. Alternation occurs between a diploid free-living coccolithophorid (Fig. 5.9), *Hymenomonas carterae,* and a haploid benthic pseudofilamentous or *Apisthonema* stage (Stosch, 1967; Leadbeater, 1970). *Hymenomonas* reproduces by cell division, but each cell may also produce four smaller non coccolith-bearing zoospores, which grow into a benthic phase. The benthic phase is capable of reproducing itself by zoospores, or it may form smaller gametes which fuse to produce a zygote, re-establishing the coccolith-bearing phase. *Ochrosphaera*, a non-motile coccolith-bearing genus, is also implicated in this life cycle (Gayral & Fresnel-Morange, 1971), suggesting pleiomorphism. Parke and Adams (1960) have described an alternation between two free-living coccolithophorids, the non-motile *Coccolithus pelagicus* and the motile *Crystallolithus hyalinus*; meiosis and syngamy were not observed, and it is unclear whether this is a sexual life cycle. In *Emiliania huxleyi*, the dominant phase is non-motile, but it may produce motile, scaly, but non coccolith-bearing cells, which appear to have less DNA than the coccolith-bearing phase (Paasche & Klaveness, 1970).

A sporic life cycle involving the alternation of a free-living haploid gametophyte generation with a morphologically similar or dissimilar free-living diploid sporophyte is characteristic of all sexually reproducing brown algae, with the exception of members of the Fucales and related orders. This was recognized by Reinke (1878) over a century ago in the Cutleriales in *Zanardinia*, with an isomorphic alternation, and in *Cutleria* (Falkenberg 1879) with a heteromorphic cycle, involving a prostrate sporophyte generation previously identified as a separate genus, *Aglaozonia*.

Meiosis occurs in unilocular sporangia borne

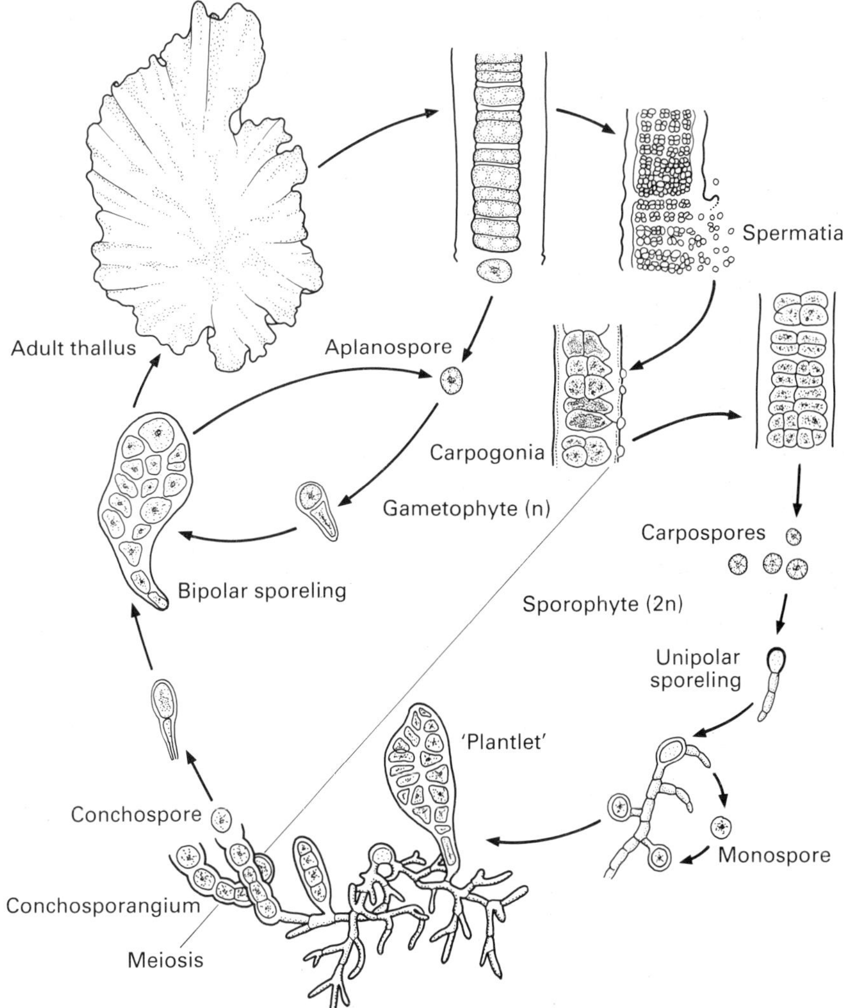

Fig. 5.26 Life history of *Porphyra* spp. (After West & Hommersand 1981.)

on the sporophyte generation, and synapsis has been detected ultrastructurally in *Pilayella* and *Chorda* (Toth & Markey, 1973). In most orders the spores produced are numerous and motile. On germination they produce gametophytes which may be isogamous, anisogamous, or oogamous. Marked anisogamy or oogamy in the Phaeophyceae is a characteristic of the orders Cutleriales, Sphacelariales, Dictyotales, Laminariales, and Desmarestiales, while the Ectocarpales, Dictyosiphonales, and Chordariales are predominantly isogamous. The orders Ectocarpales, Tilopteridales, Dictyotales, Sphacelariales, and Cutleriales (in part) show predominantly isomorphic alternations, whereas alternation of heteromorphic phases occurs in other orders. In isogamous and anisogamous species the gametangia are plurilocular structures, the compartments of which result from sequential divisions and contain at maturity one gamete each. Similar plurilocular sporangia may occur on the sporophytes of some species and produce spores rather than gametes. In oogamous plants one egg is produced in each oogonium; male gametes are produced either in abundance in plurilocular antheridia or singly in one-chambered antheridia, which may be regarded as reduced plurilocular structures (Henry & Cole, 1982b).

Deviations from this characteristic alternation of sporophyte and gametophyte are found in many orders (Wynne & Loiseaux, 1976), and most commonly involve the recycling of the haploid or diploid phase via neutral zoids from plurilocular sporangia; one of the phases may also be suppressed as result of apomeiosis or parthenogenesis. There are also a number of reports, reviewed by Caram (1972), of the fusion of the products of unilocular sporangia, i.e. the zoospores functioning as gametes to produce what would in effect be a gametic life cycle. This was first described by Knight (1929) for *Ectocarpus siliculosus*, but has since been refuted for this species (Müller, 1975), and it appears that there is no corroborative cytological evidence for its occurrence in the Phaeophyceae.

Knight (1929) showed an isomorphic alternation of generations in *Ectocarpus siliculosus*, with plurilocular structures occurring on both the haploid and diploid phases, but with the unilocular sporangia confined to the diploid sporophytes. Müller (1967, 1972) confirmed this alternation, and in addition showed that gametophytes could be haploid or diploid and sporophytes could be haploid, diploid, or tetraploid (Fig. 5.27). Haploid sporophytes developed by parthenogenetic germination of unfused male or female gametes and bore apomeiotic unilocular sporangia. Tetraploid sporophytes occasionally developed spontaneously from zoids released from the plurilocular sporangia of these haploid sporophytes. Zoids from the unilocular sporangia of these tetraploids produced both diploid sporophytes and diploid gametophytes. Sporophytes and gametophytes could be distinguished morphologically, and although isogamy occurs there is a pronounced physiological anisogamy in that the female gamete settles and produces a pheromone which attracts the male (*see* Chapter 6). In many other members of this order, particularly in the crustose forms, sexuality appears to be lacking (Wynne & Loiseaux, 1976).

In the Dictyotales, *Dictyota dichotoma* and *Dilophus ligulatus* (Gaillard, 1972) show the characteristic isomorphic pattern in which antheridia and oogonia develop from cortical cells to form sori on separate male and female gametophytes. The antheridia are plurilocular (Fig. 5.28a), and each of the thousands of locules releases a single uniflagellate sperm, whereas each oogonium produces a single egg (Fig. 5.28b).

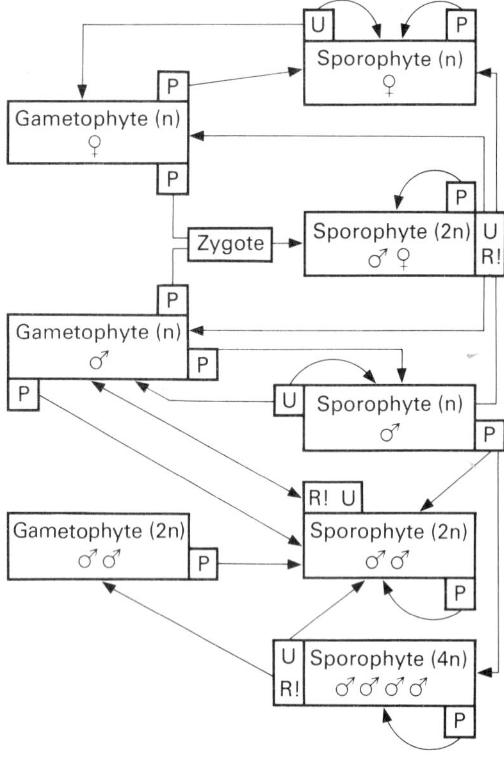

Fig. 5.27 Outline of life cycle of *Ectocarpus siliculosus* involving haploid, diploid, and tetraploid phases. U, unilocular structures; P, plurilocular structures; R!, meiosis. (After Wynne & Loiseaux 1976, from Müller 1967.)

The sporophytes bear meiotic tetrasporangia (Fig. 5.28c), although apomeiosis is reported in *Padina pavonica* and *Dilophus fasicola*. Isomorphic life cycles are also characteristic of the Sphacelariales, although *Sphacelaria rigidula* (as *S. furcigera*; Hoek & Flinterman, 1968) shows a slight heteromorphy in which the dioecious gametophytes are less robust than the sporophytes. Female gametes may develop parthenogenetically to produce female gametophytes or haploid sporophytes.

In true heteromorphic cycles the two generations differ markedly in form, and members of the Laminariales exemplify such cycles, which were first demonstrated by Sauvageau (1915) for *Saccorhiza polyschides* (Fig. 5.29). The sporophyte is macroscopic and bears meiotic unilocular sporangia in sori borne either on the vegetative blade or on specialized sporophylls. The zoospores from a single sporangium produce equal

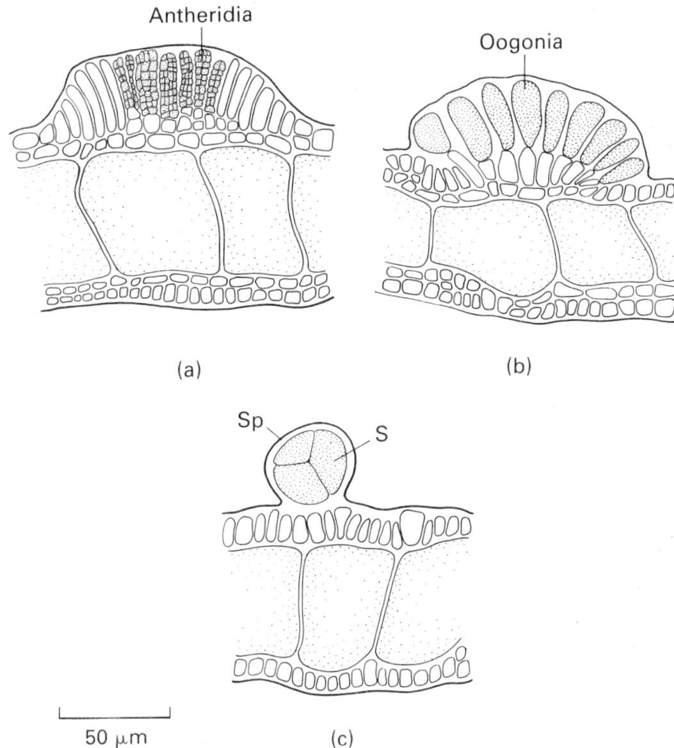

Fig. 5.28 Reproductive structures in *Dictyota*. (a) Antheridia. (b) Oogonia. (c) Tetrasporangia, Sp, sporangia; S, spore. (After Scagel *et al.* 1982.)

numbers of microscopic male and female gametophytes (Schreiber, 1930). Evans (1965) showed that sex was genotypically determined in *Saccorhiza polyschides* by the presence of a large X chromosome in the female and a small Y chromosome in the male. These chromosomes were seen to pair in meiosis. On germination the protoplasm usually moves out of the spore into a germination tube and a wall is formed to create the first cell of the gametophyte. The male has smaller cells and is more branched than the female; it bears antheridia which each produce a single sperm. In the females each oogonium liberates a single egg, and after fertilization the zygote develops into a sporophyte, which rapidly enlarges and differentiates. The eggs of *Alaria* may develop parthenogenetically into haploid sporophytes morphologically indistinguishable from their diploid counterparts (Nakahara & Nakamura, 1973). Members of the same morphological sections in the genus *Laminaria* apparently readily hybridize (Lüning *et al.*, 1978) and intergeneric hybrids are also reported between *Macrocystis* and *Pelagophycus* (Sanbonsuga & Neushul 1978). The developmental processes associated with the life cycles of members of the Laminariales are well documented in *Laminaria* (Kain, 1979).

Similar life cycles occur in the Sporochnales and Desmarestiales. In *Sporochnus pedunculatus*, a large sporophyte-bearing meiotic unilocular sporangia alternates with a microscopic gametophyte (Caram, 1965). A *Laminaria* type of alternation of generations occurs in the Desmarestiales, with monoecious gametophytes in *Desmarestia viridis* (Kornmann, 1962) and dioecious ones in *D. aculeata* (Chapman & Burrows, 1971). Monoecious gametophytes are also found in *Arthrocladia*; fertilization was not observed and similar chromosome numbers were obtained from both phases suggesting the occurrence of both parthenogenesis and apomeiosis (Müller & Meel, 1982).

A pronounced alternation of morphological phases occurs in the recently described Syringodermatales (Fig. 5.30) (Henry, 1984). In *Syringoderma phinneyi* (Henry & Müller, 1983), meiospores from the unilocular sporangia give

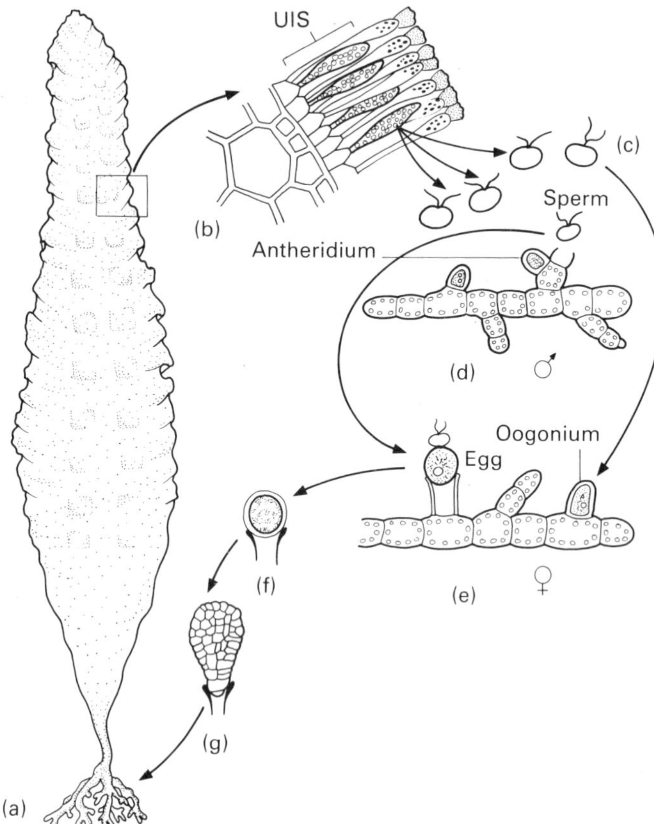

Fig. 5.29 Life cycle of *Laminaria*. (a) Macroscopic sporophyte. (b) Section of blade of sporophyte with unilocular sporangia (UIS), site of meiosis. (c) Haploid zoospores. (d) Microscopic male gametophyte with sperm-producing antheridia. (e) Microscopic female gametophyte with egg-producing oogonia. (f) Zygote. (g) Young sporophyte. (After Bold, 1973.)

rise to microscopic, dioecious, isogamous, prostrate gametophytes, whose zygotes re-establish the sporophyte generation. In *S. floridana* (Henry, 1984), the zoospores from the unilocular sporangia are scarcely mobile, and the gametophyte generation, which develops *in situ*, is reduced to two cells, both of which differentiate as isogametes. In a similar manner *S. abyssicola* is interpreted as having similar four-celled gametophytes (Walker & Henry, 1978); morphologically these are the most reduced free-living gametophytes in the Phaeophyceae.

Life cycles in the Chordariales, Dictyosiphonales, and Scytosiphonales are among the most confusing in the brown algae; basically they are heteromorphic, but in many instances pleiomorphism occurs without any evidence of a change in ploidy. The reduced phases may be filamentous 'plethysmothalli' or crusts, showing morphological and reproductive features linking these orders to the Ectocarpales *sensu lato*. Life cycles are further complicated in some species, in that they apparently vary with geographical locality. Those with a true alternation of generations include *Papenfussiella callitricha* (Peters, 1984), *Sphaeorotrichia divaricata* (Ajisaki & Umezaki, 1978), *Hecatonema streblonematoides* (Loiseaux, 1970a) *Soranthera ulvoidea* (Wynne, 1969), and *Stictyosiphon adriaticus* (Caram, 1965). In *Scytosiphon lomentaria*, Tatewaki (1966) described an alternation between an erect gametophyte phase and a crustose *Ralfsia*-like sporophyte which bore unilocular sporangia (Fig. 5.31). This alternation of morphological phases was confirmed from other localities and reported for the related *Petalonia fascia* by Wynne (1969), Hsiao (1969), and Edelstein *et al.* (1970). Others reported *Microspongium* (Lund, 1966) as well as *Streblonema* and *Compsonema* (Loiseaux, 1970b) phases in this life cycle complex. Unlike Tatewaki (1966), these latter authors found no evidence of sexuality although this was later reconfirmed for

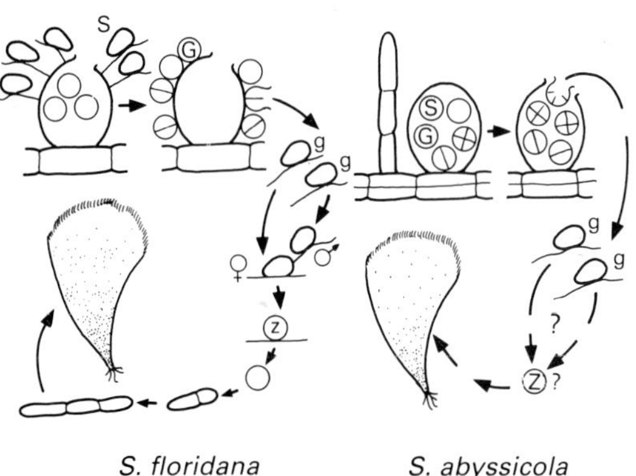

Fig. 5.30 Diagrammatic life cycles of the four species of *Syringoderma*. G, gametophyte; g, gamete; S, spore; Z, zygote. (After Henry 1984.)

Japanese plants (Nakamura & Tatewaki, 1975) and demonstrated for other members of the order *Colponema bullosa, Petalonia fascia* and *Endorachne binghamiae*. Clayton (1981) has shown that sexual activity in *Scytosiphon* is dependent on both environment and the age of the plants, and is greatest in older plants in the winter months. Heteromorphic alternations without apparent changes in ploidy are also reported in the Chordariales, *Haplogloia andersonii* (Wynne, 1969), *Splachnidium rugosum* (Price & Ducker, 1966) and in the Dictyosiphonales, *Phaeostrophion irregulare* (Mathieson, 1967), *Coliodesme californica* (Wynne, 1972), and *Striaria attenuata* (Nygren, 1975) from Sweden. The latter is of particular interest as the life cycle in the Mediterranean is reportedly sexual (Caram, 1965).

Alternation between apparent gametophytes and sporophytes, but without evidence of meiosis or syngamy, is not restricted to heteromorphic species. From a morphological perspective *Tilopteris mertensii* and *Haplospora globosa* (Tilopteridales) reproduce sexually in that they produce eggs and sperm, yet the sperm is non-functional and meiosis does not occur in the sporangia (Kuhlenkamp & Müller, 1985). In *Haplospora* (Fig. 5.32) unfertilized eggs from the monoecious gametophyte generation develop into an isomorphic spore-producing sporophyte generation. The sporophyte may be recycled via uninucleate spores or produce tetranucleate spores which give rise to the gametophyte generation. The situation is still simpler in *Tilopteris*, in that the unfertilized eggs directly recycle the gametophyte

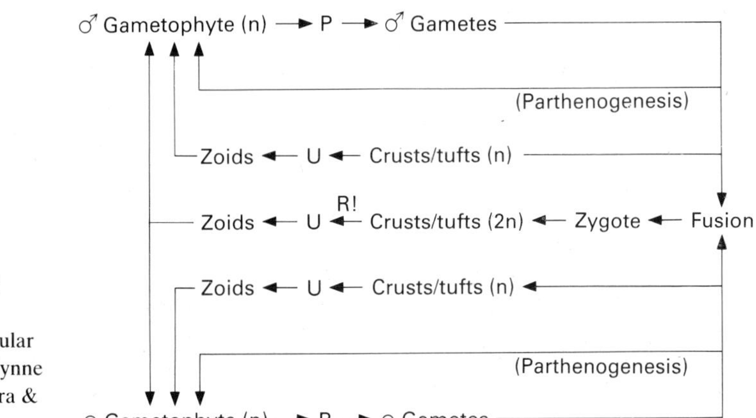

Fig. 5.31 Outline of life cycle of *Scytosiphon lomentaria*. U, unilocular structures; P, plurilocular structures; R! meiosis. (After Wynne & Loiseaux 1976, from Nakamura & Tatewaki 1975.)

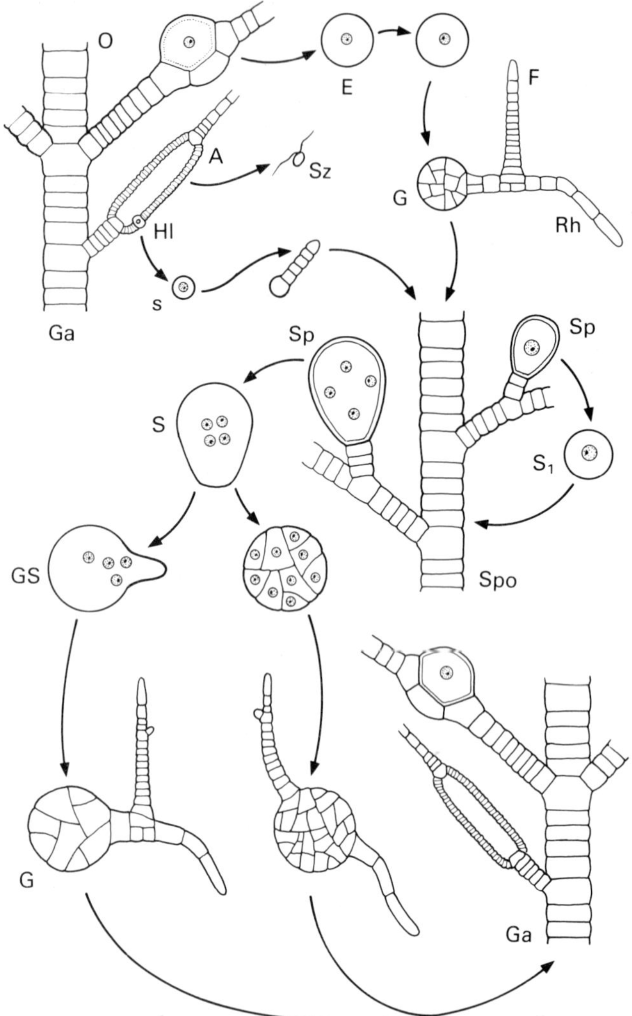

Fig. 5.32 Life cycle of *Tilopteris mertensii*. O, oogonium; A, antheridium; Hl, hypertrophied locule; E, egg; Sz, spermatozoid; s, spore from hypertrophied locule; G, germling; Ga, gametophyte; Rh, rhizoid; F, filament; Spo, sporophyte; Sp, sporangium; S, quadrinucleate spore; S_1, uninucleate spore. (After Kuhlenkamp & Müller, 1985.)

and the sporophyte generation is apparently lacking. It is particularly interesting that while no syngamy occurs the eggs still produce a pheromone which attracts the sperm. This suggests that the loss of sexuality is a derived phenomenon and that sexual populations may remain to be discovered.

Another as yet little understood phenomenon in the Phaeophyceae is heteroblasty, in which zoids, independent of external conditions or sexuality, germinate by two differing modes. Initially ovoid germlings become filamentous, while amoeboid forms typically become stellate and later develop into discoid growths. Heteroblasty is particularly common in the Myrionemataceae (Chordariales) (Pedersen, 1981).

It is now generally accepted that life cycles cannot be used as a primary criterion in the classification of the Chlorophyceae at the ordinal level, as each type has probably independently arisen on more than one occasion (Hoek, 1981). Nevertheless, isomorphic sporic life cycles are strongly associated with the Cladophorales and the Ulvales, and heteromorphic ones with the Acrosiphoniales and some members of the Caulerpales.

In the marine Cladophorales an alternation occurs between a quadriflagellate, spore-producing sporophyte generation and a monoecious gametophyte generation which produces biflagellate gametes. This is supported by both culture and cytological evidence in *Chaetomorpha* (Patel, 1971, 1972; Kornmann, 1972), *Cladophora* (Wik-Sjostedt, 1970), and *Rhizoclonium* (Bliding, 1957). There are, however, reports that the freshwater species *Cladophora crispata* has a zygotic life cycle (Siddeque & Faridi, 1977) and *C. glomerata* a gametic one (List, 1930). *Cladophora* gametes may develop parthenogenetically and undergo spontaneous diploidization, and polyploidy also occurs (Wik-Sjostedt, 1970).

In the Ulvales the genera *Percursaria, Enteromorpha, Ulva, Ulvaria,* and *Blidingia* have an isomorphic alternation of generations similar to those reported for the Cladophorales (Kornmann, 1956; Foyn, 1934; Bliding, 1963; Dube, 1967; Tatewaki & Iima, 1984). The most detailed studies are for *Ulva mutabilis* (Fig. 5.11) (Fjeld & Lovlie, 1976; Lovlie & Bryhni, 1978), in which 12 morphologically identical forms differ in cytological and reproductive details. Mating types are genotypically determined by a single allelic pair, and the sporophytes thus produce, by meiosis, equal numbers of zoospores of each type. The resulting haploid gametophytes produce gametes which undergo syngamy with a complementary mating type to return the diploid sporophyte. Parthenogenetic development of gametes occurs, of which 1–2% recycle the parent gametophyte, and the remainder produce haploid sporophytes. Diploidization of the haploid sporophytes may occur (Hoxmark & Nordby, 1974), but may be incomplete, and hence haploid, diploid, and partially haploid and diploid parthenosporophytes are known. In addition the two mating strains differ in their abilities to undergo diploidization. Both diploid and haploid parthenosporophytes produce zoospores by meiosis, and in the latter random segregation of chromosomes produces spores of very low viability. Zoospores from such parthenosporophytes are always of one mating type, further demonstrating the allelic nature of sex determination. Parthenogenetic development of gametes has also been observed in other species of *Ulva* (Bliding, 1968; Tanner, 1981), *Enteromorpha* (Kapraun, 1970), and *Percursaria* (Kornmann, 1956), with some strains apparently lacking normal sexual reproduction.

Sexual reproduction has been demonstrated in few genera of the Chaetophorales. In *Entocladia* (O'Kelly & Yarish, 1981) there is an isomorphic alternation of generations, and this is also reported for *Eugomontia* (Kornmann, 1960), although a *Codiolum*-like stage has also been reported in this genus (Wilkinson & Burrows, 1970).

Heteromorphic alternations involving a macrophytic gametophyte generation and a reduced *Codiolum* or *Chlorochytrium*-like sporophyte are reported in a number of green algae, for example *Ulothrix* (Lokhorst, 1978), *Monostroma* (Kornmann & Sahling, 1962), *Urospora* (Fig. 5.33a) (Kornmann, 1961), and *Spongomorpha* (Fig. 5.33b, c, d) (= *Acrosiphonia*;) (Kornmann, 1964). These genera have a confused taxonomic position at the ordinal level and are variously placed in the Ulotrichales, Acrosiphoniales or, in the case of the thallose forms in the Monostromataceae, of the Ulvales (Hoek, 1981; Tanner, 1981). A class, the Codiolophyceae, has also been proposed (Kornmann, 1973). The thick-walled unicellular *Codiolum*-like phases may be interpreted as stalked hypnozygotes with attachment discs (Bold & Wynne, 1985) with meiosis occurring during zoosporogenesis, hence the life cycle could be regarded as zygotic. Alternatively, they may be regarded as a reduced unicellular sporophytic phase (Tanner, 1981) and hence the life cycle is sporic. Differences in morphology between the stalked *Codiolum* phase and the coccoid *Chlorochytrium* are believed to be controlled by environmental conditions (Kornmann, 1973).

Life cycles in *Acrosiphonia* show an intergradation from heteromorphic sporic to zygotic (Kornmann, 1970b, 1973). *A. spinescens* is typically heteromorphic (Fig. 5.33b). Zoosporogenesis is lacking in *A. grandis* (Fig. 5.33c), but the *Codiolum* nucleus divides several times prior to developing into the gametophyte, whereas in *A. arcta* from Heigoland (Fig. 5.33c), the zygote undergoes two nuclear divisions, which may be meiotic, prior to germinating to form a gametophyte.

Members of the Caulerpales are generally characterized by their gametic life cycles, however species of the *Halicystis–Derbesia–Bryopsis* complex, and related genera, have a heteromor-

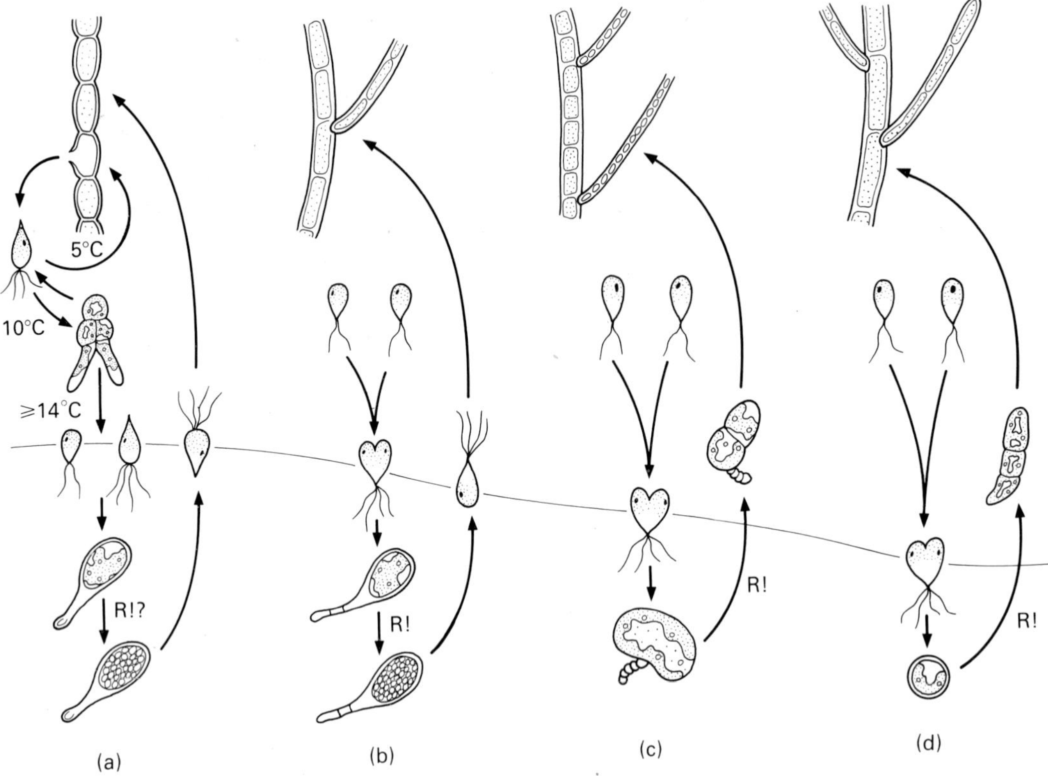

Fig. 5.33 Life cycles in (a) *Urospora* and (b–d) *Acrosiphonia*, showing progressive reduction of the *Codiolum*-like sporophyte. R!, meiosis. (After Tanner, 1981.)

phic alternation, though variation may occur even within a single species (Rietema, 1970; Tanner, 1981). The filamentous *Derbesia* phase is regarded as the sporophyte, with meiosis occurring during the production of the stephanokontic zoospores, which, depending on the species, germinate to produce either a multinucleate sac-like *Halicystis* (Fig. 5.13a) or the branched coenocytic filaments of *Bryopsis* (Fig. 5.13b). The gametophytes are anisogamous; gametes in *Bryopsis* are produced in vegetative branches which become converted into gametangia (Burr & West, 1970), and in *Halicystis* part of the vegetative protoplasm becomes converted to gametes (Hollenberg, 1935).

In some isolates of *Bryopsis* a full *Derbesia* (Fig. 5.13c) stage is lacking and gametes develop into a uninucleate protonemal stage which may produce spores or may develop directly into the gametophyte (Rietema, 1971; Bartlett & South, 1973). There are also many reports (Tanner, 1981) of apomeiotic strains and various asexual accessory reproductive processes. Conclusive cytological demonstration of levels of ploidy and the occurrence of meiosis is almost completely lacking in this group.

GAMETIC LIFE CYCLES

Noctiluca miliaris (Fig. 5.34) is a dinoflagellate which exhibits a gametic life cycle (Zingmark, 1970) and is also unusual in having an apparently eukaryotic vegetative nucleus, though its gametes have typically dinophycean nuclei. During gametogenesis the nucleus divides meiotically and then mitotically to produce upwards of 2000 uniflagellate isogametes. Pairs of gametes fuse and the zygote, after a resting period, develops directly into a vegetative cell.

In diatoms sexual reproduction is gametic. In most species in addition to genetic recombination,

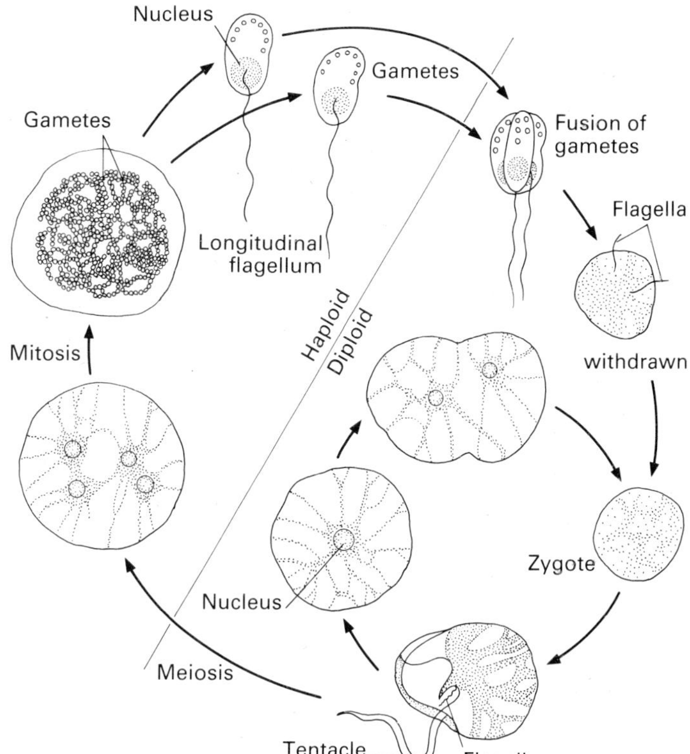

Fig. 5.34 The life cycle of the dinoflagellate *Noctiluca miliaris*. (After Lee, 1980; from Zingmark, 1970.)

sexual reproduction brings about a restoration of maximum cell size (Drebes, 1977). Pennate diatoms are isogamous (Fig. 5.35) and centric species oogamous (Fig. 5.36). In the centric forms male cells may undergo a series of mitoses, whose products, the spermatogonia, remain within the parent frustules until a final meiotic division occurs to produce uniflagellate sperms. The details of spermatogenesis vary among genera: in *Cyclotella* the vegetative cells apparently function as spermatogonia (Schultz & Trainor, 1968), in others, such as *Stephanophyxis* (Drebes, 1966), the spermatogonia develop thin silica shells prior to meiosis. The oogonia develop directly from vegetative cells, and in diatoms cytokinesis is lacking during egg formation. After each division of meiosis only one functional nucleus remains. Some, such as *Lithodesmum undulatum* and *Biddulphia mobiliensis*, produce two functional eggs (Stosch, 1954). After fertilization the zygote increases in size to form an auxospore, and inside the auxospore envelope the initial diatom cell is formed. The formation of each valve is preceded by an acytokinetic mitotic division, with the formation of supernumary nuclei. The auxospore envelope eventually ruptures, liberating the new enlarged cell (Drebes, 1977).

Autogamy is reported in the centric diatoms. In *Cyclotella meneghiniana* (Schultz & Trainor, 1968) sperm are known, but fertilization has not been observed. Instead, after meiosis in the auxospore mother cell, two of the nuclei fuse.

In contrast to the Biddulphiales sexual reproduction in the pennate diatoms involves non-flagellated isogametes which show physiological anisogamy. In the fossil record the centric diatoms predate the pennate forms (Drebes, 1977) and are considered primitive, hence the supposedly derived forms reproduce by the more primitive isogamy. It should be noted, however, that the isogametes are aflagellate and that this indicates the derivative nature of these gametes. In most pennate diatoms isomorphic, sexually

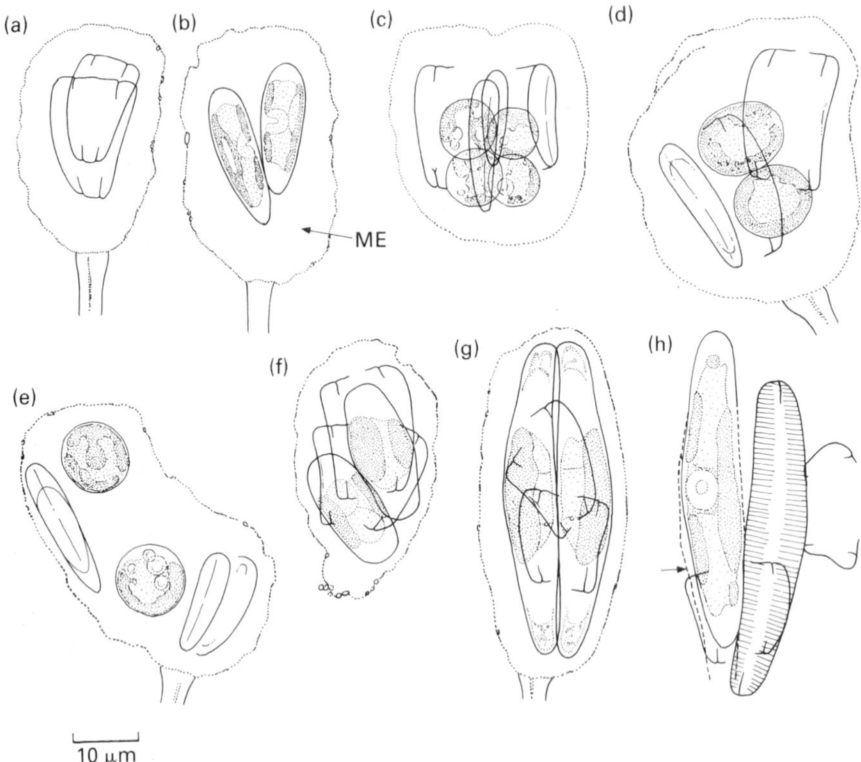

Fig. 5.35 Sexual reproduction in a pennate diatom, *Rhoicosphenia curvata*. (a & b) Gametangial cells within a mucilage envelope (ME), more active partner has the higher position. (c) Four gametes immediately prior to fusion. (d) Two zygotes lie between the empty gametangial frustules. (e) Separation of the zygotes. (f) Early stages of auxospore formation. (g) Fully expanded auxospores with plastids restricted to the central region. (h) Auxospores after valve deposition — the plastids have expanded to occupy almost the total length of the cell. The left cell has partially escaped from its perizonium. (After Mann, 1982.)

induced cells come together in pairs and secrete a copulation mucus. The first meiotic division is normal, but in the second no cytokinesis occurs and two gametes are produced, each with one haploid and one degenerate supernumary nucleus. Isogamy is known, but physiological anisogamy is more common, with each mother cell producing one active and one passive gamete. Fusion produces two zygotes which develop into auxospores. These enlarge, and their outer membrane is replaced by a silicified perizonium, and as in the centric forms mitosis and nuclear degeneration accompany valve formation. In pennate diatoms such as *Eunotia* and *Cocconeis* only one gamete is produced per mother cell; autogamy also occurs in this group (Drebes, 1977).

Gametic life cycles are also proposed for the Fucales (Phaeophyceae) (Fig. 5.37), in which the only free-living plant is diploid and meiosis is closely associated with the formation of gametes. Reproduction in the order is best known in *Fucus* in which monoecious (e.g. *F. spiralis*) or dioecious (e.g. *F. serratus*) species occur. At reproductive maturity the tips of the vegetative branches become receptacles, which contain sunken flask-shaped conceptacles in which the antheridia and oogonia are formed together with sterile paraphyses. The antheridia, which are not separated by walls (Berkaloff & Rousseau, 1979), are borne on branched paraphyses and produce numerous haploid sperm. The oogonia are borne on a stalk cell and undergo meiosis followed by mitosis to produce eight egg cells.

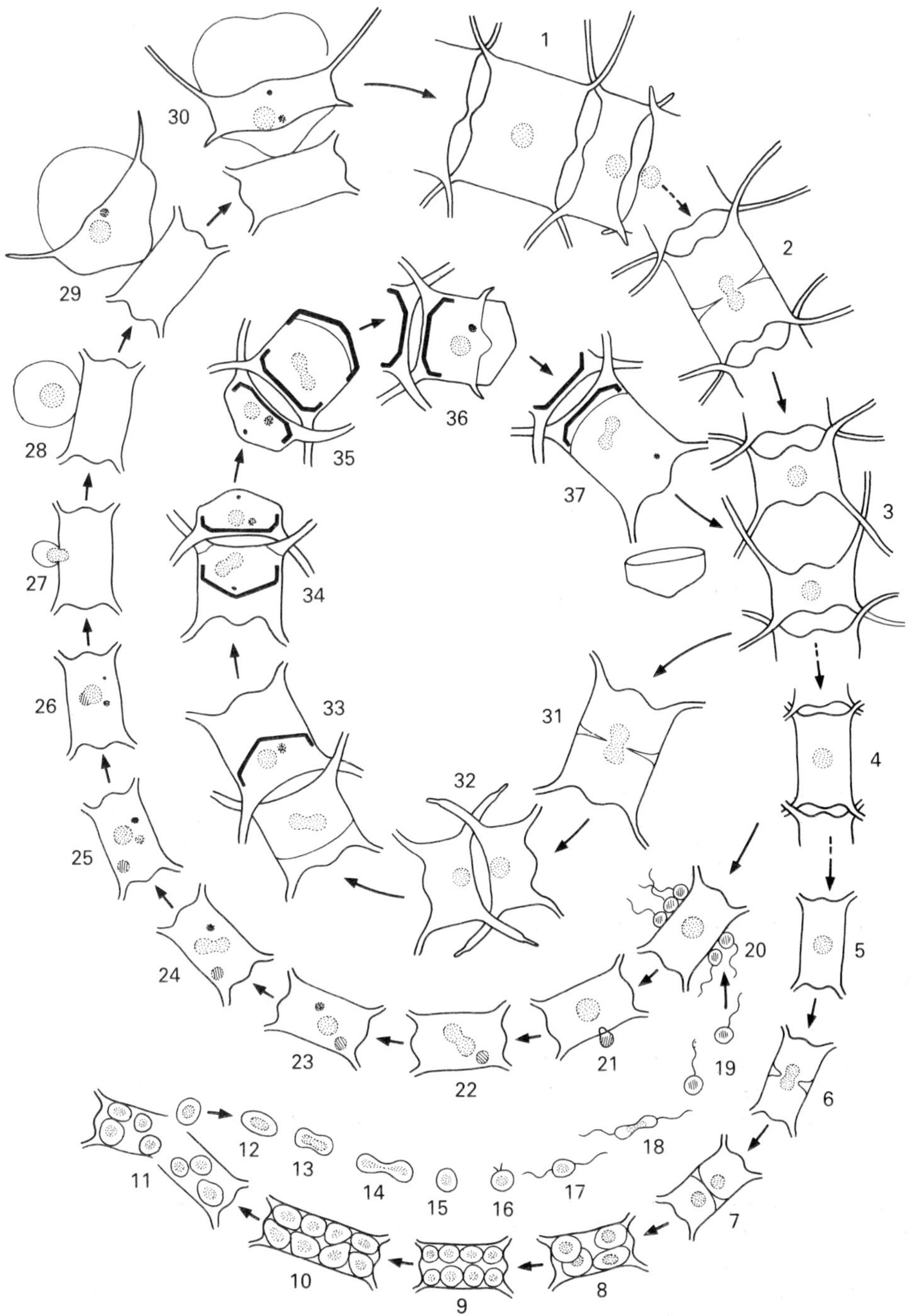

Fig. 5.36 Life cycle of a centric diatom *Chaetoceros didymum*. Vegetative cell division 1–4, sexual cycle 6–30, 5–11 sperm production, 20 oogonium, 21–27 processes associated with syngamy, 28–30 development of auxospores, 31–37 asexual resting spore cycle. (After Stosch *et al.*, 1973.)

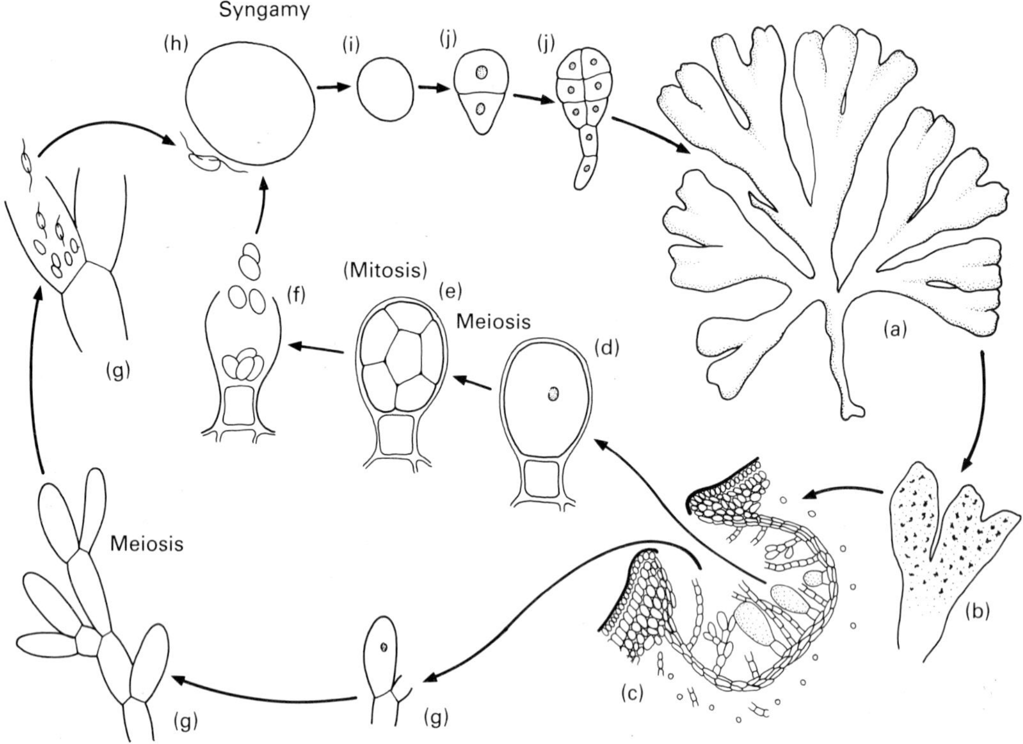

Fig. 5.37 Life cycle of a monoecious species of *Fucus*. (a) Mature sporophyte. (b) Receptacles develop at tips of branches. (c) Conceptacles within receptacles contain oogonia, antheridia, and sterile hair cells (paraphyses). (d) Oogonium. (e) Meiotic then mitotic division within the oogonium produces eggs. (f) Release of eggs. (g) Development of antheridial branch and antheridia, meiosis followed by mitosis in the antheridium produces sperm. (h) Syngamy. (i) Zygote. (j) Development of embryonic sporophyte. (After Scagel *et al.*, 1982.)

At maturity the outer wall of the oogonium ruptures and the eggs are released as a packet. Once free of the conceptacle, the inner walls disassociate to free the eggs, though the sperm are capable of penetrating the oogonial sac before egg release. Zygote germination and physiology of early embryo development have been intensively studied (Evans *et al.*, 1982). Genera differ in the numbers of eggs produced per oogonium, four in *Ascophyllum*, two in *Pelvetia*, and one in *Cystoseira* or *Sargassum*. In all cases eight nuclei are initially produced, but the supernumary ones are extruded or degenerate. In *Xiphophora* (Naylor, 1954) and *Bifurcariopsis* (Jensen, 1974), internal walls form in the developing oogonium and are presumably laid down at the eight-nucleus stage, isolating the final products of division into four binucleate chambers, of which only one nucleus remains at maturity. The young oogonium may thus be regarded as a unilocular sporangium in which meiosis produces four spores, each of which divides endosporically to produce the gametophytes. These observations lend credence to the theory first advanced by Strasburger (1906), who suggested that an alternation of generations occurs in the Fucales, but with a highly reduced gametophyte.

The reduced gametophyte theory has been used by Moe and Henry (1982) to explain the life cycle in the related order *Ascoseirales*. The free-living *Ascoseira* thallus is interpreted as a sporophyte on which chains of unilocular sporangia are developed in conceptacles. Each spore divides to form a two-celled gametophyte of which one cell degenerates while the other produces a single motile gamete. No syngamy was

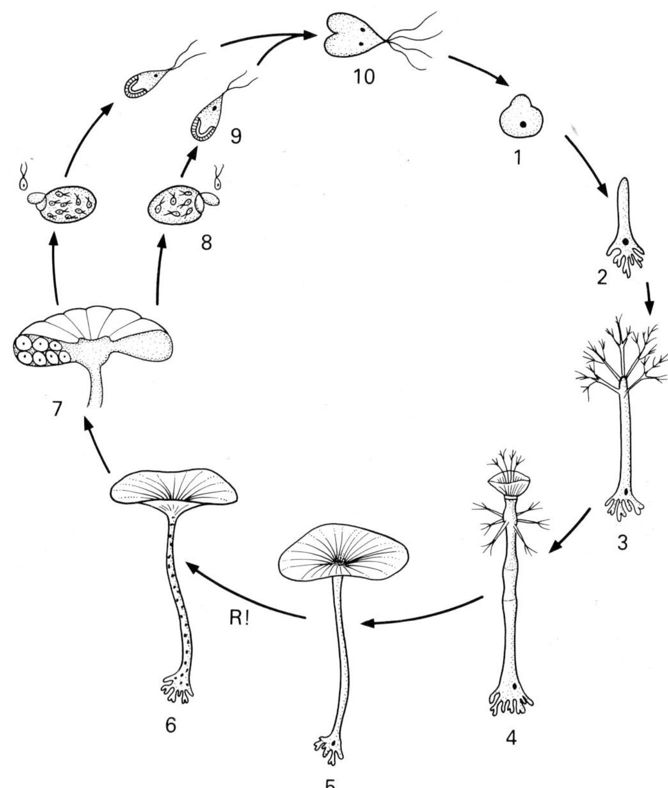

Fig. 5.38 Life cycle in *Acetabularia mediterranea*. (1) Germinating zygote. (2–5) Development of sporophyte. (6) Meiotic division of primary nucleus and migration of secondary nuclei into cap. (7) Development of cysts in cap. (8) Release of gametes from cysts. (9) Isogametes. (10) Syngamy. R!, meiosis. (After Tanner, 1981; from Bonotto, 1975.)

observed and the gametes presumably develop parthenogenetically; the occurrence of meiosis has not been observed.

Members of the Dasycladales are the only species of the Chlorophyceae for which karyological and genetic evidence exists showing them to have a gametic life cycle (Fig. 5.38) (Koop, 1979). In the vegetative state the siphonous thallus of *Acetabularia* contains a single large nucleus located in one of the basal rhizoids. Gametogenesis commences with the meiotic division of this nucleus, followed by mitosis to produce a large number of haploid daughter nuclei, many of which migrate to the cap and are incorporated singly into gametangial cysts. The resistant cysts are released from the cap, and mitotic division within cysts leads to the release of biflagellate isogametes. The zygote resulting from gamete fusion develops directly into the new *Acetabularia* thallus (Bonotto, 1975). The cyst nuclei possess the same amount of DNA as the gametes, but half the amount of the zygote, showing that meiosis occurs prior to gamete formation (Koop, 1979).

With the exception of the *Derbesia–Bryopsis* complex, members of the Caulerpales are assumed to have a gametic life cycle (Fig. 5.39), but cytological confirmation is generally lacking. In *Codium*, differentiated gametangia are cut off from the siphonous thallus by a septum (Borden & Stein, 1969), but in *Udotea* (Meinesz, 1980) and *Caulerpa* (Goldstein & Morrall, 1970) differentiated gametangia are not produced and gametes are formed by cleavage of the somatic protoplasm. Most species produce anisogametes, although isogamous forms are known. The zygote of *Caulerpa* (Price, 1972) grows directly into a new adult thallus, but in the Udoteaceae, e.g. *Udotea* and *Halimeda*, germination produces a spherical uninucleate structure, the protosphere, which in turn produces a filamentous stage from which the adult thallus develops (Meinesz, 1980). The pro-

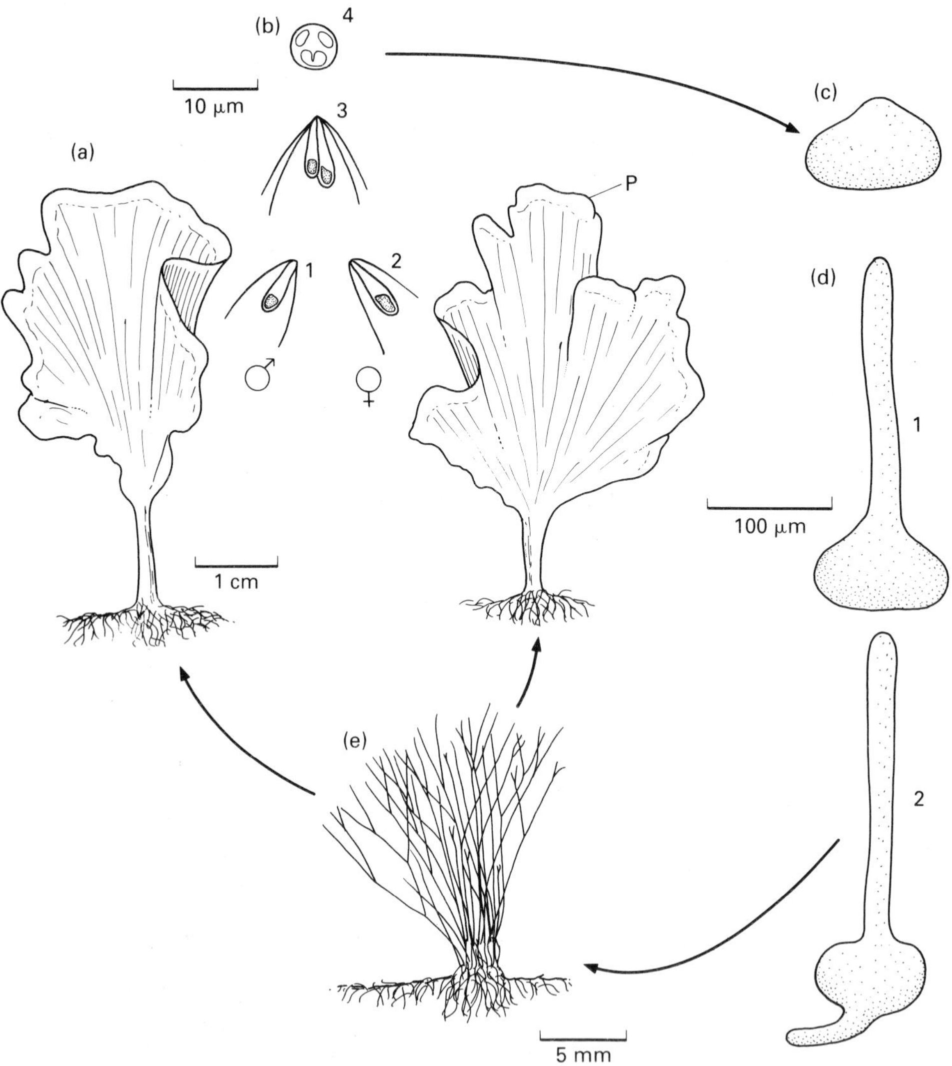

Fig. 5.39 Life cycle of *Udotea petiolata*. (a) Fertile thallus. (b) 1, male gamete; 2, female gamete; 3, syngamy; 4, zygote at 10 days. (c) Protosphere. (d) 1, protosphere with apical growth; 2, development of rhizoids. (e) Branched stage with rhizoids and erect filaments, which aggregate to produce the mature siphonous thallus. (After Meinesz, 1980.)

tosphere and filamentous stages may be analogous with the uninucleate filamentous stages in some *Bryopsis* life cycles that are capable of forming zoospores or directly producing the adult thalli (Hoek, 1981). Such an interpretation suggests the possession of a sporic life cycle with a highly reduced sporophyte and somatic meiosis. This has not been observed in the Udoteaceae (Meinesz, 1980). It is, however, interesting that Rietema (1971, 1972) suggested that at northern latitudes the sporophyte of *Bryopsis* loses the capacity to produce spores, and somatic meiosis occurs. In support of this, Sears and Wilce (1970) showed that sporangia of the sporophytic *Derbesia* failed to undergo cytokinesis and developed directly into *Bryopsis* gametophytes.

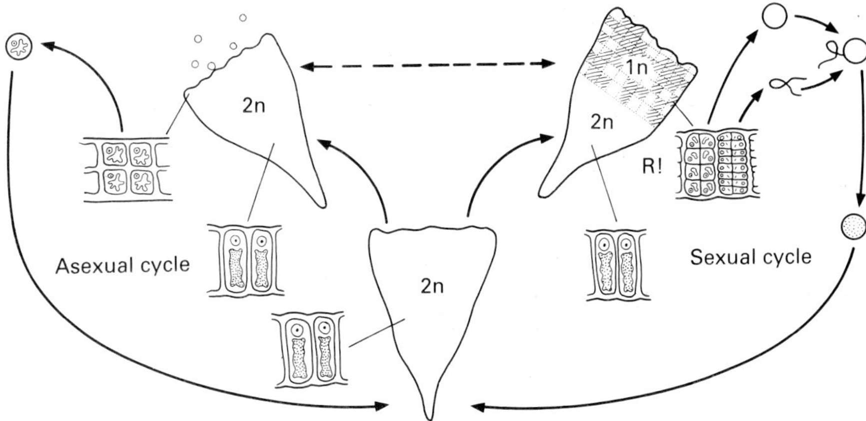

Fig. 5.40 Life history of *Prasiola* species with somatic meiosis. Vegetative areas of the thallus are monostromatic, sporogenic areas are distromatic, and gametogenic areas are polystromatic. R!, meiosis. (After Tanner, 1981; from Friedmann, 1959.)

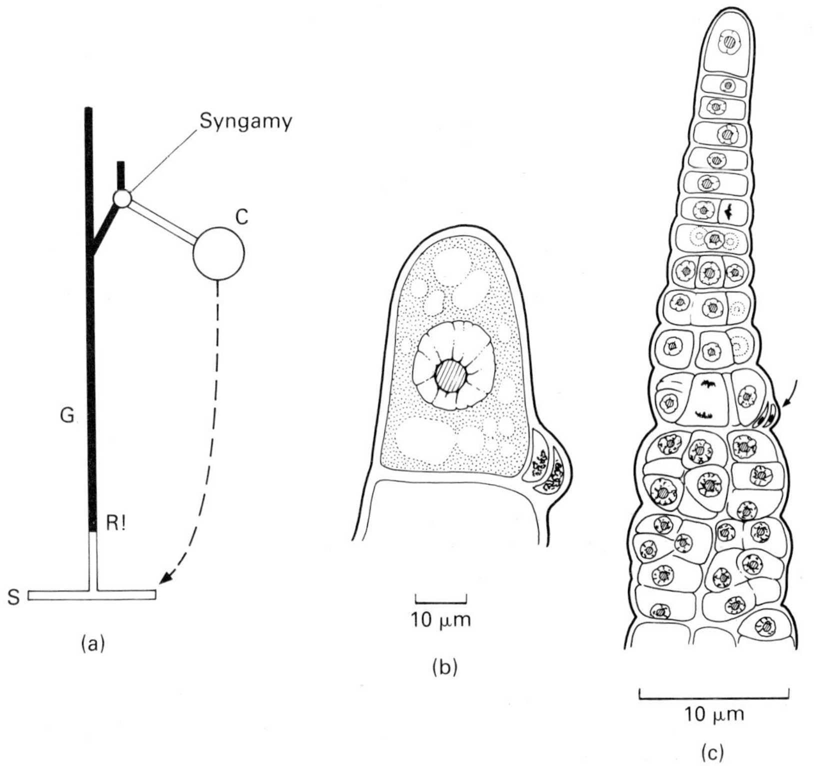

Fig. 5.41 *Lemanea mamillosa*. (a) Life cycle: C, carposporophyte; S, diploid 'sporophytic part of thallus'; G, haploid gametophytic part of thallus; R!, meiosis. (After Dixon, 1973.) (b) Meiotic division results in a haploid apical cell plus lateral residual nuclei. (c) Development of a haploid thallus, arrow indicates residual nuclei. (After Magne, 1967.)

SOMATIC LIFE CYCLES

Life cycles in which meiosis occurs in a vegetative cell and does not immediately lead to the production of spores or gametes may occur in several algal groups, but are unequivocably demonstrated in the Prasiolales of the Chlorophyceae and in some members of the Rhodophyceae.

In *Prasiola meridionalis* and *P. stipitata* (Fig. 5.40) the cells in the upper portion of the blade may divide meiotically and their offspring undergo further mitotic divisions to produce a mosaic pattern of paired male and female gametangial areas. This has been interpreted as evidence for genotypic determination of sex during meiosis (Friedmann 1959; Cole & Akintobe, 1963). However, Bravo (1965) showed in *P. meridionalis* that all of the cells from a single mitosis developed into either male or female gametangia. Gamete release is facilitated by the disintegration of the gametangial walls. Syngamy is oogamous and the biflagellate sperm is incorporated into the egg. A flagellum is lost, but the remaining one is functional and serves to propel the zygote prior to settlement and growth into a new thallus. Other life cycles reported for this genus include zygotic for the freshwater *P. japonica* (Fujiyama, 1955) and heteromorphic sporic for the marine *P. calophylla* (Kornmann & Sahling, 1974).

Freshwater red algae generally have heteromorphic life cycles (Sheath, 1984), and in the Batrachospermales these involve a macroscopic gametophyte and a reduced filamentous diploid phase termed 'chantransia'. The only spores formed are the carpospores produced by the carposporophyte as a result of fertilization of a carpogonium by a spermatium. Tetraspores are unknown and the gametophyte generation arises directly from the chantransia. In *Lemanea mamillosa* (Fig. 5.41) (Magne, 1967) and *Batrachospermum mahabaleshwarensis* (Balakrishnan & Chagule, 1980) meiosis occurs in the apical cell of the chantransia phase, with residual nuclei from each meiosis being extruded into lateral protruberances from which no further growth occurs. In some species such as *Batrachospermum* sp. (Stosch & Tiel, 1979) and *Lemanea fluviatilis* (Huth, 1981) meiosis appears to be delayed and occurs not in the chantransia but in the apical cell of what is morphologically the gametophyte phase.

Other red algae with reduced tetrasporophytes may have the potential for the suppression of tetrasporogenesis and the direct development of a gametophyte generation via somatic meiosis. Several such potential occurrences are reviewed by Hawkes (1983), who described the possible occurrence of this phenomenon in *Hummbrella hydra*, a member of the Gigartinales from New Zealand. However, unequivocal cytological evidence for a *Lemanea*-type of life cycle is lacking except in the Batrachospermales.

CHAPTER 6

Physiology and biochemistry

Introduction

While many of the physiological and biochemical processes which occur in higher plants are very similar to those occurring in the algae, there are a number of notable differences. The algal groups are evolutionarily diverse, they usually exhibit a simpler anatomy than the higher plants, and most are aquatic. The evolutionary diversity is reflected in their photosynthetic pigmentation and food storage products, while structural and ecological differences may be seen in the mechanisms of nutrient uptake and internal transportation. This chapter stresses aspects of plant physiology which are either unique to the algal groups of particularly important to their success.

Light and its measurement

Photosynthesis in the algae occurs mainly in the visible range of the spectrum between 400 nm and 700 nm, and photomorphogenic responses may extend the physiological effects of light into the far red from 700 to 760 nm. The spectral sensitivity of these reactions is very different from that of the human eye or from that of photographic emulsions. Light meters designed to measure illuminances in lux or foot-candles are therefore inappropriate for most biological purposes. Phycologists are interested principally in photon flux density (PFD), the number of quanta (= photons in the light range of the electromagnetic spectrum). The measurements are appropriate for most photobiological events which depend on the number of quanta absorbed rather than on their energy content. Such quantum irradiance measurements are reported in SI units as micromoles metre^{-2} second^{-1} (Incoll *et al.*, 1977); the identical, but non-SI, units of micro-einsteins metre^{-2} second^{-1} are also frequently utilized. Irradiance may also be measured and expressed energetically in units of watts metre^{-2}. For monochromatic light, irradiance and photon flux density may be readily interconverted as:

$$1\,\mu\text{mol}\,\text{m}^{-2}\,\text{sec}^{-1} = (119.7/\Lambda)\ \text{W}\,\text{m}^{-2}$$

where Λ = the wavelength of the light in nm.

For light of broader spectral range the conversion from irradiance to PFD requires the integration of the PFD obtained over the spectral range of interest. This is readily done using a spectroradiometer to measure the energy content of narrow spectral bands, converting this to PFD, plotting the measurements against the wavelength, and determining the area under the curve.

Although it is generally not possible to relate directly illuminances measured in lux to measures of either light energy or photon density, the relationship:

$$1\,\text{W}\,\text{m}^{-2} = 5\,\mu\text{mol}\,\text{m}^{-2}\,\text{sec}^{-1} = 250\ \text{lux}$$

is approximate for most white light sources. Conversion factors for most common light sources are also available (Lüning, 1981) and are appropriate where absolute accuracy is not required.

Photosynthetic pigments and light harvesting

Photosynthesis in algae, as in all photoautotrophic organisms, depends on the harvesting of light energy and its conversion into chemical energy, specifically into ATP and a reductant, NADPH. The first step in this process is the absorption of light by the photosynthetic pigments. The distribution of the photosynthetic pigments, chlorophylls, biliproteins, and carotenoids, within the algal phyla and the evolutionary implications of these distributions are discussed in Chapters 2 and 8.

Functional chlorophylls (Fig. 6.1a) consist of four linked pyrrole rings with a magnesium atom chelated at the centre; *in vivo* they are conjugated with proteins (Thornber & Barber, 1979). Chlorophylls *a* and *b* possess a long phytol chain, but this is lacking in chlorophyll *c*. Chlorophyll *a* is found in all eukaryotic photosynthetic algae and in the prokaryotic Cyanophycota and Protochlorophycota. Acetone extracts show two absorption peaks of approximately equal intensity, one, the Soret band, in the blue part of the spectrum at 430 nm, the other in the red at 660 nm.

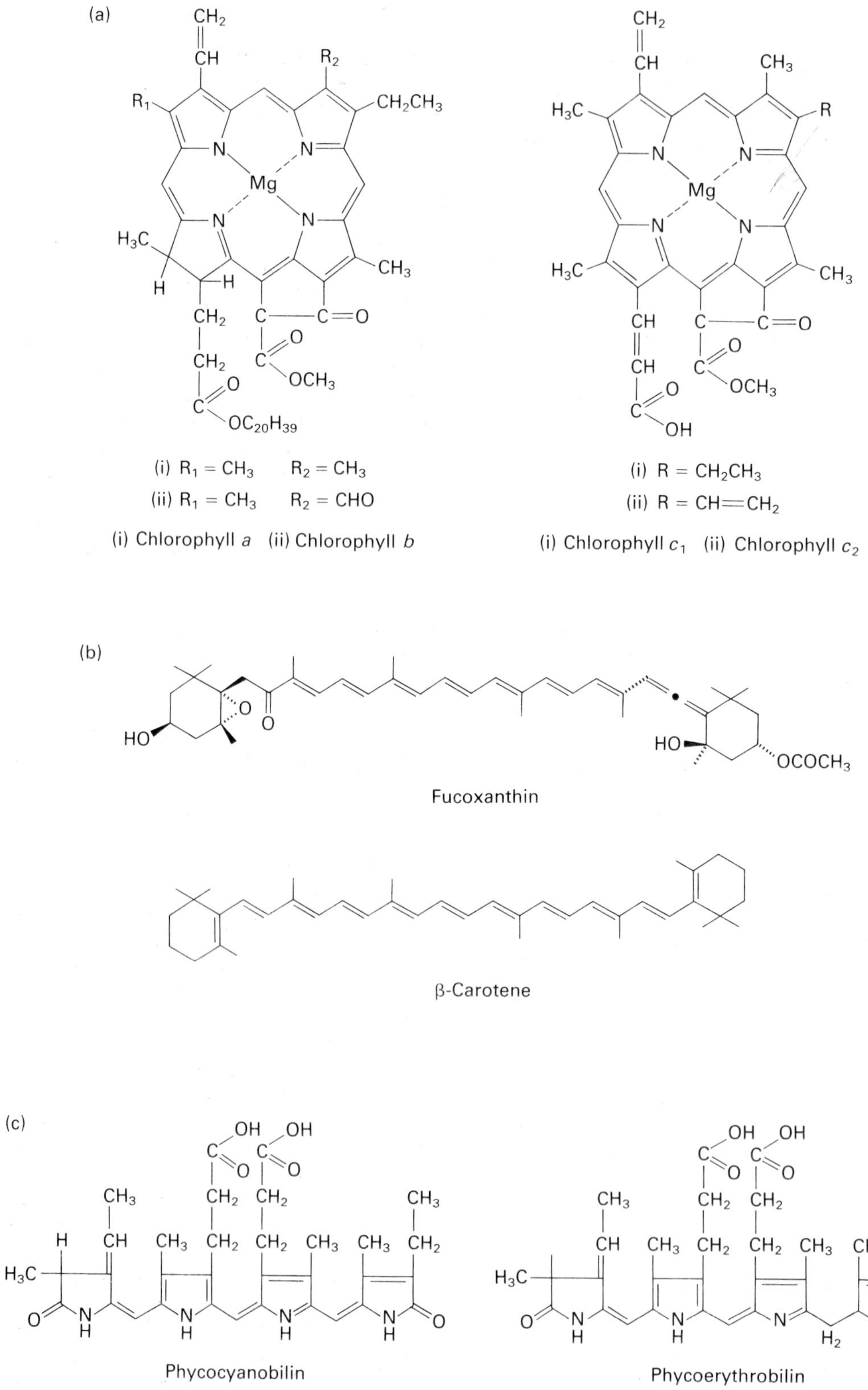

(a)

(i) $R_1 = CH_3$ $R_2 = CH_3$

(ii) $R_1 = CH_3$ $R_2 = CHO$

(i) Chlorophyll *a* (ii) Chlorophyll *b*

(i) $R = CH_2CH_3$

(ii) $R = CH{=}CH_2$

(i) Chlorophyll c_1 (ii) Chlorophyll c_2

(b)

Fucoxanthin

β-Carotene

(c)

Phycocyanobilin

Phycoerythrobilin

Fig. 6.1 Chemical structure of three main groups of photosynthetic pigments. (After Ragan, 1981.)

The Chlorophycota contain chlorophyll *b*, which differs from chlorophyll *a* in that a methyl group is replaced by an aldehyde group. Under the same extraction conditions it shows a light absorption in the Soret band at 435 nm of almost three times the intensity of the red absorption at 643 nm (Prezlin, 1981), showing its importance in light harvesting in the blue end of the spectrum. The Chromophycota lack chlorophyll *b*, but contain chlorophyll *c*, of which two principal forms are known, chlorophyll c_1 and chlorophyll c_2; as with chlorophyll *b* they are strong absorbers of blue light.

The biliproteins, the phycoerythrins, and phycocyanins contain the linearly arranged tetrapyrrolic bilin pigments phycoerythrobilin and phycocyanobilin (Fig. 6.1c), which unlike the chlorophylls are strongly covalently bonded to a protein component. The red phycoerythrins and the blue-green phycocyanins exist in a number of forms, which are found in the Cyanophyceae, Cryptophyceae, and Rhodophyceae. Differing in the arrangement of their polypeptide components, they can be distinguished by their spectral characteristics (Prezlin, 1981).

Carotenoids (Fig. 6.1b) are isoprenoid polyene pigments present in many organisms, but can be synthesized *de novo* only by plants and photosynthetic bacteria. They consist of the hydrocarbon carotenes and their dihydroxy derivatives, the xanthophylls, and are usually red, brown, or yellow in colour; they absorb light in the blue-green part of the spectrum. The occurrence and distribution of algal carotenoids are reviewed by Ragan (1981). The role of many of the minor carotenoids is photoprotective rather than light harvesting, as they effectively screen out potentially harmful blue and near-UV radiation (Prezlin 1981).

Many chlorophyte algae resemble the higher plants in carotenoid composition, with β-carotene, and the xanthophylls lutein, violaxanthin, antheraxanthin, zeaxanthin and neoxanthin. Most siphonous green algae have α-carotene, siphonoxanthin and its fatty acid ester siphonein. The chromophyte lines of algae have a predominance of xanthophylls; in the Phaeophyceae, the Bacillariophyceae, and the Chrysophyceae, fucoxanthin is the most abundant xanthophyll, but in the dinoflagellates it is replaced by peridinin. These two xanthophylls are similar in structure, and differ from most of the others in their relatively high oxygen content. Zeoxanthin and lutein are the commonest xanthophylls in the Rhodophyceae, while the Cyanophyceae contain mixoxanthin (= echinenone) and myxoxanthophyll.

LIGHT ABSORPTION

The photosynthetic conversion of light energy into chemical energy is described by the Hill and Bendall (1960) 'Z' model (Fig. 6.2). There are two photosystems, PSI and PSII, each with their light-absorbing trap, P_{680} and P_{700} respectively. The two photosystems have separate pigment systems, do not always occur in a 1:1 ratio, and can be physically separated by treatment by mild detergents. The electron donor to photosystem II (PSII) is water, and the ultimate electron acceptor in photosystem I (PSI) is oxidized nicotinamide adenine dinucleotide phosphate (NADP), with adenosine triphosphate (ATP) produced in the PSI reactions.

Many cyanophyte species have the capacity to utilize sulphide as an electron donor in anoxygenic photosynthesis (Padan & Cohen, 1982). This facultative capacity shows their links with both the strictly anaerobic phototrophic bacteria and the obligately aerobic eukaryotic algae.

The P700 complex has been isolated from several classes of algae (Phaeophyceae; Anderson & Barrett, 1979: Rhodophyceae; Gantt, 1981: Chlorophyceae; Nakamura *et al.*, 1976: Dinophyceae; Prezlin & Alberte, 1978). It consists of a complex of 40–50 molecules of chlorophyll *a*–protein for each reaction centre molecule of P_{700}, which itself is a form of chlorophyll *a*. The P_{680} trap of PSII is less well understood and appears to consist of shorter wavelength-absorbing forms of chlorophyll *a* associated with chlorophyll *b* and accessory pigments.

The analogy of an antenna–receiver–transducer system has been applied to the light capture process (Ramus, 1981), with signals (photons) of varying quality (wavelength) captured by the antenna of a receiver (pigment complex) and channelled to a trap where the signal is transduced (converted into chemical energy). When an antenna pigment molecule absorbs a photon, an electron is promoted to a higher orbital shell; this electron is more energetic, and the molecule less stable, and without coupling to another system the electron would fall back to the ground state and the released energy would be emitted

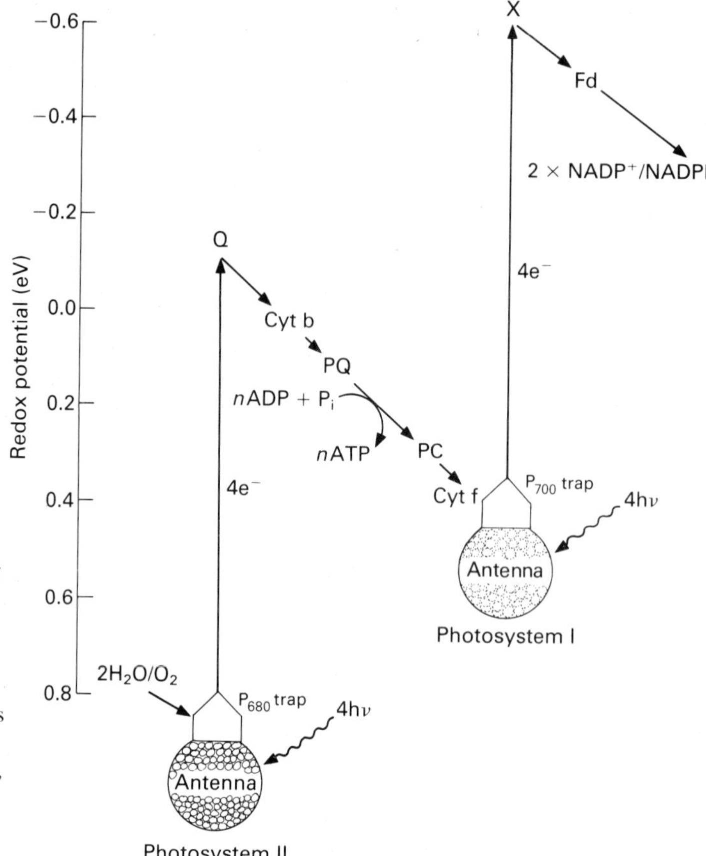

Fig. 6.2 Hill–Bendall Z scheme for electron flow in the light reactions of photosynthesis. The bold vertical arrows represent the photochemical or light reactions. Only non-cyclic photophosphorylation is shown. Q and X are primary electron acceptors for PSII and PSI respectively. Cyt, cytochrome; PQ, plastoquinone; PC, plastocyanin; Fd, ferrodoxin are electron carriers. (After Ramus, 1981.)

as fluorescence. As energy would be lost to entropy in this process, light of longer wavelength is emitted. The antenna pigment molecules are arranged in a manner such that the energy of the electron returning to its ground state is not emitted as fluorescence, but is passed on to another pigment molecule, and ultimately to the photosystem traps (Fig. 6.3). The shorter the wavelength of the light absorbed, the more accessory pigments involved. In a red alga the chain for shorter wavelength light would be phycoerythrin, phycocyanin, allophycocyanin, and chlorophyll *a* (Ramus, 1981), and this is reflected in the spatial arrangements of these molecules within the phycobilosome (Fig. 6.4) (Gantt, 1981).

ACTION SPECTRA

The contribution of pigments to the phyotosynthetic process is shown by the action spectrum of an alga, in which the *in-vivo* absorption of light, adjusted for constant photon flux density across the spectrum, and its effect measured as photosynthesis are plotted against wavelength (Fig. 6.5). In the Chlorophycota and Chromophycota, the action spectrum and the absorption spectrum show a close correlation and show the importance of both the chlorophylls and the other accessory pigments in the light harvesting process. Red algal plots are, however, significantly different, the absorption spectrum being broader than the action spectrum. This suggests that most light harvesting is confined to the biliproteins, with the chlorophylls having a reduced involvement.

CHROMATIC ADJUSTMENT

It has been widely shown that algal cells grown at low irradiances have higher pigment concentrations than those grown at high irradiances

(Ramus, 1981; Yentsch, 1980), and thus are more efficient at absorbing available light, although the energy yield per quantum absorbed is unaltered. In diatoms it is variation in irradiance in the blue end of the spectrum which brings about this effect (Vesk & Jeffrey, 1977), but the pigment ratios remain constant and thus no complementary chromatic adjustment occurs. Complementary chromatic adjustment occurs in those Cyanophyceae capable of c-phycoerythrin synthesis (Tandeau de Marsac, 1977), with the phycocyanin/phycoerythrin ratio being highest under red light and lowest under green. Similar pigment adjustments in the Rhodophyceae have not been detected (Waaland *et al.*, 1974; Ramus *et al.*, 1976), but may occur in the Cryptophyceae (Vesk & Jeffrey, 1977).

COMPLEMENTARY CHROMATIC ADAPTATION AND THE VERTICAL DISTRIBUTION OF SEAWEEDS

Engelmann (1893) produced a crude action spectrum for photosynthesis in a number of algae by using aerotactic bacteria as a bioassay for oxygen evolution. He noted the importance of green light in the Rhodophyceae and proposed that the pigmentation of the major seaweed groups was an adaptation to the wavelengths of light prevailing at the depths at which they were found in the ocean. The Chlorophyceae would therefore be commonest in the shallow waters and the Rhodophyceae be found in greatest abundance in the deep subtidal. This theory of complementary chromatic adaptation has entered many texts as an example of adaptation of plants to their environment, but it is now apparent that it is supported by neither physiological nor distributional evidence (Ramus, 1981; Dring, 1981). On the continental shelf of the Carolinas (Schneider, 1976), 25% of the green algae are found below 50 m, and in Hawaii (Doty *et al.*, 1974) there is a progressive enrichment in the green algal flora with depth: at 90 m they are in greater abundance than the red algae, while a green alga *Johnson-sea-linkia profunda* at 157 m (Littler *et al.*, 1985) holds the depth record for a fleshy alga. Some red and also green algae do occur at greater depths, but these are encrusting or otherwise reduced forms, and their success, in addition to being able to grow at low irradiances, is presumably due to their abilities to withstand grazing pressures. Experiments involving transplants and the shading of shallow water species show that any phenotypic adaptation that occurs is due to overall decrease in irradiance rather than a change in the spectral distribution (Ramus *et al.*, 1976, 1977).

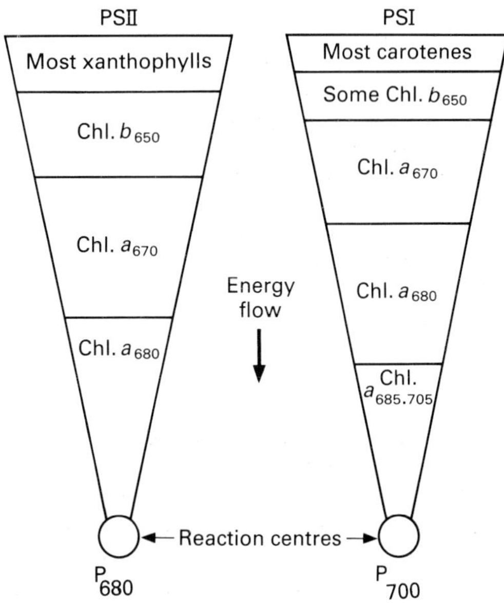

Fig. 6.3 Hypothetical arrangement of pigments in the light-harvesting antennae for PSII and PSI. (After Govindjee & Braun, 1974.)

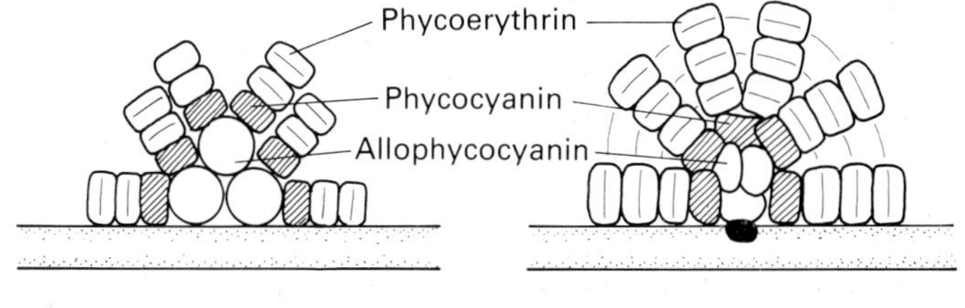

Fig. 6.4 Structure of phycobilisomes. (a) Hemidiscoidal type found in *Rhodella* and *Pseudoanabaena* (Cyanophycota). (b) Hemispherical type of *Porphyridium* (Rhodophycota); the black dot represents a protein presumed to anchor the phycobilisome to the thylakoid. (After Gannt, 1981.)

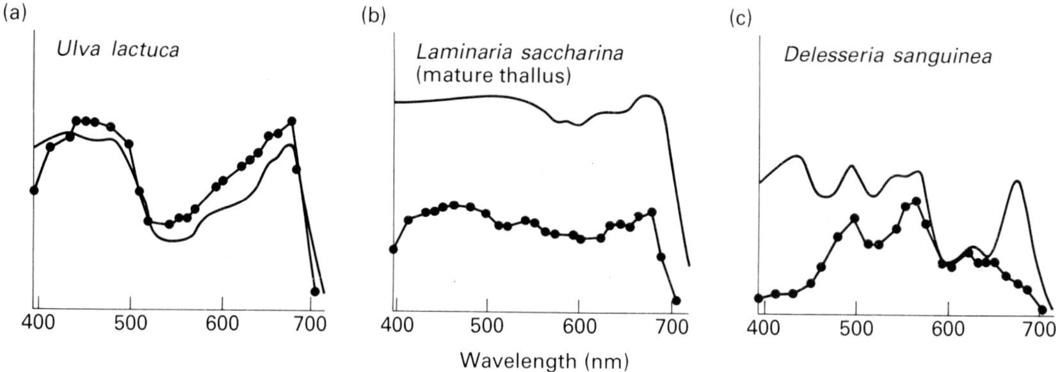

Fig. 6.5 Action spectra for photosynthesis (●) and absorption spectra (–). (a) A chlorophyte *Ulva lactuca*. (b) A chromophyte *Laminaria saccharina* (Phaeophyceae). (c) A rhodophyte *Delesseria sanguinea*. (After Dring, 1982; unpublished data of K. Lüning.)

PHOTOSYNTHESIS AND IRRADIANCE

The relationships between photosynthesis and irradiance levels are depicted in P *v*. I curves. At low irradiances photosynthetic rate is limited by the light reactions of photosynthesis, and this increases linearly with increase in irradiance (and is temperature-independent) until a plateau is reached at the saturating irradiance level ($P_{max.}$) Moderate further increases in irradiance have no effect on photosynthesis, though very high irradiances are inhibitory (Fig. 6.6). The maximum photosynthetic rate obtained at the saturating light levels is a function of the carbon-fixing processes of photosynthesis. As such it is dependent on such factors as temperature and nutrient availability. The photosynthetic rate may frequently be increased by the addition of nutrients or an increase in temperature.

The modifications in pigment concentration in response to levels of irradiance are reflected in photosynthetic performance. The increase in pigments typically results in an increase in initial slope of the P *v*. I curve and a lowering of the $P_{max.}$. The light compensation point at which photosynthesis and catabolic processes are equal is also lowered.

Daylength and photoperiodic responses

Photoperiodism is defined as 'the control of some aspect of a life cycle by the timing of light and darkness' (Dring, 1984). It is important to distinguish between responses due to the amount of light and responses due to its duration: only the latter are true photoperiodic responses (Fig. 6.7). Photoperiodic responses appear to be particularly common in those members of the Chlorophyceae, Phaeophyceae, and Rhodophyceae which have heteromorphic phases in their life cycles (Dring, 1984). A genuine photoperiodic response should be all or nothing and must satisfy the criterion that a light-break interrupted night should have the same effect as a long day (Vince-Prue, 1975). Such responses are known for many macroalgae (Dring, 1984); for example, they occur in conchospore formation in the rhodophyte *Porphyra* (Dring, 1967; Rentschler, 1967) and *Bangia* (Richardson, 1970). Daylength effects which do not satisfy the all or nothing, or light-break, criteria are also known. Tetrasporogenesis in *Acrosymphyton purpuriferum* (Rhodophyceae; Cortel-Breeman & ten Hoopen 1978), which occurs under short daylengths, is uninhibited by a light break. *Halymenia latifolia* (Rhodophyceae) shows a partial photoperiodic response which satisfies the light break requirement, but sporogenesis still occurs at a reduced rate under long days (Maggs & Guiry, 1982).

In the Phaeophyceae photoperiodic responses have been most thoroughly investigated in *Scytosiphon lomentaria*. Under long daylengths zoospores produce crusts, but subsequent exposure to short days induces the formation of upright thalli (Dring & Lüning, 1975). Light breaks of less than one minute inhibit this response, and only blue light is effective.

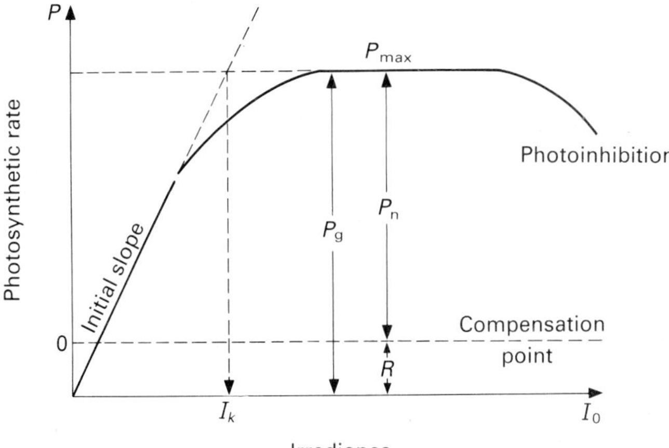

Fig. 6.6 Light saturation curve for photosynthesis. P, photosynthesis; P_{max}, maximum photosynthesis; P_g, gross photosynthesis; P_n, net photosynthesis; R, respiration; I_0, incident photon flux density; I_k, saturating photon flux density. (After Ramus, 1981.)

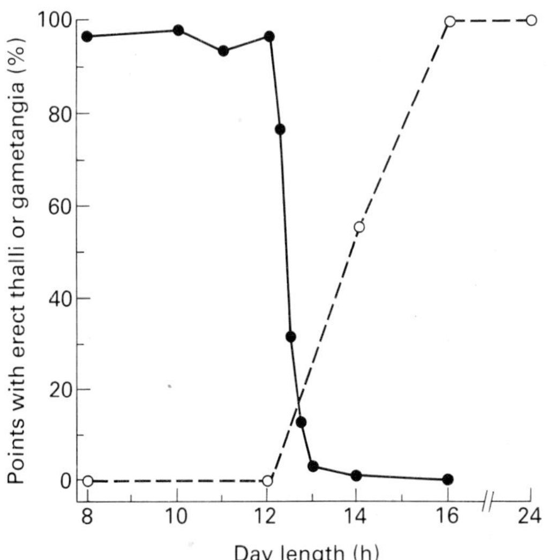

Fig. 6.7 Developmental response to various day lengths of two brown algae at constant irradiance and temperature. (○) *Scytosiphon lomentaria* — percentage forming erect thalli; (●) *Sphacelaria rigidula* — percentage forming gametangia. (After Dring, 1984.)

In contrast to the situation in higher plants the physiological basis for photoperiodic responses in the algae has received little attention. All photoperiodic reponses in higher plants are thought to be mediated by phytochrome. Red light (660 nm) is the most effective at promoting phytochrome-mediated responses, and these may be subsequently inhibited by exposure to far-red wavelengths (730 nm). All major groups of green plants are known to possess phytochrome (Dring, 1984). Phytochrome-mediated chloroplast movements are well documented in the green algae *Mougeotia* and *Mesotaenium* (Haupt, 1983). In the Rhodophycota, *Porphyra* shows the spectral responses typical of phytochrome (Dring, 1967; Rentschler, 1967) and a phytochrome-like receptor pigment has been detected in extracts of *Acrochaetium daviesii* (van der Velde & Hemrika-Wagner, 1978). Blue light is effective in the responses of several red and brown algae, and Dring (1984) has suggested that an alternative pigment, 'cryptochrome', may be responsible for daylength perception in in these instances.

Abdel-Rahman (1982) has demonstrated a link between the the overall length of the light–dark cycle and the length of the dark period in tetrasporogenesis of *Achrochaetium asparagopsis* (Fig. 6.8). This is normally a short-day event, but only occurs when the overall cycles are of approximately 24 or 48 hours' duration. These observations indicate links between circadian rhythms and photoperiodism.

Other wavelengths of light may produce photomorphogenic effects unrelated to photoperiodism. Growth of *Fucus* embryos is strongly inhibited by wavelengths of 575–625 nm, and this inhibition can be reversed by the addition of either longer or shorter wavelengths (McLachlan & Bidwell, 1983). Blue light effects are reported for the Phaeophyceae, and include the formation of hairs and two-dimensional crusts in *Scytosiphon lomentaria* (Dring & Lüning, 1975) and

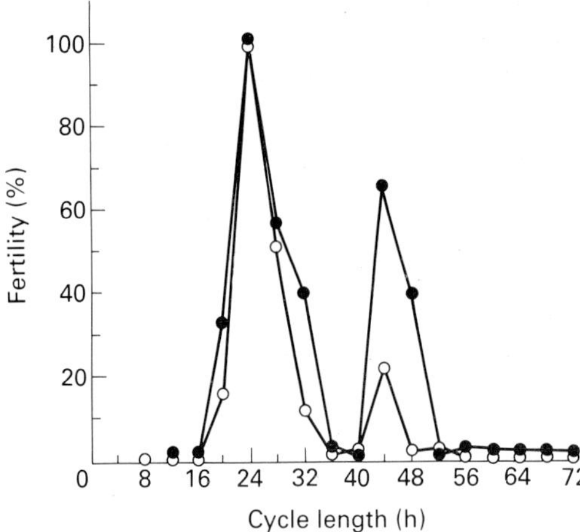

Fig. 6.8 The effect of light–dark cycle length on tetrasporangial formation in *Acrochaetium asparagopsis*. Photoperiods of six hours (open circles) or eight hours (closed circles) were followed by dark periods of various lengths. Note peaks in 24- and 44-hour cycles and almost complete inhibition in all other cycles. (After Abdel-Rahman, 1982.)

the induction of egg formation in laminarialean gametophytes (Lüning & Dring, 1975).

Carbon metabolism

CARBON FIXATION

The metabolic pathways involved in the photosynthetic fixation of carbon have been extensively studied in the unicellular green algae. Short-term exposure of *Chlorella* and *Scenedesmus* to ^{14}C-labelled carbon dioxide showed that fixation was by carboxylation of ribulose-1, 5-biphosphate (RuBP), utilizing the enzyme ribulose biphosphate carboxylase (RuBPC), and that the first stable product of these reactions was phosphoglyceric acid (PGA). Numerous experiments have confirmed the universality and importance of this photosynthetically driven carboxylation mechanism. It has been demonstrated in such diverse groups as the Cyanophyceae (Pelroy & Bassham, 1972), the Rhodophyceae (Bean & Hassid, 1955) and the Phaeophyceae (Bidwell *et al.*, 1958)

Hatch and Slack (1966) demonstrated that in sugar cane an alternative mechanism for photosynthetic carbon uptake occurred, involving the carboxylation of phosphoenolpyruvate (PEP) to produce oxaloacetate. Two enzymes are potentially involved in this carboxylation, PEP carboxylase (PEPC) and PEP carboxykinase (PEPCK); the former utilizes the bicarbonate ion and the latter CO_2. This mechanism is now known to be widespread in a variety of higher terrestrial plants. It is, however, an auxiliary mechanism which facilitates reductive carbon assimilation, and all plants in which it is found rely entirely on RuBPC for net carbon fixation. Initial studies in algae indicated the presence of several C_4 acids which became rapidly labelled in $^{14}CO_2$ uptake experiments, suggesting that a C_4 uptake pathway was important in seaweeds (Joshi *et al.*, 1974). PEPC activity is, however, extremely low in the Chlorophyceae, Phaeophyceae, Rhodophyceae (Kremer & Küppers, 1977), and Bacillariophyceae (Kremer & Berks, 1978). By contrast, PEPCK activity, while almost zero in the Chlorophyceae, accounts for 10% of the activity of RuBPC in the Rhodophyceae, and in some Phaeophyceae its activity reaches 70% (Kremer & Küppers, 1977). High values are also reported for the Bacillariophyceae (Kremer & Berks, 1978). However, phosphoglyceraldehyde (PGAL) is the first product of short-term photosynthetically driven carbon fixation in these plants, and the evidence suggests that this is utilized to produce the PEP substrate for further carboxylation (Kremer, 1981a, b). There is no evidence for a photosynthetically driven C_4 pathway similar to that operating in higher plants. PEPCK activity in the Laminariales is highest in the meristematic regions, and appears important in carbon conservation, allowing the refixation of CO_2 derived from the metabolism of mannitol (Weidner & Küppers, 1982). As such, the dark uptake of CO_2 via the action of PEPCK may be ecologically significant.

It would appear that, while algae are C_3 plants biochemically, they physiologically resemble C_4 plants in the low CO_2 compensation points reported for some species (Birmingham & Colman, 1979; Coughlan & Tattersfield, 1977). These observations may be explained if the algae are operating some CO_2-concentrating mechanism (Raven & Beardall, 1981). The exact nature of the carbon utilized in algal photosynthesis is difficult to determine as, unlike terrestrial plants which are exposed to atmospheric CO_2, aquatic plants have available carbonic acid (H_2CO_3) and its ions, bicarbonate (HCO_3^-) and carbonate (CO_3^{2-}). The overall amounts and the proportions of these carbon species in the water depends upon pH, salinity, the partial pressure of the CO_2 in the atmosphere, and the temperature. In acidic waters free carbon dioxide dominates, but rapidly declines above neutrality, in the pH 7–9 range HCO_3^- predominates, and above this level CO_3^{2-} becomes important (Fig. 6.9). All algae seem to be able to take up free CO_2 (Raven, 1974) which is exclusively utilized as a substrate for RuBPC. The ability to utilize HCO_3^- has also been demonstrated in a variety of algae (Beardall *et al.*, 1976; Jolliffe & Tregunna, 1970), and this may be due to the dehydration activity of carbonic anhydrase to produce CO_2 (Reed & Graham, 1981). Bidwell and McLachlan (1985) examined photosynthesis in a variety of seaweeds maintained in moist air rather than aqueous culture medium, and concluded that these algae normally absorb bicarbonate rather than CO_2 from seawater.

While photosynthetically driven uptake of inorganic carbon is the predominant mechanism of carbon nutrition, heterotrophic CO_2 fixation also occurs and some algae are capable of utilizing organic carbon substrates either via photosynthetically driven uptake or heterotrophically (Droop, 1974a). Species lacking photosynthetic pigments are known in most algal classes, and these must of necessity be heterotrophic for carbon; others retain their photosynthetic apparatus and may utilize simple organic substrates, such as acetate, while providing ATP and reductant via photosynthesis. Ecological aspects of algal heterotrophy are discussed in Chapter 7.

DARK RESPIRATION

The major features of dark respiration in algae are essentially identical with those found in other aerobic plants (Lloyd, 1974; Raven & Beardall, 1981). The glycolytic or EMP (Embden–Meyerhoff–Parnas) pathway runs in parallel with the hexose monophosphate shunt (HMP), while the tricarboxylic acid (TCA) or Krebs cycle is in series with these processes. The system produces ATP for the cell, and still larger quantities are produced under aerobic conditions by membrane-associated oxidative phosphorylation using reductant generated by dehydrogenases acting on organic substrates. Reductant, mainly as NADPH, is also generated from the HMP.

The location of these processes in the cell is essentially the same as in the higher plants. The enzymes of the EMP and HMP pathways lie in the cytosol, but also occur in the chloroplasts, which contain starch (Raven & Beardall, 1981). The TCA cycle enzymes are located in the mitochondrial matrix, while the oxidative phosphorylation processes are associated with the inner mitochondrial membrane. Differences of location are found in the prokaryotic Cyanophyceae, where the EMP, HMP, and TCA enzymes are found in the cytosol, and oxidative phosphorylation is associated with membranes.

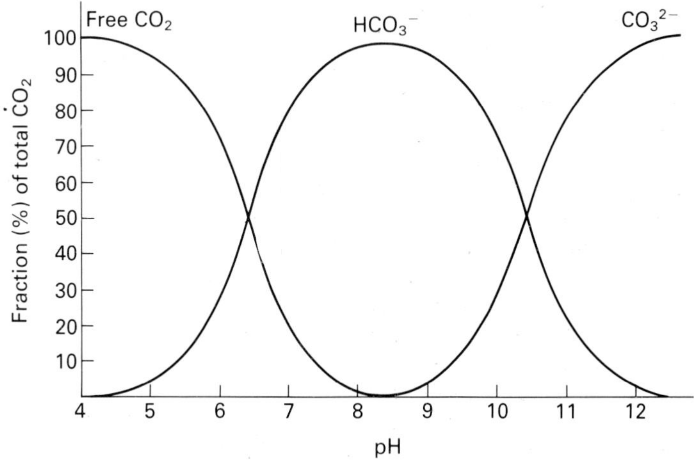

Fig. 6.9 Relation between pH and the relative proportions of inorganic carbon species in solution in fresh waters. (After Golterman *et al.*, 1978.)

Much of the ATP and NADPH produced in photosynthesis is utilized in the reduction of CO_2 to carbohydrate, but this use is not obligate and these compounds may be utilized for other cellular processes. Determination of the contribution of dark respiration to a photosynthesizing cell is not an easy task. For comparative purposes the respiratory capacity of an algal cell may be expressed as the ratio of the maximum production of carbon dioxide in the dark ($u_{r\ max.}$) to the maximum achieved growth rate (u_g). For phototrophically grown cells the ratio may be as low as 0.3–0.5, while for comparable heterotrophically grown cells the ratio is 0.5–1.2. This indicates that the respiratory capacity for a given maximum growth rate is lower for phototrophs than for heterotrophs, and that a significant portion of the ATP and reductant for growth is produced directly from the light reactions of photosynthesis. Data from *Anacystis nidulans* (Doolittle & Singer, 1974) indicate that the ratio in the Cyanophyceae may be less than 0.5, suggesting an almost total reliance on photosynthetically produced ATP and reductant. At the opposite extreme, the dinoflagellates appear to be a group of algae uniquely characterized by high rates of dark respiration ($u_{r\ max.}/u_g = 1.1$; Prezlin & Sweeney, 1978).

PHOTORESPIRATION, GLYCOLATE METABOLISM, AND LOSS OF ORGANIC CARBON

In addition to its carboxylating functions RuBPC is an oxidase, promoting the oxidation of RuBP to phosphoglycerate and phosphoglycolate (Raven & Beardall, 1981). In the vascular plants, bryophytes and charophytes, further oxygen uptake occurs and glycolate oxidase catalyses the conversion of glycolate to glycoxylate. In contrast, in the majority of algae this oxidation is due to the action of glycolate dehydrogenase (Raven & Beardall, 1981). Two molecules of glycoxylate may ultimately be converted to one of glycerate, with loss of carbon dioxide. This light-dependent oxygen uptake and carbon dioxide production in higher plants is termed photorespiration, and may consume over 50% of the carbon dioxide fixed, with no net gain in ATP (Bidwell, 1983). While it is well established that algae have a glycolate pathway, there is considerable controversy about its importance in the carbon budgets of algal populations in nature. Studies on a wide range of algae (Lloyd *et al.*, 1977) have failed to detect any photorespiratory gas exchange. Glycolate, however, is an extracellular product of many algae (Raven & Beardall, 1981; Bidwell, 1983) with amounts excreted ranging from almost zero to 70% of the carbon fixed. In the laboratory, maximum glycolate excretion occurs when algae are grown under high CO_2 concentrations (> 0.2%) and then transferred to low CO_2 conditions (Ingle & Colman, 1976). Presumably the relatively large quantities of RuBP produced under the high CO_2 regime are subsequently available for oxidation to glycolate. Under more natural conditions glycolate excretion may be stimulated by nutrient limitation and it is suggested (Harris, 1980) that this is a mechanism by which the algal cell can rid itself of excess carbon under those conditions in which the rate of carbon uptake temporarily exceeds that at which it can be metabolized or stored.

Algae may release a variety of other organic substances, including carbohydrates, amino acids, lipids, phenolics, and enzymes (Hellebust, 1974; Kremer, 1981a). The larger members of the Phaeophyceae are known to release large quantities of phenolics (Kremer 1981a) and these are thought to be responsible for the production of the yellow humic substances present in inshore seawaters. Glycolate excretion occurs in some marine macroalgae, but is low in comparison to that of phytoplankton. Dissolved carbohydrate is released from many seaweeds, but the circumstances and the amounts are controversial (Kremer, 1981a). In *Ascophyllum nodosum* very little release of carbohydrate occurs (Moebus *et al.*, 1974), whereas in *Laminaria* spp. 30–35% of the net carbon fixed is lost as carbohydrates and phenolics (Johnston *et al.*, 1977; Hatcher *et al.*, 1977).

ACCUMULATION AND STORAGE OF PHOTOSYNTHETIC PRODUCTS

The essentially identical nature of the pathway of carbon fixation in all algal groups is not reflected in their storage products. The storage products of photosynthesis in the eukaryotic algae are carbohydrates (Craigie, 1974; Kremer, 1981a), and while the early literature contains references to fats and oils as storage products the major role of algal lipids is as membrane components (Wood, 1974). In addition to storage carbohydrates the algae produce a variety of sugars and sugar alcohols (Craigie, 1974; Kremer, 1981a). One of these, the polyhydric alcohol mannitol, may account for 20–30% of the dry weight of some members of the Phaeophyceae. These low molecular weight compounds have an osmotic role (Kauss, 1978; Reed *et al.* 1985) as well as serving immediate metabolic needs. The true storage carbohydrates (Craigie, 1974; McCandless, 1981) are larger, relatively insoluble molecules, which are mainly glucose polymers. The Chlorophyceae produce typical higher plant starches consisting of two polyglucans, a straight chain α-1, 4-linked amylose and a branched α-1, 4; 1, 6 amylopectin; they give the characteristic blue-black staining reaction with iodine. The starches of the Rhodophyceae and Cryptophyceae are similar, with a higher amylopectin content; they produce a red iodine staining reaction. Cyanophycean starch in its molecular size and hydrolysis products more closely resembles glycogen than higher plant starch. Laminarin (Phaeophyceae) is a linear β-1, 3-linked polyglucan of 16–32 monomers and which may have terminal mannitol units and some 1, 6 branching. Chrysolaminarin (Chrysophyceae) is similar in structure, but lacks mannitol, whereas paramylum (Euglenophyceae), also reported in the Prymnesiophyceae (Kreger & van der Veer, 1971), is a larger molecule.

In most algae, and especially in the unicells, the translocation of photoassimilates is not a problem as most will be metabolized *in situ*. Translocation does occur between various symbiotic partnerships which involve algae and heterotrophs (Chapter 7). Translocation within algae has been mostly studied in Phaeophyceae, particularly in the Fucales and the Laminariales (Fig. 6.10) (Buggeln, 1983). Transportation in virtually all members of the Laminariales is via sieve tubes, and in *Macrocystis* these resemble the sieve tubes of higher plants. In *Laminaria* they are more occluded with cytoplasmic organelles, and the pores in the sieve plates are smaller and more numerous. *Saccorhiza* lacks sieve tubes and their transport function is carried out by solenocytes (Fig. 6.11) (Emerson *et al.*, 1982). These are elongate cells occuring in the medulla and are cross-connected by other cells, the allelocysts. The observations on the movement of photoassimilates in Laminariales are consistent with the mass flow (Munch) hypothesis, though the evidence is largely circumstantial (Buggeln, 1983). Buggeln *et al.* (1985) have shown that the flow rate of photoassimilate is not uniform and that pulses precede, and are superimposed upon, the mass flow (Fig. 6.12). Similar events are known to occur in higher plants and possibly indicate the occurrence of an energy-assisted transport mechanism.

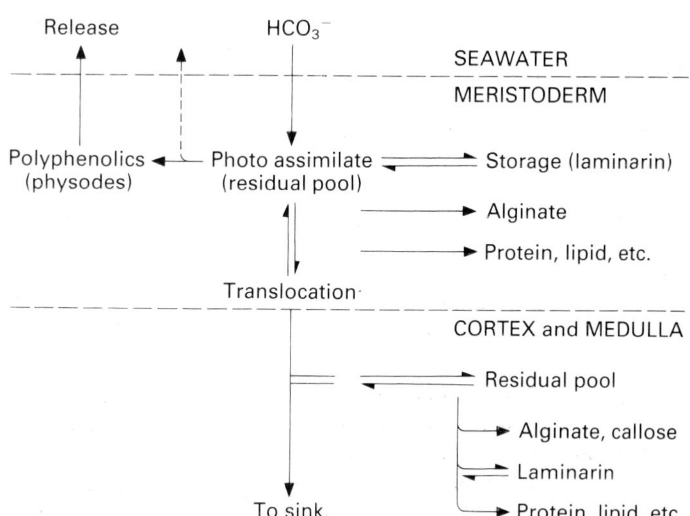

Fig. 6.10 Presumptive fates of photoassimilates in tissues of a kelp blade. (After Buggeln, 1983.)

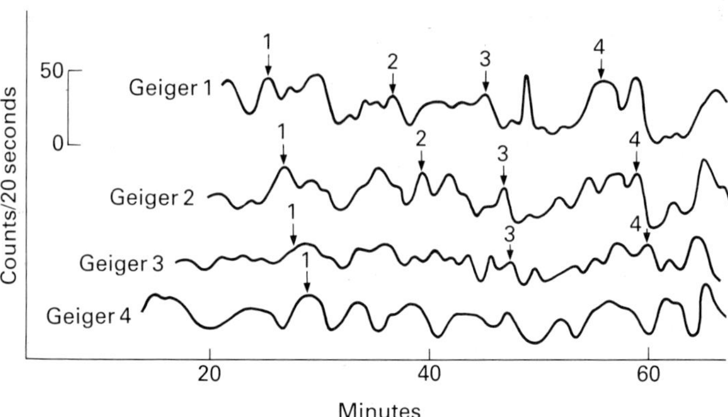

Fig. 6.11 Pulses of ^{11}C-labelled photoassimilate passing successive detectors arranged linearly at 1 cm intervals on a blade of *Macrocystis pyrifera*. Examples of peaks of radioactivity, thought to be associated, are indicated by numbered arrows. The radioactive pulse was withdrawn at T = 0 minutes. (After Buggeln *et al.*, 1985.)

Inorganic nutrients

The elements carbon, hydrogen, oxgyen, phosphorus, nitrogen, magnesium, iron, copper, manganese, zinc, molybdenum, sulphur, potassium, and calcium are required by all algae, and others require sodium, iodine, bromine, boron, vanadium, chlorine and silicon (Table 6.1) (O'Kelly, 1974). Inorganic nutrients, such as nitrogen and phosphorus, are required in relatively large quantity and are termed macronutrients. Others such as copper and molybdenum are required as co-factors in enzyme systems or are utilized in electron transport systems; these are usually required in relatively small amounts and are referred to as micronutrients.

NITROGEN

After carbon, oxygen, and hydrogen, nitrogen is the most abundant element in algae without mineralized walls. In algae growing without nitrogen limitation it may constitute 10% of the dry weight (Syrett, 1981). Nitrogen is available to algae in three basic forms: free nitrogen gas, and as combined inorganic or organic compounds.

Nitrogen fixation

The ability to reduce, or fix, nitrogen gas is found only among the prokaryotes, and in the algae exclusively in the Cyanophyceae. Nitrogen fixation requires nitrogenase, an oxygen-sensitive iron-, sulphur- and molybdenum-containing enzyme complex (Bothe, 1982) which also brings about the reduction of other substrates containing triple covalent bonds (nitrous oxide, cyanides, isocyanides, cyclopropene, and acetylene). Acetylene reduction to ethylene and the simple separation of substrate and product by gas chromatography has been adopted as the standard assay for nitrogen fixation.

Most nitrogen-fixing cyanophytes are filamentous and contain large, pale, thick-walled cells called heterocysts (Fig. 6.13), which are formed in the absence of utilizable combined nitrogen. They lack the oxygen-evolving photosystem II apparatus, RuBP carboxylase, and may lack or have reduced amounts of the photosynthetic biliproteins (Wolk, 1982). The heterocyst walls contain oxygen-binding glycolipids (Lambein & Wolk, 1973), and these together with respiratory oxygen consumption maintain the anaerobic conditions necessary for nitrogen fixation (Bothe, 1982).

Nitrogenase activity requires ATP and reduced ferrodoxin (Bothe, 1982). ATP is produced in the heterocyst primarily via photosystem I cyclic photophosphorylation, as is some reduced ferrodoxin. Most ferrodoxin, however, appears to be reduced by NADPH obtained principally from the hexose monophosphate shunt pathway using carbohydrate imported from adjacent cells. Nitrogen is reduced to NH_4^+, which is converted to glutamine before export from the heterocyst to adjacent cells of the filament.

Nitrogen fixation has been reported in planktonic non-heterocystous colonial species of *Oscillatoria (Trichodesmium)* in cells located at the centre of the colony. The cells did not produce O_2 photosynthetically and did not produce thickened specialized cell walls, as in heterocysts (Carpenter & Price, 1976). In a similar manner

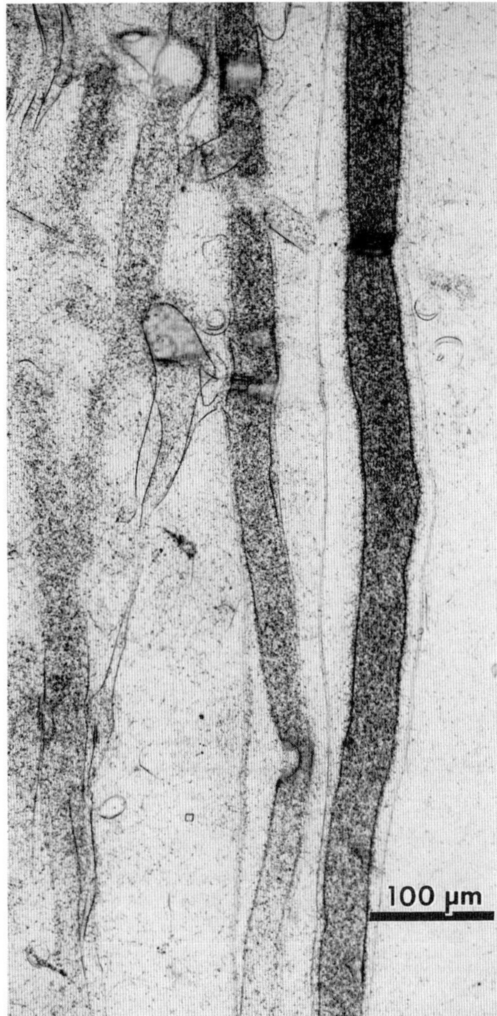

Fig. 6.12 Autoradiograph of a section through the medulla of ^{14}C bicarbonate-pulsed blade of *Saccorhiza dermatodea* showing heavily labelled solenocysts, presumptive long distance translocating elements. (Courtesy of Carolyn Emerson.)

nitrogen fixation activity in the marine planktonic *Oscillatoria erythraea* is related to the ability of the filaments to aggregate into bundles (Bryceson & Fay, 1981).

Inorganic combined nitrogen

Almost all chlorophyll-containing algae will grow on nitrate, NO_3^-, nitrite, NO_2^-, or ammonium, NH_4^+, ions and, in most instances, comparable growth occurs on all three sources. In non nitrogen-starved cells the uptake of NO_3^- ceases when NH_4^+ is added, although in severe nitrogen deficiency simultaneous uptake occurs (Syrett, 1981).

Whatever the source of inorganic nitrogen it is generally agreed that conversion to NH_4^+ occurs before incorporation into cellular organic compounds (Syrett, 1981). The pathway of conversion is a two-stage process:

$$NO_3^- \xrightarrow[2e^-]{\text{Nitrate reductase}} NO_2^- \xrightarrow[6e^-]{\text{Nitrite reductase}} NH_4^+$$

Eukaryotic algae have a nitrate reductase complex similar to that found in higher plants, consisting of molybdenum, flavin, and cytochrome components; it uses NADH (and sometimes NADPH) as a source of reductant. The prokaryotic nitrate reductase is simpler; it also contains molybdenum, but uses reduced ferrodoxin as an electron source. The second reduction to ammonium is catalysed by ferrodoxin nitrite reductase, and both NADH and NADPH are inactive as electron donors (Syrett, 1981).

The most widely accepted pathway for the incorporation of the ammonium ion into cellular organic compounds (Syrett, 1981; DeBoer, 1981) is that glutamine synthetase catalyses the addition of NH_4^+ to glutamate to produce glutamine. One amino group is then transferred to oxyglutamate to produce two molecules of glutamate. One molecule feeds back into the cycle and the other is available for the production of amino acids via transamination reactions (Figs 6.14 & 6.15).

The overall process of uptake and conversion of nitrate to amino acids requires 10 electrons per molecule, in contrast to the reduction of CO_2 to carbohydrate, which requires only four. It is therefore not surprising that it has been repeatedly shown that assimilation of both NH_4^+ and NO_3^- in normally grown algae is dependent on photosynthesis, with little or no uptake occurring in the dark (Syrett, 1981). However, this is not an obligate process, and nitrogen-starved *Chlamydomonas* with accumulated carbohydrate reserves assimilates both NO_3^- and NH_4^+ in the dark, but does so at even higher rates in the light (Thacker & Syrett, 1972).

Table 6.1 Role of inorganic nutrients in algal metabolism. (After DeBoer, 1981.)

Element	Probable functions	Examples of compounds
Nitrogen	Major metabolic importance as compounds	Amino acids, purines, pyrimidines, porphyrins, amino sugars, amines
Phosphorus	Structural, energy transfer	ATP, GTP, nucleic acids, phospholipids, coenzymes, coenzyme A, phosphoenol pyruvate
Potassium	Osmotic regulation. pH control, protein conformation and stability	Probably occurs predominantly in the ionic form.
Calcium	Structural, enzyme activation, ion transport	Calcium pectate, calcium carbonate
Magnesium	Photosynthetic pigments, enzyme activation, ion transport, ribosomal stability	Chlorophyll
Sulphur	Active groups in enzymes and coenzymes, structural	Methionine, cystine, cysteine, glutathionine, agar, carrageenan, sulpholipids, coenzyme A
Iron	Active groups of porphyrin molecules and enzymes	Ferredoxin, cytochromes, nitrate reductase, nitrite reductase, ferritin, catalase
Manganese	Electron transport in photosystem II, maintenance of chloroplast membrane structure	Manganin
Copper	Electron transport (photosynthesis), enzymes	Plastocyanin, amine oxidase
Zinc	Enzymes, auxin metabolism(?), ribosome structure(?)	Carbonic anhydrase
Molybdenum	Nitrate reduction, ion absorption	Nitrate reductase
Sodium	Enzyme activation, water balance, enzymes	Nitrate reductase
Chlorine	Photosystem II	Terpenes
Boron	Regulation of carbon utilization(?), RNA metabolism(?)	
Cobalt	Component of vitamin B_{12}, C_4 photosynthesis pathway	Vitamin B_{12}
Bromine*	?	Wide range of halogenated compounds in red algae
Iodine*	?	

* Possibly an essential element in some seaweeds.

Organic nitrogen sources

Many algae are capable of using organically combined nitrogen, especially amino acids, urea, and purines, as their sole nitrogen source (Neilson & Larsson, 1980). Antia *et al.* (1975) found that 23 of 26 species of marine phytoplankton they studied could use urea, and 17 could use the purine hypoxanthine. Others use amino acids, particularly glycine, serine, alanine, and glutamic and aspartic acids (Wheeler *et al.*, 1974).

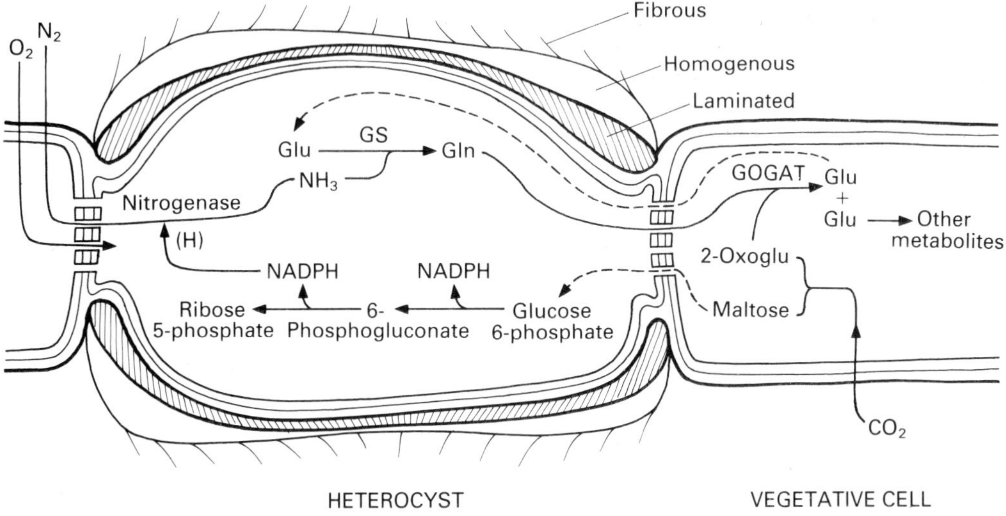

Fig. 6.13 Diagram of the flow of carbon and nitrogen between heterocysts and vegetative cells in nitrogen-fixing Cyanophycota. GS, glutamine synthetase; GOGAT, glutamine oxoglutarate amino transferase; Glu, glutamic acid; Gln, glutamine. The layered wall of the heterocyst is impermeable to nitrogen, CO_2 and oxygen. (After Haselkorn, 1978.)

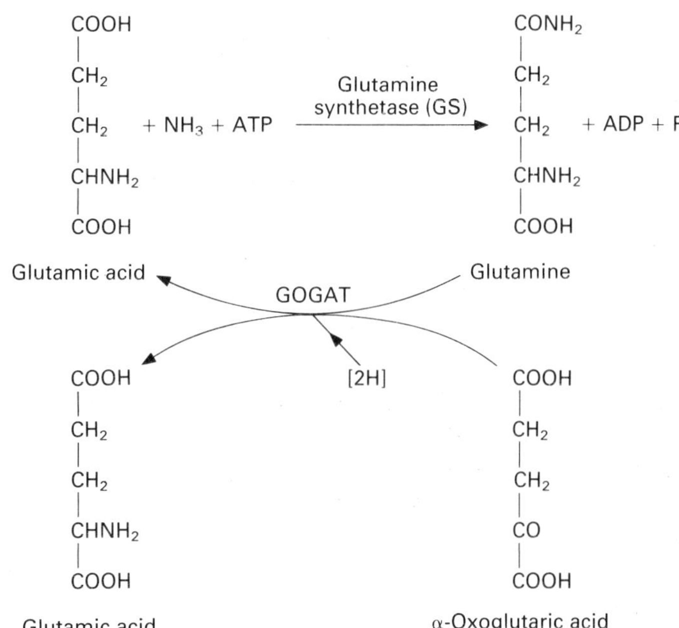

Fig. 6.14 Pathway of incorporation of ammonia utilizing GOGAT (glutamine oxoglutarate amino transferase). (After Syrett, 1981.)

Nitrogen storage

Assimilated nitrogen may be stored as nitrate, ammonium, or low molecular weight organic compounds and used for growth at some future time. Several seaweeds, including *Ulva lactuca* (Chlorophyceae), *Gracilaria folicera* (Rhodophyceae; Rosenberg & Ramus, 1982), and the kelps *Alaria esculenta* (Buggeln, 1978) and *Laminaria longicruris* (Chapman & Craigie, 1977), store nitrogen, which enables them to continue growth

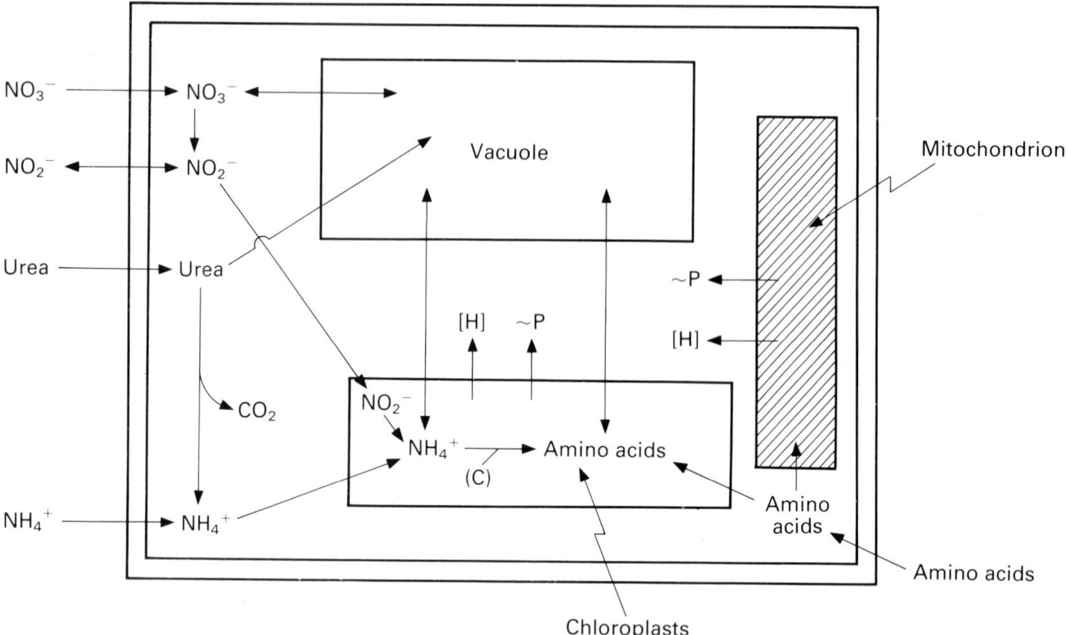

Fig. 6.15 Main features of nitrogen assimilation and partitioning in an algal cell. (After Syrett, 1981.)

when external concentrations in the ocean have become limiting (Chapter 7). The Cyanophyceae appear unique in their ability to store nitrogen as structured granules of cyanophycin (Lang *et al.*, 1972), a polypeptide polymer of arginine and aspartic acids. Synthesis of this peptide is unusual as it appears to be non-ribosomal (Simon, 1973).

PHOSPHORUS

Phosphorus is required in algal cells to form nucleic acid, phospholipids, and various ester phosphates such as phosphorylated sugars, ATP, and NADP.

Orthophosphate phosphorus is the only important inorganic phosphorus source for algae, although the ability to utilize polyphosphates and various organic phosphates appears to be widespread (Kuenzler, 1965). Such organic phosphates are hydrolysed by phosphatases located on the cell membrane. Alkaline phosphatase is induced by phosphorus limitation (Healey, 1973) and may be liberated from the cells into the surrounding medium (Reichardt *et al.*, 1967; Perry, 1972). The ability to utilize organically bound phosphates is advantageous in aquatic habitats in which such combined phosphate sources are present in concentrations higher than those of free phosphate.

Most algae store excess phosphate as polyphosphate granules of 30–500 nm diameter (Harold 1966), though some may store ionic phosphate in vacuoles (Healey, 1973). The uptake and storage of a nutrient beyond the immediate metabolic needs of the cells is termed luxury consumption and presumably provides the alga with a supply of phosphorus when external levels might otherwise be limiting. It has been suggested, however, that the storage function of polyphosphates is only secondary to their role in regulating the concentration of free phosphate ions in the cell (Kulaev, 1975).

NITROGEN/PHOSPHORUS RATIOS

Phytoplankton cells growing in a nutrient-unlimited environment typically show a C:N:P atomic ratio of 106:16:1, the so-called Redfield ratio (Redfield *et al.*, 1963). Analyses in which the plankton have been separated from associated detrital material show that N:P ratios for 'normal phytoplankton' range from 5:1 to 15:1 (Ryther & Dunstan, 1971). Chemostat studies with *Scenedesmus* (Rhee, 1978), however, suggest that cells are nitrogen-limited up to an N:P ratio of about 30, and are phosphorus-limited above this level (Fig. 6.16). Benthic marine algae are depleted more in phosphorus and less in nitrogen relative to carbon than phytoplankton, and show an average ratio of 550:30:1 (Atkinson & Smith, 1983). Large departures from these ratios suggest nutrient limitation, and the general response is a decrease in the cell quota of the limiting nutrient. The effects of both phosphorus and nitrogen limitation have been extensively studied in the marine diatom *Thalassiosira pseudonana* (Eppley & Renger, 1974; Perry, 1976; McCarthy & Goldman, 1979). Nitrogen limitation causes a decrease in growth rate expressed as generation time and photosynthetic rate. The cell quotas for nitrogen, phosphorus, and chlorophyll *a* decline at a faster rate than that of carbon, bringing about an increase in the ratio of carbon to the other cell components. Phosphorus deficiency also increases the generation time, but while the cell quota of phosphorus drops nitrogen remains constant and the carbon per cell increases.

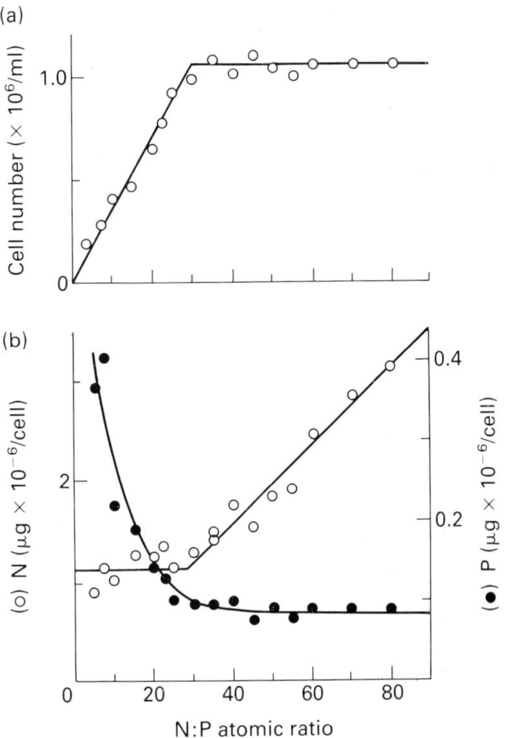

Fig. 6.16 Response of *Scenedesmus* to N:P ratios in chemostat cultures. Phosphorus concentration was held constant while nitrogen concentration was varied. (a) Growth curve. (b) Cellular concentration of nitrogen (open circle) and phosphorus (solid circle). (After Rhee, 1978.)

CALCIUM

Calcification in algae involves the precipitation of $CaCO_3$ around or within algal cells. It occurs in the Cyanophyceae, Rhodophyceae, Dinophyceae, Prymnesiophyceae, Phaeophyceae, Chlorophyceae, and Charophyceae (Borowitzka, 1982). Photosynthetic removal of CO_2 results in a rise in pH and an increase in the abundance of CO_3^{2-} ions; these combine with Ca^{2+} and $CaCO_3$ is deposited. Two crystalline forms of $CaCO_3$ are found in algae, calcite and aragonite, although each family of calcareous algae contains only one crystalline form. Aragonite is most frequently found in the marine algae, and calcite in the freshwater species, although the the marine coralline red algae are notable exceptions in their production of calcite.

The calcification process is best understood in the tropical green alga *Halimeda*, which deposits needle-like crystals of aragonite outside the cell walls, but within an intercellular space separated from the external seawater by a layer of tightly adpressed utricles (Fig. 6.17). This compartmentation allows the rapid utilization of CO_2 and HCO_3^-, the release of OH^-, and an increase in pH causing precipitation of $CaCO_3$. The model can be applied to other aragonite-depositing marine algae in which the presence of densely interwoven filaments allows the development of a long diffusion path from the external seawater to the site of precipitation (Borowitzka, 1982).

Members of the Corallinaceae (Rhodophyceae) deposit calcite among the fibrils of the cell wall (Giraud & Cabioch, 1979). Again, deposition is linked to photosynthesis (Borowitzka 1982). The mechanisms of deposition are not well understood, and an electrochemical model, which does not rely on CO_2 utilization, has been

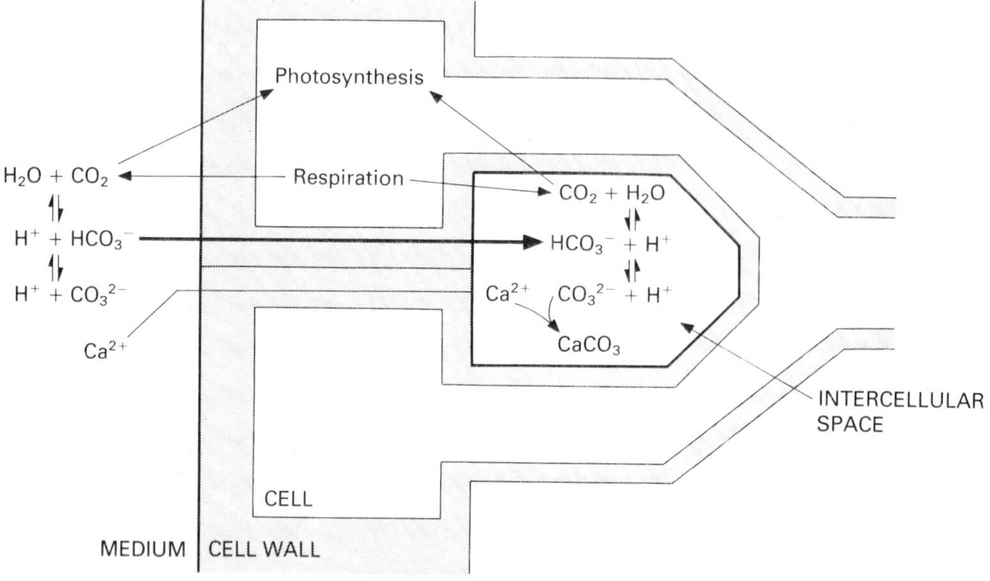

Fig. 6.17 Diagrammatic representation of the ionic carbon fluxes leading to calcium carbonate deposition in the walls of a calcareous green alga *Halimeda*. (After Borowitzka, 1982.)

proposed (Digby, 1979), but has been rejected by Borowitzka (1982).

Those members of the Prymnesiophyceae commonly known as the coccolithophorids are the only algae in which intracellular calcification occurs (Fig. 6.18). In most species the deposition of calcite occurs on an organic base within the cisternae of the Golgi apparatus (Klaveness & Paasche, 1979) to form delicately sculptured calcareous plates, or coccoliths. Coccoliths contain polysaccharides and associated proteinaceous materials which are capable of binding Ca^{2+} ions and hence presumably act as nucleating agents in the precipitation process (Borowitzka, 1982). Calcification proceeds in the light, and in *Emiliania huxleyi* CO_2 is utilized in photosynthesis while HCO_3^- provides the carbon for calcification (Sikes *et al.*, 1980). When this species is subject to nitrogen starvation it precipitates aragonite rather than calcite (Wibur & Watanabe, 1963), suggesting that it is the proteinaceous component at the site of calcification that determines the crystalline form deposited.

One important question is why algae remain uncalcified under conditions which support calcification in other species. One theory is that they produce on their cell surfaces substances, such as polyphenols, which inhibit $CaCO_3$ crystal formation (Reynolds, 1978).

SILICON

The diatoms with their silica frustules have a nutritional requirement for silicon in the form of orthosilicic acid. Silicon has also been detected in the Phaeophyceae and Chlorophyceae (Parker, 1969), in the chrysophytes, which produce siliceous scales or cysts (Paasche, 1980), and the prasinophyte *Platymonas* (Fuhrman *et al.*, 1978). Silicon is transported into an intercellular pool via a carrier enzyme located on the cell membrane (Sullivan, 1976, 1977). Uptake is an energy-requiring process, but the requirement appears to be satisfied by respiration, as uptake rate does not significantly increase in the light (Nelson *et al.*, 1976).

Experiments with synchronized cultures of diatoms have shown that silicic acid uptake occurs only during cytokinesis and new wall formation (Darley *et al.*, 1976; Chisholm *et al.*, 1978). Some of the large marine plankton diatoms may be capable of continuous uptake and storage of silicon in their large vacuoles (Chisholm *et al.*, 1978). Prolonged silicon starvation leads to an

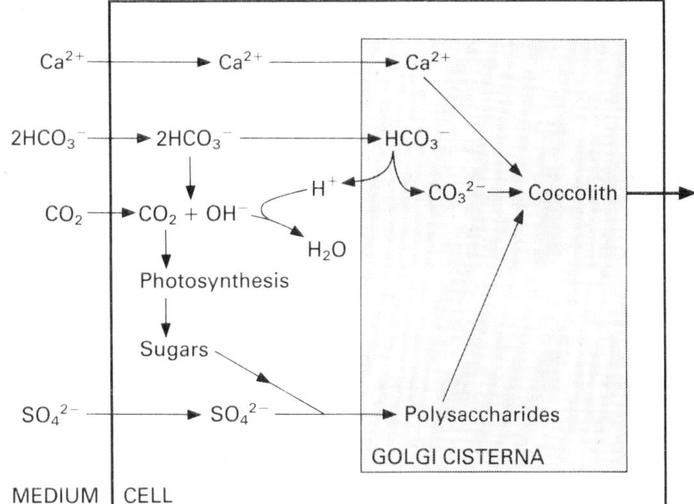

Fig. 6.18 Diagrammatic representation of the ionic carbon fluxes leading to calcification and coccolith formation in *Emiliania huxleyi*. Calcium carbonate deposition occurs internally in the cell within the Golgi cisternae. (After Borowitzka, 1982.)

accumulation of daughter cells, which remain united as no walls can be formed (Coombs *et al.*, 1967).

Diatoms are frequent contaminants in macroalgal cultures, and their absolute requirement for silicon may be utilized in their control by using germanium as a competitive inhibitor of silicon uptake (Lewin, 1966). Germanium dioxide may retard the growth of some members of the Phaeophyceae (Parker, 1969; Markham & Hagmeir, 1982), which may indicate a possible silicon requirement. However, the addition of extra silicates to the medium does not overcome germanium toxicity to *Fucus* (McLachlan *et al.*, 1971), suggesting that the nature of the toxicity does not involve competitive inhibition of silicate uptake.

POTASSIUM AND SULPHUR

Potassium and sulphur are inorganic nutrients which are required in greater than trace amounts. Potassium is present in many algae in high concentrations relative to the external medium. Its functions include osmotic regulation (Kauss, 1978) and the maintenance of the electrochemical environment of the algal cells. In addition to its occurrence in the amino acids methionine and cysteine, sulphur is incorporated into several other compounds, some of which appear to be of unique occurrence in the algae. Sulphated polysaccharides are important wall components of the seaweeds (McCandless, 1981, *see* Chapter 9), and sulphur-containing compounds are utilized in osmoregulation by some marine macrophytes (Reed, 1983).

TRACE ELEMENTS

The unequivocal demonstration of the need for many trace elements has been demonstrated in relatively few species of algae. Enrichment experiments show that productivity is increased with the addition of trace metals (Axler *et al.*, 1980) and trace metals are included in most artificial culture media (Provasoli *et al.*, 1957).

Quantitatively, iron is the most important trace metal for phytoplankton (Huntsman & Sunda, 1980). It is required in numerous redox reactions and in the synthesis of chlorophyll (O'Kelly, 1974). Many algae take up zinc in high concentration relative to the surrounding environment (Chapter 9), and much of the phycological interest in this element has concerned the uptake mechanisms (O'Kelly, 1974). The disappearance of cytoplasmic ribosomes correlates with zinc deficiencies in *Chlorella* (Prask & Plocke, 1971) and with a decrease in chlorophyll and protein synthesis in *Porphyra tenera* (Noda & Horiguchi, 1971). Molybdenum is required in nitrogen metabolism and is associated with the reduction of nitrate. In the Cyanophyceae molybdenum, along with iron, is a constituent of the nitrogenase enzyme complex (Bothe, 1982). Copper is an essential micronutrient, as a constituent of plastocyanin (a protein involved in photosynthetic electron transport; Katoh, 1960), and as a co-factor for several enzymes. Many algae are extremely sensitive to excess copper (Chapter 9), and this property has found use in the formulation of algicides and antifouling paints. Cobalt, as a constituent element of vitamin B_{12}, is also an essential micronutrient (Bunt, 1970), and there is a specific requirement for manganese in the photosystem II reactions of photosynthesis (O'Kelly, 1974). It has been suggested that levels of manganese and cobalt may be among the factors controlling the growth of *Macrocystis pyrifera* in southern California (Kuwabara, 1982). A requirement for boron has been demonstrated in *Fucus* embryos (McLachlan, 1977) and in several microalgae (Lewin, 1976). Stosch (1964) reported a requirement for arsenic in *Asparagopsis armata* (Rhodophyceae).

Many marine algae are strong concentrators of halogens. Chloride appears to be essential to photophosphorylation reactions (O'Kelly, 1974), and there is an apparent bromine requirement in *Polysiphonia urceolata* (Rhodophyceae; Fries, 1975) and *Fucus edentatus* (Phaeophyceae; McLachlan, 1977). Absolute requirements for iodine have been demonstrated in *Polysiphonia urceolata* (Fries, 1966) and *Ectocarpus fasciculatus* (Phaeophyceae; Pedersen, 1969); iodine also appears necessary for the growth and morphogenesis of *Petalonia fascia* (Phaeophyceae; Hsiao, 1969). Halogenated organic compounds are present in many marine macrophytes (Chapter 9) and may be important deterrents to herbivores.

Vitamins and growth factors

In addition to inorganic nutrients, many algae require exogenously supplied vitamins: cyanocobalamin (B_{12}), thiamin, and biotin (Swift, 1980). Of 388 algal species tested (Provasoli & Carlucci, 1974), 203 were heterotrophic for these vitamins; 85% required B_{12}, 40% thiamin, and 7% biotin. Although there are exceptions, some algal groups do not require biotin (Xanthophyceae, Phaeophyceae, Cyanophyceae, and Chlorophyceae), and in addition most members of the Cyanophyceae and Rhodophyceae are autotrophic for thiamin. Members of the Chrysophyceae, Prymnesiophyceae, Euglenophyceae, and the Dinophyceae show a much greater tendency to vitamin auxotrophy. Many algal species which do not require exogenous vitamins may synthesize and liberate vitamins in sufficient amounts to support the growth of other vitamin-requiring species (Carlucci & Bowes, 1970). Some, however, produce extracellular products that chelate free B_{12}, making it unavailable to other species (Pintner & Altmyer, 1979).

Other organic compounds may be necessary to algal growth, or may be inhibitory. Multicellular algae frequently fail to develop normal morphologies when grown under axenic conditions (Provasoli & Carlucci, 1974; Provasoli *et al.*, 1977), but normal growth can be restored by the addition of living or autoclaved marine bacteria. Numerous allelopathic relations of phytoplankton algae are known (Maestrini & Bonin, 1981a). The growth of a crustose red alga *Rhodophysema elegans* is inhibited by the proximity of a crustose brown alga *Ralfsia spongiocarpa* (Fletcher, 1975), suggesting that diffusible compounds produced by benthic algae may also have growth-regulating properties of ecological importance.

Uptake of nutrients

Oxygen and carbon dioxide uptake, except at high pH when active transport is necessary for the uptake of bicarbonate at rates sufficient to support net photosynthesis, is by passive diffusion (Raven, 1970). Other inorganic nutrients are generally taken up in ionic forms, and this uptake usually involves active transport linked to photosynthesis, which provides the energy necessary for the uptake.

The mechanisms and kinetics of active nutrient uptake in algae have been the subject of considerable research over the past two decades (DeBoer, 1981; McCarthy, 1981). The main stimulation for this work came from two papers (Dugdale, 1967; Eppley & Coatsworth, 1968) which examined the role of nutrient uptake kinetics in the growth of phytoplankton. The rate of nutrient uptake was shown to be related to the extracellular nutrient concentration (Fig. 6.19):

$$V = V_{\max}\left(\frac{S}{K_s + S}\right)$$

where V = the nutrient uptake rate, $V_{\max}$ = the

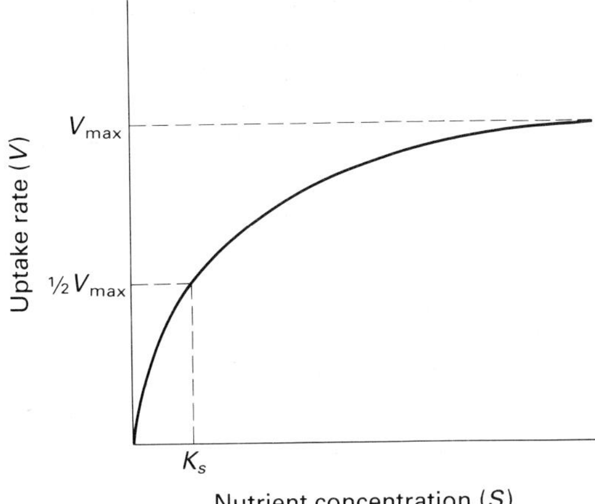

Fig. 6.19 Relationship between substrate concentration (S) of a nutrient and the uptake rate. V_{max} maximum rate of uptake; K_s, substrate concentration at which uptake rate is one half the maximum rate (half saturation) constant. (After Darley, 1982.)

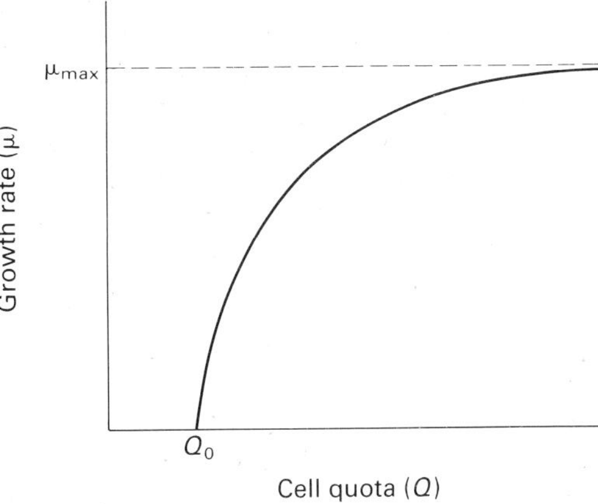

Fig. 6.20 Relationship between the growth rate, μ, and the internal cellular concentration of nutrient (Q), the cell quota. Q_0, minimum cell quota; μ_{max}, maximum growth rate. (After Rhee, 1980.)

maximum uptake rate, S = the concentration of the nutrient, and K_s = the half saturation constant or substrate concentration at which $V = V_{max}/2$.

This implies that at low nutrient concentrations the rate of uptake increases rapidly with increase in external nutrient concentration, but at higher concentrations increase in concentration adds progressively less to uptake rate until a high point is reached (V_{max}) at which uptake is virtually constant and thus independent of the external concentration of the nutrients. The similarity between nutrient uptake and the Michaelis–Menten equation for enzymatic catalysis suggests comparable mechanisms, in which nutrient uptake is by specific energy-requiring membrane-bound enzyme systems, with the nutrient as the substrate.

The application of this equation to algal nutrient uptake assumes that there is only a single nutrient or that the concentrations of the other nutrients remain constant, that only the initial rates of the uptake are measured for each nutrient concentration, and that the medium is sufficiently well mixed to prevent nutrient depletion at the surface of the algal cells.

The growth rate of an alga will decline if the concentration of the nutrients in the external medium drops below that which will support an uptake rate sufficient to maintain the current growth rate. Under such conditions the growth rate of the alga and the nutrient concentration show a relationship analogous to the Michaelis–Menten equation for nutrient uptake. This is the Monod equation (Monod, 1942) which was originally applied to the growth of bacteria in chemostat culture, but also describes nutrient-limited, steady-state growth of algae (Droop, 1974a):

$$\mu = \mu_{max}\left(\frac{S}{K_s + S}\right)$$

where μ is the specific growth rate, μ_{max} is the maximum specific growth rate, S is the substrate concentration, and K_s is the half saturation constant for growth.

Other studies (Droop, 1968) have shown that growth rate is also related to the cellular concentration of the limiting nutrient rather than to its concentration in the external environment (Fig. 6.20). This relationship is expressed in the Droop (1968) equation:

$$\mu = \mu'_{max}\left(1-\frac{Q_0}{Q}\right)$$

where Q is the amount of the nutrient in the algal cells, Q_0 is the lowest level of Q at which the algae can grow and μ'_{max} is the growth rate at infinite Q.

Under steady-state conditions both the Monod

and Droop equations are applicable to the description of nutrient-limited growth. The Droop approach is often the more practical, as the higher nutrient levels of the cell quota are more readily measured than the very low external nutrient concentrations. The Droop equation is a theoretical description, in that Q is never infinite and the true maximum growth rate μ_{max} at Q_{max} will be less than the theoretical maximum μ'_{max}; thus both Q_{max} and μ_{max} must be determinated experimentally (Rhee, 1980).

The theoretical basis of nutrient uptake by macroalgae is essentially the same as that for microalgae (DeBoer, 1981). It is, however, complicated by the size of the plants and the increase in the amount of apoplastic free space, which must be considered in any uptake experiments, and in the measurement of the cell quota.

Growth and reproduction

PHEROMONES

Sexual attractants are known in many animals, in the Oomycota and some freshwater green algae, including *Chlamydomonas* (Tsubo, 1961) and *Oedogonium* (van den Ende, 1976). The control of gametogenesis in a number of species of *Volvox* is also under pheromonal control (Starr *et al.*, 1980); in dioecious species a glycoprotein induces the development of sexual spheroids, but in a monoecious species critical levels of L-glutamic acid, which arise due to increasing population density, induce sexuality.

In the algae the identity and actions of such pheromones are best known in the Phaeophyceae (Table 6.2). The pheromones are functional in 10^{-7} mol/l concentrations and have an attractive range of 0.5–1 mm (Maier & Müller, 1986).

Ectocarpus siliculosus is nominally isogamous, but gametes which may be classified as female show a shorter period of motility and settle on a substratum and withdraw their flagella. Soon thereafter they become attractive to male gametes which respond to a highly volatile compound produced by the female gamete. It was the scent of this compound, vaguely reminiscent of oranges, which was detected in clonal cultures of the female gametophytes and which led ultimately to the purification, identification and synthesis of a cyclic C_{11} olefine which was termed ectocarpene. Other low molecular weight olefinic pheromones of similar structure are known from Cutleriales, Syringodermatales, Chordariales, Dictyotales, Laminariales, Desmarestiales, Sporochnales, Sphacelariales, and Tilopteridales (Müller, 1981; Maier & Müller, 1986). The same pheromones have also been shown to initiate sperm release in members of the Laminariales, Desmarestiales, and Sporochnales (Maier, 1982; Maier & Müller, 1986). Members of the Fucales and Durvillaeales produce linear, or cyclopropane-containing, olefinic pheromones. Fucoserratine is a C_8 compound found in *Fucus*, while others have C_{11} compounds. The C_{11} pheromone finavarrene, produced by *Ascophyllum nodosum*, is also found in the Dictyosiphonales, while the C_{11} homosirene, typical of several Australasian species, occurs in the Scytosiphonales (Maier & Müller, 1986).

FERTILIZATION AND GERMINATION

Although there is great deal of information on the morphological and anatomical processes which characterize the germination of algal spores and zygotes, nearly all of our physiological information has been obtained from a single source: *Fucus* eggs. Two attributes make them attractive to developmental physiologists: they are large (70–80 μm diameter), and when newly released they lack cell walls, although the secretion of alginic acid for wall formation begins immediately after fertilization. They have been used in cell–cell recognition studies, in research on sex attractants, and in polarity studies (Evans *et al.*, 1982), and the *Fucus* egg/zygote is clearly an important model system in developmental biology with the potential for providing answers to fundamental developmental questions far removed from phycology.

In *Fucus* the process of gamete binding, unlike that of pheromonal attraction, is highly species-specific (Bolwell *et al.*, 1977). Only a few sperm bind to eggs (Evans *et al.*, 1982), possibly due to changes in the egg membrane induced by fertilization, or perhaps to a paucity of receptor sites. Specific sugar-binding proteins, lectins, have been implicated in this recognition, and have been shown to bind to fucose, mannose, galactose, and glucose residues (Bolwell *et al.*, 1979). Lectins inhibit fertilization, presumably by blocking sperm receptor sites on the surface of the *Fucus* egg. Proteins which cause species-specific inhibition of fertilization when added exogenously have been isolated from the egg membrane, and

Table 6.2. Chemical structure and distribution of pheromones in the Phaeophyceae. (After Maier & Müller, 1986.)

Structure	Trivial name	Taxa
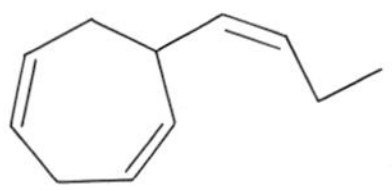	Ectocarpene	*Ectocarpus siliculosus* *E. fasciculatus* *Sphacelaria rigidula* *Adenocystis utricularis*
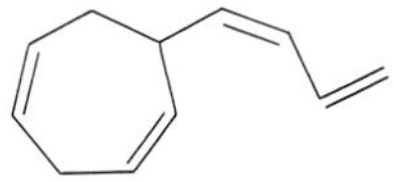	Desmarestene	*Desmarestia aculeata* *D. viridis*
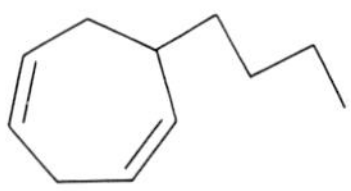	Dictyopterene	*Dictyota dichotoma*
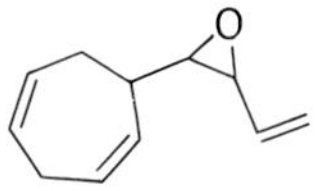 O	Lamoxirene	Laminariaceae Alariaceae Lessoniaceae (29 species)
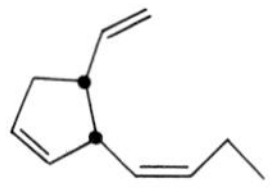	Multifidene	*Cutleria multifida* *Chorda tomentosa*
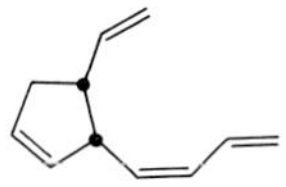	Viridiene	*Syringoderma phinneyi*
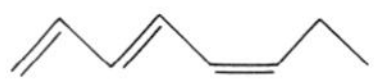	Fucoserratene	*Fucus serratus* *F. vesiculosus* *F. spiralis*
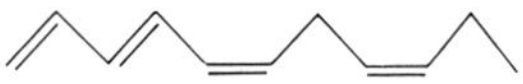	Finavarrene	*Ascophyllum nodosum* *Dictyosiphon foeniculaceus*
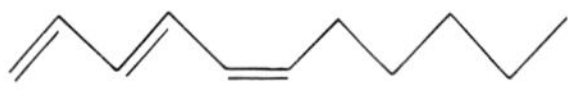	Cystophorene	*Cystophora siliquosa*
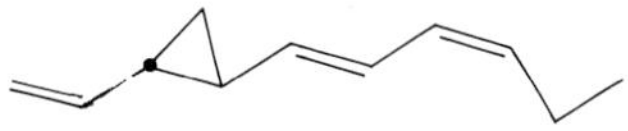	Hormosirene	*Hormosira banksii* *Xiphophora chondrophylla* *X. gladiata* *Durvillaea potatorum* *D. antarctica* *D. willana* *Scytosiphon lomentaria* *Colpomenia peregrina*

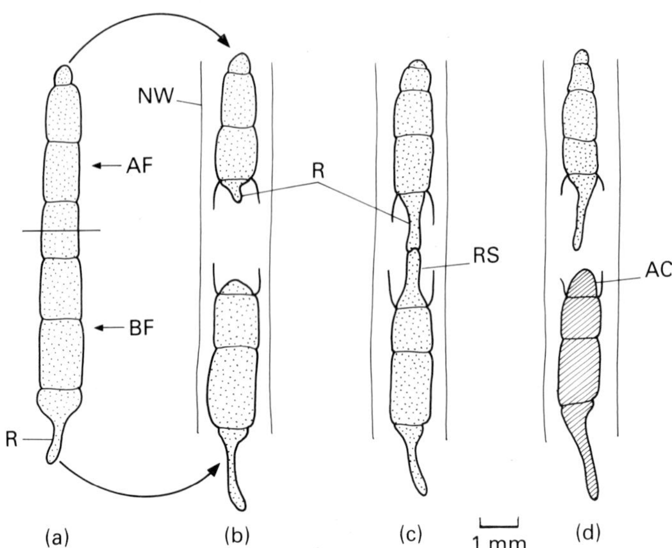

Fig. 6.21 Experimental repair system using *Griffithsia* spp. (a) *Griffithsia* filaments of the same species are decapitated. Basal (BF) and apical (AF) segments are inserted into a cytoplasm-free *Nitella* (NW) cell wall cylinder in (b). Regenerating rhizoid (R) in apical segment induces formation of a repair shoot cell (RS) from the basal segment and fusion occurs in (c). When the apical and basal segments are derived from different species (d) the basal segment regenerates an apical cell (AC) rather than a repair shoot cell and fusion does not occur. (After Buggeln, 1981; from Waaland, 1975.)

carbohydrate-binding proteins have been isolated from *Fucus* sperm (Bolwell *et al.*, 1980).

POLARITY

Polarity is the occurrence of any asymmetrical state within a living system and is usually expressed in terms of differences between two ends of a cell or organ. Newly released *Fucus* eggs are apolar and the first overt sign of polarization occurs several hours after fertilization with the development of a protuberance which elongates to form an attachment rhizoid. At this stage the embryo has become highly polar, with an unequal distribution of organelles and distinct chemical gradients. Strong unilateral irradiance induces the formation of rhizoidal emergence from the shaded side, while weaker irradiance causes the protrusion to arise from 90–135° away from the source (Jaffe, 1958).

Respiration rates increase following fertilization, presumably correlating with the decrease in stored food reserves (Quatrano & Stevens, 1976). The most striking of the post-fertilizational changes are those associated with cell wall formation. Alginic acid secretion occurs within minutes of fertilization, after 60 minutes the cell walls contain equal amounts of alginates and cellulose (Quatrano & Stevens, 1976). During this period internal organelles become asymmetrically distributed prior to nuclear division and cytokinesis (Brawley *et al.*, 1977).

The mature thallus of all multicellular algae displays some kind of apicobasal polarity, presumably initiated by the first division of the spore or zygote. Polarity may be obvious in the morphology or anatomy of the plant, or more subtly expressed in physiological gradients such as chlorophyll content, photosynthetic capacity, respiration, and dark CO_2 fixation along, for example, the blade of *Laminaria saccharina* (Johnston *et al.*, 1977). A clear demonstration that such polarity occurs in single cells is seen in the regeneration of *Griffithsia* (Rhodophyceae; Duffield *et al.*, 1972), in which the basal poles of isolated axial cells produce rhizoids and the apical poles produce axial cells (Fig. 6.21).

The polarity of individual cells is perhaps one manifestation of totipotency whereby a cell retains the capacity to divide, differentiate, and ultimately develop into a complete organism. Isolated protoplasts of the coenocytic *Bryopsis plumosa* (Chlorophyceae; Tatewaki & Nagata, 1970), rhizoidal cells of *Pelvetia wrightii* (Phaeophyceae; Saga *et al.*, 1978), and medullary cells of *Chondrus crispus* (Rhodophyceae; Chen & Taylor, 1978) all show such totipotency. Perhaps the most spectacular example of totipotency is shown in *Laminaria angustata*, where excised single cells form a callus tissue which, on addition of a bacterized extract of the parent plant, develop into sporophytes (Saga *et al.*, 1978).

There are many other instances of apico-basal gradients which occur in multicellular fragments

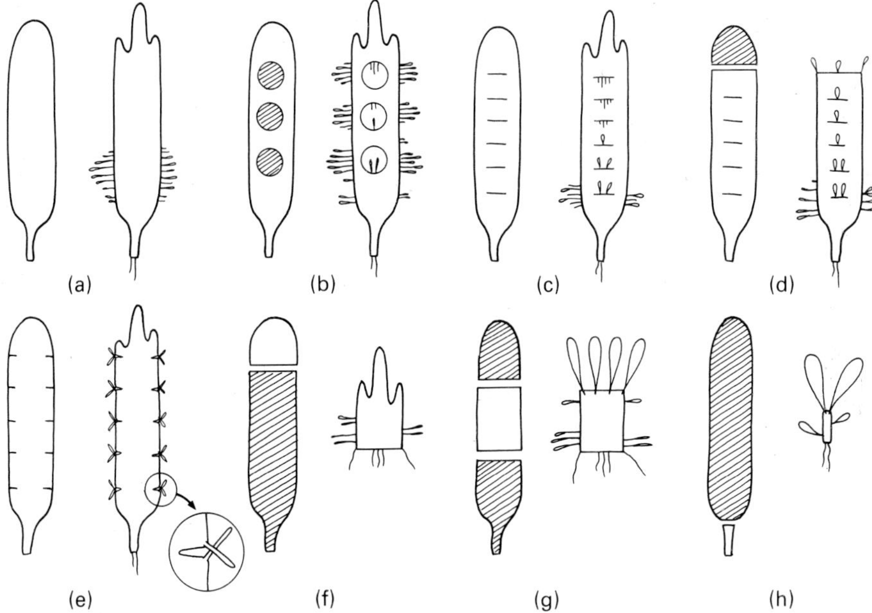

Fig. 6.22 Regeneration effects resulting from experimental wounding or removal of portions of the blade of the red alga *Schottera nicaeense*. The left member of each pair shows the initial condition; lines represent cuts, shaded portions were removed. On the right is the adventive embryony which had developed after 30–90 days. (After Perrone & Felicini, 1972.)

of macroalgae, indicating that physiological differences exist along the thalli from which they were derived. In *Enteromorpha* (Chlorophyceae) more rhizoids are formed from basal than from apical segments (Eaton *et al.*, 1966), while in *Caulerpa prolifera* (Chlorophyceae) excised horizontal shoots or 'rhizomes' regenerate rhizoids with a greater frequency at the apical end (Jacobs, 1964).

APICAL DOMINANCE

Apical dominance effects have been reported in a number of algae, especially in those which have well-developed apical cells or apical meristem regions. The effects primarily involve the control of the growth of lateral branches. Adventive embyrony occurs in *Fucus*. Young plants arise from a wounded algal thallus, and their number and orientation are correlated with position along the apico-basal gradient of the parent thallus (Moss, 1964, 1966). Similar polarity effects are seen in the red alga *Schottera nicaeensis* (Fig. 6.22) (Perrone & Felicini, 1972). Removal of the apical cells increases the growth of the lateral branches in *Apoglossum ruscifolium* (Rhodophyceae; Abelard & L'Hardy-Halos, 1975), *Antithamnion plumula* (Rhodophyceae; L'Hardy-Halos, 1971a); and *Sphacelaria cirrhosa* (Phaeophyceae; Ducreux, 1977). In the Fucales (Phaeophyceae) apical excision leads to the development of lateral branch primordia in *Ascophyllum nodosum* (Moss, 1970), and in *Sargassum muticum* to the development of secondary laterals (Chamberlain, *et al.* 1979), the presence of which in turn controls the development of tertiary laterals.

INTERACTIONS BETWEEN THALLUS CELLS

Morphogenic interactions among the cells of intact algae are also recognized, particularly among the rhodophytes, whose apical growth and filamentous, pseudoparenchymatous construction makes them ideal systems for such studies. Cells of simple filamentous red algae may undergo enormous enlargement during their development (as much as 40 000 times in *Antithamnion plumula;* Dixon 1973), which compares

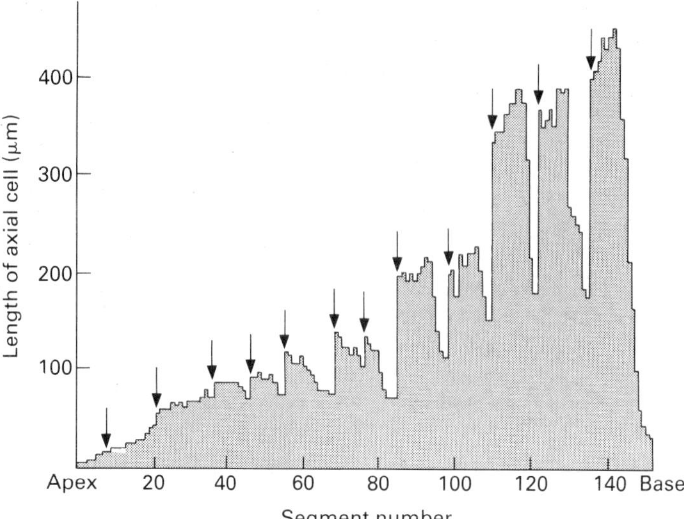

Fig. 6.23 Histogram showing the length of every cell along a principal axis of *Ceramium rubrum*. The arrows indicate the positions of cells at which major branching dichotomies occurred. (After Dixon, 1971.)

with an average cell volume increase in angiosperms of 100–200 fold. Such cells can be easily stained with vital fluorescent dyes to determine the rate of cell enlargement, and this may be coupled with simple observational determination of the rate of cell division (Waaland & Waaland, 1975). These two components, the rate of division and the rate of enlargement, can be readily related to the age of the cell and its position in the plant. In several species of the Florideophyceae the final size of the cell appears to depend not only on age and distance from the apical cell, but also on its location in relation to lateral branches (Dixon, 1971). Axial cells at the bases of lateral branches are frequently much smaller than those closer to the apex of the plant.

HORMONES AND ALGAL DEVELOPMENT

Five groups of hormones, auxins, gibberellins, cytokinins, abscissic acid, and ethylene, are well documented in vascular plants, but the evidence for their occurrence and actions in the algae is largely circumstantial (Buggeln, 1981).

Auxin (indole-3-acetic acid, IAA) is an endogenous regulator of cell expansion in higher plants. In the cell membrane of *Valonia* (Chlorophyceae) it appears to regulate K^+ transport (Zimmermann *et al.*, 1976), but it does not apparently affect wall elasticity (Zimmermann, 1977). It has been detected in natural collections of *Undaria pinnatifida* (Phaeophyceae; Abe *et al.*, 1972), but may well have been produced by epiphytic micro-organisms. In *Alaria esculenta* growth responses due to auxin and non-active analogues could not be distinguished (Buggeln, 1976). Blade growth was inhibited, and was corrected with the inhibition of photosynthesis and structural alterations of chloroplasts (Buggeln & Bal, 1977). Apical dominance in decapitated algae was reported to be restored by the exogenous application of auxin (Moss, 1965; Chamberlain *et al.*, 1979). Growth of bacterized cultures of *Codium fragile* (Chlorophyceae) was enhanced by the addition of auxin, but not by synthetic analogues (Hanisak, 1979). IAA has also been detected in *Caulerpa* (Chlorophyceae) by high-performance liquid chromatography and mass spectroscopy; its concentration of *c.* 1 p.p.m. of fresh material is similar to that reported for higher plants (Jacobs *et al.*, 1985).

Gibberellin-like activity has been demonstrated in extracts from seaweeds (Augier, 1976) and from seawater (Fries, 1973). Inhibition of lateral growths of *Sargassum muticum* (Phaeophyceae) was partially overcome by the addition of exogenous gibberellic acid (Gorham, 1977), but other species so inhibited showed no recovery (Buggeln, 1981). Treatment of the large coenocytic green alga *Caulerpa prolifera* with hormonal levels of gibberellic acid resulted in increased elongation of the rhizome and more frequent intiation of rhizoids, but no effects were observed on the blades (Jacobs & Davis, 1983).

Cytokinin-like activity has been demonstrated in seawater extracts (Pedersen, 1973), and exogenous kinetin was required for the production of upright filaments in axenic cultures of *Ectocarpus fasciculatus* (Phaeophyceae; Pedersen, 1968). Ethylene is reported to be produced by auxin-treated *Porphyra tenera* (Rhodophyceae), *Codium latum* (Chlorophyceae) and *Padina arborescens* (Phaeophyceae; Watanabe & Kondo, 1976), though this could have been produced by the epiphytic micro-organisms. Ethylene has no effect on the elongation of the blades of *Alaria esculenta* (Buggeln, 1981).

The best evidence for the production of an endogenous growth regulator specific to algae is the occurrence of the diffusible glycoprotein 'rhodomorphin' (Watson & Waaland, 1983). This is produced by rhizoids of *Griffithsia* spp. (Rhodophyceae; Waaland, 1975) and induces the apical end of a wounded cell to produce a new shoot which grows towards, and ultimately fuses with, the rhizoid cell.

TROPISMS

Tropisms are directional responses to directional stimuli; while numerous tropic responses are recognized in higher plants, phototropism and geotropism are best known in the algae. Phototropic responses in filamentous algae are reported in the Xanthophyceae, Chlorophyceae, Phaeophyceae, and the Rhodophyceae (Buggeln, 1981). Blue light causes the maximum phototropic response, which is usually positive in the shoots and negative in rhizoids and haptera.

Kataoka (1975a, b) showed that the positive phototropic response in the shoot apices of the coenocytic *Vaucheria geminata* (Xanthophyceae) is a multistage process, inhibited by high light intensities. The process is a growth phenomenon, with the growing point moving to the illuminated side. Branching could also be induced by unilaterally illuminating part of the thallus. Kataoka (1977) concluded that while cyclic AMP was involved in the curvature process auxin was not. The separation of the reception of the stimulus and reponse occurs in the rhizoids of *Boergesenia forbesii* (Chlorophyceae; Ishizawa & Wada, 1979a, b) at 15°C and there is no growth and hence no response to unilateral illumination. When stimulated plants are transferred to darkness at 25°C, growth resumes, and curvature of the rhizoids away from the initial direction of the stimulus occurs. Larger multicellular structures such as the haptera of the kelp *Alaria esculenta* also show negative phototropic responses (Buggeln, 1974). All phototropic responses show the maximum stimulation by blue light of wavelength 430–450 nm, with little effect by light above 500 nm. It is suggested that the receptor pigment may be a carotenoid or riboflavin (Buggeln, 1981).

Geotropism in the algae has been unequivocably demonstrated to occur in the positive response of the rhizoids of *Chara* (Charophyceae), which possess statoliths (Sievers *et al.*, 1979). Experiments involving bilateral illumination suggest that gravity may play a role in the orientation of the rhizoids of *Caulerpa racemosa*

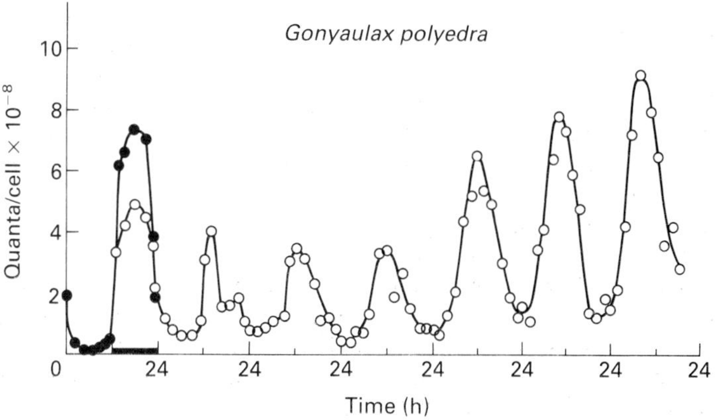

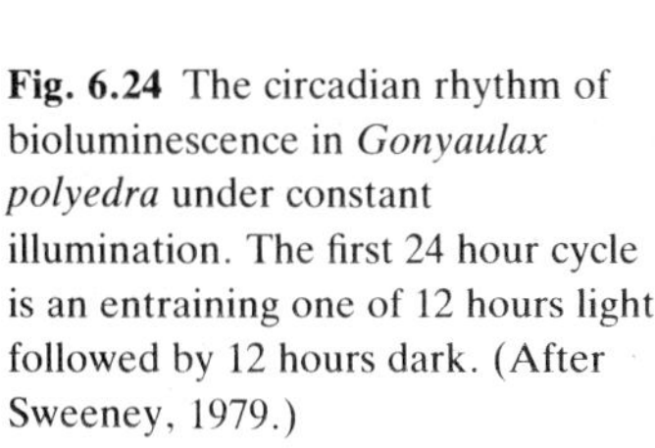
Fig. 6.24 The circadian rhythm of bioluminescence in *Gonyaulax polyedra* under constant illumination. The first 24 hour cycle is an entraining one of 12 hours light followed by 12 hours dark. (After Sweeney, 1979.)

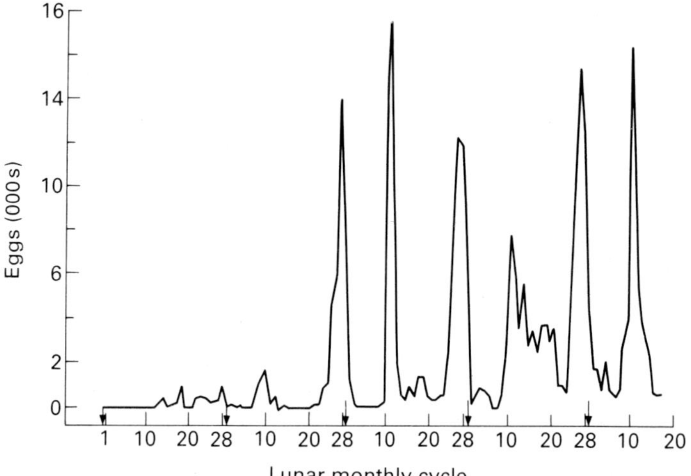

Fig. 6.25 Egg production in laboratory cultures of *Dictyota dichotoma* growing under 14-hour days. The dark period was replaced by dim light on day 1 and day 28 of the first lunar period to simulate conditions of the full moon. (After Müller, 1962.)

(Chlorophyceae), and that its detection is via the accumulation of starch-containing amyloplasts on the lower surfaces of the rhizoids (Matilsky & Jacobs, 1983). It is also suggested that the geotropic response may be mediated by endogenous gibberellin (Jacobs & Davis, 1983).

Contact phenomena have been observed in red algal thalli (L'Hardy-Halos, 1971b) with the production of attachment rhizoids suggesting a thigmatropic response.

ENDOGENOUS RHYTHMS

The algae have evolved in an environment with a diurnal periodicity of energy supply, and the regularity of this light–dark cycle imposes a need for co-ordination among the synthetic and reproductive events of their cells (Chisholm, 1981). Circadian rhythms having a period of approximately 24 hours are absent in the Cyanophyceae, but present in all the eukaryotic algae (Fig. 6.24) (Sweeney, 1983). The precise length of such cycles is dependent on an endogenous self-sustaining clock, which in nature is set by the 24-hour light–dark photoregime. When an alga is released from this entrainment by exposure to constant light the rhythms persist, showing that the circadian clock is independent of the stimulus. The disappearance of rhythms in algae held under constant conditions is due to the loss of synchrony between individuals rather than to a loss of the circadian clock (Chisholm, 1981).

In the algae circadian rhythms are manifested in numerous physiological and ecological events. The two most studied species are *Euglena gracilis* and *Gonyaulax polyedra* (Chisholm, 1981). Circadian rhythms occurring in multicellular algae include cell division in *Giffithsia pacifica* (Rhodophyceae; Waaland & Cleland, 1972), photosynthesis and enzymatic activity in *Spatoglossum pacificum* (Phaeophyceae; Yamada *et al.*, 1979), and reproductive behaviour in *Oedogonium cardiacum* (Chlorophyceae; Ruddat, 1961), *Pseudobryopsis* sp. (Chlorophyceae; Okuda & Tatewaki, 1982) and *Acrochaetium asparagopsis* (Rhodophyceae; Abdel-Rahman, 1982).

In *Gonyaulax* there is evidence for a single clock mechanism controlling all rhythmic processes (Sweeney, 1983). The periodicity of luminescence is due to a rhythm in the concentration of the luciferase protein (Johnson *et al.*, 1984), and protein turnover rates may be implicated in the clock mechanism.

Field observations on intertidal marine algae have shown that some exhibit a periodicity of reproductive activity which coincides with specific periods of the lunar tidal cycle (Tanner, 1981). An endogenous lunar rhythm has been demonstrated in the brown alga *Dictyota dichotoma* (Müller, 1962), which in nature releases gametes at *c.* 14-day intervals to coincide with spring tides (Fig. 6.25). In culture the rhythm can be set by exposing the plants to low-intensity light during a normally dark period. Gamete release occurs at approximately two-week intervals thereafter, suggesting that in nature the entraining factor is the light of the full moon.

CHAPTER 7

Ecology

Introduction

Algae are found in virtually all habitats where photosynthesis can be conducted at a rate sufficient for net production, i.e. growth after losses due to metabolism and predation. They are found from polar regions to hot deserts, from mountain tops to the limits of the photic zone in oceans and lakes. They occur on almost all inorganic substrates, form associations with varying degrees of intimacy with other organisms, and are found free-floating, as plankton, in all but the most temporary water bodies.

Round (1981) was the first phycologist to undertake the daunting task of attempting to provide a unified account of algal ecology in a single book, a labour which can only be compared to that undertaken almost 50 years earlier by Fritsch (1935, 1945), for the structure and reproduction of these organisms. In this chapter the approach that has been taken is functional, and some of the processes involved in ecological interactions are described. These are illustrated by reference to three major ecological groups of algae: (1) phytoplankton, (2) benthic algae, specifically the epilithic seaweeds, and those found associated with unconsolidated sediments and soils, and (3) symbiotic associations. In addition, algae which occur at extremes of temperature, in hot springs and associated with snow and ice, are also mentioned. Many algal ecologists have taken a floristic approach to their subject and have provided detailed lists of species found in a wide variety of habitats and locations. Space does not permit the inclusion of such data in this work, but this should not be regarded as a disapproval of such methods; on the contrary, as stated by Round (1981), to ignore lists of species is to deny the complexity of algal communities. Both experimental and descriptive approaches are necessary for a greater understanding of algal ecology, and works which combine both approaches are to be especially welcomed.

One unifying and overwhelmingly important feature of the vast majority of algae is that they are primary producers; they convert the physical energy of the sun's radiation into chemical energy by the process of photosynthesis. As primary producers they provide much of the basis for the food web in aquatic environments, either for direct consumption or via detrital systems. The importance of this production can be seen when it is realized that over 70% of the surface of the planet is covered by water. Many algal communities are extremely productive and rival many terrestrial systems (Fig. 7.1).

Phytoplankton

The phytoplankton are the most widespread and extensively studied of all ecological groups of algae. Phytoplankton occur in virtually all bodies of water, where they float freely and largely involuntarily. While most are independent of shores and benthos, some shallow-water species have adopted life history strategies involving benthic resting stages (Fryxell, 1983). All algal divisions except the Rhodophyceae, Charophyceae, and Phaeophyceae contribute species to the phytoplankton flora, and in coastal zones the zoospores produced from benthic phaeophytes, and chlorophytes, may at least temporarily add to the plankton. A number of excellent books and reviews pertaining to the ecology of phytoplankton have recently appeared; these include Sournia (1978), Morris (1980), Platt (1981), Round (1981), and Reynolds (1984).

Phytoplanktonic algae are mainly unicellular, though many colonial and filamentous forms occur, especially in fresh waters. In size they range from small flagellates and cocci < 5 μm in diameter to colonial forms such as *Volvox* at 500 μm. They are classified by size into two groups, the net plankton, retained by nets of approximately 64 μm mesh size, and the smaller nanoplankton. There is no universal agreement on the size used to separate the two groups. Zeitzschel (1978) defines nanoplankton as < 20 μm diameter, and Manny (1972) as < 10 μm. Cell size and the ratio of surface area to volume have important implications for buoyancy, nutrient

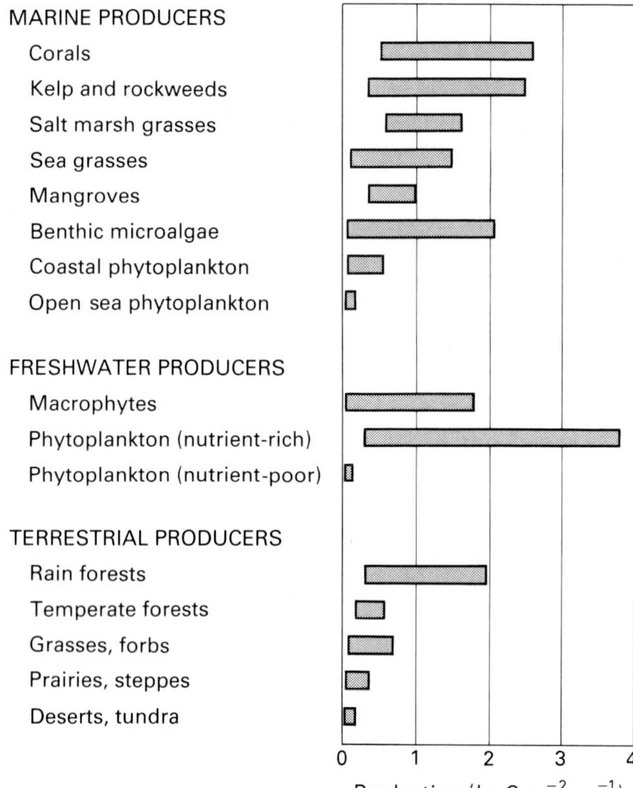

Fig. 7.1 Comparison of annual net primary production rates in marine, freshwater, and terrestrial environments. (After Valiela, 1984.)

absorption, and resistance to predation. Large free-floating macroalgae, such as the *Sargassum* of the Sargasso Sea, are by definition also planktonic, but are usually excluded due to their size and affinities with the benthic algal flora.

In the oceans the net algae are primarily diatoms and dinoflagellates, while the nanoplankton is composed of small cyanophytes and small flagellates, principally prymnesiophytes, many of which are coccolithophorids. Diatoms and dinoflagellates are also important in fresh waters, but members of the other algal classes, particularly Cyanophyceae, Chlorophyceae, and Chrysophyceae, may become dominant. In oligotrophic (nutrient-poor) waters, chrysophytes and desmids predominate, while in eutrophic (nutrient rich) areas cyanophytes and diatoms are usually found, and euglenophytes may be especially abundant in small nutrient-enriched pools (Round, 1981)

METHODOLOGIES

Studies of phytoplankton generally fall into two categories, the first where the whole assemblage is treated together, and the second where individual species are identified and treated separately in subsequent analyses. The first is particularly useful in determining standing stocks of phytoplankton, which may be reported as biomass, total carbon, organic nitrogen, chlorophyll, ATP, or a plethora of other values. Production of the total phytoplankton assemblage may be estimated by measurements based on changes in the above values, or by methods based on the determination of photosynthetic rates utilizing ^{14}C methods (Parsons *et al.*, 1984).

When details of the contributions of individual species are required, microscopic examination and counting methods must be employed (Sournia, 1978). Autoradiographic methods allow the assessment of productivity of individual cells and thus the determination of the role of separate species in the productivity of phytoplankton populations (Knoechel & Kalff, 1976).

The use of these wide varieties of methods has contributed greatly to our understanding of phytoplankton ecology, but there has been a price to pay in that the lack of standardization frequently makes comparison between studies difficult if not impossible.

FLOTATION AND BUOYANCY

For a population to survive phytoplankton must maintain themselves in the water column where irradiances will support photosynthetic levels sufficient for net growth and reproduction. The tendency to sink should not, however, be regarded as necessarily disadvantageous, as movement in the water column enhances nutrient uptake and may also be a mechanism for avoiding predators and high, potentially damaging, levels of irradiance at the water surface.

All algae, with the exception of the gas vesicle-forming Cyanophyceae, have densities greater than water and thus must inevitably sink. For the motile species this is not a problem as they can actively swim towards the light, but many others such as diatoms and coccoid forms of various divisions cannot. Most cytoplasmic components have densities higher than water (nominally 1.0008 g/ml for fresh water or 1.027 g/ml for sea water). The silica walls of diatoms, one of the most successful phytoplankton groups, have a density of 2.6. Only lipids at 0.86 are less dense than water and, while they may contribute to the reduction of cell density, cannot produce neutral buoyancy. Smayda (1970) calculated that the increase in the lipid content of a marine diatom from 9 to 40% would reduce the overall density of the cell by 3.5% to 1.15 (still considerably above the density of sea water), but would result in a 25% reduction in the sinking rate. Usually lipids account for 2–20% of algal dry weight, but they are also accumulated in similar concentrations in benthic species, suggesting that their aid in the reduction of sinking may be incidental (Reynolds, 1984). Ionic regulation, in which heavier ions are replaced by lighter ones, may have some effect in reducing the density of marine species. Anderson and Sweeney (1978) showed that the density of the vacuolar sap of the diatom *Ditylum brightwelli* may be reduced by the selective accumulation of K^+ and Na^+ ions, which replace the heavier divalent ions. The dinoflagellate *Pyrocystis noctiluca* also showed an ability to reduce the concentration of the heavy SO_4^{2-}, Ca^{2+} and Mg^{2+} ions sufficiently to become positively buoyant (Kahn & Swift, 1978).

Planktonic members of the Cyanophyceae possess flotation devices termed gas vacuoles (Walsby, 1978), which consist of a number of closely packed cylinders, or vesicles, up to 1 μm in length by 70 nm diameter and bounded by a proteinaceous membrane. When exposed to sufficient external pressure, in nature brought about by the turgor of the cell, the vesicles collapse and the cell looses buoyancy.

When *Anabaena flos-aquae* is grown under low irradiance it produces gas vacuoles and becomes buoyant, but when transferred to high irradiances the vacuolation decreases and the buoyancy is lost. During photosynthesis cell turgor pressure may rise by several atmospheres due to the accumulation of photosynthate and the uptake of K^+ ions (Dinsdale & Walsby, 1972; Allison & Walsby, 1981). In this manner planktonic cyanophytes may regulate their position in the water column to maintain themselves at depths most favourable for growth (Reynolds & Walsby, 1975). They thus have considerable advantages over other planktonic algae in being able to form blooms in a stable water column. Gas vacuolate species are not common in the sea, where the usual planktonic adaptations are to a turbulent environment (Margalef, 1978), which would reduce the effectiveness of such flotation devices. One notable exception is the colonial filamentous *Trichodesmium*, which forms blooms in tropical seas (Fogg, 1982). The pressure necessary for the collapse of its gas vesicles is of the order of 37 atmospheres, 8–10 times that required for *Anabaena* and far beyond any turgor changes the cell may experience. *Trichodesmium* thus maintains a positive buoyancy, but does not possess a regulatory capacity. It passively ascends in the water column and relies on turbulent mixing to take it down into nutrient-richer deeper waters, where its strong gas vesicles can withstand the external hydrostatic pressure. This strategy is essentially the opposite of that shown by non-buoyant phytoplankton, which rely on turbulent mixing to maintain themselves in the upper part of the water column.

Thus, most phytoplankton cells are maintained in the water column by currents. Convection cells, wind-induced rotations in surface waters (Fig. 7.2) (Langmuir, 1938), are perhaps the most important of these movements, where downwelling occurs at the lines of convergence of the cells and is replaced by upwelling at the divergence between the cells.

The sinking rate of a phytoplankton cell is directly dependent on size and, if all other variables remain constant, a large cell will sink faster than a small one (Fig. 7.3). Microscopic size, coupled with mechanisms to reduce cell density, is the most important adaptation to a reduction in sinking rate. If the volume and mass of a cell remain constant, any morphological changes which increase its surface area will increase friction and enhance the potential for entrainment in a turbulent water mass. Walsby and Xypolyta (1977) experimentally demonstrated this by enzymatically removing the chitinous fibres from the diatom *Thalassiosira fluviatilis* and showing that the sinking rate was doubled. The physiological status of diatoms can also affect their sinking rate, and dead cells, or even senescent ones, sink up to twice as fast as viable cells. This may be an adaptation allowing nutrient-exhausted cells to enter richer, deeper waters, but the mechanism is not understood (Reynolds, 1984). It does not depend on the treatment of the cells prior to death, and there is no evidence for the formation of vital flotation devices such as gas vacuoles.

LIGHT

Many species of phytoplankton appear to be able to modify their photosynthetic response to ambient irradiances. Thus, a cell which is in the lower part of the photic zone will increase the concentration of its photosynthetic pigments and reduce the level of light at which photosynthesis becomes saturated. This increase in chlorophyll concentration is associated with an increase in the number of thylakoids per cell, and the size of the photosynthetic unit increases with decreasing irradiance (Perry *et al.*, 1981). Although shade-adapted algae may absorb light more efficiently, they cannot utilize it with any increased efficiency, and the yield per photon of light absorbed remains constant. In contrast, in surface or sun-adapted plankton, chlorophyll levels are reduced and P_{max} is increased, as evidenced by an increase in the levels of the carboxylating enzymes. The changes in chlorophyll concentration may be brought about by the increase in the proportion of blue light which occurs with increasing depth, rather than simply be due to a decrease in the total irradiance (Vesk & Jeffrey, 1977). Except for some members of the Cyanophyceae, and possibly some cryptomonads, no complementary chromatic adaptation has been observed (Chapter 6). Many cyanophytes form extensive surface blooms, and in *Microcystis* the carotenoid concentration increases to enhance light harvesting and to filter out the potentially dangerous levels of ultraviolet light (Paerl *et al.*, 1983).

One interesting question in phytoplankton ecology is whether a permanent-shade flora

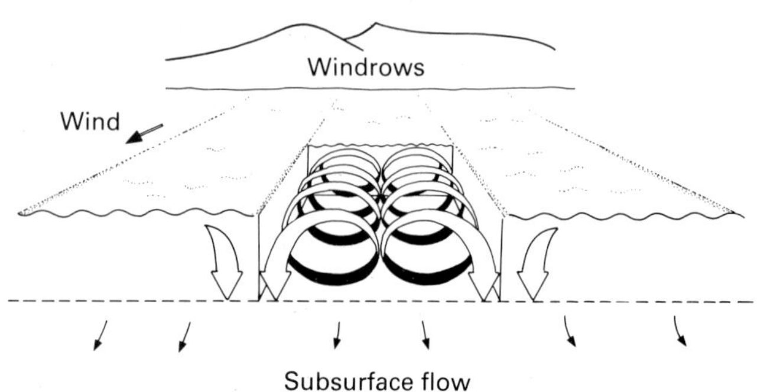

Fig. 7.2 Diagram of wind-induced surface flow resulting in Langmuir circulation. (After Reynolds, 1984.)

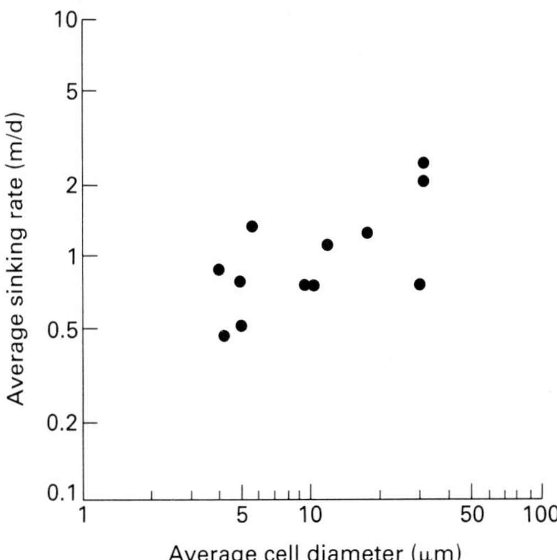

Fig. 7.3 Relationship between sinking rate and cell diameters in chain-forming marine diatoms. (After Reynolds, 1984.)

occurs in the deeper layers of the oceans. Many species have been reported as occurring preferentially or exclusively at depths of 100 m or below, and Sournia (1982) has collated the available data to produce a list of some 20 species in 12 genera and five classes, covering virtually the whole spectrum of phytoplankton taxa, size, and shapes. Although a definite algal community seems to occur under these low-light conditions, no common morphological adaptation is evident. Still to be answered is the question of whether the species actually grow at these depths or merely survive there. The survival of microalgae without light, in the laboratory, is well documented (Antia, 1976), and physiological adaptation to low light levels also occurs (Falkowski, 1981; Geider *et al.*, 1986). *Phaeodactylum tricornutum*, a non-shade diatom, can survive and maintain a basic metabolism, at irradiances below 1 μmol cm^{-2} sec^{-1}, but is capable of rapid increases in photosynthetic rate when there is an increase in the amount of available light.

In addition to the photosynthetically active region of the spectrum, sunlight contains wavelengths about 700 nm, the infrared region, which are absorbed by water and have a heating effect. The magnitude of the warming depends upon the size of the body of water and its ambient temperature. Continued surface heating decreases the water density and results in the horizontal stratification of the water body into two (or more) layers, a warm surface layer and a deeper, cooler layer (Fig. 7.4). The zone between layers is known as the thermocline, in which the temperature decreases abruptly and the subsequent density differences prevent exchange of waters between the deep (hypolimnion) and shallow (epilimnion) layers, although circulation occurs within the layers (Fig. 7.5). In coastal zones, density stratification may also be caused by reduction of surface salinity brought about by freshwater run-off from the land, or from the melting of sea ice. The boundaries between such density- or salinity-induced water layers are known as pycnoclines or haloclines. Stratification has important ecological implications, as the surface waters may become depleted in nutrients, while the deeper waters may remain nutrient-rich but out of the reach of the algal cells above the thermocline. Shallow water thermoclines develop only under calm conditions, when turbulent mixing of the surface by wind action does not occur. Once they are at a depth beyond the effects of wind action they may remain until the surface waters cool sufficiently to sink and mix, and in temperate regions they are largely seasonal, occurring in the summer. In higher latitudes, at least in the oceans, the surface waters may remain cold throughout the year and very little thermocline development is found. In contrast, in the tropics stratification is a year-round phenomenon.

NUTRIENTS

Phytoplankton require a variety of both inorganic and organic nutrients, which are present in dilute solution at spatially and temporally varying concentrations. Given the universal importance of such elements in algal metabolism, their availability has profound ecological importance, as all algae require the same macronutrients (the requirement for silicate by diatoms and some flagellates is a notable exception).

Phosphates, inorganic nitrogen compounds, and silicates (for diatoms) are the nutrients required in the greatest abundance (Chapter 6), and are also those whose effects on phytoplankton populations have received the greatest attention. In the marine environment, nitrogen seems to be the macronutrient which most often limits

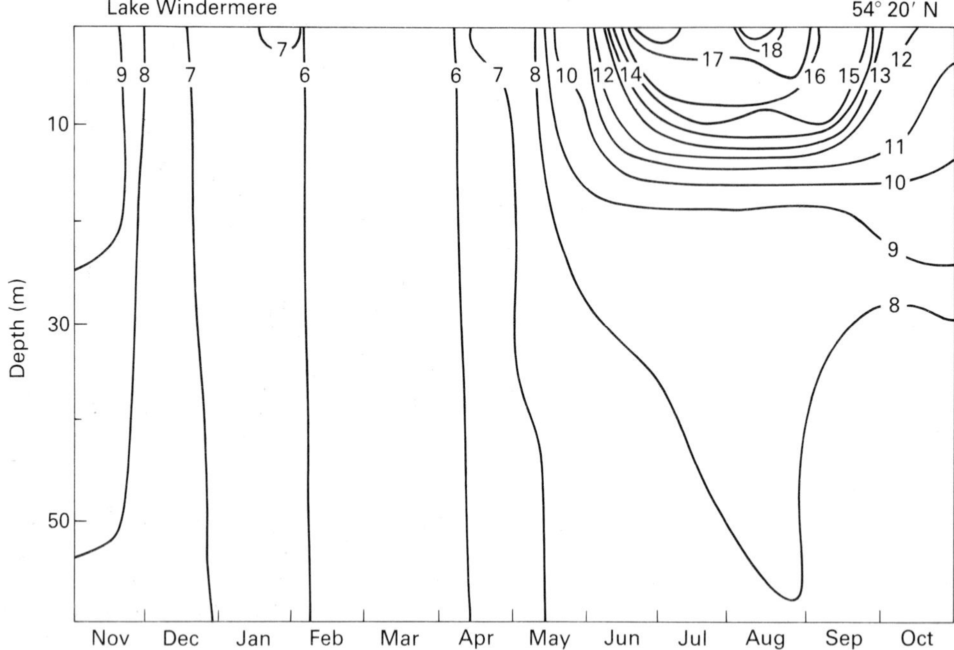

Fig. 7.4 Depth–time diagram of thermal stratification in a temperate lake showing effects of summer thermocline and temperature stratification in the summer months. (After Jenkin, 1942.)

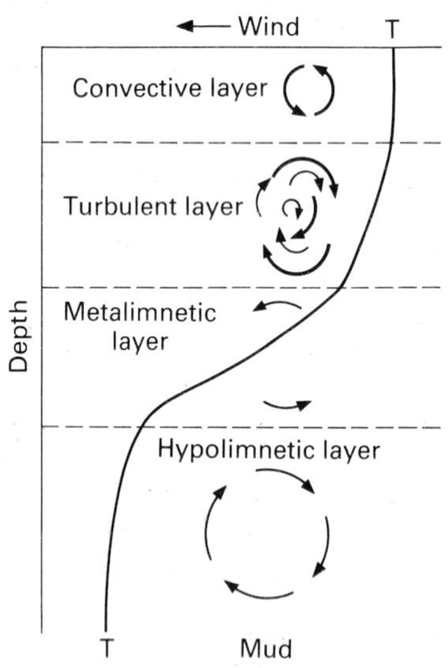

Fig. 7.5 Diagram of vertical circulation patterns showing separation above and below the thermocline. T, temperature profile with depth. (After Reynolds, 1984.)

growth (Ryther & Dunstan, 1971), while whole-lake enrichment experiments implicate phosphorus in this role in fresh waters (Fig. 7.6) (Schindler, 1974, 1977; *see also* Chapter 9).

It is the surface of the phytoplankton cell which permits the exchange of energy and nutrients, and for any given shape of cell the surface to volume ratio (S/V) will decrease with increasing cell volume. It would be expected therefore that phytoplankton growth rate and S/V ratios would be positively correlated, and this has been shown in both nature and laboratory cultures (Sournia, 1981). Parsons and Takahashi (1973) compared the effects of cell size on growth rates using the diatom *Ditylum brightwelli* and the nanoplanktonic coccolithophorid *Emiliania huxleyi*. Growth rates of the large cells exceeded those of the small cells only at high light levels, high nutrient concentrations, and during water column instability. They suggest that this explains why nanoplankton quantitatively overwhelms net plankton in most marine systems. The S/V ratio has also been shown to increase in nutrient-defi-

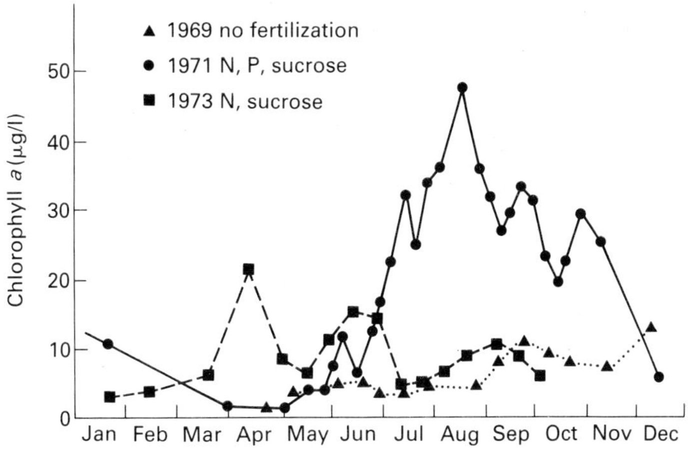

Fig. 7.6 Effects of experimental fertilization of a lake on the phytoplankton biomass (measured as chlorophyll). In 1969 there was no fertilization; in 1970 NH_4, PO_4 and sucrose were added; and in 1973 NH_4 and sucrose were added. The peak in 1971 shows that of the three fertilizers phosphate was limiting. (After Schindler, 1974.)

cient media (Harrison *et al.*, 1977), suggesting that small algae with high S/V ratios would be favoured in oligotrophic waters. In fresh waters this is supported by the observations of Watson and Kalff (1981), who showed an increasing preponderance of nanoplankton along a decreasing gradient of nutrient concentration. Large algae have, however, the potential of storing more nutrients. Plankton such as dinoflagellates, capable of migrating at night to the nutrient-rich waters of the hypolimnion and returning to the shallow photic waters by day, have obvious ecological advantages (Dugdale & Goering, 1967).

While the addition of nutrients may increase both the growth rates and the final biomass of a phytoplankton population, there is some evidence that nutrients are not always limiting to the growth of phytoplankton, even when they are at low concentration and the algal biomass is also low. It has been found that natural populations of phytoplankton do not show the extreme departures from the idealized Redfield ratio of C:N:P typical of severe nutrient limitation (Chapter 6) and may be near to their maximum potential growth (Goldman *et al.* 1979). Yentsch *et al.* (1977) showed that phytoplankton grown under conditions of nitrogen deprivation showed rapid uptake of bicarbonate ions in the dark when provided with a supply of ammonium ions. They used this technique to test the nitrogen limitation of inshore phytoplankton populations in the Gulf of Maine. Three blooms occurred, *Chaetoceros* in April, *Skeletonema* in July, and *Gonyaulax* in September; nitrogen deficiency was only found during the bloom periods, suggesting that populations occurring between blooms were in equilibrium with recycled nitrogen, and that blooms occurred during periods of upwelling, which fetched additional nitrogen into the surface waters.

There is thus a paradox in which it appears that the population as a whole, judged in terms of final biomass, is nutrient-limited, but the individual cells, judged on the concentrations of nutrients in their internal pools, are not. Maestrini and Bonin (1981b) suggest that the concept of nutrient-unlimited growth should be restricted to those species which have the ability of rapid assimilation and are able to take advantage of transient pulses of nutrients. Such nutrient-rich microniches may occur due to animal excretion or to the remineralization of dead organisms (McCarthy & Goldman, 1979).

Many phytoplankton possess the ability to take up organic compounds in the dark (Droop, 1974). Algae from the undersurface of Antarctic sea ice are reported to assimilate amino acids at ambient concentrations, but these provide less than 1% of the carbon obtained by photosynthetic fixation (Palmisano *et al.*, 1985.) The most convincing evidence for ecologically significant uptake of organic substrates, however, is for photoheterotrophy rather than dark uptake. Light stimulated the uptake of acetate by phytoplankton in the lower regions of the photic zone of Lake Tahoe (California/Nevada), but had no effect on uptake by algae higher in the water column. In this photoheterotrophic community acetate accounted for 0.25–2.5 × the carbon uptake from inorganic sources (Vincent & Goldman, 1980). Uptake of particulate organics by potentially photosynthetic algae has also been reported, and some species of the freshwater chrysophytes *Dinobryon* and *Uroglena* are capable of ingesting bacteria at rates comparable to those recorded for non-photosynthetic flagellates (Bird & Kalff, 1986). In the deeper waters the *Dinobryon* obtained at least 50% of its carbon by bacterivory, and removed more bacteria from the water column than the crustacean, rotifer, and ciliate communities combined.

GRAZING

Losses from phytoplankton populations are even more difficult to quantify than increases. Cells may permanently sink from the water column, and others may suffer from parasitism or microbial attack (Shilo, 1971; Youngman *et al.*, 1976), but it seems generally agreed that losses to predation (Fig. 7.7) are quantitatively the most important (Frost, 1980, Crumpton & Wetzel 1982). The most obvious of these predators are the zooplankton, and in the oceans copepod grazing has been the most extensively studied. Other groups of marine grazers include the protozoa, mainly ciliates (Fenchel, 1980), and tintinnids (Capriulo & Carpenter, 1980), as well as appendicularians (Alldredge, 1981), salps, and pteropods (Silver & Bruland, 1981). Freshwater grazers are principally crustaceans and rotifers, though protozoa (Canter, 1973) and members of the Chytridomycota (Canter & Jaworski, 1978) may have significant effects on some diatom populations.

Any generalizations on the effect of algal cell size on predation appear hazardous. Kalff and Knoechel (1978) suggest the best minimizers of grazing loss in oligotrophic fresh waters will be large, such as dinoflagellates and colonial species. It is frequently reported that smaller species are especially susceptible to grazing. Porter (1973) showed that zooplankton grazing decreased the small cells in a population and allowed larger algae such as desmids, diatoms, and dinoflagellates to develop. Both Frost (1980) and Sournia (1981) stress that no size of class or group of algae should be considered immune from grazing, and a large proportion of any phytoplankton biomass appears available to grazers, though

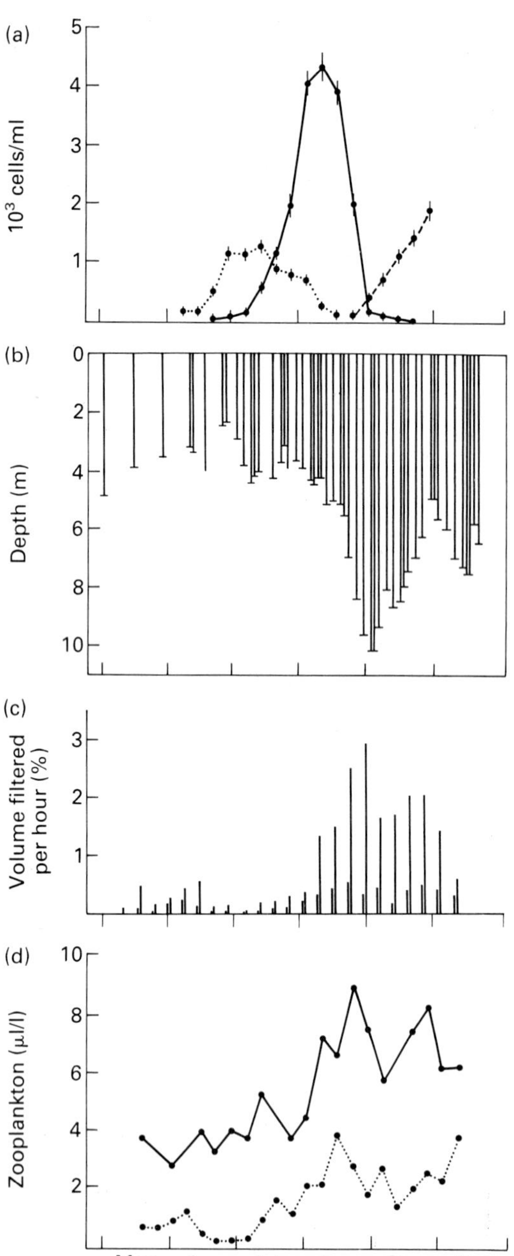

Fig. 7.7 (a) Population curves for *Cyclotella michiganiana* (...), *Cyclotella comensis* (–), and *Sphaerocystis schroeteri* (__). (b) Transparency (Sechi depth in metres). (c) Grazing rates. The first of each pair of vertical bars is daytime rate, the second night-time. (d) Daytime (...) and night-time (–) zooplankton biomass. Note mid-August data for correlations of decrease in algal biomass, increase in transparency, increase in filtering rates, and increase in zooplankton biomass. (After Crumpton & Wetzel, 1982.)

certain cyanophytes may be less palatable than other groups (Arnold, 1971; Porter, 1973). It has long been known that there is a negative correlation between phytoplankton and zooplankton density and that the distribution of plankton species is uneven; grazers are one of the major factors responsible for this patchiness; for example, Cushing and Vucetic (1963) followed the development of a patch of the copepod *Calanus finmarchicus* for three months and found that the phytoplankton declined while inorganic nutrients remained plentiful.

It should not be assumed that all the organic material ingested is incorporated by the grazers. In calanoid copepods about 20–30% of the organic carbon of the phytoplankton is passed in the faeces (Eppley & Peterson, 1979). Some tropical, algivorous, freshwater fish assimilated on average only 43% of the carbon ingested (Moriarty & Moriarty, 1973). In addition to allowing organic material to enter the higher trophic levels, grazing is a first step in the remineralization of inorganic nutrients to support the growth of other algae. Grazing may actually be beneficial to a species. Colonies of the freshwater green alga *Sphaerocystis schroeteri* are only partially disrupted by passage through the gut of *Daphnia magna*, and most emerge intact and in viable condition. During passage, they take up phosphates from the gut, and these stimulate photosynthesis and division in the defecated cells. This enhanced algal growth can compensate for minor losses to the population caused by the grazing damage (Porter, 1976).

POPULATION DYNAMICS

Phytoplankton are rarely found uniformly distributed in any water body; considerable horizontal patchiness occurs as the rule rather than the exception, and this may occur on a variety of scales, from less than millimetres to tens or even hundreds of kilometres (Platt & Denham, 1980). Differences in vertical distribution also occur and both vertical and horizontal distributions may change with time. These temporal changes are also affected by the scale of measurement; they may be diel, seasonal, or long-term changes occurring over many years. A phytoplankton assemblage must therefore be thought of as a three-dimensional patch, whose boundaries and species composition change with time.

Diel fluctuations in algal cell biology are well documented (Chisholm, 1981; Sweeney, 1983) and many, while entrained by the light–dark cycle, are endogenous (Chapter 6). Such cycles include cell division, vertical migration, photosynthetic capacity, bioluminescence, and nutrient uptake. These circadian rhythms ensure that the cell is at the optimum physiological state to deal with the changing environmental conditions that are encountered during a 24-hour period, and as not all species show the same rhythms this could be a mechanism for reducing interspecific competition.

Happey-Wood (1976) has investigated vertical migrations of several phytoflagellates (including *Ochromonas, Chroomonas, Mallomonas, Cryptomonas*) over a 24-hour period in a freshwater pond (Fig. 7.8). The species moved down the water column in the dark and returned to the surface during the day. Under clear skies they occurred in the subsurface waters, but were found on the surface under cloudy conditions. The non-motile green alga *Oocystis* served as a control, showing that the movements were active and not merely due to circulation in the water column.

Seasonal changes, particularly in temperate lakes, are among the most studied aspects of phytoplankton ecology. Temperature and light are both low in winter, and though nutrients may be present in abundance the standing crop of phytoplankton and its productivity are also low. Increasing irradiance in the spring brings about a surge in growth and, depending on the stability of the water column, may lead to a bloom. This is further enhanced in the surface layers by the development of a thermocline. The bloom continues until predation and the depletion of nutrients brings about a summer decline in abundance. The species composition at any time depends on many factors, perhaps most importantly the basal nutrient status of the waters. In general, there is a succession from small to large species, which is predicted on theoretical grounds (Margalef, 1978). Spring blooms are dominated by diatoms, and as these are consumed by the grazers they are replaced in the summer by larger dinoflagellates and green flagellates. In eutrophic waters, cyanophyte blooms may develop in the warmer periods of the late summer. A second bloom of diatoms may appear in the autumn as the surface waters cool and the disappearence of the thermocline allows mixing of the surface and nutrient-rich deeper waters (Fig. 7.9).

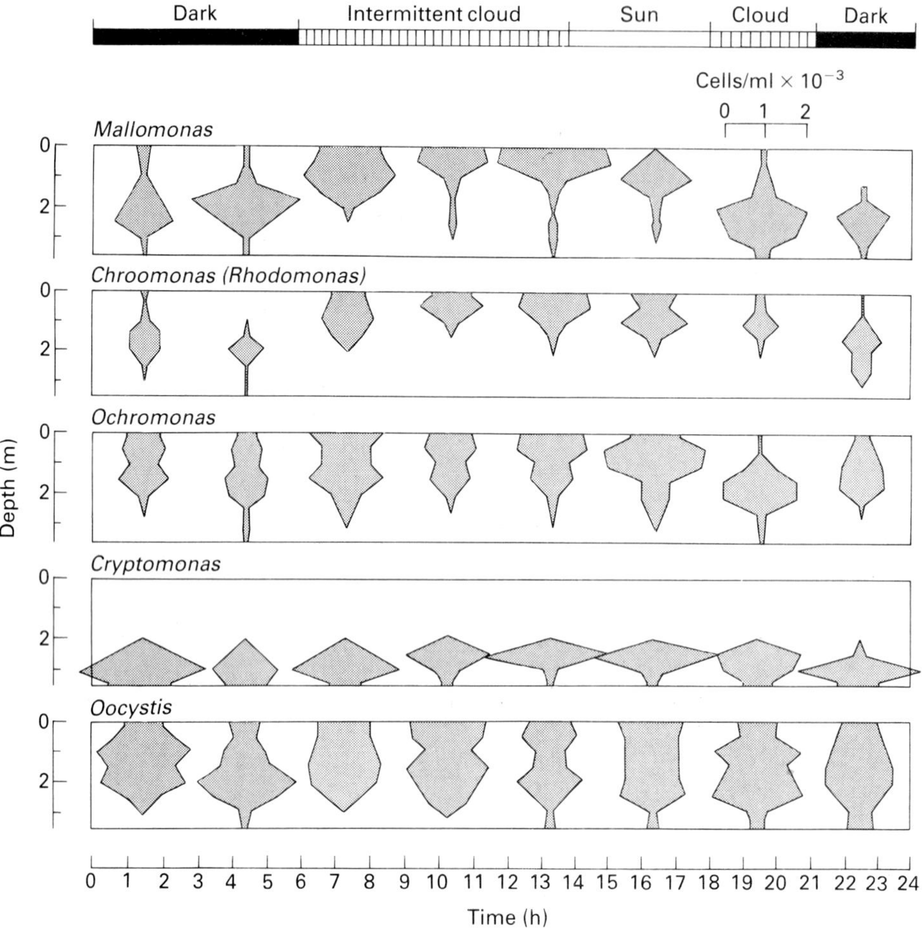

Fig. 7.8 Vertical distribution of phytoplankton in a pond over a 24-hour period. Top bar shows sky conditions. (After Happey-Wood, 1976.)

High-latitude lakes differ from temperate lakes in their extended period of ice cover. There are lakes in polar regions which are permanently frozen but support an algal flora (Baker, 1967). In most, however, maximum growth occurs immediately following the thaw, and the species composition of such lakes is extremely variable (Round, 1981). Species succession appears correlated with the length of the ice-free period, and there may be but a single summer bloom (Fig. 7.10).

In contrast, tropical lakes show less seasonal variation than temperate ones. Many are eutrophic due to the rapid remineralization occurring at the higher water temperatures. One of the interesting features of the plankton of such lakes is the occurrence of pennate diatom genera such as *Nitzschia*, which are predominantly benthic in more temperate waters (Richardson, 1968). Tropical lakes in general appear to support a higher biomass of algae, but show less species diversity than temperate ones. Lewis (1978) reported little seasonal change and a dominance of members of the Chlorophyceae in lakes in the Philippines. In contrast, Lake George, Uganda, one of the most intensively studied of tropical lakes, is dominated by cyanophytes, which make up 70–80% of the biomass. Green algae are

Fig. 7.9 Seasonal changes in algal biomass measured as chlorophyll concentration in a temperate lake. Thin line, 1961; thick line, 1963. Note spring and autumn peaks. (After Szczepansky, 1966.)

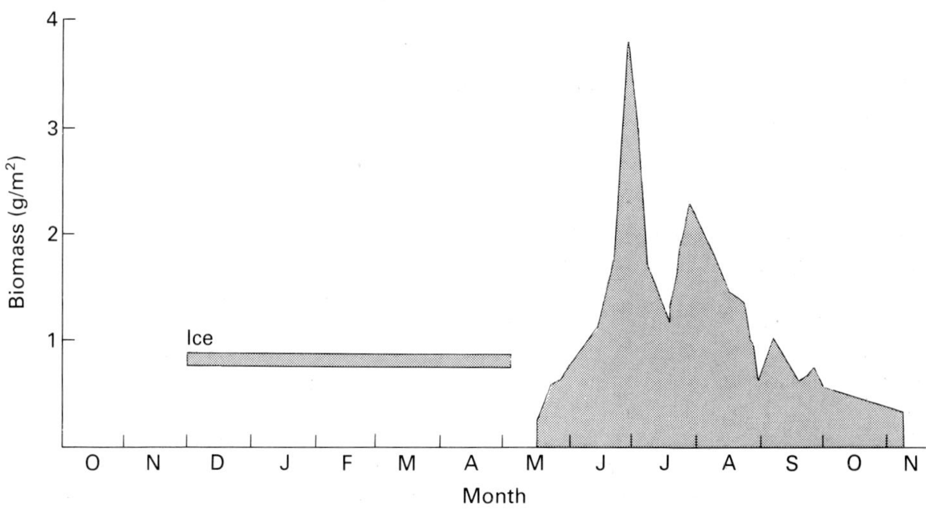

Fig. 7.10 Seasonal changes in biomass in a high-latitude lake which is ice-covered for five months. (After Ilmavirta & Kotimaa, 1974.)

well represented in the flora, but diatoms are relatively rare. There is only a twofold annual variation in biomass, as determined by chlorophyll levels, with slight peaks occurring during the periods of maximum rainfall (Fig. 7.11) (Ganf, 1974).

Smayda (1980) has reviewed phytoplankton distribution and species succession in the marine environment. As with the freshwater environment, the largest number of studies have been made in temperate regions, particularly in the inshore. In general, diatoms dominate in the colder, nutrient-rich waters, and coccolithophorids in the warmer oligotrophic waters, while dinoflagellates appear to be somewhat intermediate in environmental preference. A major ecological distinction can usually be made between the phytoplankton flora of coastal (neritic) and oceanic waters, though the environmental conditions responsible for these differences remain unclear. Coastal waters are undoubtedly influenced by run-off from the land, which may be richer in nutrients. In addition, coastal plankton are more likely than those of the open seas to have benthic stages in their life cycles.

With large bodies of water it is important to distinguish between a true species succession, which is the change in species composition within a given water mass, and changes due to the replacement of one water mass, with its characteristic flora, by another supporting a different floristic assemblage. A good example of a true species succession is reported for Norwegian coastal waters (Braarud *et al.*, 1958) in which a spring diatom bloom was replaced by a less diverse and smaller population of euglenoids, then replaced, in turn, by the summer maximum of dinoflagellates and coccolithophorids (Fig. 7.12). Coastal regions, however, show great variability in the specific composition of their phytoplankton flora. Long Island Sound and Narragansett Bay on the eastern seaboard of the U.S.A. show seasonal successions different from those reported from Norway or even from the adjacent Gulf of Maine (Smayda, 1981). The phytoplankton population was larger and bloomed more frequently, rather than exhibiting distinct spring and summer–autumn blooms (Fig. 7.13). There are fewer important species, and diatoms, particularly *Skeletonema costatum*, are continuously domi-

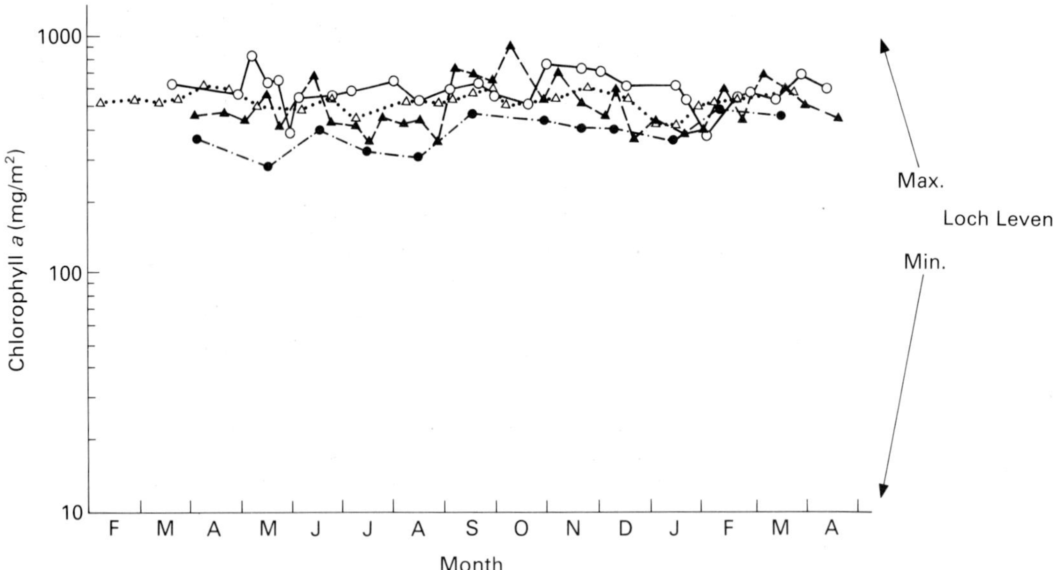

Fig. 7.11 Seasonal changes in algal biomass in a tropical lake (Lake George) measured as chlorophyll concentration. Each line is for a year's data between 1967 and 1971; there is little seasonal change in biomass. For contrast the maximum and minimum chlorophyll values for the same period for a temperate lake (Loch Leven) over the same period are also presented. (After Ganf, 1974.)

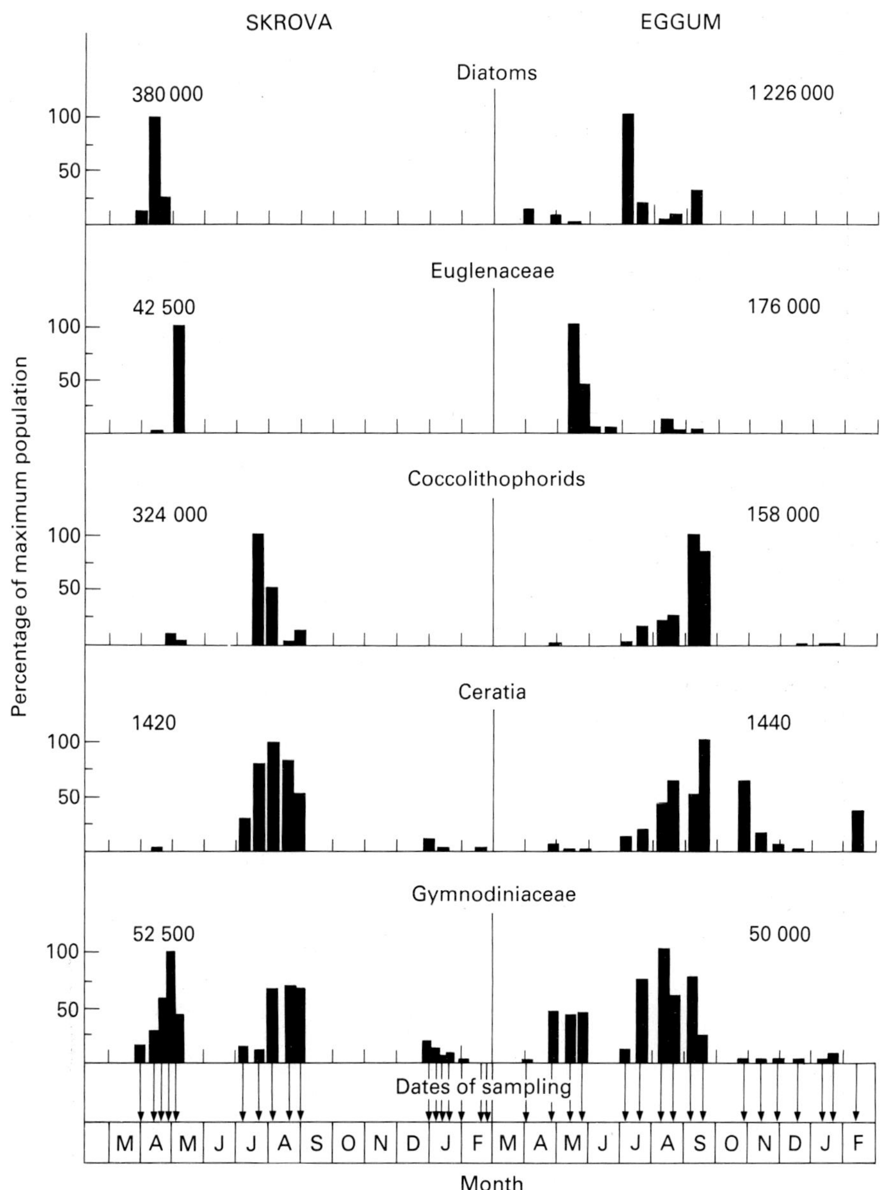

Fig. 7.12 Seasonal changes in the main components of the phytoplankton at two stations (Skrova and Eggum) in Norwegian coastal waters. For each date the maximum number recorded is presented as the percentage of the maximum population recorded for the group during the year. (After Smayda, 1980; data from Braarud *et al.*, 1958.)

nant, while dinoflagellates are important in the summer. However, unlike coastal Norway, dinoflagellates such as *Ceratium* and coccolithophorids are almost absent from the summer flora. Succession is thus reduced to primarily a series of seasonal oscillations in abundance of the commoner species.

The open oceans show considerable latitudinal differences in species composition and seasonal changes. In the low temperatures of the

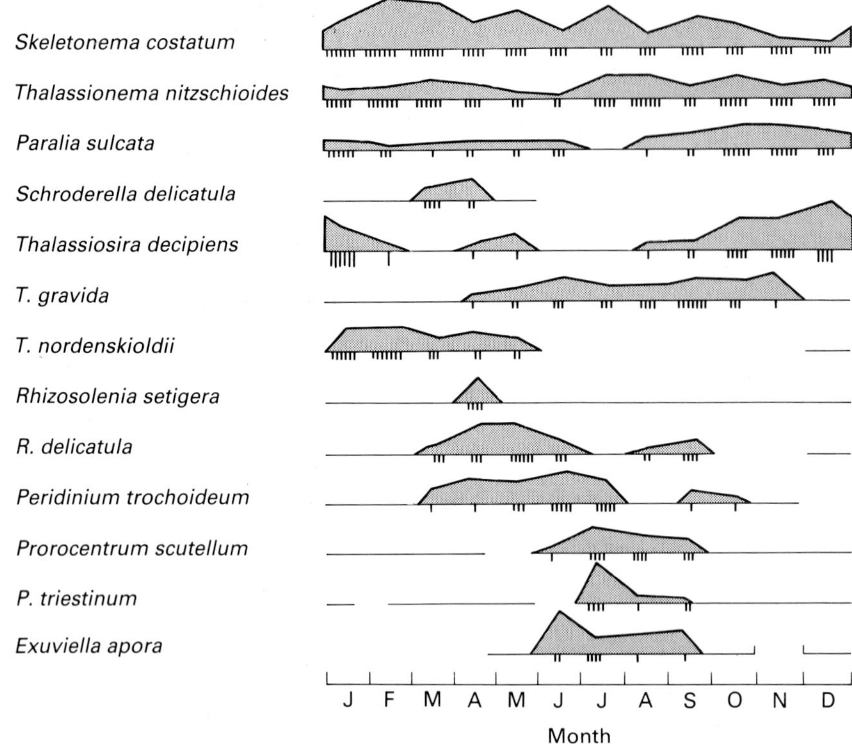

Fig. 7.13 Seasonal occurrence and frequency of the most important species of phytoplankton in Long Island Sound. (After Smayda, 1980; data from Riley, 1967.)

Labrador Sea (summer maximum < 10° C) the annual cycle is dominated by diatoms, with *Fragilaria nana* exhibiting a bimodal dominance in June and in September–October (Holmes, 1956). The spring bloom was not followed by a predominance of dinoflagellates, though *Exuviaella baltica* was relatively abundant from June through August. In more temperate waters off Iceland (summer maximum *c.* 15°C), a March bloom of the coccolithophorid *Emiliania huxleyi* was succeeded in May by large diatoms dominated by *Rhizosolenia* spp. By July this had been replaced by a dinoflagellate maximum of several species of *Ceratium* and *Peridinium*, which in turn was replaced in September by several species of the dinoflagellate *Exuviaella* and the diatom genus *Thalassiosira*.

On many tropical coasts there are areas of upwelling of deep water, which is both cooler and more nutrient-rich than the surface waters it replaces. In these areas there are considerable differences in the phytoplankton populations inhabiting the coastal and adjacent oceans, and as the upwellings may be seasonal there are marked annual cycles of the plankton. Off the coast of West Africa a cool-water flora dominated by diatoms, principally *Chaetoceros, Rhizosolenia, Coscinodiscus*, develops between July and October. For the remainder of the year the flora is more warm-water and is dominated by dinoflagellates (Reyssac & Roux, 1972).

Tropical oceanic water shows low levels of nutrients, and their limiting nature is supported by enrichment experiments (Glooschenko & Curl, 1971). Phytoplankton cell numbers are generally low, and diatoms and coccolithophorids are the most abundant groups (Round, 1981). In some tropical seas the cyanophyte *Trichodesmium thiebaudii* is fairly common and may form extensive blooms (Fogg, 1982). Hulbert *et al.* (1960) recorded a seasonal succession in the subtropical waters of the Sargasso Sea (Fig. 7.14). Coccoli-

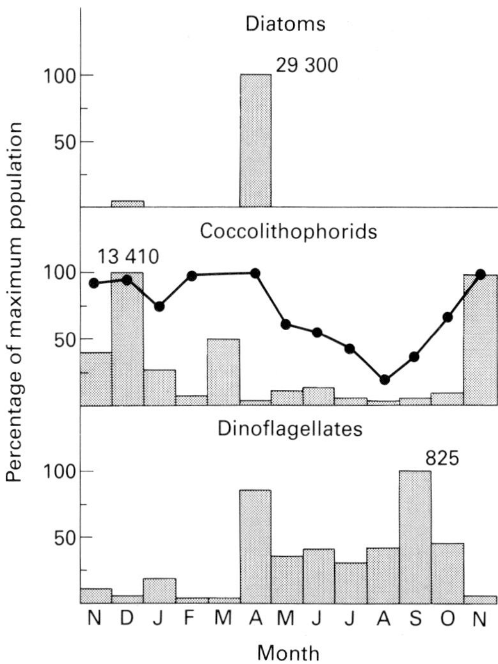

Fig. 7.14 Seasonal changes in the principal phytoplankton groups in the surface waters of the Sargasso Sea. Concentrations are shown as the maximum observed concentration recorded for each group. Solid line on coccolithophorid data is for percentage represented by *Emiliania huxleyi*. (After Smayda, 1980; data from Hulbert *et al.*, 1960.)

thophorids occurred throughout the year, dominated by a single species, *Emiliania huxleyi*. In the summer, from May to October, the flora was sparse, with dominance passing from coccolithophorids to dinoflagellates. The summer thermocline disappeared in November, and *Emiliania huxleyi* became dominant. There was a short period of domination by the diatom *Rhizosolenia* in April which was coincident with the onset of thermal stratification.

THE PARADOX OF THE PLANKTON

Hutchinson (1961) commented on a phenomenon which he termed the 'paradox of the plankton', where 10–50 species of algae appear to coexist within apparently uniform bodies of water. This would seem to contradict the competitive exclusion principle, which prohibits two or more species from occupying the same ecological niche (Hardin, 1960). Hutchinson and others (Richerson *et al.*, 1970; Platt & Denham, 1980) suggest that no violation occurs as the habitat is heterogenous due to temporal and spatial fluctuations in environmental conditions.

An alternative explanation is that several species may coexist in equilibrium, in a stable environment, providing that each is limited by a different resource. This has been verified in culture by Tilman (1977), who grew two freshwater diatom species with similar maximum growth rates, *Asterionella formosa* and *Cyclotella meneghiniana,* in semi-continuous culture at different ratios of silicate and phosphate concentration (Fig. 7.15). *Asterionella* can take up phosphorus more efficiently than *Cyclotella* at low phosphate concentrations, while the reverse is true for silicates. Thus at high Si:P ratios *Asterionella* has the competitive advantage, whereas at high P:Si ratios *Cyclotella* is favoured. At intermediate ratios the species will stably coexist because they are limited by different nutrients. This limiting nutrient explanation has been extended to multispecies environments (Sommer, 1983). Sommer (1984) has also shown, however, that oscillations in the level of a single nutrient increase species diversity by allowing the coexistence of species competing for the this nutrient.

Epilithic algae — the seaweeds

Epilithic algae occur on solid substrates in habitats where water movement prevents the accumulation of sediments; such habitats include river beds and the shorelines of lakes and oceans. The marine epilithic macrophytes, colloquially termed seaweeds, have received the most attention from ecologists. The comparative structural simplicity of seaweed-dominated communities, coupled with the rapid growth of the constituent species, make them ideal for manipulative experiments designed to test ecological hypotheses. Over the past two decades such studies have had considerable influence on general ecological theories, especially those concerning competition and predation.

LIFE FORMS AND ADAPTIVE MORPHOLOGIES

The seaweeds are comprised of three major taxonomic groups, Chlorophyceae, Phaeophyceae, and Rhodophyceae (Lobban & Wynne,

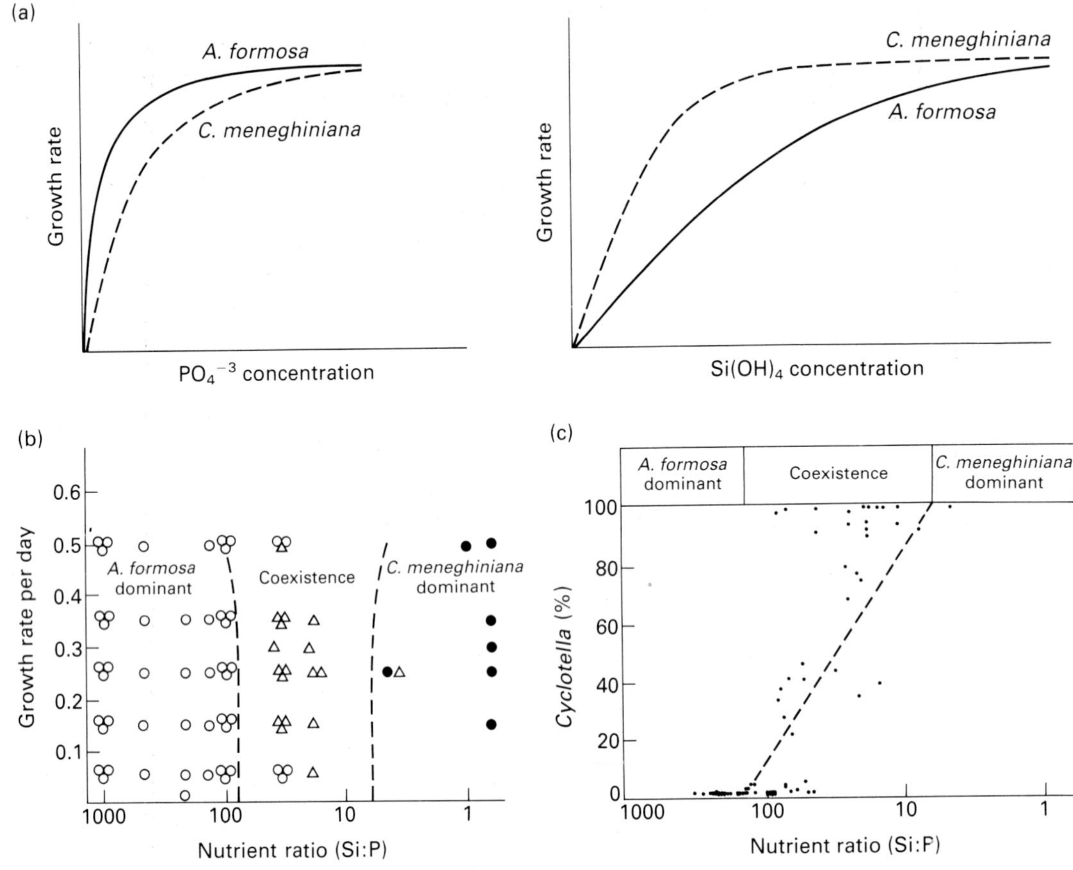

Fig. 7.15 Competition between *Asterionella formosa* and *Cyclotella meneghiniana* along a resource gradient of Si:P concentration. (a) The nutrient-related growth kinetics for each species in axenic culture. (b) The results of competition experiments as a function of growth rate and Si:P ratio. *Asterionella* as a dominant, open circle; *Cyclotella* dominant, solid circle; both species co-existing, open triangle. (c) The relative abundance of *Cyclotella* as a percentage of the total number of *Asterionella* and *Cyclotella* cells in samples from Lake Michigan plotted against the Si:P ratios measured in the same samples. (Adapted from Darley, 1982; data from Tilman, 1977.)

1981), though colonial or filamentous forms of other divisions (Cyanophyceae; Xanthophyceae; Chrysophyceae; Bacillariophyceae) may be locally abundant and ecologically important. The seaweeds are the largest and most structurally complex of the algae, and the three divisions show a remarkable parallelism of morphologies (Chapter 4), but from an evolutionary perspective they represent distinct lines of development (Chapter 8). It would thus seem likely that this parallelism of form is adaptive, conferring fitness on phyletically diverse organisms growing in a common habitat. This suggestion is further strengthened by the observation that the overall morphologies of seaweeds may be influenced by environmental conditions in similar ways, irrespective of their systematic affiliations (Norton *et al.*, 1981).

The recognition of the importance of morphology has led to attempts at ecological classifications of seaweeds based on life forms, in a manner similar to that proposed by Raunkiaer (1934) for terrestrial vegetation. The attempt is to unite under a single name thallus morphology, longevity, and life history; such an approach is best suited to describing perennation strategies in a biogeographical context (Garbary, 1976). Other life form classifications include those of Feldmann (1951), Chapman and Chapman (1976), and Russell (1977). Many other workers use terms such as annual, perennial, and pseudoperennial (e.g. Sears & Wilce, 1975), which have some counterparts in the life form schemes.

Littler and Littler (1980) have proposed a functional form hypothesis, under which algal morphologies are classified in an ecological context, and adaptive features of structure and function are assessed by cost–benefit analyses (Tables 7.1 & 7.2). They distinguished between opportunistic forms with rapid growth but low biomass and high reproductive potential (r-selected) and late successional forms with slower growth, higher final biomass, and reduced reproductive potential (K-selected). Littler and Arnold (1982) assessed photosynthetic performance in six such functional form groups; it was greatest in thin tubular and sheet-like forms, and lowest in encrusting forms. This was reflected ecologically in that the rapidly growing r-selected species were characteristically early successional stages found in fluctuating or unstable environments, while K-selected species occurred in more stable habitats. K-selected species would be expected to show greater resistance to predation. There is considerable evidence that slow-growing crusts may become overwhelmed by epiphytes unless these are removed by grazers or other physical disturbance (Jackson, 1977; Wanders, 1977). It has been suggested that heteromorphic algal life cycles involving a crustose thallus may have evolved in response to grazing pressure (Lubchenco & Cubit, 1980; Slocum, 1980).

Table 7.1 Hypothetical (*a priori*) survival strategies available to opportunistic macroalgae representative of stressed* communities versus macroalgae characteristic of non-stressed† communities. (After Littler & Littler, 1980.)

Opportunistic forms	Late successional forms
1 Rapid colonizers on newly cleared surfaces	1 Not rapid colonizers (present mostly in late seral stages); invade pioneer communities on a predictable seasonal basis
2 Ephemerals, annuals, or perennials with vegetative short-cuts to life history	2 More complex and longer life histories; reproduction optimally timed seasonally
3 Thallus form relatively simple (undifferentiated); small with little biomass per thallus; high thallus area to volume ratio	3 Thallus form differentiated structurally and functionally with much structural tissue (large thalli high in biomass); low thallus area to volume ratio
4 Rapid growth potential and high net primary productivity per entire thallus; nearly all tissue photosynthetic	4 Slow growth and low net productivity per entire thallus unit due to respiration of non-photosynthetic tissue and reduced protoplasm per algal unit
5 High total reproductive capacity with nearly all cells potentially reproductive; many reproductive bodies with little energy invested in each propagule; released throughout the year	5 Low total reproductive capacity; specialized reproductive tissue with relatively high energy contained in individual propagules
6 Calorific value high and uniform throughout the thallus	6 Calorific value low in some structural components and distributed differentially in thallus parts; may store high-energy compounds for predictable harsh seasons
7 Different parts of life history have similar opportunistic strategies; isomorphic alternation; young thalli just smaller versions of old	7 Different parts of life history may have evolved markedly different strategies; heteromorphic alternation; young thalli may possess strategies paralleling opportunistic forms
8 Escape predation by nature of their temporal and spatial unpredictability or by rapid growth (satiating herbivores)	8 Reduce palatability to predators by complex structural and chemical defences

* Young or temporally fluctuating.
† Mature, temporally constant.

Table 7.2 Hypothetical costs and benefits of the survival strategies proposed in Table 7.1 for opportunistic (inconspicuous) and late successional (conspicuous) species of macroalgae. (After Littler & Littler, 1980.)

Opportunistic forms	Late successional forms
Costs	
1 Reproductive bodies have high mortality	1 Slow growth; low net productivity per entire thallus unit results in long establishment times
2 Small and simple thalli are easily outcompeted for light by tall canopy-formers	2 Low and infrequent output of reproductive bodies
3 Delicate thalli are more easily crowded out and damaged by less delicate forms	3 Low surface area to volume ratios relatively ineffective for uptake of low nutrient concentrations
4 Thallus relatively accessible and susceptible to grazing	4 Overall mortality effects more disastrous because of slow replacement times and overall lower densities
5 Delicate thalli are easily torn away by shearing forces of waves and abraded by sedimentary particles	5 Must commit a relatively large amount of energy and materials to protecting long-lived structures (energy that is thereby unavailable for growth and reproduction)
6 High surface area to volume ratio results in greater desiccation when exposed to air	6 Specialized physiologically and thus tend to be stenotopic
7 Limited survival options due to less heterogeneity of life history phases	7 Respiration costs high due to maintenance of structural tissues (especially during unfavourable growth conditions)
Benefits	
1 High productivity and rapid growth permits rapid invasion of primary substrates	1 High quality of reproductive bodies (more energy per propagule) reduces mortality
2 High and continuous output of reproductive bodies	2 Differentiated structure (e.g. stipe) and large size increases competitive ability for light
3 High surface area to volume ratio favours rapid uptake of nutrients	3 Structural specialization increases toughness and competitive ability for space
4 Rapid replacement of tissues can minimize predation and overcome mortality effects	4 Photosynthetic and reproductive structures relatively inaccessible and resistant to grazing by epilithic herbivores
5 Escape from predation by nature of their temporal and spatial unpredictability	5 Resistant to physical stresses such as shearing and abrasion
6 Not physiologically specialized and tend to be more eurytopic	6 Low surface area to volume ratio decreases water loss during exposure to air
	7 More available survival options due to complex (heteromorphic) life history strategies
	8 Mechanisms for storing nutritive compounds, dropping costly parts, or shifting physiological patterns permit survival during unfavourable but predictable seasons

Succession as understood by terrestrial plant ecologists involves the modification of the environment by each successive stage in such a manner as to facilitate the development of the sere, until, ultimately, a climax community develops. Many apparent successional patterns have been noted in the marine benthos. Lodge (1948) reported four successional stages on a cleared intertidal strip in the Isle of Man (U.K.): (1) diatom slime; (2) filamentous ephemerals such as *Ulothrix;* (3) a dense carpet of *Enteromorpha;* (4) a climax sward of *Fucus*. Other studies have shown that successional stages may be less predictable and that several routes may lead to the climax (Foster, 1975; Sutherland & Karlson, 1977). There is little conclusive evidence to support the theory that early successional stages are necessary to condition the environment for those that follow (Connell & Slayter, 1977) and the apparent succession may simply be due to the rapid growth characteristics and reproductive abilities of the ephemeral pioneering algae.

NUTRIENTS

In contrast to the situation in phytoplankton there is little known about the importance of nutrients in controlling the growth and structure of seaweed communities. The most detailed information is on the nitrogen budgets of members of the Laminariales. Chapman and Craigie (1977) related the growth of *Laminaria longicruris* in St Margaret's Bay, Nova Scotia, Canada, to both ambient nitrogen levels and internal storage (Fig. 7.16). The onset of rapid growth occurs in December and is correlated with an increase in the concentration of nitrate in the sea water. The growth rate continues to increase and peaks in early May–June, some two months after a precipitous decline in nitrate concentrations in the sea. This continued growth is sustained by nitrate stored in the *Laminaria* tissue. Peak photosynthesis occurs in July, at a time when growth is lowest, and results in a build-up of carbohydrate reserves (laminarin), which presumably are utilized to support growth when light levels are unfavourable in the early winter. Thus, *Laminaria longicruris*, by its ability to store both inorganic nitrate and carbohydrate reserves, is able to take up nitrogen when it is most abundant in the autumn and winter, and utilize this for growth during the spring, under light conditions which are more favourable for photosynthesis. At another site in Nova Scotia, where upwelling maintained relatively high ambient levels of nitrogen all year round, this growth strategy was not observed; there was little storage of either carbohydrate or inorganic nitrogen, and growth followed the seasonal pattern of irradiance (Gagne *et al.*, 1982). In other genera the limiting effects of nitrogen are not as clear; the summer growth rates of *Alaria* (Buggeln, 1978) or *Laminaria hyperborea* and *L. saccharina* (Lüning, 1979) were unaffected by the supply of additional nitrate.

Similar carbon–nitrogen patterns have been demonstrated for smaller seaweeds (e.g. *Gracilaria,* Rhodophyceae; *Ulva,* Chlorophyceae; Rosenberg & Ramus, 1982.) Peaks in tissue nitrogen follow peaks in the nitrogen concentration of the ambient sea water, and both accumulate nitrogen in the winter, and show depletion in the spring proportional to the growth rate. Neither over-wintered with significant carbohydrate reserves, and the onset of growth in the spring does not depend on starch reserves from the previous growing season.

Heterotrophy in marine macrophytes is not as well demonstrated as in phytoplankton. There is some evidence from laboratory studies that uptake of acetate, sugars and amino acids may occur (Chapter 6), but its importance in nature remains to be demonstrated. Wilce (1967) hypothesized that heterotrophy may occur in some arctic seaweed species, as these were found in deep water and would have no light during the winter. However, Chapman and Lindley (1980) showed ambient irradiances were sufficient for growth of the arctic kelp *Laminaria solidungula* and that stored carbohydrate reserves were utilized to support growth during the winter.

COMPETITION

In the benthic marine environment there is competition for space not only between algal species, but also with sessile invertebrates. Attachment space, in many instances, appears to be the primary limiting resource, and the subsequent crowding only secondarily leads to shading and nutrient competition. The outcome of space competition is also dependent on the presence,

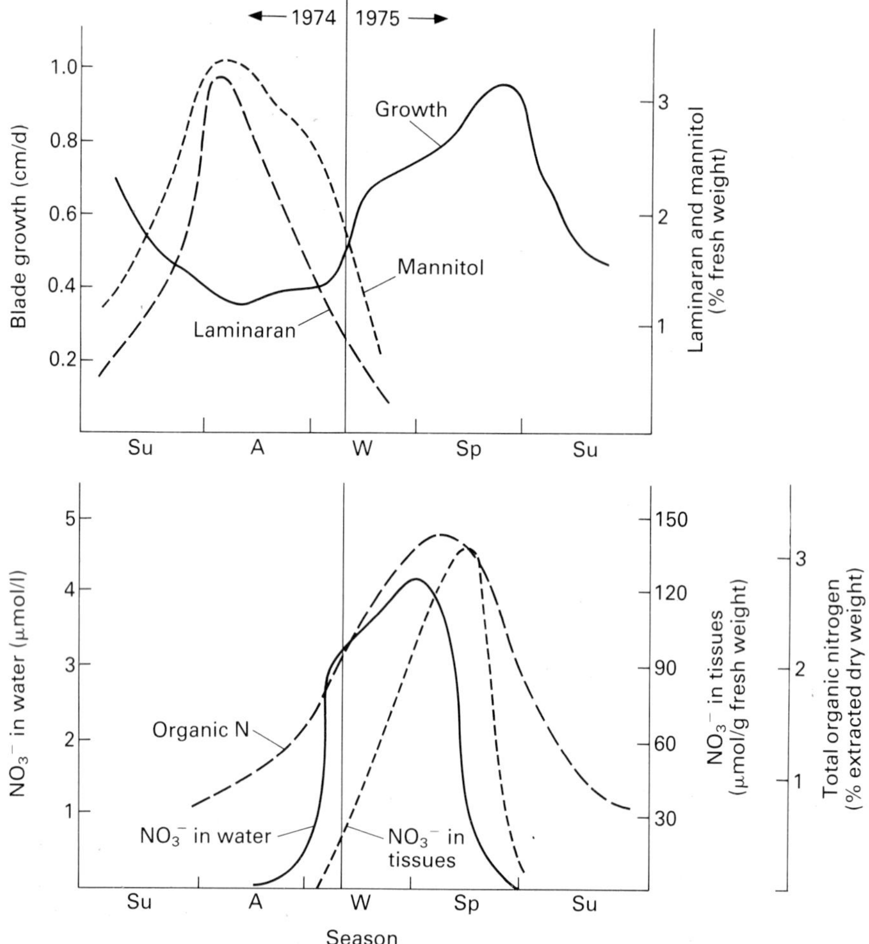

Fig. 7.16 Seasonal relationship between the rate of blade growth, internal carbon and nitrogen reserves, and nitrate concentration in seawater for *Laminaria longicruris*. (After Darley, 1982; data from Chapman & Craigie, 1977.)

or absence, of predators on both the seaweeds and the sessile animals.

Heavy settlement of algal spores may lead to competition between siblings, and there is an inverse relationship between density and individual size (Fig. 7.17) (Cousens & Hutchings, 1983). Reproduction may also be affected: *Fucus vesiculosus* under crowded conditions allocates more of its resources to vegetative growth and less to reproduction (Russell, 1976).

Interspecific competition between seaweeds has been studied in culture by Russell and Fielding (1974). Three species, a chlorophyte, *Ulothrix flacca*, a phaeophyte, *Ectocarpus siliculosus*, and a rhodophyte, *Erythrotrichia carnea*, were grown separately, and in pairs, under a variety of temperature, salinity, and irradiance regimes. All the species showed a wider tolerance of conditions when grown in monoculture than in pairs, and species extinction occurred under conditions which would permit growth in isolation.

Most experiments on competition have been made in the field and have involved manipulation of the densities of putative competitors. The kelp *Egregia laevigata* in southern California is outcompeted by the understory turf of red algae, primarily *Gigartina, Laurencia*, and *Gastrocolonium*. These red algae may propagate vegetatively

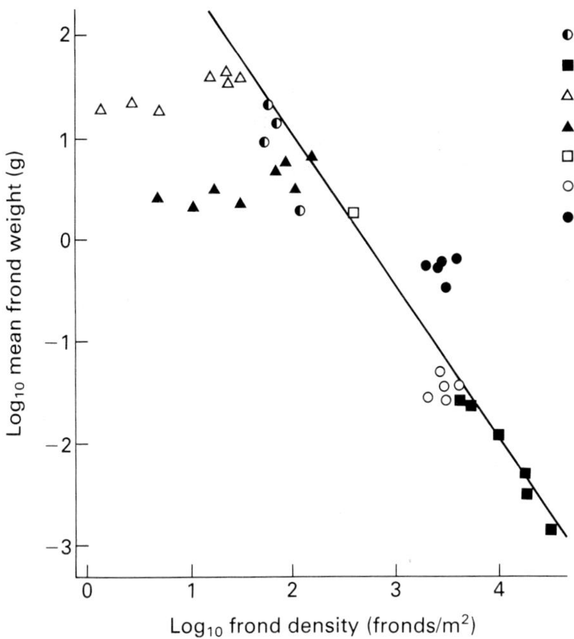

Fig. 7.17 Relationship between mean frond weight and frond density for natural monospecific populations of some marine macrophytic algae. (After Cousens & Hutchings, 1983.)

to fill space which becomes available, and sand trapped between their thalli prevents the settlement of *Egregia* spores (Sousa *et al.*, 1981). It is only if space is created by external disturbance that *Egregia* recruitment can occur. A similar competitive situation exists between the kelp *Postelsia palmaeformis* and the mussel *Mytilus californianus* in exposed locations in the U.S. Pacific northwest (Dayton, 1973).

In a series of experiments involving the removal of the kelp *Hedophyllum sessile*, Dayton (1975) showed the importance of large, dominant algal species in controlling the associated flora by mechanisms other than competitive exclusion. Removal precipitated a decline in the normal understory species and promoted the growth of ephemeral algae. Similar changes were noted by Hawkins and Harkin (1985), who removed subtidal canopy plants in the Isle of Man.

The inadvertent introduction of new species into an area also provides an opportunity to study competitive exclusion. Competition between the kelp *Macrocystis pyrifera* and the Japanese immigrant fucoid *Sargassum muticum* has been reported from Santa Catalina Island, California (Ambrose & Nelson, 1982). The *Sargassum* arrived in the area in 1971, and following a dieback of *M. pyrifera* in 1976 invaded some of the vacated space, preventing re-establishment of the kelp. Thus, it appears that local distribution of *Sargassum* is restricted by competition with *M. pyrifera*, but that once established it is capable of excluding the kelp.

GRAZING AND PREDATION

Lubchenco and Gaines (1981) provide an excellent review of much of our knowledge of seaweed–grazer interactions, and it is clear that grazing and predation are powerful biotic forces in the control of species composition and biomass in seaweed communities. The major animal groups which eat algae are the molluscs, particularly the gastropods and chitons; crustaceans, particularly amphipods; and the sea urchins and fish. Herbivorous fish are rare in temperate waters, but in the tropics 15–25% of fish are algivorous, and the high proportion of toxic algae in the tropics (Chapter 9) would appear to be an adaptation to high levels of grazing predation.

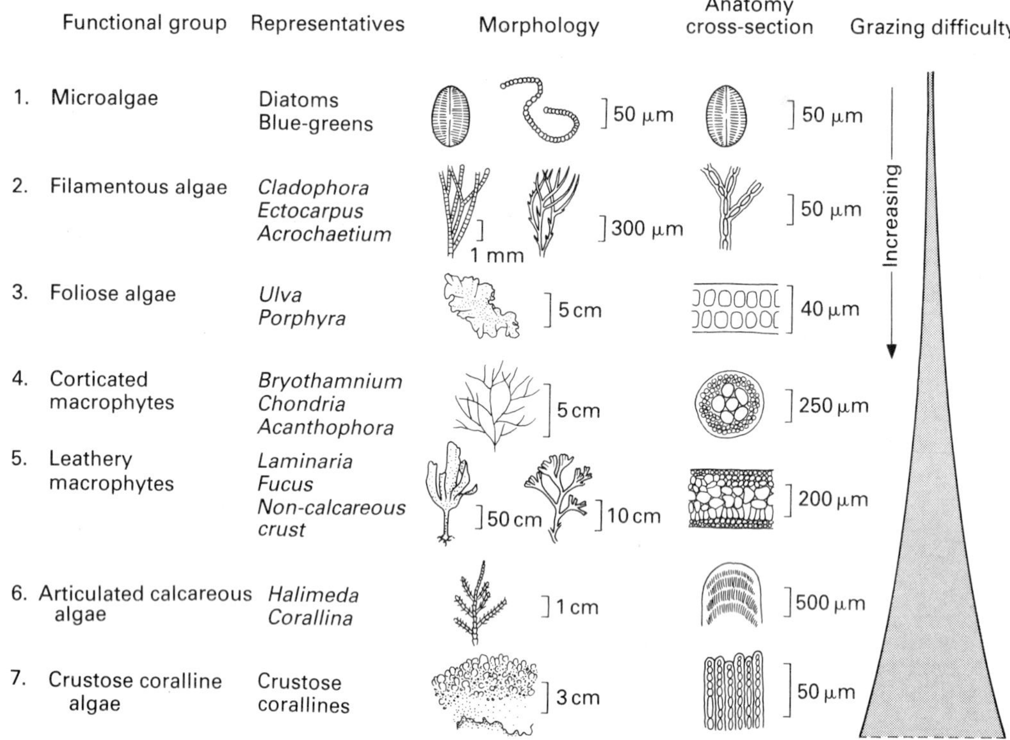

Fig. 7.18 Functional groups of benthic marine algae. Grazing difficulty refers to the structural toughness of the species. (After Steneck & Watling, 1982.)

Many grazers show considerable species selectivity, and this may be related to algal morphology. Steneck and Watling (1982) describe seven functional groups, ranging from microalgae, through filamentous and corticated forms, to cartilaginous and ultimately crustose coralline algae, presenting an increasing gradient of grazing difficulty (Fig. 7.18). The size or developmental stage of the alga may also be important: the periwinkle, *Littorina*, eats the sporelings of *Chondrus crispus*, but does not readily consume the adult plants (Cheney, 1982). The nutritional quality of the algae consumed can affect the overall fitness of the herbivore (Vadas, 1977). The amount consumed does not necessarily reflect the importance of the species to the grazer. *Littorina littorea* preferentially consumed *Enteromorpha*, but did best on a mixed diet which included *Chondrus* and *Fucus* (Cheney, 1982).

Lubchenco (1978) altered the density of *Littorina littorea* in rock pools in the New England intertidal (Fig. 7.19). All the snails were removed from pools where *Chondrus* was dominant and were added to pools dominated by *Enteromorpha*. Under increased grazing pressure the abundance of *Enteromorpha* declined and other ephemeral species which replaced it were also ultimately consumed. In the *Chondrus*-dominated pools the reduction in grazing allowed the development of *Enteromorpha*, which rapidly displaced the other species. Control pools with intermediate levels of grazing remained unchanged. Thus, under high grazing pressures all but the most resistant species are consumed, while under low pressure a competitive dominant, which would normally be eaten, occupies all available space. Algal diversity is therefore maximized at intermediate levels of grazing.

Similar grazing effects have been noted in kelp beds, where the most important herbivores are sea urchins (Lawrence, 1975). Paine and Vadas (1969) removed *Strongylocentrotus franciscanus* at Friday Harbor, and found an increase in

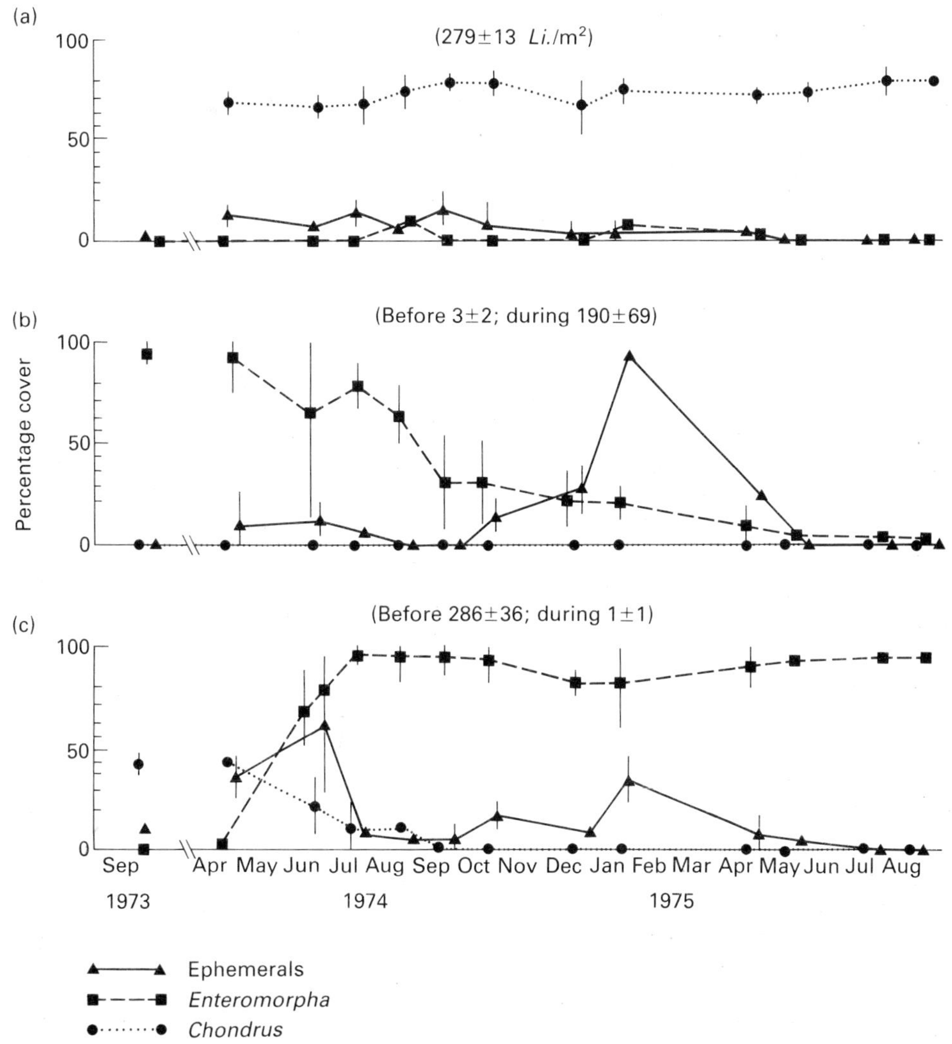

Fig. 7.19 Effect of *Littorina littorea* on algal composition in New England tide pools. (a) Control. (b) Addition of *Littorina*. (c) Removal of *Littorina*. Mean number of *Littorina* in pools is presented. (After Lubchenco, 1978.)

species diversity and a domination by *Laminaria*, whereas in control areas only calcareous corallines and some ephemeral algae were found. Removal of *Echinus esculentus* in Britain (Jones & Kain, 1967) and *Strongylocentrotus droebachiensis* in eastern Canada (Breen & Mann, 1976) led to similar increases in algal biomass and diversity.

The algal grazers and the invertebrate space competitors are in turn subject to interactions with predators, which have significant effects on the associated algal communities. The mussel *Mytilus californianus* is competitively dominant in parts of the intertidal of the Pacific coast of North America. Its major predator is the starfish *Pisaster ochraceus*. Removal of *Pisaster* caused a change from a diverse intertidal community to one dominated by mussels (Paine, 1974). Paine (1969) termed such predators which control competitive dominants 'keystone species'.

Lubchenco and Menge (1978) have reported competitive interactions between barnacles (*Balanus balanoides*), mussels (*Mytilus edulis*), and the red alga, *Chondrus*, in the New England

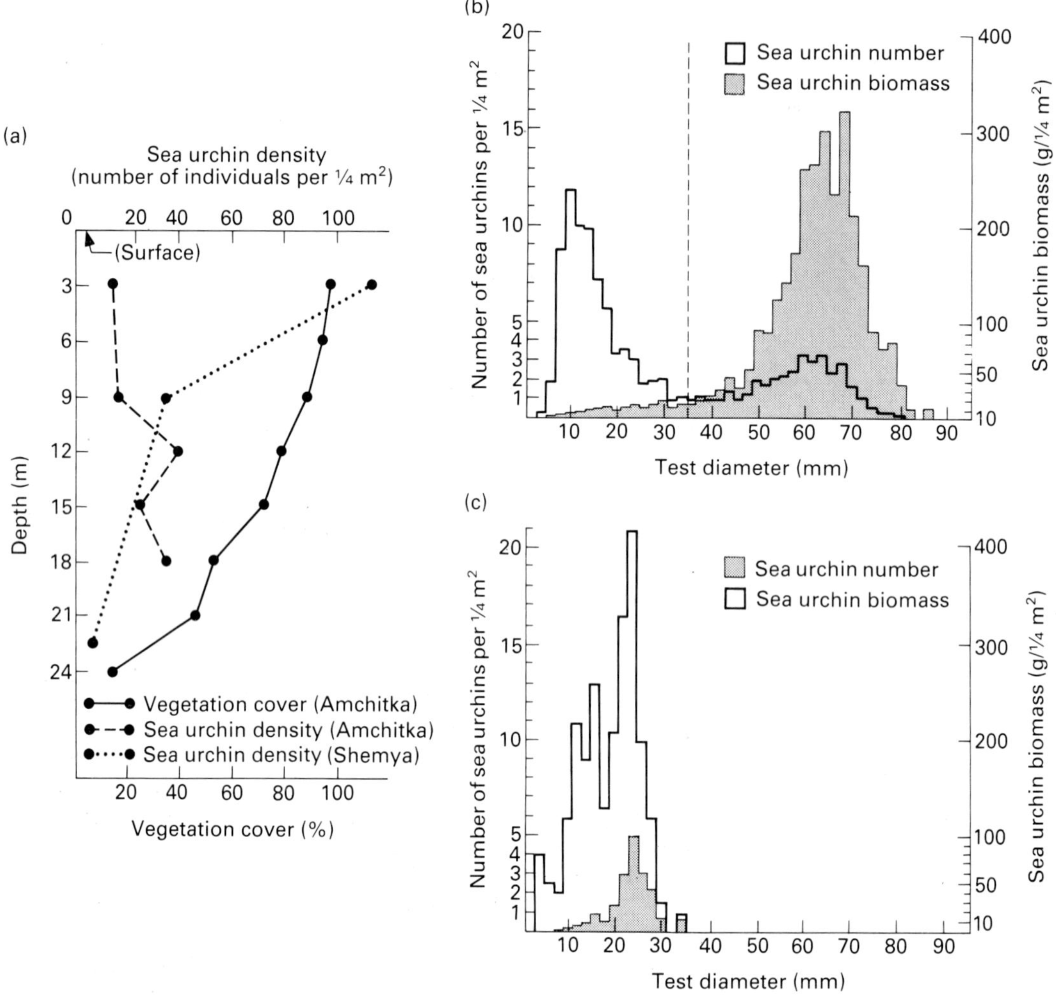

Fig. 7.20 Sea otters, sea urchins and benthic algal vegetation. (a) Vegetation coverage and sea urchin density *v.* depth; vegetation cover for Shemya Island is coincident with the ordinate. (b) Shemya Island (sea otters absent). (c) Amchitka Island (sea otters present). Sea urchin size classes and biomass. Dotted line in (b) is for comparison with (c) and represents the largest sea urchin size class observed in Amchitka. (After Estes & Palmisano, 1974.)

intertidal, which were moderated by the presence of predatory whelks (*Thais*) and starfish. Predator exclusion experiments showed that the invertebrates were competitively dominant, but the activities of the predators created space, which could then be occupied by *Chondrus*. In exposed areas the predators were removed by wave action, and *Mytilus* dominated.

The sea otter *Enhydra lutris* once occupied a range around the Pacific rim from Northern Japan to Baja, California. It was nearly extinguished by hunting, and remained in abundance mainly in parts of the Aleutian archipelago. It eats benthic invertebrates, particularly sea urchins, and the increase in urchin numbers and the concomitant decrease in kelp biomass have been attributed in part to its regional extinction. In Amchitka Island, otters are present, and there is abundant algal cover in the subtidal, whereas in Shemya Island, 400 km to the west, they are

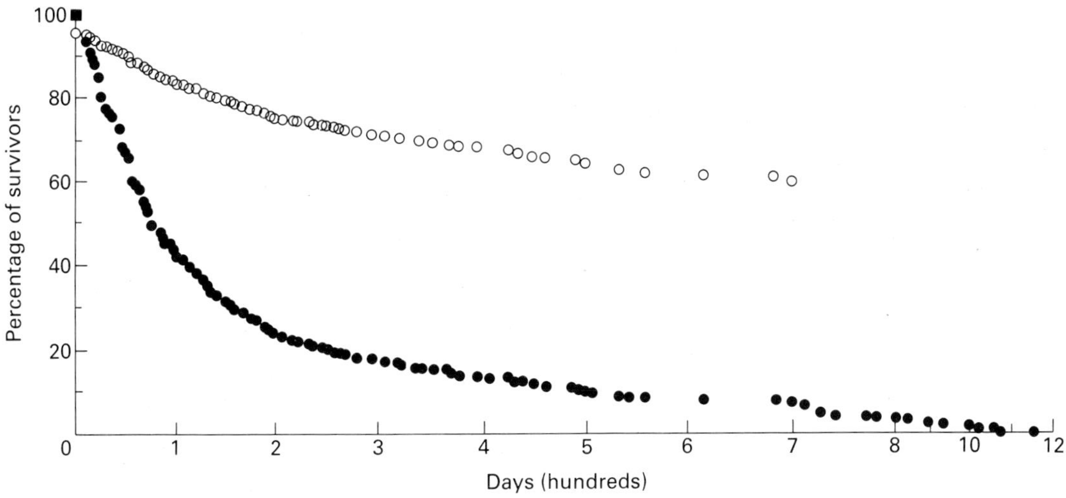

Fig. 7.21 Survivorship curves for recruits to a population of *Pelvetia fastigiata* in southern California. Filled circle is percentage survivorship of 314 individuals; open circle is the $\log_{10}$ survivorship for the 314 individuals scaled to a population of 1000. (After Gunnill, 1980.)

absent; there is no subtidal algal cover and urchins are both larger and more abundant (Fig. 7.20) (Estes & Palmisano, 1974). In areas of Alaska where sea otters have been reintroduced, kelp beds have also become re-established (Duggins, 1980). A similar keystone role has been suggested for lobsters in Nova Scotia by Breen and Mann (1976), who noted the correlation between declining lobster abundance, increases in urchin numbers and destruction of kelp beds. The evidence in support of this hypothesis is not conclusive, as other invertebrates and fish are also important urchin predators (Miller, 1985).

DEMOGRAPHIC AND PHENOLOGICAL STUDIES

In contrast to the situation for benthic animals, there are relatively few demographic studies on marine algae and few data are available on their life expectancy. The most detailed study is that of Gunnill (1980) on the perennial fucalean *Pelvetia fastigiata* from California; maximum life expectancy was 2.5 years with 80% of plants lost in the first year and and a further 10–15% in the second year (Fig. 7.21). Rice and Chapman (1982) describe changes in biomass and survivorship over the growing season of two cohorts of the annual brown seaweed *Chordaria flagelliformis* and use this information to calculate primary production. The alga appeared in June–July and disappeared in mid-September, and for most of the growing season there was a decrease in number of individuals correlated with an increase in the biomass of individual plants.

Most seaweeds exhibit seasonal differences in growth which are responses to environmental changes such as irradiance, photoperiod, temperature, nutrient availability, and predation. These phenological differences may be expressed morphologically, or in reproductive status, and for those which are heteromorphic, or have cryptic phases in their life cycles, by their apparent disappearance for part of the year. Hooper *et al.* (1980) provide a detailed analysis of the flora of Bonne Bay, a fjord in western Newfoundland, with cluster analyses being used to illustrate both seasonal occurrences and reproductive status. The vegetative data show that the seasonal ephemerals can be classified into a winter–spring flora and a summer–autumn flora (Fig. 7.22), whereas the reproductive data show three major groups, with plants which are fertile in winter, spring, and summer–fall (Fig. 7.23). Such studies suggest hypotheses to be tested in culture concerning the seasonal factors which might control the growth and reproduction of species.

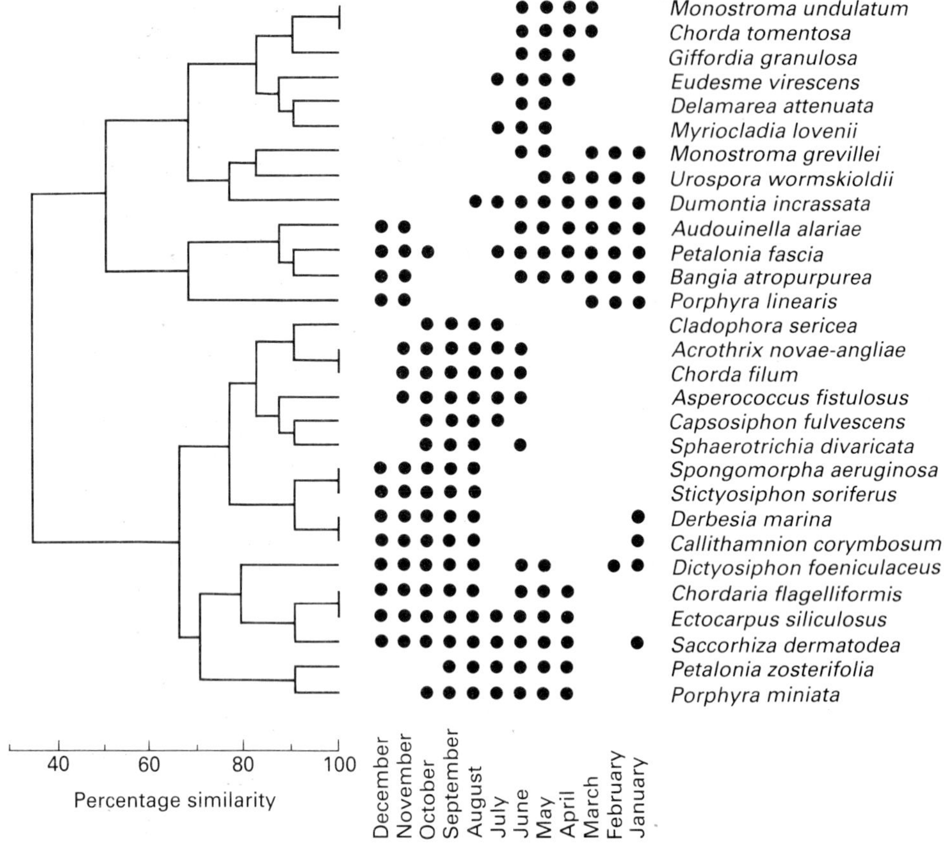

Fig. 7.22 Cluster analysis showing occurrence of species with pronounced vegetative seasonality in Bonne Bay, Newfoundland. Filled circle indicates presence in a given month. (After Hooper *et al.*, 1980.)

VERTICAL DISTRIBUTION OF SEAWEEDS

Most seaweeds are restricted to the fringes of the world's oceans in areas where the substrate is sufficiently stable to permit their attachment. They occupy a narrow coastal fringe extending vertically from the upper limits to which wave splash reaches, downwards through the intertidal, to the subtidal and the lower limits of the photic zone. The current record depth for attached seaweeds is for an undescribed coralline red alga from 268 m depth in the Bahamas (Littler *et al.*, 1985). Desiccation at the upper limits, and low irradiance at the lower, would appear to be the major abiotic factors limiting vertical distribution, but these factors are markedly modified by the biological effects of competition and predation.

Algae are not randomly distributed along this vertical gradient; many occupy discrete zones with abrupt upper and lower boundaries. While these zones extend into the subtidal, they are particularly well marked in the intertidal region.

The simplest classification methods divide the shore into a number of horizontal zones on the basis of the limits of occurrence of certain diagnostic species. One of the most widely accepted schemes is that of Stephenson and Stephenson (1949) as modified by Lewis (1961) and further expounded by Stephenson and Stephenson (1972), in which is proposed a system of three major zones (Fig. 7.24). On moderately sheltered shores the eulittoral zone comprises most of the intertidal, from the limits of the extreme high-water spring tides to the extreme low-water spring

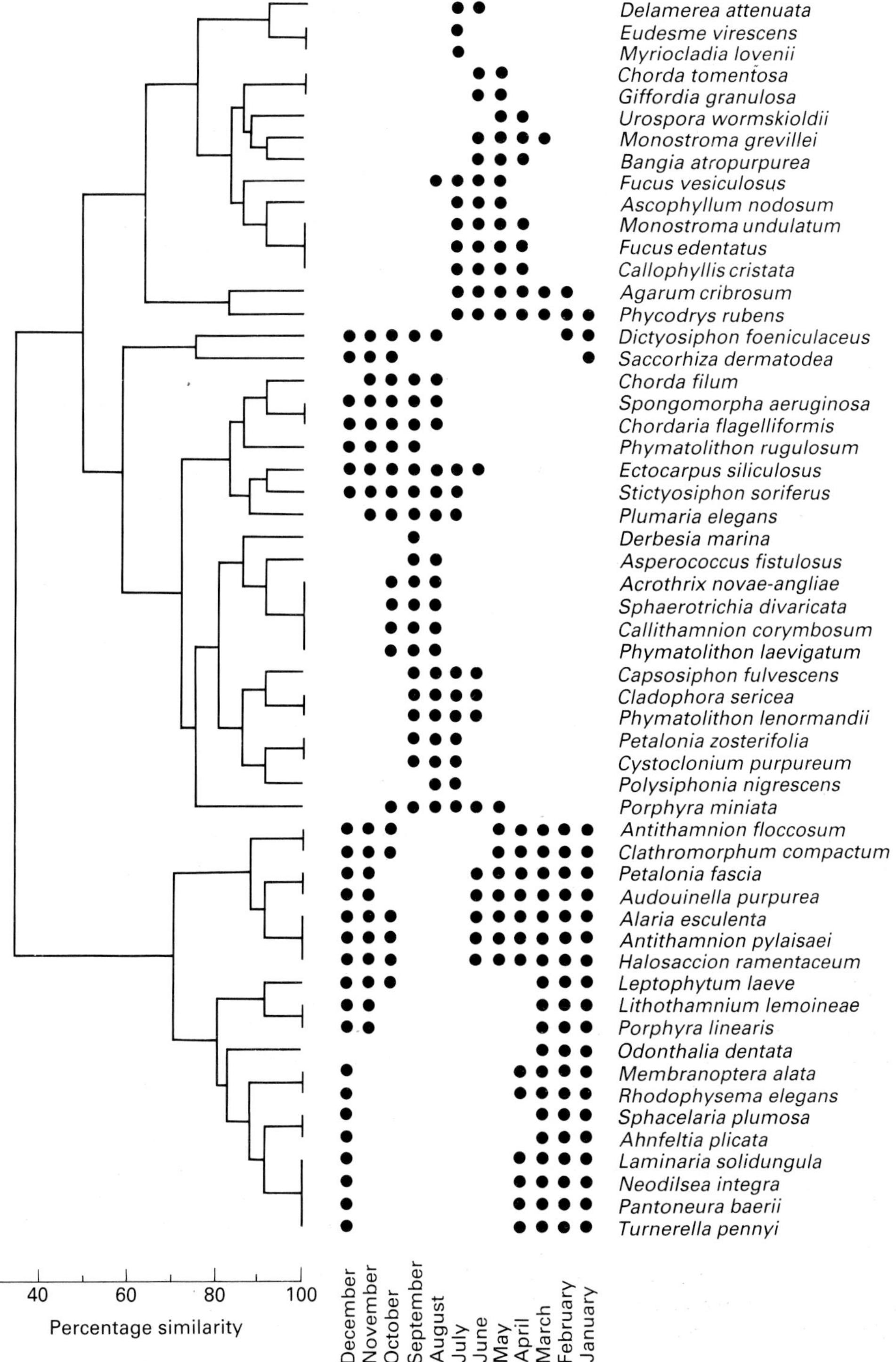

Fig. 7.23 Cluster analysis showing the reproductive phenology of selected benthic algae from Bonne Bay, Newfoundland. Filled circle indicates the presence of functional reproductive structures. (After Hooper *et al.*, 1980.)

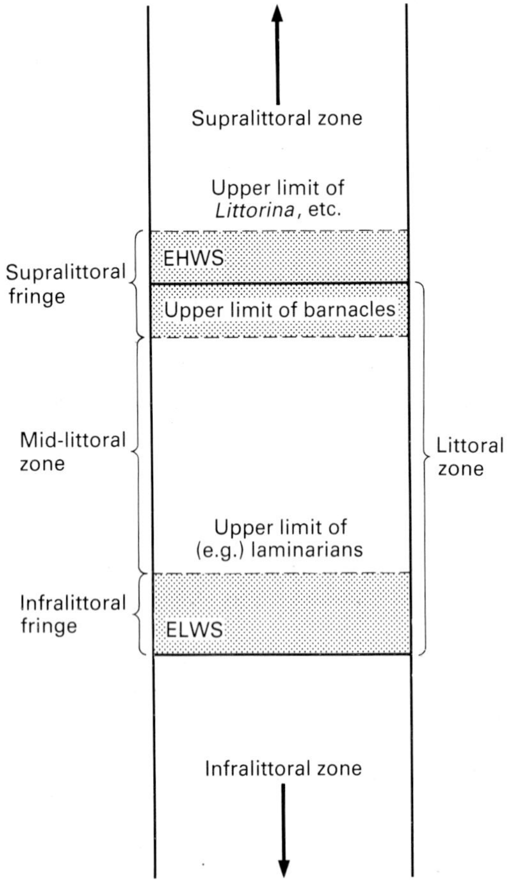

Fig. 7.24 Universal features of intertidal zonation on a rocky shore. EHWS, extreme high-water spring tides; ELWS, extreme low-water spring tides. (After Stephenson & Stephenson, 1949.)

tides. Above this is the littoral fringe and below it is the sublittoral zone. On northern temperate shores the upper limit of the laminarians characterizes the sublittoral–eulittoral boundary, while the eulittoral–littoral fringe is marked by the upper limit of barnacles. As shores become more exposed these zones are raised and become wider (Fig. 7.25) (Burrows *et al.*, 1954; Lewis, 1961; Seapy & Littler, 1979). Although the delimiting species change, this 'universal zonation system' has been widely accepted and applied to coastlines of all the world's oceans (Stephenson & Stephenson, 1972). Within these broad zones the seaweeds may also occupy discrete bands of distribution. On many northern European shores there is an upper band of ephemeral green and blue-green algae, often dominated by *Enteromorpha*: below this the dominants are fucoids, a common sequence being *Pelvetia canaliculata, Fucus spiralis, F. vesiculosus* and *F. serratus*; below is a narrow band of *Laminaria digitata*, followed by the subtidal *L. hyperborea*. Descriptions of shorelines from other parts of the world show that similar horizontal bands of algae occur (Stephenson & Stephenson, 1972; Round, 1981).

The most obvious of the factors which might effect zonation is the tidal range and periodicity. At Plymouth, U.K., Colman (1933) suggested that tides do not cover and uncover the intertidal at a uniform rate and that more abrupt changes would occur at three points in the cycle, correlating with boundary changes in the intertidal vegetation. This theory of critical tide levels was further developed by Doty (1946) who, under the more complex tidal cycles of the U.S. Pacific northwest (two daily tides of unequal height), recognized seven critical tide levels. The implication from these studies is that the seaweeds that occupy zones between critical tide levels are precisely physiologically adapted to withstand the periods of emersion and submersion that they experience, but cannot survive in the adjacent zones. There is considerable controversy as to whether critical tidal levels occur at all, let alone influence intertidal zonation. Underwood (1978) made accurate measurements of the tide emersion curve at Colman's site and showed a smooth uninterrupted change. In contrast, Druehl and Green (1982) measured uneven periods of submersion and emersion along three transects in the British Columbia intertidal. Their measurements, which accounted for both wave action and the tidal cycle, were used to generate emersion–submersion curves for the transects. Wave action caused considerable changes in the cycle of emersion–submersion which would have been predicted from tidal data alone. Algal distributions on the transects showed some significant correlations with these abrupt emersion–submersion events.

Chapman (1973) presents the other extreme view, that the existence of sharp boundaries between intertidal zones, along a continuous environmental gradient, can be explained entirely by interspecific competition, without invoking any appeal to tidal cycles.

Over the past decade a number of experiments have been conducted to determine the factors

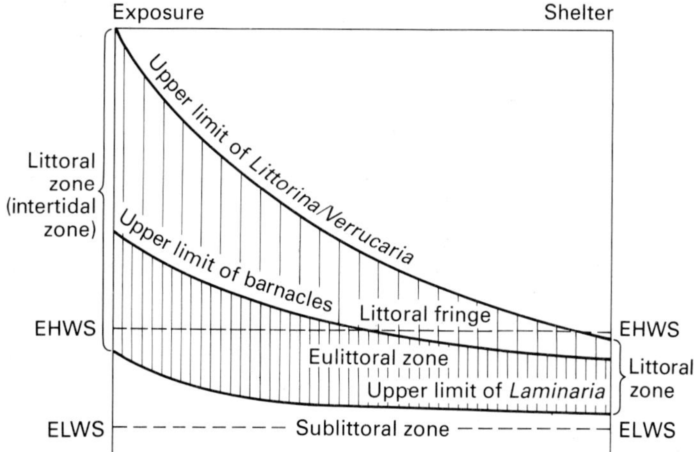

Fig. 7.25 The effects of wave exposure in broadening and raising intertidal zones. EHWS, extreme high-water spring tides; ELWS, extreme low-water spring tides. (After Lewis, 1964.)

affecting vertical distribution of algae. Schonbeck and Norton (1978) showed that low tides in combination with drying conditions caused the greatest damage to intertidal fucoids and that the low-shore fucoids, *Fucus vesiculosus* and *F. serratus*, were less resistant to desiccation than *F. spiralis* and *Pelvetia canaliculata*, located on the upper shore. Dring and Brown (1982) also provide evidence in support of a physiological basis for zonation. Their results indicated that water loss was a function of surface to volume ratio and not position on the shore. Recovery from severe desiccation took about two hours in each species, but the extent of recovery from a given degree of desiccation was greater in upper shore species. *Pelvetia* and *F. spiralis* showed complete recovery from 80–90% water loss, *F. vesiculosus* from about 70% and *F. serratus* from 60% water loss. It is thus the extent of the recovery, measured as photosynthesis, which shows the clearest correlation with zonation pattern.

In the low shore of California, Hodgson (1980) showed the importance of interspecific competition in maintaining zonation. The red alga *Gastroclonium coulteri* occupies a band about 0.3 m above mean low tide. Above is a zone of other red algae, principally *Gigartina papillata* and *Rodoglossum affine*, with the zone below occupied by a marine angiosperm, the surf grass *Phyllospadix*. When the *Gastroclonium* zone was cleared the *Gigartina* grew down to occupy the available space; when the *Gigartina* zone was cleared there was no corresponding upward movement of *Gastroclonium*. Similarly, when the *Phyllospadix* was removed, the *Gastroclonium* zone moved down the shore, but reciprocal clearing did not result in the upward movement of *Phyllospadix*. It was determined that desiccation losses prevented the movement up the shore, while the lower limit was controlled by competition. Similar experiments between upper intertidal fucoids in Britain also indicate that in the absence of competition species may extend their ranges downshore (Schonbeck & Norton, 1980; Hawkins & Hartnoll, 1985). Lubchenco (1980) also shows competition and predation to be limiting factors in establishing lower limits of algal species in the New England intertidal.

It has thus become generally accepted that upper limits are determined by physiological responses to desiccation and the lower ones by biological interactions. Underwood and Denley (1984) play devil's advocate and suggest that the evidence is still too incomplete, or inconclusive, to support such generalizations. Cubit (1984) points out that herbivores are abundant in the uppermost levels of rocky shores, the zone where algae show a distinct seasonality. Fast-growing algae are abundant in the cool winter months, but disappear in the hotter summer months, a correlation consistent with the upper shore desiccation hypothesis. Exclusion of limpets allowed the algae to persist through the summer months, and produced increased abundance and diversity in the winter. Cubit (1984) suggests that the establishment of perennial algae in this habitat is limited by the intensity of herbivory in the summer and by competition with transient algae in the winter.

Most ecological experiments in the subtidal have been concerned with establishing the effect of light, competition, and predation on the structure of algal communities, rather than determining which, if any, of these factors controls zonation. A recent exception is a series of algal removal experiments conducted in the low shore (*Fucus serratus*) and subtidal (*Laminaria digitata* and *L. hyperborea*) zones in the Isle of Man, U.K. (Hawkins & Harkin, 1985). They concluded that zonation was largely determined by space competition and that interactions between macrophytes were more important than predation in delimiting zones. Whittick (1983) examined the distributions of the four dominant epiphytes on the stipes of the kelp *Laminaria hyperborea*. Distinct zones occurred, and there was evidence of space competition between the species.

SEAWEED BIOGEOGRAPHY

Some taxonomic groups of algae are particularly well represented in certain geographic regions, for example the Laminariales in the North Pacific (Druehl, 1981), and the Fucales in Australasia (Nizamuddin, 1962). The Dictyotales and siphonalean green algae are particularly common in tropical waters, and in general the tropical flora lacks large species and has many calcified forms. Lubchenco and Gaines (1981) argue that the impact of grazing in the tropics is responsible for this flora, and the changes in herbivory with increasing latitude, such as the disappearance of herbivorous fishes, accounts for the change in seaweed form. The North Atlantic appears depauperate in kelp genera, for example *Macrocystis pyrifera* is absent but occurs in many other areas of cold water and is capable of growth in the region (Boalch, 1981). The Laminariales do not occur in Antarctica, and their niche is occupied by members of the Dersmarestiales (Moe & Silva, 1977). There is also a general agreement that brown algae increase in relative abundance compared to red algae with increase in latitude in the northern hemisphere. This is not seen in the southern hemisphere, and red algae appear particularly well represented in the subantarctic.

It is generally accepted that the occurrence of large numbers of species of a taxonomic group, including endemics, is indicative of a centre of origin, but without clear delimitations of ancestral forms, or a fossil record, it could equally be argued that these represent rapid speciation, which occurred as species moved into previously uncolonized habitats. There is clearly a need for further palaeological work, especially to relate modern distributions, earlier marine climates, and geological events.

Phytogeographic analyses depend for their information on the availability of local floras and checklists. In this regard the floras of the mid- and North Atlantic are particularly well known, and numerical analyses of these two regions have been made respectively by van den Hoek (1975) and Lawson (1978) (Fig. 7.26). Van den Hoek's analysis delimits five distinct phytogeographic regions which can be attributed to surface water temperatures. In 1915 Setchell divided the oceans into nine biogeographic zones with respect to algal distribution. These were defined by 5° C ranges of surface water temperature, except for the polar regions, which had 10° C ranges. Hutchins (1947) suggested that there are four critical temperatures which affect distribution of seaweeds: (1) the minimum for survival; (2) the minimum for reproduction; (3) the maximum for reproduction; and (4) the maximum for survival. Van den Hoek (1982b) has added two more temperatures, upper and lower limits for growth, and has reported on the distribution of benthic marine algae in relation to temperature and daylength regulation of their life histories (van den Hoek 1982b, c). Such analyses require detailed information from cultures of the factors responsible for growth and reproduction, e.g. Yarish *et al.* (1984), but are rewarding in their ability to explain the potential limits of geographic distributions. For example the gametophyte phase of the heteromorphic *Bonnemaisonia hamifera* is absent in Newfoundland, while the vegetatively reproducing *Trailliella* phase is common. Tetrasporogenesis in this species requires a combination of high water temperatures and short daylengths (Lüning, 1980), and these conditions are not met in Newfoundland, though isolates are genetically capable of tetrasporogenesis in culture. Similarly, species of *Ceramium* (Rhodophyceae) in Britain are fertile in the south, but remain sterile in Scotland (Edwards, 1973).

Physiological races are known for some algae.

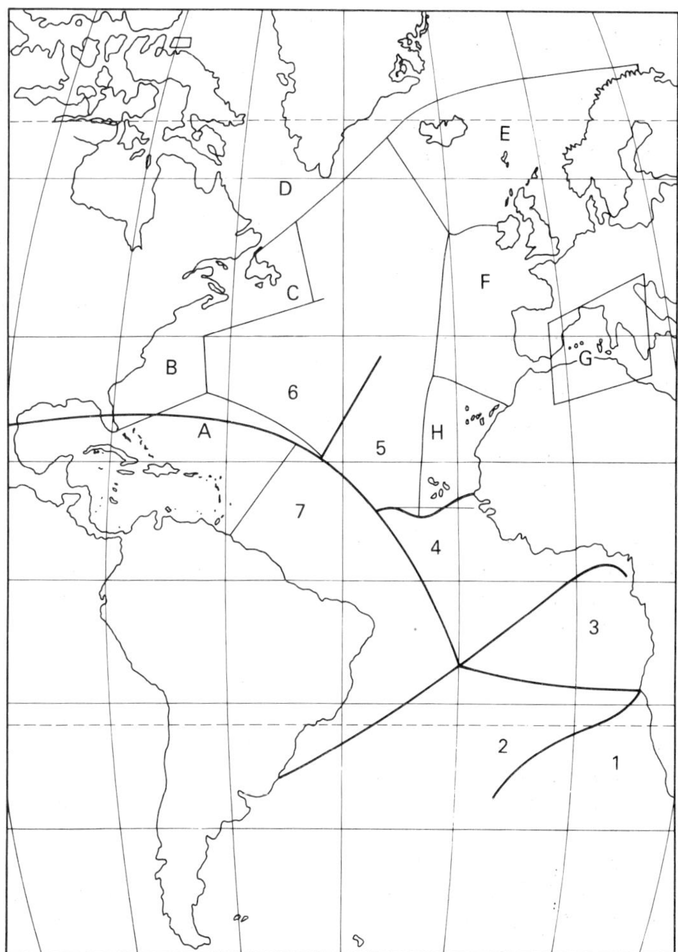

Fig. 7.26 The distribution of distinct seaweed floras in the North Atlantic (A–H) and central and South Atlantic (1–7). (After Druehl, 1981; data from van den Hoek [1975] and Lawson [1978].)

West (1972) showed the existence of genotypically determined strains of the red alga *Rhodochorton purpureum* with respect to sporulation, and Lüning (1980) showed latitudinal differences in the daylength requirement for production of the upright phase in the phaeophyte *Scytosiphon lomentaria* (Fig. 7.27). Bolton (1983) ascertained the optimal temperatures for growth, and maximum temperatures for long-term survival, of several isolates of *Ectocarpus siliculosus* collected from habitats ranging from Arctic to warm temperate. These had been maintained in culture for up to 20 years, and many had previously been shown capable of interbreeding. There was both genotypic and phenotypic variation with respect to the temperature effects, but no clear ecotypes emerged.

Temperature and daylength thus appear to be the major abiotic factors determining the distribution of seaweeds on a geographic scale. Factors such as salinity have some regional significance. Druehl (1978) determined that *Macrocystis integrifolia* had a reduced tolerance of low salinities at higher temperatures, and this caused its exclusion from parts of British Columbia. The effects of biological interactions, so important on a local ecological scale, are almost unknown at the biogeographic level. Two notable exceptions compare the distributions of congeneric species over latitudinal spans on the eastern and western seaboards of North America (Pielou, 1977, 1978). The species ranges overlapped to a considerable degree, and the closely related congeneric species overlapped to a greater degree

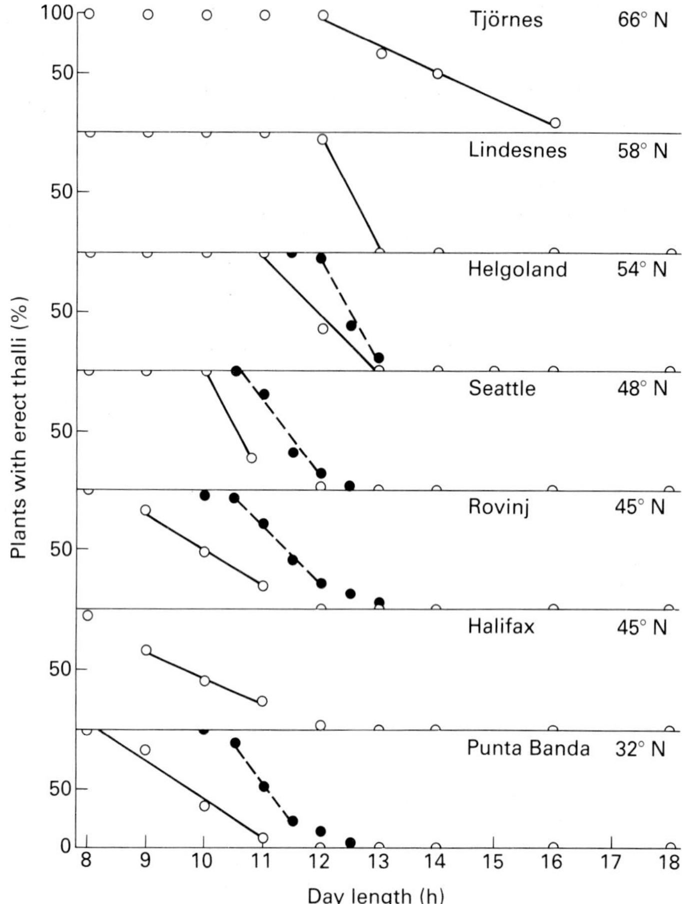

Fig. 7.27 Effect of daylength on erect thallus formation by different geographical strains of *Scytosiphon lomentaria* at 10°C (open circles) and 15°C (filled circles). Each value is based on a count of 250 plants. (After Lüning, 1980.)

than more distantly related species. It was concluded, therefore, that the geographic ranges of seaweeds are unaffected by interspecific competition.

Algae of sediments and sands

EPIPELIC ALGAE

The epipelic algae are found in both marine and fresh waters on the surface of sediments ranging from fine muds to coarse sands. The flora is dominated by pennate diatoms, cyanophytes, and motile algae of other groups; in fresh waters desmids may also be important. There is a well-documented vertical migration in the upper few millimetres of sediments in tidal environments, and these may be sampled by placing microscope slide cover glasses or lens tissue on the surface of the sediment.

Generally the cells move to the surface in the light and retreat into the sediment at night, though some may enter the sediment to avoid the high irradiances of the midday. Those from intertidal or estuarine marine sediments may adopt a rhythm based on tidal cycles. Palmer and Round (1965) suggested that the tidal rhythm showed by diatoms and *Euglena* in the Avon estuary was converted from a diurnal one by the periodic darkening of the sediments due to the high-turbidity waters. In some clearer estuarine waters the algae remain on the surface in the light in all but the flood tides (Perkins, 1960). This vertical migration is an endogenous rhythm, and is retained in the laboratory for several days under constant light conditions. *Euglena* retained a diurnal rhythm based on a light–dark cycle determined by the date of collection, while the movements of the diatom *Hantzschia virgata* remained correlated with the tides in nature (Fig. 7.28). This correlation was manifested by the occurrence of a daily advance of 50–60 min until low tide reached the evening, at which time the cycle was re-phased to pick up the morning low tide (Palmer & Round, 1967). In stream diatoms the upward migration commences prior to daylight, suggesting that, while the light–dark cycle may be important in entraining the rhythm, light *per se* does not trigger the migration.

The ecological significance of these vertical movements is not well understood; they may be mechanisms to avoid predation or to enable the algae to avail themselves of the higher nutrient levels found in the sediments. Nutrient enrichment experiments have shown that the growth of epipelic algae in salt marshes may be limited by the availability of nitrogen (Darley *et al.*, 1981)

EPISAMMIC ALGAE

In contrast to the epipelic algae the episammic flora is attached to the sediment particles. In both marine and fresh waters the flora consists in part of small coccoid chlorophytes and cyanophytes, which occupy depressions in the sediment particles. The most conspicuous components, however, are the diatoms, which may be closely adherent to the particles, or attached by short mucilage stalks; in contrast to the epipelic flora these are predominantly non-motile genera (Round, 1981). Moss (1977) reported that the episammic species were more resistant to extended periods of darkness and anaerobic conditions than those of the epipelon. In streams episammic algae may become detached and appear in the drift with peaks around midday. Mueller-Haekel (1973) suggests that this is due to a periodicity of cell division, where one cell remains attached to the substrate while the other is set free to reattach elsewhere.

MACROPHYTES AND UNCONSOLIDATED SEDIMENTS

Members of the Charophyceae are the most important rhizobenthic algae in fresh waters; they may form substantial communities in shallow waters of alkaline lakes and slow-flowing rivers. Their anchoring rhizoids consolidate particles and increase sediment stability (Round, 1981).

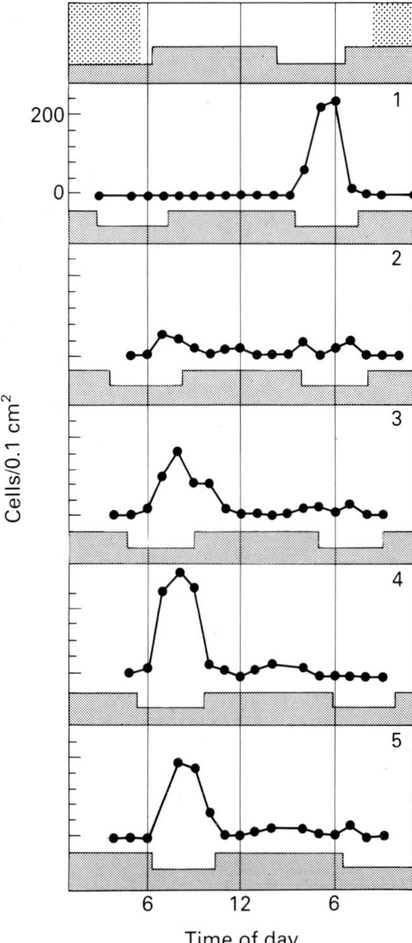

Fig. 7.28 Persistence of vertical migration, in constant light in the laboratory, of the diatom *Hantzschia virgata*. Top panel indicates tide state on day of collection from the field; 1–5 indicates subsequent consecutive days. Stippling indicates dark periods. Bar graph at the bottom of each panel indicates the time of high and low tide in the field. Note vertical migration phase shift between days 1 and 3. (After Palmer & Round, 1967.)

In the marine environment the rhizobenthic species are mainly tropical, siphonous green algae such as *Caulerpa spp*, *Halimeda* and *Udotea*,

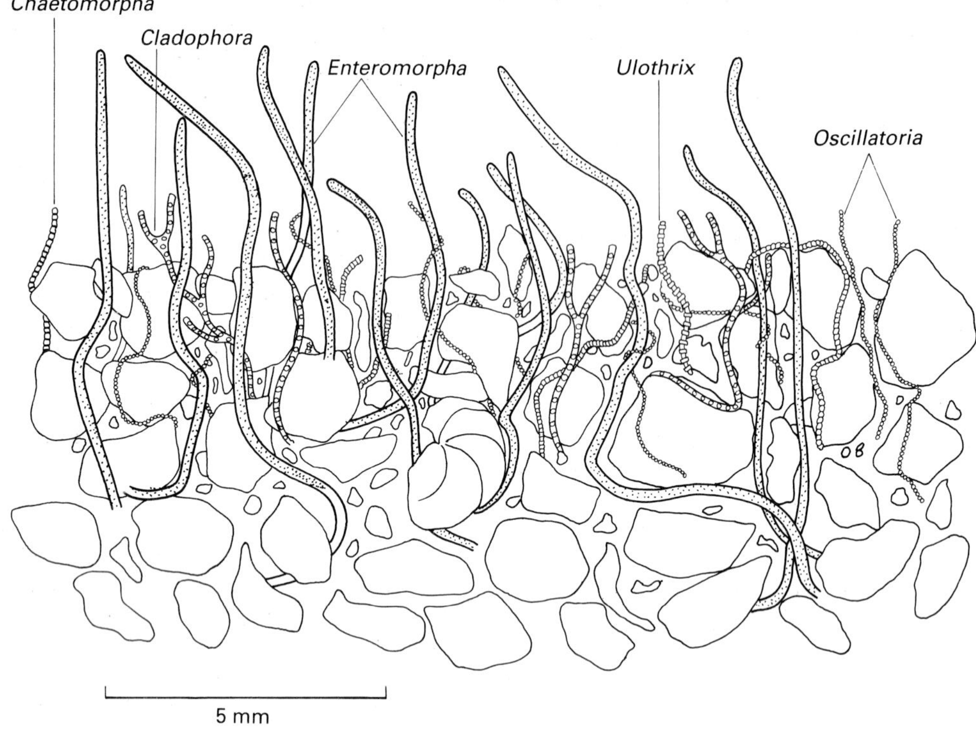

Fig. 7.29 Mat-forming filamentous algae binding sand particles. (After Scoffin, 1970.)

which produce extensive sediment-trapping rhizoidal systems, enabling them to resist moderate water movement (Fig. 7.29) (Scoffin, 1970; van den Hoek *et al.*, 1972).

Other filamentous marine algae, such as *Audouinella* spp., may also have important ecological roles in mineral deposition and sand-binding. They thus have some similarities with stromatolites (Fig. 8.4), which are rock-like structures of tropical distribution formed by cyanophytes. The classic site for modern stromatolites is Shark Bay in Western Australia, where they form in sheltered hypersaline waters (Logan, 1961).

Unattached or loosely attached seaweeds are also found in sheltered locations. Burrows (1958) listed 87 species in Port Erin Bay, Isle of Man, while in sheltered inlets in Florida about 60 species were reported (Hamm & Humm, 1976). Other species appear to be capable of persisting in more exposed sandy sites, where they may be found floating free in the water column, e.g. the brown alga *Pilayella* (Wilce *et al.*,. 1982) and the red alga *Amoenothamnion planktonicum* (Womersley & Norris, 1959).

Salt marshes, which are principally dominated by vascular plants, are another habitat in which loose-lying or unattached algae may be found (Chapman, 1974; Ranwell, 1972). Some filamentous forms, such as *Enteromorpha prolifera* and *Rhizoclonium* sp., grow entangled around the base of the stems of the vascular plants. Others (e.g. *Catenella repens; Bostrychia scorpoides*) are epiphytes, while some, such as *Vaucheria* spp. (Xanthophyceae), grow attached to the sediments. Cyanophytes (*Lyngbya; Oscillatoria; Phormidium*) are particularly common on salt marsh sediments and, together with the rhizophyllous bacteria, fix considerable amounts of nitrogen (Patriquin & McLung, 1978). Similar associations of algae, with the addition of such tropical genera as *Caulerpa*, are also found in the mangrove associations, which replace salt marshes in frost-free regions.

In the northern temperate salt marshes the

fucoids are an important component of the flora. Some are unattached, while others may be loosely embedded in the sediment. These are the same species which occur as epiliths on rocky shores, but they show considerable morphological and reproductive modification (Norton & Mathieson, 1983), and are frequently designated as varieties, or more properly as ecads. Usually they have dwarfed, twisted thalli, which lack holdfasts, and in extreme instances they form spherical balls, of which the best known examples are of *Ascophyllum nodosum* ecad *makaii* (South & Hill, 1970). Such fucoids seldom become fertile, receptacles are rare, and if formed produce few if any viable gametes; reproduction is thus totally vegetative. This absence of reproduction, other than by vegetative means, is a characteristic phenomenon of unattached, or loosely attached, species, and occurs whether their reproductive structures are elaborate, such as the fucoid receptacles, or little-modified vegetative cells, as in green algae such as *Cladophora* spp.

Soil algae

Algae are of widespread occurrence in the soils of all continents, from the polar regions to hot deserts (Starks *et al.*, 1981). The soil algal flora is drawn principally from members of the Chlorophyceae, Cyanophyceae, Bacillariophyceae, and Xanthophyceae, and is sufficiently distinct that it should not merely be regarded as a part of the aquatic flora living in an unfavourable environment (Round, 1981).

They are a taxonomically difficult group, and identification requires their isolation in culture. Such methods are undoubtedly species-selective, and a variety of media and culture conditions must be utilized in order to minimize any bias. These methods have been pioneered largely by Harold Bold and his students, and reviewed, together with the systematics and ecology of soil algae, by Metting (1981) and Starks *et al.* (1981).

It is even more difficult to obtain quantitative data. Sharabi and Pramer (1973) compared methods based on direct observation, culture, and chlorophyll levels, and concluded that the last-mentioned gave the highest estimates of algal abundance. It is clear that there is a need to standardize methods of collection and culturing of soil algae before valid comparisons of the soil flora can be made.

Feher (1948) identified 685 taxa of soil algae reported in the literature and concluded that it was not possible to determine any geographic distributions or correlations with soil types. It would appear, however, that alkaline soils support an abundance of cyanophytes, and that these are not found on soils with a pH of less than 5 (Brock, 1973; Starks *et al.*, 1981). In contrast, members of the Chlorophyceae are often found on acid soils. Shields and Durrel (1964) point out that many arid sites are alkaline and many wet sites acid, therefore soil moisture may also have to be taken into consideration. Filamentous cyanophytes are frequently reported as stabilizers of bare and eroded soils, while green algae may also produce mucilages, which contribute to the humus. There is evidence of a positive correlation between the diversity and abundance of algae and the age of soils and levels of organic matter and nutrients in dated volcanic soils (Carson & Brown, 1978). Floristic diversity is also greatest in undisturbed soils (King & Ward, 1977). Soil algae grow better in partially dry soils (Hunt *et al.*, 1979), and as many species produce thick-walled akinetes or hypnozygotes they are thus well adapted to withstand drought conditions. In the dry state some may withstand temperatures of 140° C for 1 hour and up to three months at 40° C (Trainor, 1983).

While many members of the flora occur on the surface of the soils others, presumably washed in, exist in the non-photic zone below the surface. Such algae may be present as resting stages, but others are presumably capable of heterotrophic growth (Droop, 1974a). Sheath and Hellebust (1974) showed that *Bracteococcus minor* (Chlorophyceae) could grow heterotrophically on glucose, and suggested that the evolution of such a carbohydrate uptake mechanism in a soil alga would not have occurred if it did not have survival value. It suggests that such species may be capable of at least a maintenance metabolism when washed into the soil below the photic zone.

Symbiotic associations

Algae are frequently found growing in close association with a variety of other organisms including algae, fungi, vascular plants, or invertebrates. A current definition of symbiosis is 'an association for significant portions of their life cycles of individuals that are members of different species'

(Margulis, 1981). Thus symbionts must include all degrees of parasitism as well as mutualistic and commensalistic associations (Goff, 1982a).

ALGAE–INVERTEBRATE ASSOCIATIONS

These associations have attracted both zoologists and phycologists, and there are many literature reviews, e.g. Glider and Pardy (1982), Goff (1983), Taylor (1973), and Trench (1979).

Trench (1975, 1982) has reviewed chloroplast- and cyanelle-based associations. Among the best known are those which occur in saccoglossan molluscs which feed by sucking out the cytoplasm of such siphonous green algae as *Codium* and *Caulerpa*. The chloroplasts accrue in the cells of the hepatopancreas, where they initially remain intact (Trench, 1980; Hinde, 1983) and continue to photosynthesize, but are unable to divide, or to synthesize chlorophyll or RuBP carboxylase, actions which in part are under the control of the nuclear genome. *Codium fragile* plastids, *in vitro*, normally leak 15–20% of freshly fixed carbon, mostly as glycolate. This may be increased to 74% by exposure to homogenates of the hepatopancreas (Gallop, 1974). The released photosynthate is metabolized by the host (Trench *et al.*, 1973), and *Elysia* deprived of both light and food loose weight at twice the rate of those deprived of food alone (Hinde & Smith, 1974). Eventually the chloroplasts are digested, and in well-fed *Elysia viridis* one-third to one-half are replaced each week (Gallop *et al.*, 1980).

A mutalistic association is clearly not established. The mollusc feeds on the alga and perhaps the only benefit to the plant is from a reduction in grazing pressure, which may occur due to the photosynthetic supplement obtained from the chloroplasts (Hinde, 1983).

Members of the Cyanophyceae, Prochlorophyceae, Chlorophyceae, Prasinophyceae, Bacillariophyceae, Prymnesiophyceae, Dinophyceae, and Rhodophyceae have been reported as endosymbionts from a variety of both freshwater and marine hosts. Three types of association have been most widely studied.

1 The prasinophyte *Tetraselmis* (= *Platymonas*) which occurs in the marine flatworm *Convoluta*.
2 The predominantly freshwater associations involving green algae, which ultrastructurally resemble *Chlorella*, known, particularly in the zoological literature, as zoochlorellae.
3 The marine dinophyceaen bionts termed zooxanthellae, the most widespread of which is *Symbiodinium microadriaticum*, which shows considerable morphological variation. In animal tissue it is coccoid, but when free-living may be motile with *Gymnodinium*-like characteristics (Loeblich & Sherley, 1979). Taylor (1974) considered it to inhabit as many as 80 species of marine invertebrates in three phyla (Protozoa; Coelenterata; Mollusca). However, biochemical and infectivity characters suggest that a number of genetically separate entities occur (Schoenberg & Trench 1980a, b, c).

Algae–invertebrate associations may be regarded as stable composite organisms (Glider & Pardy, 1982) with characteristics which enable them to occupy niches from which their component bionts are excluded. These characteristics include mechanisms which assure the continuation of the association through reproduction and mechanisms for the translocation of materials between the bionts.

Symbionts may be transmitted to the offspring during reproduction; in *Hydra* they are incorporated into the fertilized ovum (Muscatine & McAuley, 1982), or may be acquired *de novo* from the environment by the larvae of the host. Free-swimming *Platymonas convolutae* are chemotactically attracted to the egg cases of *Convoluta* (Holligan & Gooday, 1975). Some nudibranchs acquire their dinoflagellate symbionts by grazing on infected coelenterates and subsequently incorporate the ingested cells into their tissues.

Considerable nutrition is derived from the algal symbionts. In the light aposymbiotic anemones utilize lipid reserves, while carbohydrates, presumably derived from the algae, are metabolized in symbiotic specimens (Fitt & Pardy, 1981). Adult *Convoluta* are completely dependent on their symbionts, and will starve if they are kept in darkness. In contrast, *Hydra viridis* is less dependent on its *Chlorella* symbiont, and aposymbiotic individuals, while not found in nature, can be produced and maintained in the laboratory (Pardy, 1981). In reef-forming corals, calcification and growth are correlated with the abundance of symbionts, and over 85% of the

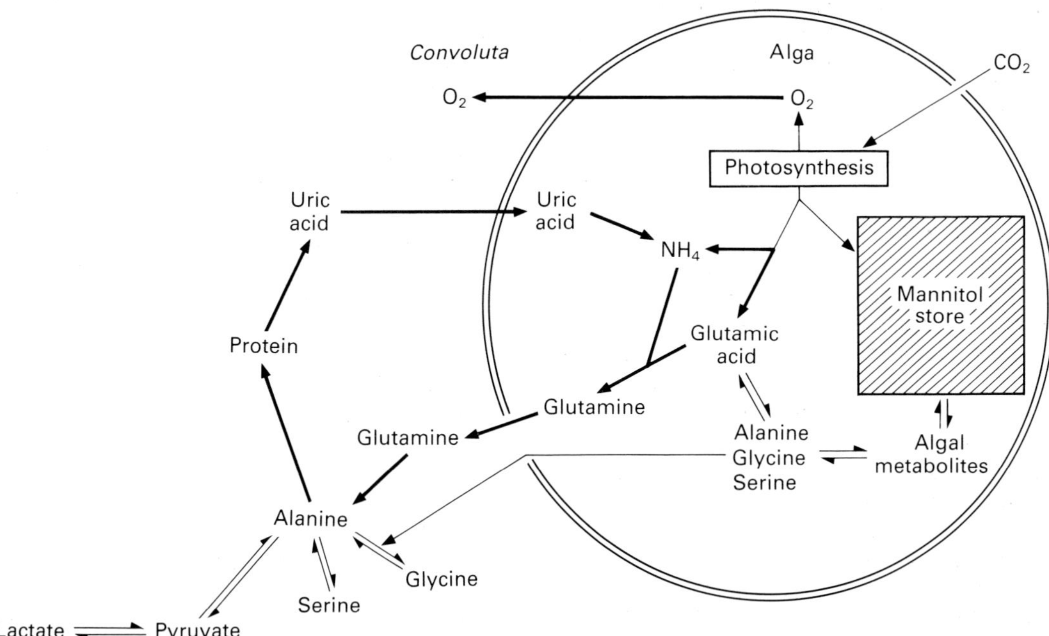

Fig. 7.30 Postulated interchanges of nutrients between the flatworm *Convoluta* and its algal symbiont. (After Boyle & Smith, 1975.)

organic carbon of the coral is obtained from the algae (Muscatine & Porter, 1977). The photosynthetic pigment concentration of these zooxanthellae increases with depth (Dunstan, 1979), which presumably compensates for the decrease in irradiance in a similar manner to that of free-living sun and shade plants. The exchange of metabolites between algae and host has been demonstrated in many associations, with most studies on the transfer of carbon from the alga to the animal. In corals this may amount to 50% of the carbon fixed, which in the animal cell is found mainly as protein or lipid, though it is transferred largely as glycerol (Muscatine, 1973). In *Chlorella*-based associations the translocated metabolite is maltose (Ziesenisz *et al.*, 1981), while *Platymonas convolutae* produces amino acids (Fig. 7.30) (Boyle & Smith, 1975).

Internal carbon cycling has been demonstrated in the coral *Porillopora capitata*, where the zooxanthellae photochemically metabolize acetate derived from the host to produce fatty acids. In addition, corals with symbionts show active uptake of inorganic nitrogen and phosphorus, whereas aposymbiotic forms show a net loss to the environment. The difference is due to the uptake by the zooxanthellae from the host, showing a symbiotic adaptation to nutrient conservation in nutrient-poor tropical waters (Muscatine, 1980). *Platymonas* obtains its nitrogen from *Convoluta* as uric acid, and this accumulates in the host when photosynthesis is inhibited.

From an evolutionary perspective one of the more interesting algal invertebrate associations is between the prokaryotic *Prochloron* and various tropical ascidians (Lewin, 1984). The alga occurs within the tunicine matrix of these invertebrates but penetrates neither the cells nor organs. *Prochloron* is capable of photosynthesis in host and *in vitro* (Fisher & Trench, 1980), but is auxotrophic for tryptophan (Patterson & Withers, 1982). Our knowledge of the biology of *Prochloron* is handicapped by the inability to grow this alga for extended periods separate from the animal host. The discovery of a free-living protochlorophyte from a eutrophic lagoon in the Netherlands is thus of considerable importance (Burger-Wiersma *et al.*, 1986).

ALGAE–PROTOZOA INTERACTIONS

The interactions of photosynthetic bodies of otherwise heterotrophic protozoans have been largely studied from an evolutionary perspective. In some instances the bodies are clearly related to free-living algae, the endosymbionts of *Peridinium balticum* (Thomas & Cox, 1973) for example, while others, such as the cyanelles of the crytomonad-like *Cyanophora*, have a plastid-like genome (Herdman & Stanier, 1977), with some of their activities, such as the synthesis of photosynthetic pigments, under the partial control of the host nuclear genome (Trench & Siebens, 1978).

The modern foraminiferans are hosts to a variety of algae, including members of the Chlorophyceae, Dinophyceae, Bacillariophyceae, and Rhodophyceae (Lee & McEnery, 1983). The common dinoflagellate symbiont of corals, *Symbiodinium microadriaticum,* is reported from a number of species. These protozoans are particularly abundant in shallow tropical and subtropical seas. Foraminiferal sands may be 50 times more productive than the plankton in the same location (Sournia, 1976) and also provide a tight recycling of nitrogen and phosphorus between the alga and its host (Lee, 1980). Light is essential for the growth of some species. *Amphistegina lessonnii* grows at a rate proportional to light intensity and does not grow in the dark even when provided with food (Röttger *et al.*, 1980), however starved *Globigerinoides sacculifer* grows only slowly and may fail to reproduce (Bé *et al.*, 1981). The algae also play a role in the calcification, with about 20% of the carbon fixed photosynthetically subsequently appearing in the foraminiferal shell (Smith & Wiebe, 1977). The mechanism whereby the animal receives carbon compounds from the alga is not clear. Kremer *et al.* (1980) concluded that photosynthate was released from the living algal cells rather than by digestion of the algal symbionts, but Lee and McEnery (1983) are sceptical and and suggest on ultrastructural grounds that algal-fixed carbon may pass to the host after the death of the algae.

In addition to symbioses with animals, algae also form intimate and frequently obligate associations with other algae, with green 'higher' plants and with fungi.

ALGAE–FUNGAE ASSOCIATIONS

The fungal associations which make up the lichens are unique among algal symbioses in that they have a morphology, physiology, and ecology which is unlike that of either the mycobiont or phycobiont component (Ahmadjian, 1981). All but the simplest lichens display a distinct internal structure, with the algae restricted to zones in the subsurface layers of the thallus (Fig. 7.31a, b). Recent descriptions of their biology can be found in Lawrey (1984) and Kershaw (1985).

Both phycobionts and mycobionts are capable of independent growth in axenic culture and there is, therefore, no obligate relationship (Ahmadjian, 1981). Lichens are frequently cited as examples of mutualism, but, while they may occupy niches unavailable to the phycobiont in the free-living state, it is difficult to see any advantage conferred on the individual algal cells.

In most lichens the mycobiont component constitutes more than 90% of the lichen biomass, and almost 30% of all fungal species are found as mycobionts. Mostly these are ascomycetes, though some basidiomycetes occur (Smith, 1978). The algal components are chlorophytes or cyanophytes; approximately 30 genera are known, of which the chlorophytes *Trebouxia* and *Pseudotrebouxia* account for more than 50% of phycobionts (Ahmadjian, 1981).

Chimeroid lichens, such as the chlorophyte-containing lichen *Peltigera*, have two algal components, and produce cephalodia, outgrowths on the surface, which contain the cyanophytes.

All the available evidence points to a one-way flow of nutrients from the alga to the fungus, comprising of carbohydrates in the primary thallus and ammonia in the cephalodia. Up to 90% of the total carbon fixed in photosynthesis moves to the fungus. Green algae excrete polyols, erythritol, ribitol, or sorbitol (Farrar, 1978), and the cyanophytes produce glucose (Hill, 1976); these algal photosynthates are converted to mannitol in the fungal thallus (Fig. 7.32). Excretion of carbohydrates from the phycobiont ceases almost immediately on their isolation into culture (Smith, 1980). The supply of carbohydrate is many times more than is required for the growth of the fungus. The additional amount may be utilized in dealing with such stress-related phenomena as the enhanced respiration which occurs

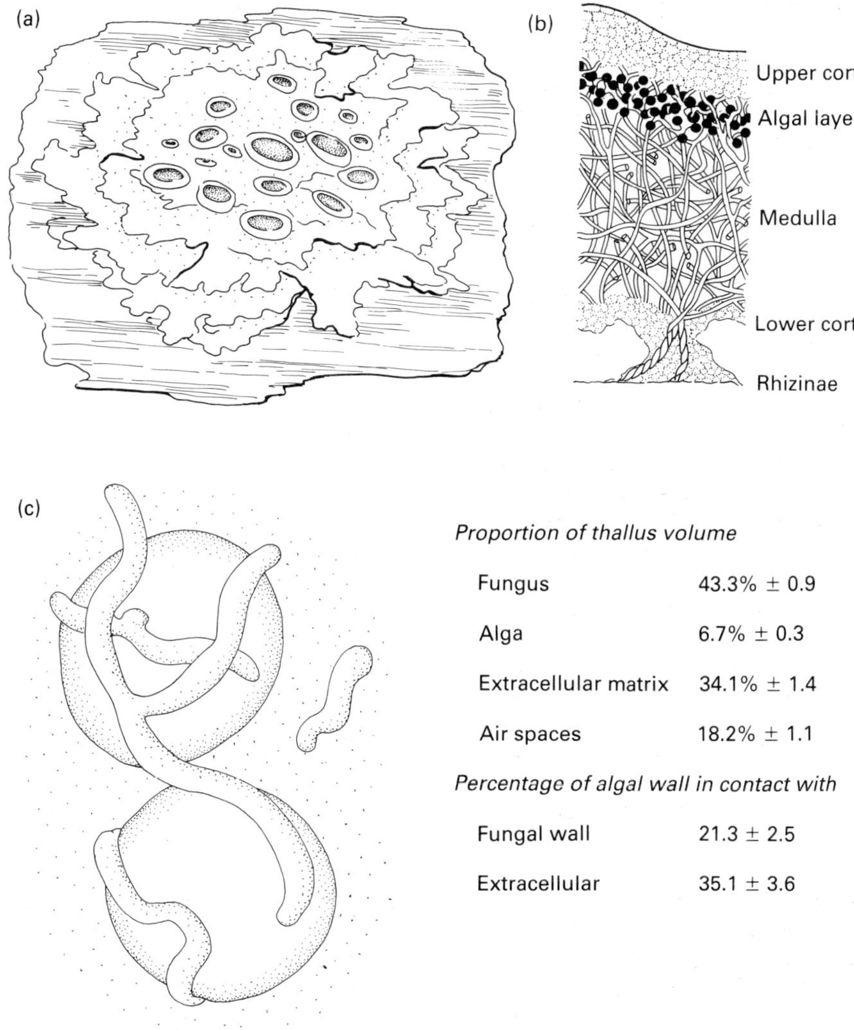

Proportion of thallus volume

Fungus	43.3% ± 0.9
Alga	6.7% ± 0.3
Extracellular matrix	34.1% ± 1.4
Air spaces	18.2% ± 1.1

Percentage of algal wall in contact with

Fungal wall	21.3 ± 2.5
Extracellular	35.1 ± 3.6

Fig. 7.31 Morphology and anatomy of the lichen *Xanthoria parietina*. (a) Habit. (b) Cross-section of thallus; note algae located immediately beneath the lower cortex. (c) Fungal hyphae in close contact with the surface of the algal cells. (After Ahmadjian, 1981.)

during the rehydration of a lichen thallus and the osmotic problems related to desiccation (Farrar & Smith, 1976). Although the symbionts may be intimately linked through either fungal haustoria or specialized contact of their cell walls in appressoria (Fig. 7.31c), there is no evidence that these are the routes of carbohydrate transfer.

The ability of lichens to fix atmospheric nitrogen is dependent upon their possession of cyanophytes, either as the primary phycobiont or as part of the cephalodium (Ahmadjian, 1981). Nitrogen fixation occurs in the heterocysts, which occur more frequently (21% *v*. 4% of cells) in the phycobiont cyanophytes than in the free-living state. Relatively little photosynthesis appears to occur in the cephalodia, and Hitch and Millbank (1975) suggest that the primary green algae provide carbohydrate to the cyanophytes. Glutamate synthetase activity, by which ammonia is converted to glutamate, is suppressed in these cyanophytes; the ammonia produced is excreted to the mycobiont. Ironically the alga may show symptoms of nitrogen starvation (Stewart & Rowell, 1977). The mycobiont appears to have regulatory control over glutamine synthetase, as normal activity of this enzyme resumes when the phycobiont is isolated from the fungus (Sampaio *et al.*, 1979).

Many species of lichen reproduce by vegetative means, involving structures which ensure the co-dispersion of the mycobiont and phycobiont. These are vegetative diaspores of which the simplest, the soredia, consist of 25–100 μm diameter balls of a few phycobiont cells wrapped in mycobiont hyphae. When the fungal component reproduces sexually the spores must locate an algal partner to enter into the lichenized state. Some common phycobionts, especially members of the Cyanophyceae, are frequently found in lichen habitats, but the commonest, the trebouxoid species, are only occasionally found free-living in nature (Tschermak-Woess, 1978). Although usually reproducing in the lichenized state by aplanospores, the *Trebouxia* phycobiont may produce, and liberate, zoospores (Slocum *et al.*, 1980). These could serve as a source of phycobiont for the germinating fungal spores.

Ahmadjian and Jacobs (1981) have been successful in axenically synthesizing the lichen *Cladonia cristatella* in the laboratory. The technique involved the mixing of clumps of algae and the mycobiont hyphae on freshly cleaved strips of mica, which had been soaked in the algal culture medium and incubated at a relative humidity of *c.* 95%. They showed that the relationship of the fungus to the alga is an aggressive one; several species of alga could be lichenized, but others were parasitized and destroyed.

Symbiotic associations of marine algae and marine fungi exist in which the fungal component grows intercellularly in the alga, but does not modify the morphology of the alga, nor lead to the production of chemical compounds unique to the association. Such symbioses, even while apparently obligate, cannot be regarded as true lichens, and have been termed mycophycobioses (Kohlmeyer & Kohlmeyer, 1972). The best known of these is *Mycosphaerella ascophylli*, which occurs in the phaeophytes *Ascophyllum nodosum* and *Pelvetia canaliculata*. It grows intercellularly in the algae, without penetrating the cells, and forms ascocarps and spermagonia in the host receptacles which may appear on the their surfaces as black dots. In axenic culture the fungus has been shown to utilize mannitol and laminarin, both of which are produced by the host (Fries, 1979).

ASSOCIATIONS WITH BRYOPHYTES AND VASCULAR PLANTS

Heterocystous members of the Nostocaceae are found in association with several bryophytes, such as the hornwort *Anthoceros* and the liverwort *Blasia*. In the vascular plants they occur in the aquatic, heterosporous fern *Azolla*, several genera of cycads, and the angiosperm *Gunnera* (Peters & Calvert, 1983).

Anthoceros and *Blasia* have cavities in their gametophyte thalli containing *Nostoc*, in which heterocyst frequency may approach 60% of the vegetative cells. Nitrogen is excreted as ammonia and taken up by the bryophyte host, which presumably supplies fixed carbon to the alga (Stewart & Rodgers, 1977). As in the cephalodial lichen associations, the activity of glutamine synthetase is arrested but resumes on isolation of the alga.

Azolla contains *Anabaena azollae* in cavities in the upper lobes of its floating leaves (Fig. 7.33). These differ from other cyanophyte components of vascular plant symbioses in the difficulty of their isolation, with the majority of the cells apparently incapable of a free-living existence (Peters & Calvert, 1983). Newton and Herman (1979) have isolated some strains which are capable of dark heterotrophic growth and which show a reduced nitrogen fixation compared to those in symbiotic association.

There is considerable synchrony in the development of the association. The *Anabaena* filaments come into contact with the leaf primordia in the apical meristem of the *Azolla*. An initial pocket develops on the underside of the leaf, and this closes to form a cavity, trapping the alga; algal division and heterocyst formation then occur within this cavity. At maturity the heterocyst frequency ranges from 20 to 60%; the symbiont accounts for 15% of the biomass of the

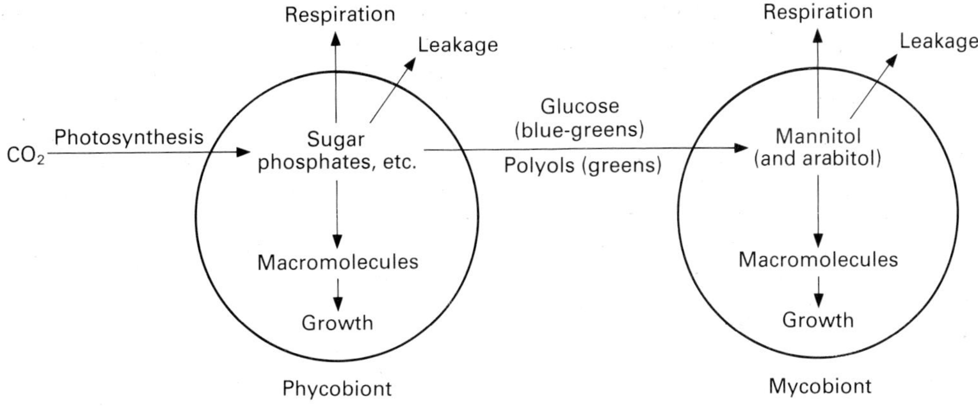

Fig. 7.32 Carbon flow through lichens. (After Darley, 1982.)

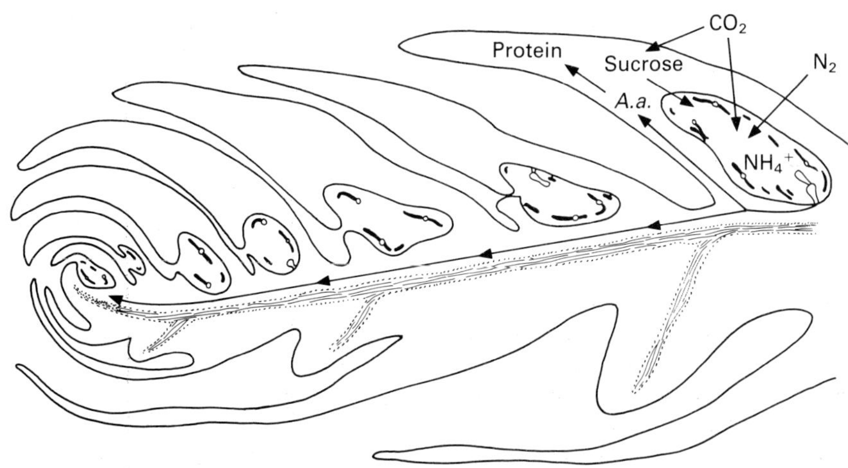

Fig. 7.33 Longitudinal section through the axis of the fern *Azolla* showing stages in the formation of the leaf cavity in the leaves and the entrapment of the symbiotic cyanophyte *Anabaena azollae*. *A. a.*, amino acids. (After Peters, 1978.)

association and is capable of providing all the nitrogen requirements of the *Azolla* (Fig. 7.34) (Peters & Calvert, 1983).

Among the higher vascular plants the cyanophyte associations of the cycads are found in specialized coralloid root nodules, while those of *Gunnera* may occur in the leaf bases (Whitton, 1973). The association of *Nostoc* with *Gunnera* is exceptional among the photosynthetic plants in that the filaments may become intercellular (Schaede, 1951)

PARASITIC ALGAE

Two major groups of algae, the Chroolepidaceae (Chlorophyceae) and the Rhodophyceae, provide examples of situations where algae become the parasites and are, to a varying degree, nutritionally dependent on another host plant.

The best known parasitic members of the Chroolepidaceae are species of *Cephaleuros*, which are tropical or subtropical in distribution. The algae grow beneath the cuticle of the host,

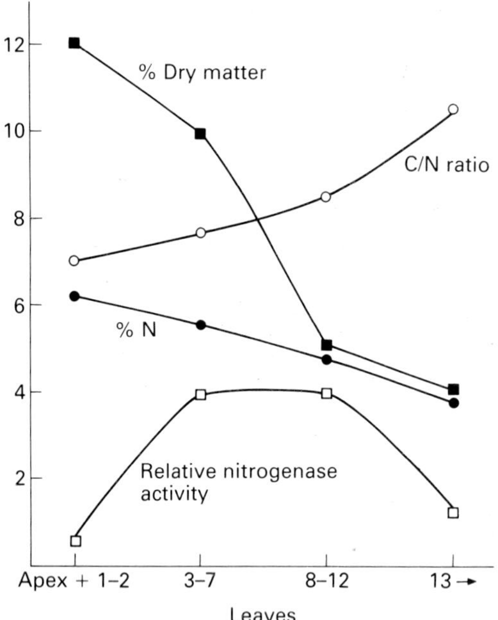

Fig. 7.34 The relative nitrogenase activity, percentage dry matter, ratio of carbon to nitrogen, and the percentage nitrogen on a dry weight basis in segments of the main stems of *Azolla* bearing sequential groups of leaves. (After Peters & Calvert, 1983.)

on the stems, leaves, flowers, and fruit. They may produce orange-red haematochrome pigments in their emergent reproductive structures, giving a fungal-like, red-rust appearance. A wide variety of vascular plants may be infected including tea, coffee, citrus, and many other commercial species. In Louisiana alone, Holcomb (1975) reported 115 species susceptible to *C. virescens*. While a wide variety of host species appear capable of infection, there is considerable variation in the degree of susceptibility even within a single species.

Zoosporangia develop on special fertile branches which are elevated above the surface of the leaf. They are thick-walled, but early in their development an escape pore forms which is blocked by a plug of proteinaceous material. The zoosporangia absciss and are distributed by wind, water, insects, and arachnids. When there is sufficient water the plug dissolves and the zoospores are released. To survive they must rapidly penetrate the cuticle of the host plant, and they are aided in this by the animal dispersal agents, which damage the surface of the leaves during their feeding activities.

Culture experiments show that *Cephaleuros* is potentially capable of obtaining nutrition from the host, as the addition of glucose or yeast extract to cultures caused a marked stimulation in growth (Joubert *et al.*, 1975). Heterotrophic growth was also obtained in complete darkness on a range of hexose sugars, although reproduction did not occur until the addition of low concentrations of auxins (Jose & Chowdray, 1978). This suggests that *in vivo* reproduction of the alga may be dependent on hormones produced by the host.

The red algal parasites are a taxonomically diverse assemblage of over 100 species in 50 genera. They fall into two distinct groups: most are adelphoparasites, which are taxonomically related to their hosts, i.e. same family or tribe; the remainder are alloparasites, which are not. Parasitic algae are generally small, < 0.5 cm, with reduced or absent pigmentation, and are morphologically simple, composed of branching filaments of cells which penetrate between the host cells. The reproductive structures are borne in a mass of emergent tissue on the surface of the host (Evans *et al.*, 1978; Goff, 1982b).

Red algal parasites only occur on other red algae and appear to be highly host-specific. *Janczewskia morimotoi* in nature appears confined to *Laurencia nipponica*, but it is reported to be capable of infecting other *Laurencia* spp., but not other rhodomelacean hosts (Nonomura & West, 1981). In contrast cross-inoculation tests on *Asterocolax gardneri*, which is found on the delesserian hosts *Nienburgia andersonniana, Phycodrys setchellii* and *Polyneura latissima*, show that the parasite, while putatively a single species, is apparently incapable of infecting species other than the one from which it was derived (Goff, 1982b).

The nutritional status of only a few red algal parasites has been examined, but most, especially those in which pigmentation is reduced or absent, are assumed to be at least partially dependent upon their hosts. In *Holmsella pachyderma*, a parasite of *Gracilaria verrucosa*, floridoside is translocated to the host, where it is converted to mannitol and then starch, setting up a source–sink concentration gradient and allowing photosynthate movement by diffusion (Evans *et al.*,

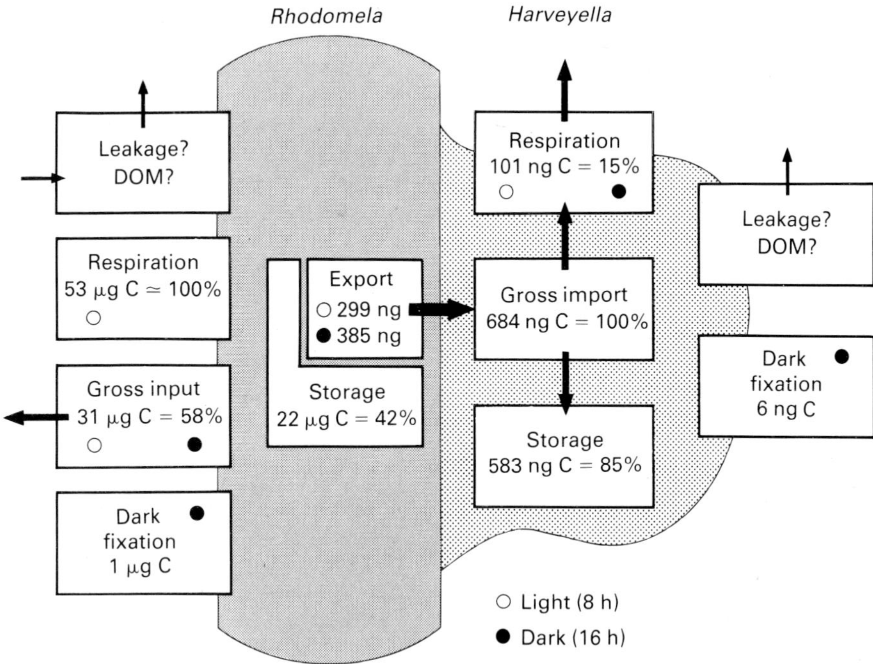

Fig. 7.35 Daily carbon budget for the nutritional relationship between *Rhodomela confervoides* and the parasite *Harveyella mirabilis*. It is assumed that the host has a mass of 200 mg and the parasite 0.05 mg fresh weight. DOM, dissolved organic matter. (After Kremer, 1983.)

1973). *Harveyella mirabilis* receives its carbon as digeneaside and perhaps as some amino acids (Fig. 7.35) (Kremer, 1983). In contrast, the larger, more highly pigmented parasites are capable of photosynthetic production of sugars and amino acids, with no evidence of host–parasite translocation (Court, 1980). Other such photosynthetic parasites, e.g. *Erythrocystis saccata*, develop normally only in the presence of the host, which suggests the involvement of growth regulatory substances (Goff, 1982b).

Secondary pit connections appear to be an inherent feature of all red algal host–parasite associations. In *Choreocolax polysiphoniae* and its host *Polysiphonia confusa*, nuclei and other cytoplasmic components of the parasite cell pass through this connection into the host cell. The transferred nuclei remain intact in the host cells and stimulate host nuclear DNA synthesis. There is also an accumulation of starch and a proliferation of host cell organelles. Thus, genetic information from the parasite may be transferred to the host, and may redirect its physiology for the benefit of the parasite (Goff & Coleman, 1985).

EPIPHYTES

Epiphytes, found on other algae or vascular plants, are the least specialized of symbiotic algae. Many species are equally likely to be found on adjacent inorganic substrates and show no special specificity or relationship with the host plant. Tokida (1960) listed almost 300 species of macroalgae occurring epiphytically on members of the Laminariales. Harlin (1980) lists over 400 macroalgal epiphytes and over 150 microalgal epiphytes (mainly diatoms) on seagrasses. Epiphytism for most species appears to be a consequence of competition for substrate, light, and nutrients in a crowded environment.

Other species, at least in nature, are obligate epiphytes found only on at best a narrow range of host species. Unlike parasites, however, there is no obvious movement of nutrients from the host. The best studied of these apparently obligate associations occurs between the red alga *Polysiphonia lanosa* and the fucoid *Ascophyllum nodosum* (Turner & Evans, 1977). The epiphyte produces rhizoids, which penetrate the host tissue, apparently by means of enzymatic digestion (Rawlence, 1972), but no transfer of metabolites has been detected (Harlin & Craigie, 1975; Turner & Evans, 1978). In fresh waters Cattaneo and Kalff (1979) found no difference in biomass or production between epiphytes on *Potamogeton* and those grown on plastic strips. The epiphytes did, however, show reduced acid phosphatase activity, suggesting organic phosphorus compounds might be obtained from the vascular plant (Fig. 7.36).

The deleterious effects of epiphytes on their hosts are due to shading and mechanical loading. D'Antonio (1985) reported a reduction in growth and an increase in thallus breakage in heavily epiphytized plants of *Rhodomela larix* on the Oregon coast (Fig. 7.37).

Some host plants have developed mechanisms which may limit the growth of epiphytes. Several brown algae are known to slough off the outer surface of the epidermis, e.g. *Halidrys siliquosa* (Moss, 1982) and *Himanthalia* (Russell & Veltkamp, 1982), while polyphenols excreted by seaweeds may inhibit epiphyte growth (McLachlan & Craigie, 1964).

Algae at temperature extremes

SNOW ALGAE

Snow algae occur worldwide in areas in which permanent or semi-permanent snow banks are found. The commonest are species of chlamydomonads, but euglenoids, chrysophytes, dinoflagellates, cryptophytes, cyanophytes, xanthophytes, and diatoms have also been reported (Hoham, 1980). There is considerable confusion in their taxonomy; many may be stages in the life cycles of other species and many supposed snow chlorococcalean species are known to be zygotic stages of *Chloromonas* spp. (Fig 5.18). Hoham (1975a, b) has suggested that true snow algae have an optimum growth below 10 °C. *Chloromonas pichinchae* would not grow above 10 °C and was palmelloid at this temperature; the motile phase only occurred between 1 and 5 °C.

Chlamydomonad blooms are initiated when the air temperature remains above freezing for several consecutive days and allows melt water to develop in the snow bank. During this period light penetrates the snow, and nutrients and dissolved gases become available to the algae. Resting zygotes may germinate in the snow, or at the snow–soil interface, and motile cells then swim to the surface and reproduce via zoospores. The movement is phototactic, *Chloromonas pichinchae* moves to the surface at dawn and dusk, but may avoid the high light of midday by moving back 10–15 cm into the snow. After a few days, gametes are formed, and the resulting zygotes may be buried under more snow, or may enter the soil as the snow bank melts. In any snow field the vegetative phase usually only lasts for about one week in each year (Hoham, 1980).

Snow algae may produce accessory pigments which, under bloom conditions, may impart a red or yellow colour to the snow. The red pigment in *Chlamydomonas nivalis* is the xanthophyll astaxanthin, and its formation may be due to nitrogen deficiency (Czygan, 1970). Other secondary carotenoids — yellows, pinks, and oranges — may be found in the zygotes. The distribution of snow algal species is related to exposure, shading, elevation, and nutrient availability. In the southwest U.S.A. green algae may be found several centimetres beneath the snow surface of snow banks in forests; orange snow may be found at or near the surface adjacent to the trees, while red snows are characteristic of open exposures above the tree line (Hoham & Blinn, 1979).

MARINE ICE ALGAE

Ice algae occur in the marine pack ice of the polar regions, where they grow in the brine channels which develop between the ice crystals and in the loose layer that occurs on the underside of the ice (Horner & Alexander, 1972). Diatoms are the commonest species, but dinoflagellates, green flagellates, and cryptophytes are also reported. Most of these algal species are non-planktonic and the assemblage is distinct from the plankton (Bunt & Wood, 1963). Biomass estimates for bottom ice communities in Antarctica are reported to be as high as 300 mg chlorophyll/m^2 (Palmisano & Sullivan, 1983).

The species are adapted to growth at low light intensities and low temperatures; *Fragillaria sublinearis* is reported capable of growth below 0 °C in light intensities of 50–100 lux (Bunt *et al.*, 1966). It has been suggested that facultative heterotrophy may occur during the winter

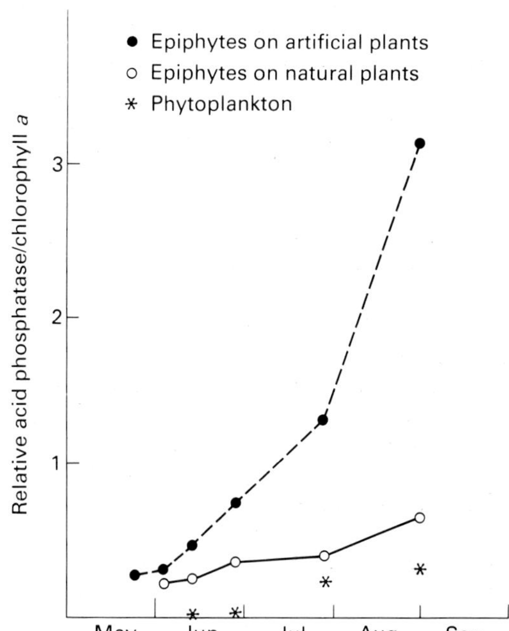

Fig. 7.36 Alkaline phosphatase activity of loosely attached epiphyte communities growing on natural and artificial substrates, the pondweed *Potamogeton*, plastic strips, and, for comparison, the adjacent phytoplankton. (After Cattaneo & Kalff, 1979.)

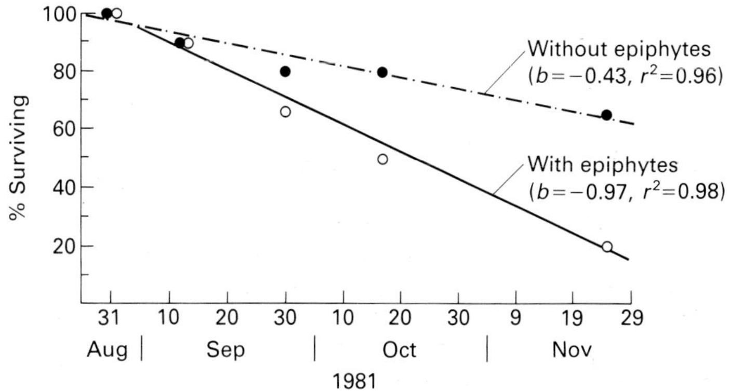

Fig. 7.37 Survival of individually marked axes of *Rhodomela larix* with and without epiphytes. (After d'Antonio, 1985.)

months (Allen, 1971), although Horner and Alexander (1972) concluded that heterotrophic metabolism of glycine, glucose, and acetate by Arctic sea-ice microalgae was negligible. In contrast, Palmisano and Sullivan (1983) showed that dark heterotrophy occurred in three species of sea-ice diatoms in Antarctica, and Palmisano *et al.* (1985) found uptake of serine by sea-ice diatoms at substrate concentrations close to ambient, but this represented less than 0.3% of the photosynthetic fixation of carbon.

ALGAE OF HOT SPRINGS

At the other end of the biotic temperature spectrum are the algae of hot springs, which are common in active volcanic regions. In addition to elevated temperatures, such waters have 10–15 times higher concentrations of dissolved salts, and are particularly rich in sodium, potassium, carbonate, silicate, sulphate, chloride, and sulphides. Others have low pH and may also have high concentrations of heavy metals (Chapter 9). While these conditions may be extreme, they are seasonally constant, and the waters are usually shallow and clear. Associated algae therefore do not have to adapt to annual environmental changes other than those associated with irradiance. Brock (1978) has summarized much of the available information on these thermophilic algae.

There appears to be an inverse relationship between cellular complexity and the ability to survive at elevated temperatures. Some heterotrophic bacteria are capable of surviving boiling waters; photosynthetic Cyanophyceae, such as *Synechococcus lividus* (Brock, 1967), may occur at +70°C, while the phylogenetically enigmatic *Cyanidium caldarium* grows at up to 57 °C (Doemel & Brock, 1970). Between 45 and 40 °C the first unequivocal eukaryotes, diatoms (*Achnanthes; Navicula*), and green algae (e.g. *Ulothrix*) appear (Kullberg, 1971). Fairchild and Sheridan (1971) found *Synechococcus lividus* along a temperature gradient of 53–73 °C and isolated four genetically distinct strains, all of which had characteristic temperature optima. Sheridan (1979) discovered that there are genetically distinct sun and shade forms of *Plectonema notatum*, which occur at different depths in the mat.

Many species of hot spring algae (e.g. *Cyanidium caldareum*; *Mastigocladus* spp.) have wide geographic distributions and it is difficult to see how these could be explained by any long-distance dispersal hypothesis. Brock (1969) suggests that the thermophilic species are relict populations from periods early in the earth's history when volcanic conditions were more widespread. It is interesting that in Iceland, a comparatively recent volcanic island, Castenholz (1969) found only 6–8 species of hot spring cyanophytes compared to several times the number that would be expected in continental hot springs.

CHAPTER 8

Evolution and phylogeny

Major events of algal evolution

The unravelling of the origins of the algae is a central theme in the study of evolution. Starting from the earliest prokaryotes and proceeding to the present time, the process of algal evolution spans most of the period of known life on earth, and embraces some of the main events in botanical evolution: the emergence of photoautotrophic prokaryotes, the origin of eukaryotes, mitosis, syngamy and meiosis, alternation of generations, and adaptations to terrestrial life.

The origins of the prokaryotes lie remarkably close in time to the beginning of the geological record at *c.* 3.8 Ga (= years $\times 10^9$) ago (Fig. 8.1). The first autotrophic organisms, the cyanophytes, made their appearance about halfway through the Archaean (3.8–2.6 Ga BP), the earlier part of the Precambrian (Schopf & Walter, 1982a). In the latter part of the Precambrian, the Proterozoic (2.6–0.58 Ga), they were ubiquitous (Fig. 8.2), and representatives of most present-day cyanophycotan orders have been recorded from the fossil record of this period, often referred to as the age of the cyanobacteria (= Cyanophycota) (Schopf & Walter, 1982a).

The development of the nucleated eukaryotic cell type took place probably about 1.4 Ga ago (Schopf, 1978). Meiosis and eukaryotic sexuality originated at least 800 Ma (= years $\times 10^6$) before present (Hofmann & Aitken, 1979), but may have come into existence well before then, possibly as early as 1330 Ma ago (Walter *et al.*, 1976). The subsequent development of life histories with an alternation between diploid and haploid generations in a meiosis–syngamy cycle evolved in four main patterns (Chapter 5); there is no clear record of the precise time of initiation of each of these patterns, although the sequence in which they developed can be deduced from a

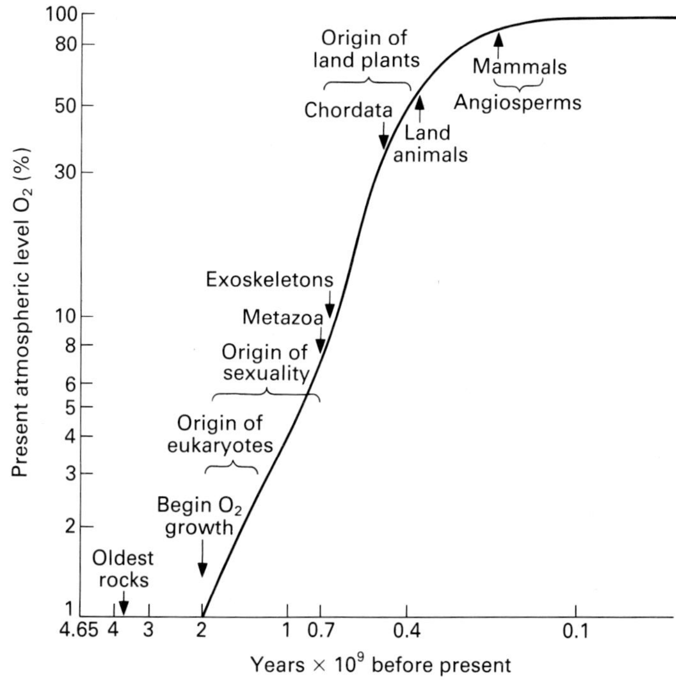

Fig. 8.1 Approximate timing of events in the evolution of the biosphere compared with hypothetical levels of oxygen (based on % of present oxygen). Scales are logarithmic and the curve is highly generalized; departures from the mean probably occurred. (Modified from Cloud, 1976.)

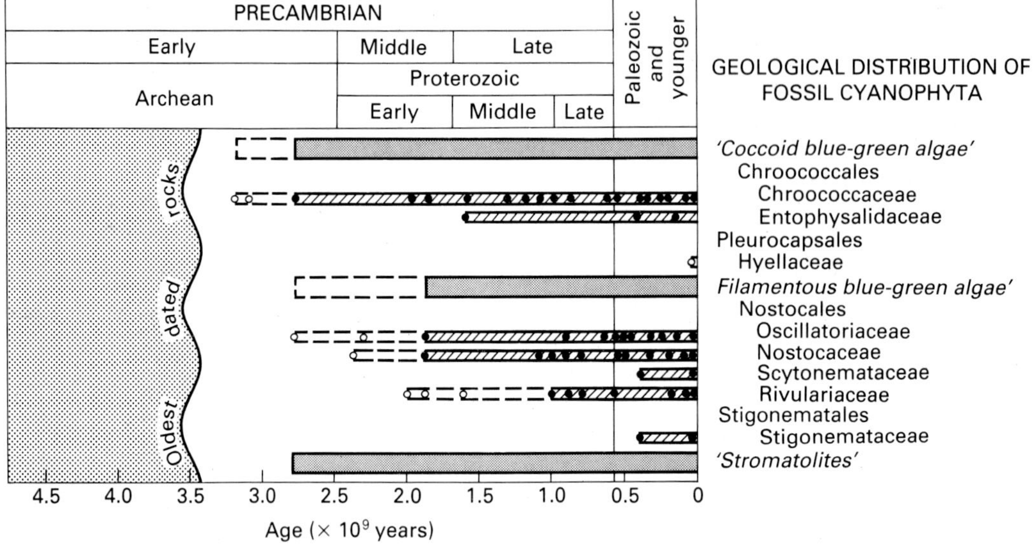

Fig. 8.2 Histogram showing the known occurrence of cyanophyte families, based on fossils. (Modified from Schopf, 1970.)

knowledge of life histories in modern plants (Scagel *et al.*, 1982). Alternation of generations was one of the adaptations essential for a terrestrial existence that eventually played a significant role in the evolution of the land plants (embryophytes). It has been widely accepted that it was among the ancestors of the green algae that this major advance took place, probably somewhere between 700 and 400 Ma ago (Cloud, 1976).

Origins of the prokaryotes and photoautotrophs

The major events in the evolution of the primitive earth and life forms are summarized in Fig. 8.3. The age of the earth has been proposed as between 4 and 6 Ga, and for approximately the first half of its existence the earth's atmosphere lacked molecular oxygen (Cloud, 1976). The prokaryotes evolved during the biogenic period, between 4.5 and 3.5 Ga ago, when the atmosphere contained gases such as hydrogen, carbon dioxide, nitrogen, sulphur dioxide, and hydrogen chloride, and virtually no free oxygen (Walker *et al.*, 1982). The earliest prokaryotes were thus presumably aquatic anaerobes. Their fossil remains are found in cherts, shales, and in stromatolites. Stromatolites are flat, mound-like, columnar or dome-shaped internally laminated sedimentary structures built up, layer by layer, by the sediment-trapping, binding and/or precipitating activities of prostrate microbial communities (principally cyanophytes) called algal mats. Similar communities build stromatolites or stromatolite-like structures at the present time (Fig. 8.4), but since the end of the Precambrian such activity has been on a reduced scale. Thus stromatolites were most widely distributed and abundant during the late Archaean and much of the Proterozoic when, over a period of some 2.5 Ga, prokaryotes were the dominant form of life on earth (the age of prokaryotes).

It is not certain when the first of the prokaryotes evolved, or in what sequence; both bacteria and cyanophytes occur in the earliest dated stromatolites. Of great importance in this consideration, however, is the suggestion that the cyanophytes arose from the eubacteria, one of three ancient prokaryote evolutionary lines (archaebacteria, urkaryotes, and eubacteria) derived

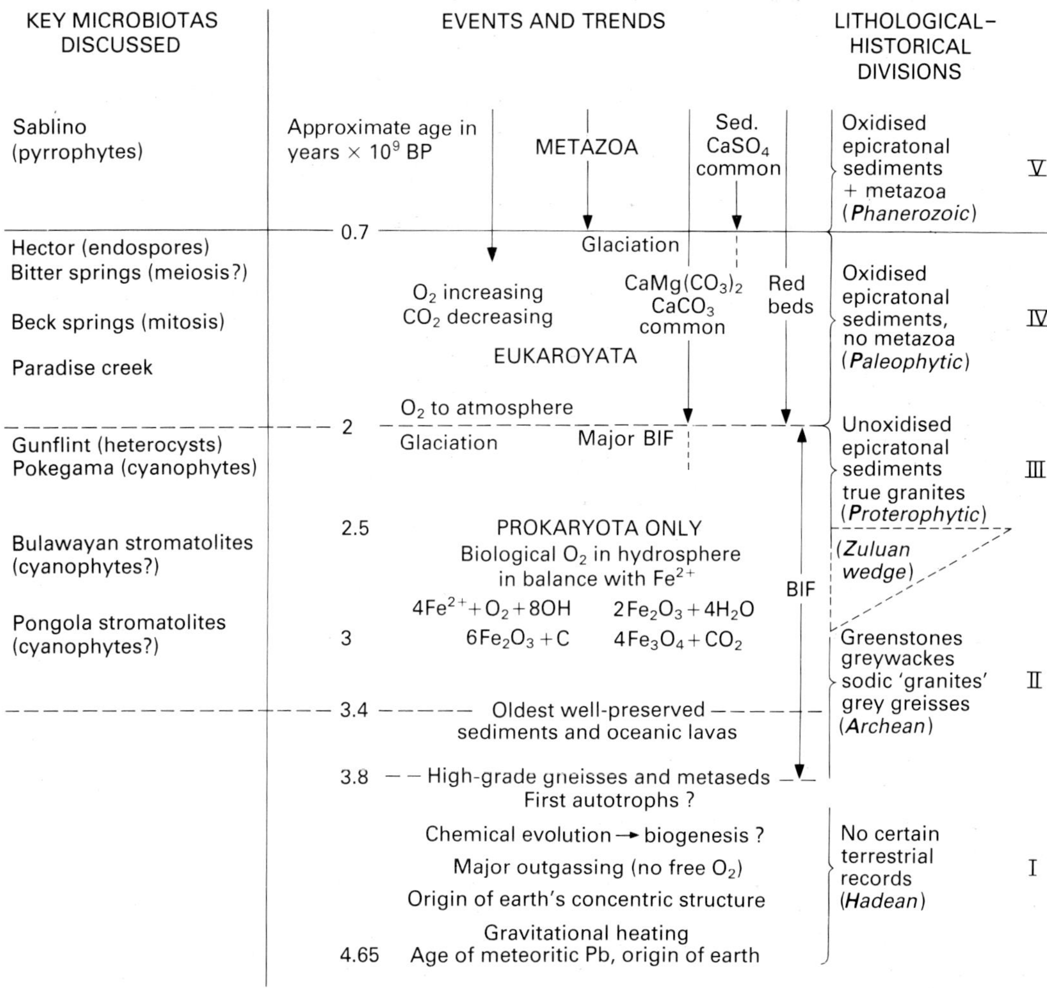

Fig. 8.3 Major divisions (I–V) of earth history with related aspects of biological and geological evolution. (Modified from Cloud, 1976.)

from a universal ancestor (Woese, 1981; Fig. 8.5). Even the best-preserved Archaean microfossils provide insufficient information on their cellular morphology to distinguish between primitive cyanophytes and other prokaryotes (such as beggiatoaceans; Schizomycophycota). It is also not possible on the basis of current information to determine whether or not the earliest cyanophytes were autotrophs, and if autotrophic whether they used both photosystems I and II, and hence released oxygen as a byproduct of photosynthesis (Schopf & Walter, 1982a). Until this question is resolved, it will not be possible to pinpoint accurately the origin of the cyanophytes from their presumably photosynthetic but non oxygen-evolving precursors. Fox *et al.* (1980), in a detailed analysis of 16S ribosome RNA sequence characterization of bacteria, indicated that bacterial photosynthetic phenotypes are extremely ancient, and that many non-photosynthetic groups appear to have arisen from photosynthetic ancestry; this questions the long-held view that the earliest bacteria were anaerobic heterotrophs. Possible ancestors of the cyano-

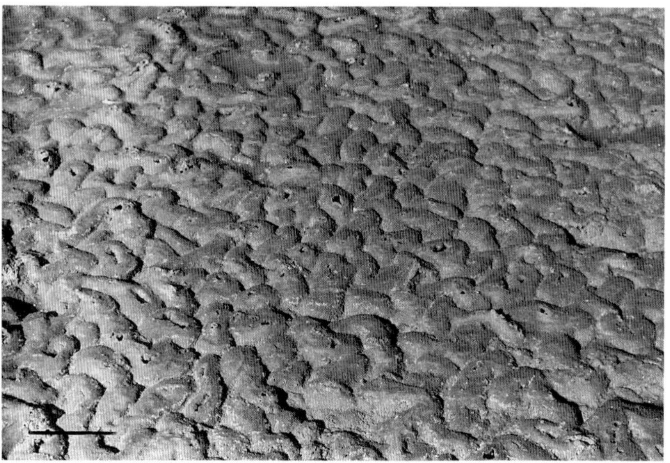

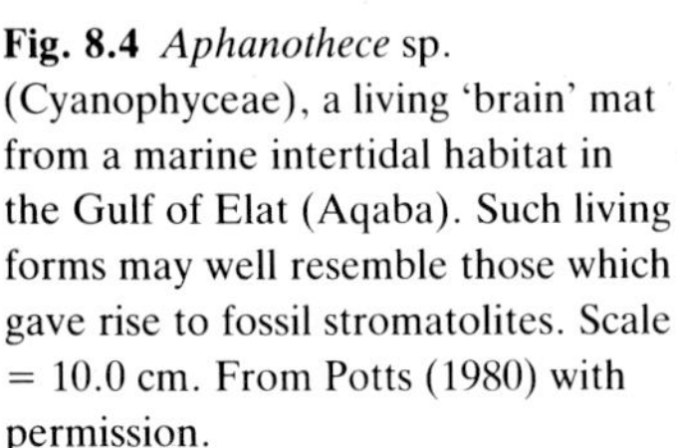

Fig. 8.4 *Aphanothece* sp. (Cyanophyceae), a living 'brain' mat from a marine intertidal habitat in the Gulf of Elat (Aqaba). Such living forms may well resemble those which gave rise to fossil stromatolites. Scale = 10.0 cm. From Potts (1980) with permission.

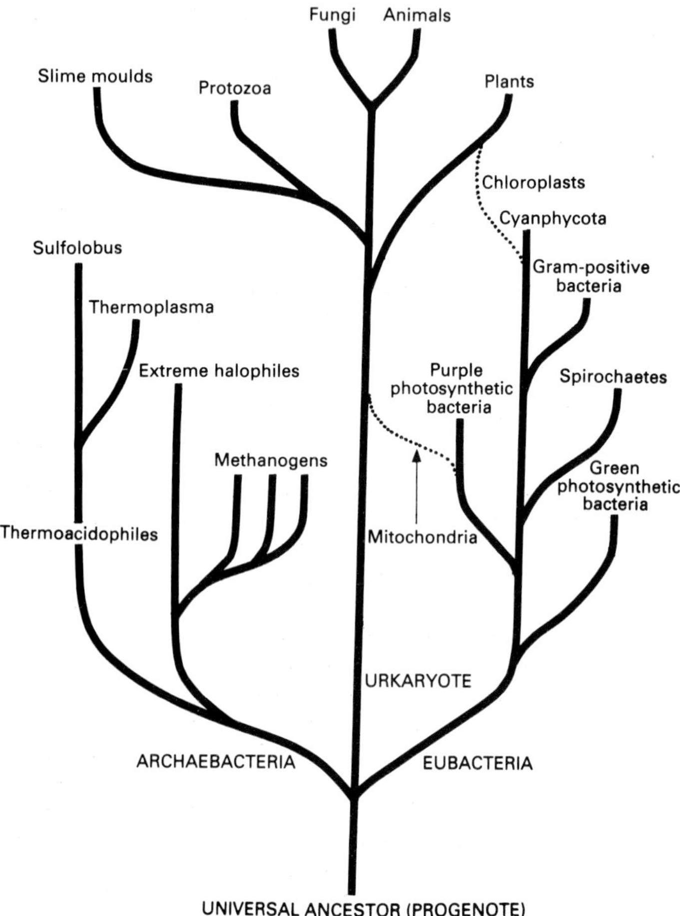

Fig. 8.5 Hypothetical phyletic tree incorporating the concept of three primary kingdoms, separating the archaebacteria from all other bacteria. The archaebacteria, eubacteria and urkaryotes are presumed to have stemmed from a common ancestor (the progenote) that was much simpler than the simplest present-day prokaryotes. Eukaryotes derived from the urkaryote became 'hosts' for bacterial endosymbionts that developed into mitochondria and chloroplasts during eukaryogenesis. (Modified from Woese, 1981.)

phytes would have included the anaerobic, photosynthetic bacteria, of which there are three major living groups: green and purple sulphur bacteria and purple non-sulphur bacteria (Margulis & Schwartz, 1982). These prokaryotes do not utilize water as a hydrogen donor in photosynthesis, and evolve not oxygen, but rather a variety of other products, especially sulphur.

Indirect evidence can, however, be used to deduce approximately when oxygenic photosynthesis began during the Archaean. Oxygen would, at first, have been released into the hydrosphere, where oxidation processes could affect bottom deposits. During the period from 3.5–2.0 Ga ago there was widespread deposition of banded iron formations. They consist of thick sequences of cherts, the layers or bands of which are alternately iron-rich and iron-poor, the iron in the former being in the oxidized or ferric state. This oxidation and precipitation of ferrous iron originally dissolved in sea water must have been brought about by oxygen liberated by photosynthesis. Consequently the origin of oxygenic photosynthesis either preceded or was coincident with the time, over 3.5 Ga ago, when banded iron formations first began to be deposited. Red bed deposition succeeded formation of the banded iron deposits about 2.0 Ga ago, which indicates that by that time oxygen was being released into the atmosphere in sufficient quantity to bring about oxidation during weathering of rocks exposed at the surface. When the atmosphere's oxygen content reached about 1% of its present level, a layer rich in ozone formed in the atmosphere. This event heralded the arrival of the eukaryotes. According to Schopf and Walter (1982a), the build-up of atmospheric oxygen could have been a relatively rapid process in geological terms.

From the indirect evidence it is thus possible to place the origins of the cyanophytes somewhere about 3.5 Ga ago. Unravelling the fossil evidence is, however, fraught with difficulties. The earliest microfossils (3.5 Ga old) do not include recognizable cyanophytes (Schopf & Walter, 1982a). The Insuzi stromatolites from South Africa (Mason & Van Brunn, 1977), aged at 3.1 Ga, are remarkably like those presently being formed at the Hot Springs at Yellowstone National Park (U.S.A.); they were probably laid down by phototactic, filamentous photoautotrophs, but no micro-organisms have been found in them. The first fairly convincing fossilized cyanophytes are not found until the late Archaean, *c.* 2.8–2.5 Ga ago (the Fortescue Group of Western Australia; Schopf & Walter, 1982b). Their morphology is comparable with that of *Oscillatoria* and *Lyngbya*.

The 'age of cyanobacteria' (cyanophytes)

The cyanophytes were the predominant organisms on earth for more than 2 billion years, from *c.* 3.0 to 1.0 Ga ago (Schopf, 1974a, b). During that period they had a profound influence on the evolution of the earth's atmosphere, which progressed from an anoxic to an oxic condition, as a result of photosynthesis; in response the cyanophytes evolved photoautotrophic forms. In addition, filamentous species developed the heterocyst which 'protected' their oxygen-sensitive nitrogenase enzyme complex (Schopf, 1974b).

Proterozoic cyanophytes were ubiquitous; they commonly occurred in mat-forming communities that gave rise to stromatolites, although they occupied every available photic habitat, especially in the later, oxic period. The emergence of eukaryotic calcareous and non-calcified algae, and of boring and grazing metazoans (Schopf & Walter, 1982a), at the end of the Precambrian and early in the Cambrian brought the age of cyanophytes to a close. Their remarkable evolutionary conservativism and their highly adaptable nature have, however, aided in their survival until modern times.

Origin of the eukaryotes

The greatest discontinuity among living organisms is that between prokaryotes, which lack membrane-bounded organelles and have no 9 + 2 flagella and no nuclear envelope, and the eukaryotes which (except for the Rhodophycota which lack 9 + 2 structures) possess all of these characters. It was not, however, until the 1960s that the fundamental differences between pro- and eukaryotes could be rigorously defined (Stanier, 1961). The discontinuity is not only morphological, but biochemical (Ragan & Chapman, 1978), and there are no obvious intermediate forms except, perhaps, the Prochlorophycota (Lewin, 1976, 1977). As stated by Chapman and Trench (1982), however: 'With the outstanding exception of the presence of chlorophyll *b*, *Prochloron* is a "good" cyanobacterium.'

In the last 20 years major advances have been made in attempts to answer the questions of when and from where the eukaryotes evolved. Cloud (1976) has suggested that they could have arisen any time after *c.* 2.0 Ga ago, or whenever the earth's atmosphere contained sufficient free oxygen. They appear to have been well established prior to 1.3 Ga ago according to Walter *et al.* (1976), who described a variety of megascopic, possibly eukaryotic, algal fossils from the Greystone Shale of the Belt Supergroup in Montana (e.g. *Proterotainia; Lanceoforma*). Steps in their evolution involved the development of membrane-bounded organelles, including the nucleus, chloroplasts, mitochondria, and 9 + 2 flagella (except in the Rhodophycota), and the evolution of the linear chromosomes and spindle apparatus associated with mitosing cells.

There are two greatly different schools of thought as to how these steps occurred. The more exciting one proposes an exogenous origin of membrane-bound organelles through a series of endosymbioses (Mereschkowsky, 1905; Margulis, 1970, 1981; e.g. mitochondria were derived from bacterial endosymbionts, chloroplasts from cyanophytes, and flagella from spirochaetes) (Fig. 8.6). The theory has considerable attraction in that the symbiotic process is well demonstrated in living algae/algae (e.g. the Glaucophytes; *see* p.45) and algae/animal relationships. It requires, however, provision for up to five separate endosymbiotic events and 27 independent transitions from mitotic to meiotic cell division (Margulis, 1970; Cloud, 1976). It also has the drawback of requiring a series of coincidences for the process to have been successful. McQuade (1983) has postulated that mitosis, phagotrophy, the mitochondrion, flagella, sexual reproduction, and the chloroplast are so complex that it is unlikely they evolved *de novo* more than once. The exogenous theory has yet to explain satisfactorily the origin of the nucleus, and the proposed endosymbiotic process in the acquisition of 9 + 2 flagella has not received general acceptance.

The other theory suggests a gradual evolution of the membrane-bounded organelles from the non membrane-bounded systems of the prokaryote ancestor via an autogenous process (Allsopp, 1969; Cavalier-Smith, 1975). Cavalier-Smith (1975, 1978) suggested this most plausible alternative to endosymbiosis: the most important step was the evolution of endocytosis (phagocytosis and pinocytosis; Fig. 8.7), which permitted the compartmentation of the cell by intracellular membranes. He argued that endosymbiosis would be one of the inevitable consequences of phagocytosis, but is not the cause of the eukaryote condition. The phagocytic capacity gave the prealga (or uralga) a selective advantage over other cyanophytes (from which it evolved) in that it could photosynthesize in the light and phagocytose in the dark. The autogenous theory proposes mechanisms for the evolution of the nucleus, mitosis, meiosis, the spindle, mitochondria, and the microtubules without the need for endosymbiotic events. Cavalier-Smith (1975) allowed, however, that the origin of mitochondria and chloroplasts via symbiosis is compatible with autogenous origins of nuclei and mitotic apparatus, and proposed (1983a) that mitochondria and chloroplasts were coevolved much later (*c.* 700 Ma BP) than the origins of eukaryotes (*c.* 1400 Ma BP). He incorporated this concept of coevolution in his proposed scheme of unified phylogeny (Cavalier-Smith, 1983b; Fig. 8.8). Cavalier-Smith argued strongly that the exogenous theory does not offer a satisfactory explanation for the origin of the nucleus, mitotic apparatus, and flagella. The last-mentioned he suggested (Cavalier-Smith, 1978) arose by a complex stepwise process beginning with microtubule-containing filopodial extensions from simple planktonic prokaryotes. There are, however, as indicated by Taylor (1974) and Ragan and Chapman (1978), virtues in both schools of thought. In any event, it is unlikely that we shall ever know with absolute certainty how the eukaryotes arose.

The evolution of mitosis occurred in conjunction with that of the nucleus, although there are fundamental differences between the exogenous versus autogenous schemes in accounting for how this occurred. The exogenous scheme (Margulis, 1981) proposes that mitosis preceded meiosis; the origin of the nucleus was conceivably dependent on a prior symbiotic event leading to the evolution of the mitochondria. A sequence of 11, not necessarily sequential, steps in the evolution of mitosis is proposed by Heath (1980). Cavalier-Smith (1975, 1978), however, favours a simultaneous development of mitosis, cell divi-

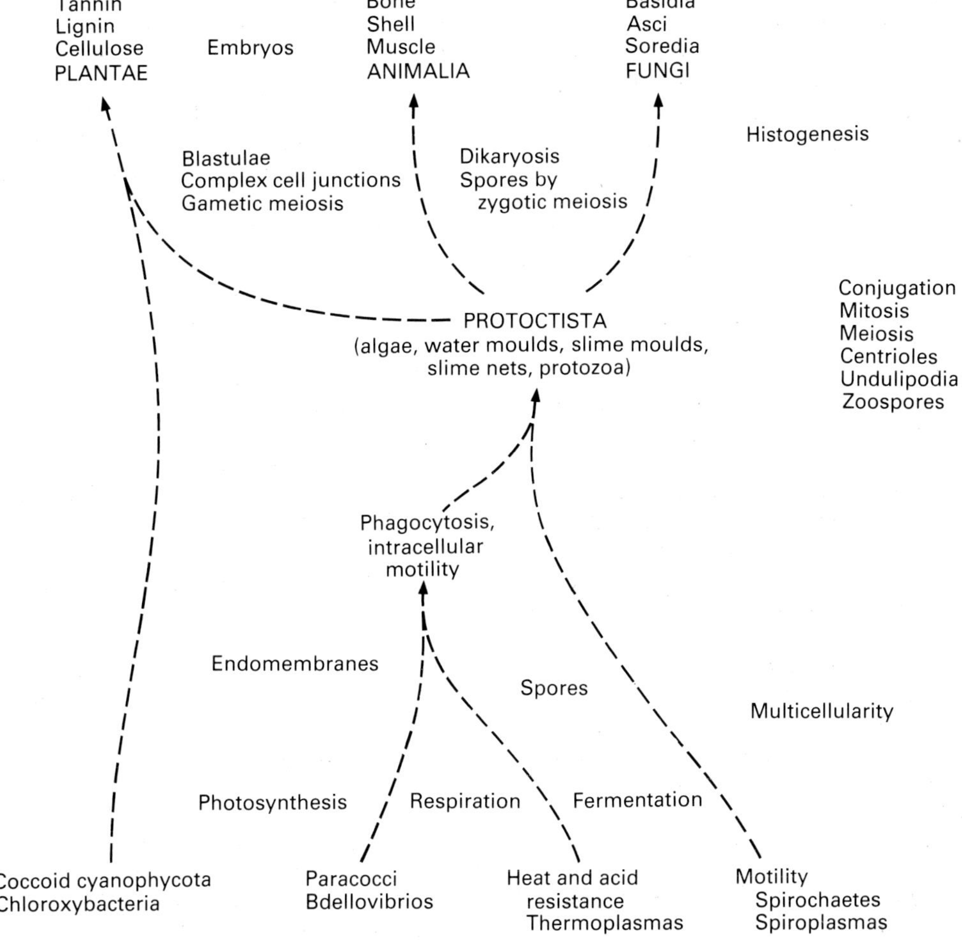

Fig. 8.6 A model for the exogenous origin of eukaryotic cells through symbiosis. The origins of plants, animals, and fungi are traced. The acquisition of flagella from spirochaetes proposed here has not received general acceptance. (Based on Margulis, 1981.)

sion, and meiosis. He speculated (1975) that the development of the spindle microtubules, the nuclear envelope, and the cleavage furrow for cell division occurred in parallel, and that the evolution of the sexual process started very early on in the evolution of eukaryotic cells, and before the development of the nucleus was completed.

The early eukaryotes were, from all accounts, probably phagocytic autotrophs, lacking a cell wall, and possessing an amoeboid or planktonic form. Whether they achieved a eukaryotic condition by exogenous or autogenous means, or by a combination of both, it is accepted that the process was a relatively rapid one when considered in the context of evolution, and may have taken no more than 100 Ma.

Evolution of meiosis, syngamy, and the alternation of generations

Whether or not sexuality (meiosis and syngamy) developed gradually or more or less simultaneously with the eukaryotic condition (Cavalier-Smith, 1975), its arrival was a major step on the

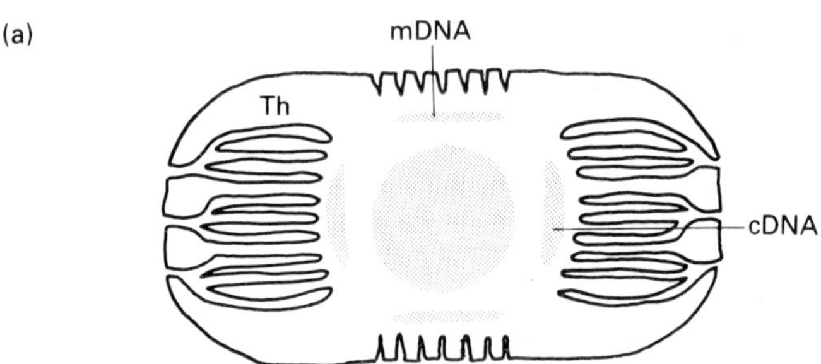

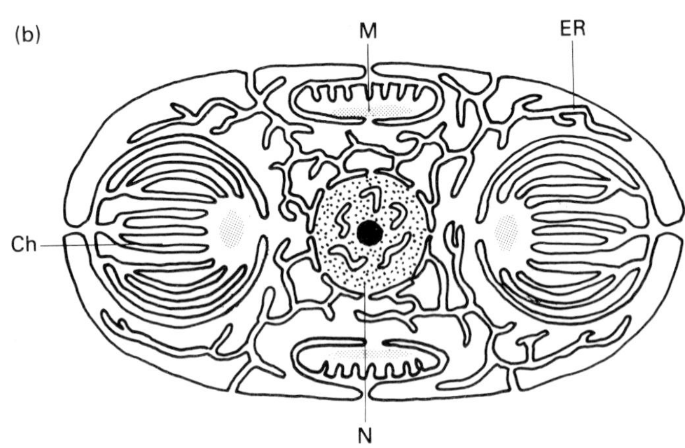

Fig. 8.7 The proposed evolution of eukaryotic features from an autogenous process involving two stages of endocytosis (or phagocytosis and pinocytosis), giving rise to the chloroplast, mitochondria, nucleus and endoplasmic reticulum. The model does not, however, account for the origin of flagella. (a) Primary invaginations. (b) secondary invaginations. Ch, chloroplast; DNA, deoxyribonucleic acid; cDNA, plastid DNA; mDNA, mitochondrial DNA; ER, endoplasmic reticulum; N, nucleus; M, mitochondrion; Th, thylakoids. From Chadefaud (1974) with permission.

evolutionary pathway to the vascular plants. The evolution of diverse eukaryotes became possible through the processes of segregation, recombination, and natural selection. Precisely when sexuality arose is unclear, but the biogeological evidence shows that it must have predated 700 Ma ago, when the metazoans first appear in the fossil record (Cloud, 1976). The debate as regards timing returns to a consideration of the exogenous versus the autogenous theories of eukaryote evolution. Taking the fossil evidence together with the currently available cellular, molecular, and physiological evidence, Cloud (1976) presented arguments which tend to support Cavalier-Smith's (1975, 1978) views: viz. that mitosis and sexuality arose more or less simultaneously. Such a conclusion is consistent with the biogeological evidence that mitosis and meiosis were both extant 1.3 Ga ago (*see* Walter *et al.*, 1976). It is also consistent with Schopf's (1974b) conclusion that sexuality may well have arisen prior to 1000 Ma ago.

A consequence of eukaryotic sexuality was that it permitted the advance from the cellular to the tissue and organ levels of somatic complexity (Cloud, 1976). While there is an inadequate fossil record for the majority of present-day algal classes, it is likely that they had their origins during the evolution of eukaryotic cells; it is not known, however, whether the main groups segregated before or after the development of sexuality. Groups such as the Euglenophycota, which entirely lack sexual reproduction, might have arisen before sexual reproduction evolved, or they may subsequently have lost it. Sexual reproduction certainly resulted in an increasing diversity of organisms, and generally speeded up the process of evolution.

Among the earliest multicellular fossils, probably sexual algae from the late Precambrian, are the dasycladalean-like *Papillomembrana* (Spjeldnaes, 1963), the presumed brown algae *Vendotaenia* and *Tryrasotaenia* (Gnilovskaja, 1971), and the macroalgal impressions from Australia estimated to be *c.* 760 Ma old (Milton, 1966).

A consequence of the development of sexuality

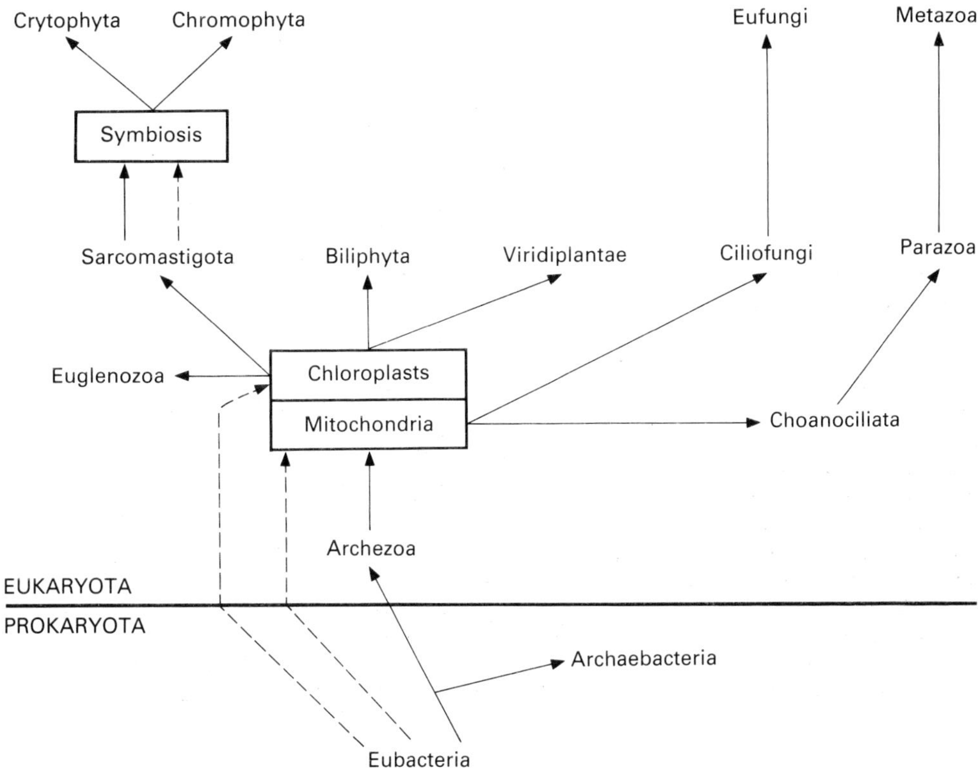

Fig. 8.8 Possible phylogenetic relationships among the 14 subkingdoms recognized by Cavalier-Smith (1983). Central to the pattern of eukaryotic diversification is the acquisition of chloroplasts and mitochondria. From Cavalier-Smith (1983) with permission.

was the persistence of both diploid and haploid phases in the early algae; it is likely that initially both phases were unicellular (Scagel *et al.*, 1982). The subsequent evolution towards the elaboration of either or both phases led to the highly varied life cyles of algae and, in the most advanced forms, an alternation of generations, between a haploid gamete-producing gametophyte generation and a diploid, spore-producing sporophyte generation (Chapter 5). There is no precise fossil record to indicate how or when these developments occurred, although it is evident from an analysis of living examples that four principal elaborations occurred, based on the timing of meiosis in the life cycle (Chapter 5).

A possible evolutionary sequence is suggested by Scagel *et al.* (1982; Fig. 8.9); all of the stages survive in present-day algae to varying degrees. The earliest to evolve may have been zygotic meiosis. From this condition two separate patterns evolved in the algae: the gametic and sporic meiosis types. The gametic type resulted in diploid organisms, and the sporic type in an alternation of generations. Further evolution of the alternation of generations proceeded in two directions: to isomorphic and heteromorphic types. In the latter there were possibilities for elaboration of either the gametophyte or sporophyte generation. Both led to the evolution of terrestrial plants, but it was the diploid-dominant algae (and metazoans) that managed to evolve the relatively complex tissues and organs and physiological processes necessary for adaptation to the rigorous ecological requirements of a terrestrial existence.

A fourth pattern of alternation, the dikaryon, did not occur in the algae and is limited to some

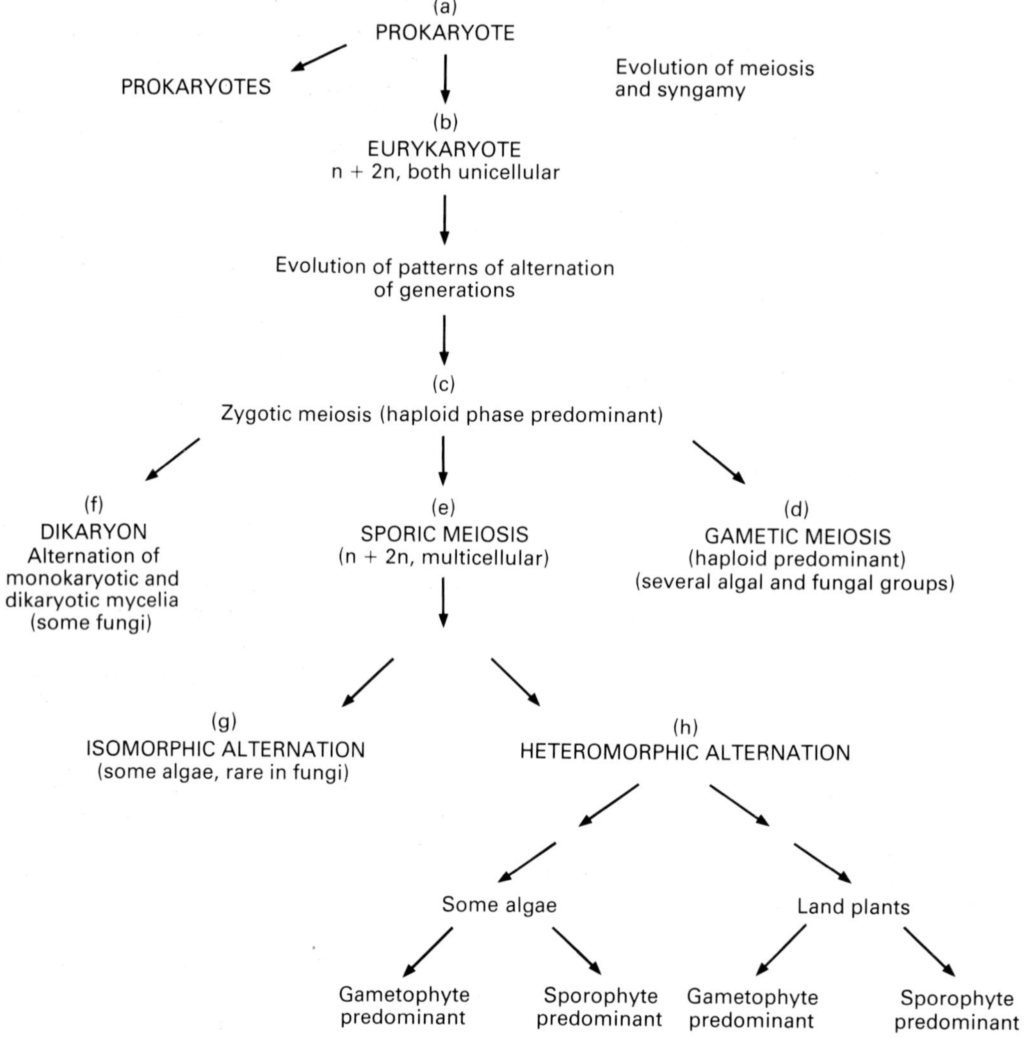

Fig. 8.9 Steps in the evolution of the main types of alternation of generations. (Based on Scagel *et al.*, 1982.)

fungi. It consists of a delay in karyogamy following plasmogamy or somatogamy (fusion between thalli), and results in diploidization in which paired, compatible nuclei occur in the cytoplasm, delaying their fusion until the formation of spores. Its origins are obscure.

Evolution of land plants

The evolution of land plants (embryophytes) is briefly considered here, since the search for evolutionary links between the algae and land plants is a major preoccupation of modern evolutionary biology.

While their precise origins are obscure, it is generally agreed that the land plants were derived from phragmoplastic chlorophytes (Scagel *et al.*, 1982). Of the several groups suggested as possible ancestors, the most likely are among the Ulotrichales. The steps necessary for the transition to the hostile terrestrial environment were summarized by Jeffrey (1962). As mentioned earlier, a major split occurred in the process, between those in which the haploid gametophyte generation, and those in which the diploid sporophyte, became the elaborated phase. The former pathway led to the bryophytes and an essentially dead-end in that they were restricted to moist habitats because of the need for water to allow fertilization. The latter pathway led to the ancestors of the remainder of the embryophytes, which successfully adapted to a terrestrial existence.

The hypothetical ancestral alga is most likely to have been filamentous and heterotrichous, and to have had an alternation of two multicellular generations. Sexual reproduction was probably oogamous, with fertilization occurring in the oogonium. In the earliest adaptations the zygote would have been retained on the haploid plant (as in *Coleochaete*; Graham, 1985), protected by sterile cells. The bryophyte line was derived from those forms in which the diploid sporophyte generation developed while remaining attached to the gametophyte; the embryophyte line evolved from those ancestral forms in which the zygota ultimately separated from the gametophyte and continued as an independent, spore-producing generation. In adapting to terrestrial existence the sporophyte developed cutinization of emergent parts, water absorption and conducting tissues, specialized photosynthetic organs (leaves), and stomata; the gametophyte generation became progressively reduced.

There is no sharp dividing line between the chlorophytes and the bryophytes and embryophtes. The land plants simply represent the most advanced of the green plants, and in many schemes (*see* Edwards [1976] for a review) they are grouped together in a single kingdom.

Phylogeny of the algae

Tracing the evolutionary history of the algae is an exercise in phylogeny or phylogenetics, requiring a consideration of both living (extant) and fossil (extinct) representatives. The evolutionary history can be represented in a phylogenetic tree, the first example of which was that of Haeckel (1896). A natural classification of extant species (Chapter 2) might represent the final branches and twigs of such a tree, in which the groupings of taxa are arranged on the basis of commonly derived traits. Such a classification is said to be based on phylogenetic systematics, or cladistics. An identical or very similar classification could, however, be arrived at by classifying the extant species on the basis of their overall similarities. This process is called phenetics. The classification of the algae adopted in this text and many recent accounts is a result of a combination of both phylogenetic and phenetic approaches. There is no ideal classification, and a great deal is left to speculation because of the lack of an adequate fossil record for most groups. It is generally agreed that some of the divisions are polyphyletic, being derived from several ancestral sources, while others are monophyletic (e.g. Euglenophycota).

Numerous phylogenies involving the algae have been proposed, and among the more notable ones are those of Copeland (1938), Cronquist (1960), Christensen (1964), Sagan (1967), Whittaker (1969), Leedale (1974), Cavalier-Smith (1975) and Scagel *et al.* (1982). Virtually all utilize biochemical characters (e.g. pigmentation) in combination with cytological and morphological features, although some (e.g. Dodge, 1974) consider only morphological and ultrastructural characteristics and others (e.g. Ragan & Chapman, 1978) only biochemical features. Klein and Cronquist (1967) provided an exhaustive comparative account of biochemical, micromorphological, and physiological aspects of algal phylogeny, and constructed a whole array of phyletic schemes based on individual characteristics. Phylogenies may also be constructed taking only a single feature into consideration (e.g. mitosis; Heath, 1980: flagellar apparatus; Moestrup, 1982; Melkonian, 1982a: photosensory apparatus; Kivic & Walne, 1983). Alternatively, they may concentrate on a single class (e.g. Euglenophycota; Leedale, 1978).

While many of the schemes differ in detail and emphasis, there is general agreement on significant aspects of algal evolution such as the distinction of the Rhodophycota from the remaining eukaryotes, the isolation of the euglenophytes, dinophytes and cryptophytes, the demarcation between the chromophyte and chlorophyte classes, and the derivation of the land plants from ancestral chlorophytes. By and large, biochemical and molecular information tends to confirm conclusions resulting from largely morphological, cytological, and ultrastructural studies. As new information is added it gradually becomes incorporated into phylogenetic schemes.

The general relations among the classes of algae are shown in Fig. 8.10. These are based

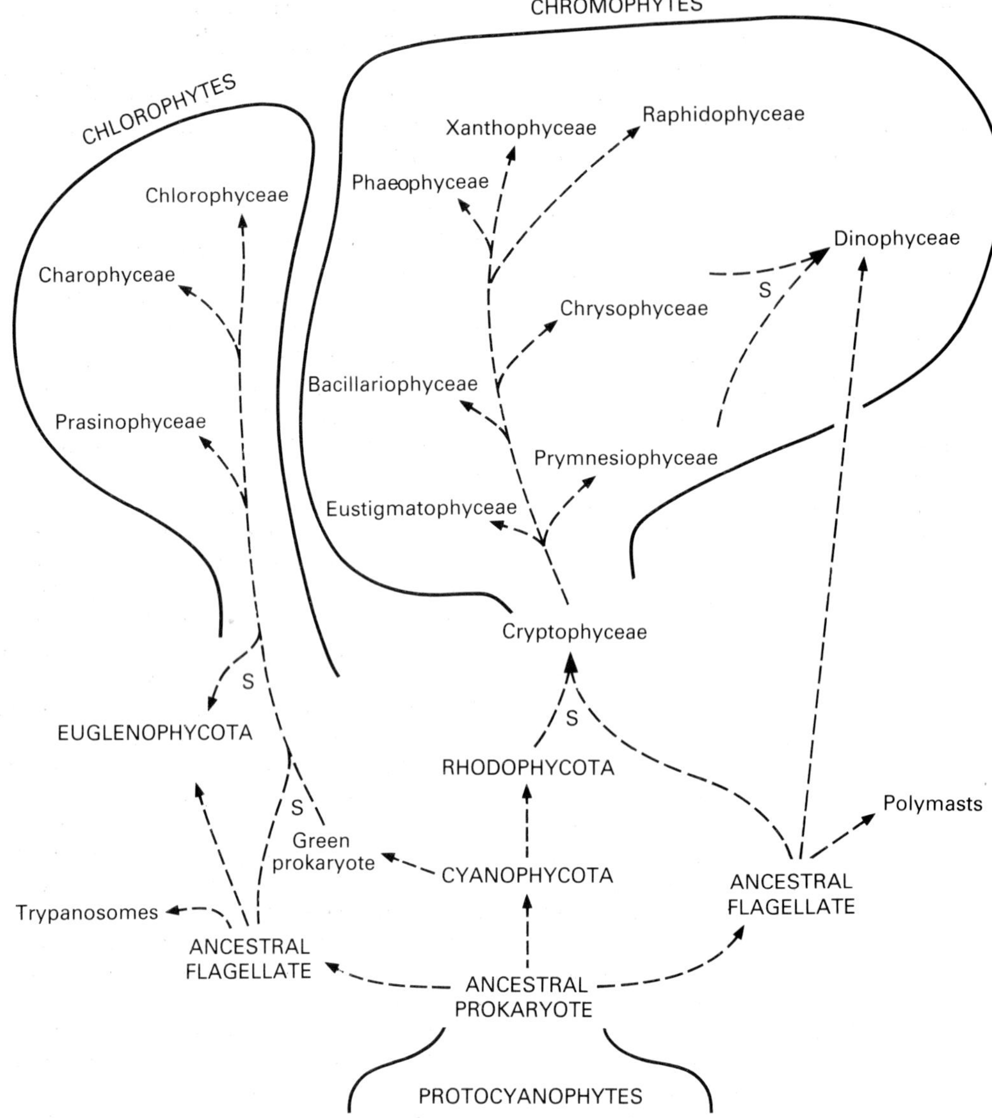

Fig. 8.10 Generalized phyletic tree showing hypothetical relations among the classes of algae. Five symbiotic events (S) are assumed. Two ancestral flagellate lines are suggested, one giving rise to the chlorophyte series following endosymbiosis with a green prokaryote (perhaps resembling *Prochloron*), the other giving rise to the chromophyte series following two endosymbiotic events involving cyanophyte and rhodophyte progenitors. The origins of the Dinophyceae are problematical. (Based on Dodge, 1979.)

primarily on similarities in pigments, flagellation, the cell covering, and the principal storage products. All of the classes except for the Chrysophyceae appear to represent distinct evolutionary lines, although it should be remembered that this is a largely phenetic series. It is evident from preceding discussions (Chapters 2, 3, and 6) that, in addition to the accepted demarcations based on pigments, there are many other morphological, cytological, and physiological distinctions among the classes which strongly suggest that they do indeed represent significant points of divergence in algal evolution. Caution must be applied, however, in interpreting an essentially phenetic scheme, since it cannot be taken as evidence of how the process occurred through time. The general directions in which algal evolution proceeded, however, are fairly firmly established.

The known fossil record of the algae is discussed below (p.253). For most classes there are many shortcomings in the fossil record, and there is a good representation only of calcareous (e.g. the Dasycladales, Chlorophyceae) or siliceous (e.g. diatoms) forms. For the majority of the remainder there are only fragmentary fossils of a very few species. Construction of a phylogenetic tree for the algae in which proper account is taken of their past evolutionary record is thus extremely difficult. It is useful, nonetheless, to review the state of our knowledge of the phylogeny of the principal algal classes.

PROKARYOTES: THE CYANOPHYCEAE AND PROCHLOROPHYCEAE

It has been widely hypothesized that the early, primitive coccoid cyanophytes arose concurrently with 'modern' bacteria from archaic bacterial ancestors (Klein & Cronquist, 1967). The cyanophytes do have many features in common with bacteria. The relationship between them was first suggested by Cohn (1853) and has been discussed in detail by many recent workers (e.g. Echlin & Morris, 1965; Hall, 1971; Fogg *et al*, 1973; Ragan & Chapman, 1978; Fox *et al.*, 1980; Starr *et al.*, 1981; Woese, 1981; Schopf & Walter, 1982a). Of considerable interest is the conclusion (Fox *et al.*, 1980; Woese, 1981) that there are three lines of descent of living organisms from a universal ancestor, or progenote (Woese, 1981): the archaebacteria, true bacteria, and eukaryotes. The cyanophytes, according to this scheme, were derived from the true bacteria (eubacteria); they possess, however, some fundamental differences from the bacteria which suggest that their evolutionary lines diverged very early on. The cyanophytes are structurally more diverse than bacteria (e.g. multicellular thalli; branched filaments; colonies), they have different cytological features (e.g. no bacterial endospores; possession of the heterocyst), and they never form bacteria-like flagella. There are also basic differences in the photochemical process and in their fatty acid composition (Fogg *et al.*, 1973).

The evolution of the cyanophytes can be traced with reasonable certainty from the fossil record (Fig. 8.2). The earliest forms were coccoid unicells, which had achieved some diversity by 1.7 Ga ago, when a number of extant families were recognizable. The filamentous types were probably derived from the coccoid, and representatives of several modern families were evident before 2.0 Ga ago (Schopf, 1974b). Hofmann (1976) described 18 genera and 24 species of coccoid and filamentous taxa from the Precambrian (1.9 Ga) microflora of the Belcher Islands, Canada. Fritsch (1945) concluded that because of the diversity of the filamentous forms, they represent several separate evolutionary lines in the cyanophytes which probably arose from an early filamentous ancestor. The timing and sequence of the evolutionary events leading to the diversification of the cyanophytes are unlikely to be precisely determined, although it does appear that the Stigonematales are the most recently evolved members of the filamentous line (Schopf & Walter, 1982a), as they are the only order not represented in the Precambrian fossil record.

The prochlorophytes, possessing features intermediate between the cyanophytes and chlorophytes (Lewin, 1981a, b; Fig. 8.11), provoke some interesting phylogenetic speculation (Lewin, 1983a). Since it is unlikely that recognizable ancestral prochlorophytes will be retrieved from the fossil record, their origins will necessarily be based on phenetic data. If an exogenous origin of the eukaryotes is assumed, then the prochlorophytes may be phylogenetically related to ancestral chloroplasts. Assuming an autogenous process, however, it could be argued (Lewin, 1983a) that the prochlorophytes may be related to ancestral eukaryotic chlorophytes. Lewin (1983a) has summarized these considerations in some simple phylogenetic tree diagrams (Fig. 8.12). He suggests that in order for the prochlorophytes to have evolved from the cyanophytes, it would have been necessary for them to lose phycocyanin and gain chlorophyll *b* synthesis.

RHODOPHYCEAE

The Rhodophyceae are undoubtedly an ancient, natural algal group, with unmistakable fossils known from the Cambrian (Schopf, 1970). On

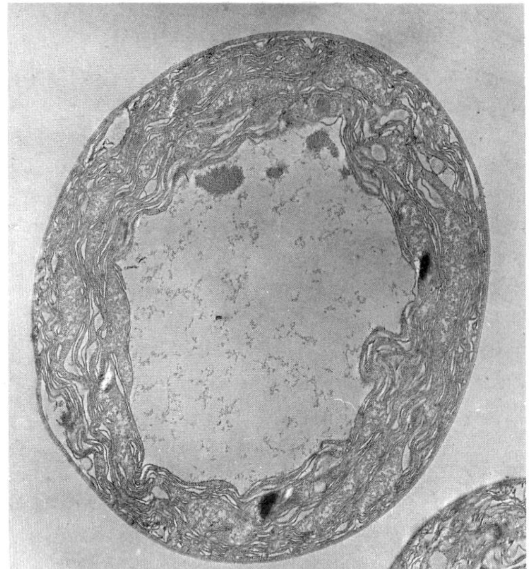

Fig. 8.11 Thin section micrograph of *Prochloron* (Prochlorophycota). The central region appears to be delimited by membranes, while the thylakoids and dense cytoplasm are concentrated in the peripheral region (× 7500). From Giddings *et al.* (1980) with permission.

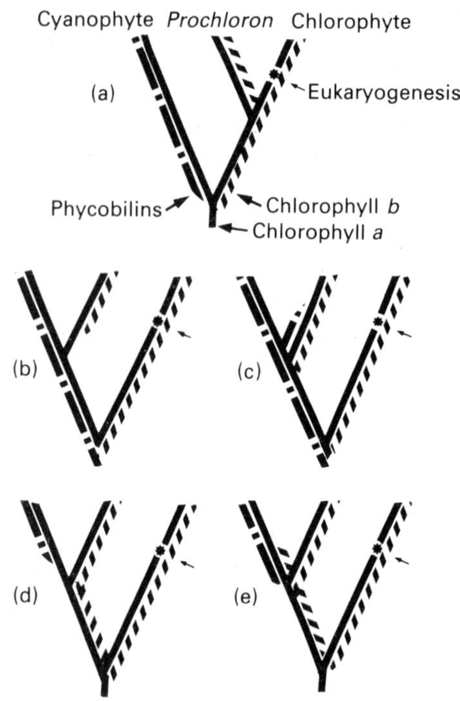

Fig. 8.12 Phylogenetic tree diagrams indicating the possible origins of *Prochloron* and chlorophytes from prokaryotic algae. The chlorophytes and *Prochloron* share chlorophylls *a* and *b* synthesis and a lack of phycobilins. *Prochloron* may have evolved from cyanophytes through the loss of phycobilin and the gain of chlorophyll *b* synthesis (trees b–e), or it may be phylogenetically related to ancestral chlorophytes (tree a). From Lewin (1983a) with permission.

biochemical, physiological, and structural grounds there is substantial evidence to suggest that they were probably derived from the cyanophytes (Klein & Cronquist, 1967). A *Cyanidium*-like alga has been suggested as a possible intermediate form. There is disagreement, however, as to whether *Cyanidium caldarium* is a cyanelle-containing organism homologous with, for example, *Cyanophora* (Kremer, 1982) or whether it is a true rhodophyte (Bisalputra, 1974; Ford, 1984). The difficulty with the cyanophyte origin of the rhodophytes lies in the question of the evolution of the eukaryotes: the unique features of the red algae clearly place them at a very early and separate point of divergence from the remaining algae, as reflected in various phyletic schemes.

The first rhodophytes were probably unicells, from which simple filamentous forms evolved at an early stage. In the absence of a good fossil record, any schemes attempting to derive the various ensuing evolutionary lines are largely speculative. It is not agreed whether or not the major subdivisions of the class, the Bangiophycidae and Florideophycidae, represent natural groups derived from two separate lines of evolution (as suggested by Lee, 1980). Recent discoveries have revealed much less distinction between them than proposed in earlier accounts (e.g. Fritsch, 1945). There may be considerable merit to the suggestion (Scagel *et al.*, 1982) that the conchocelis stage of *Porphyra* and *Bangia* (Bangiophycidae) represents a prototype in the red algal line from which the Florideophycidae emerged, leaving the Bangiophycidae as a dead-end.

Klein and Cronquist (1967), Cavalier-Smith (1978) and Kohlmeyer and Kohlmeyer (1979) are among various workers who suggest that the red algae gave rise to the higher fungi, through loss of the plastid. The lack of motile cells, and similarities in the fertilization process and subsequent development, are often cited as reasons for aligning these two groups (Scagel *et al.*, 1982). The contrary view is that the fungi arose from simple protozoan ancestors, with the higher fungi being derived from a zoomycetous ancestor, any similarities with the rhodophytes being the result of parallel evolution. These suggestions must remain highly speculative, since so little is known about the origins of either group.

THE CHROMOPHYTE AND CHLOROPHYTE LINES

As outlined earlier, classification (and phyletic) schemes for the algae have depended on the weight placed on differences in pigmentation or flagellar features. The scheme adopted here presumes an early divergence between those algae that synthesized chlorophyll *c* (the chromophytes) and those that synthesized chlorophyll *b* (the chlorophytes). This divergence must, however, have occurred after the origin of flagella, a key event in the early evolution of eukaryotes. As pointed out by Cavalier-Smith (1978), most of the adaptive radiations that led to the evolution of the main groups of eukaryotes (plants and animals) occurred in unicellular flagellates. The flagellum is sufficiently complex yet stable (especially in the features of the basal body and 9 + 2 axoneme) that it can serve as a reliable indicator of relatedness (Manton, 1965; Taylor, 1976; Moestrup, 1982). Other flagellar characteristics (e.g. roots, external features, arrangement, and number) often straddle several major groups (Moestrup, 1982) and indicate close relationships where common modifications are shared.

By contrast, Cavalier-Smith (1978) suggested that nutrition-related characters (such as chloroplasts, with accompanying cytological modifications) are often more useful in showing differences between groups. Both pigments and flagellar features appear to have been little altered in the transition from unicellular to multicellular plants, and there is extensive evidence to suggest that multicellular organisms arose many times during the course of evolution (Cavalier-Smith, 1978). In considering the phylogeny of the algae (apart from the rhodophytes), the features of flagella and pigments (with their accompanying cytological and biochemical features) are therefore overwhelmingly important in the determination of relationships and differences at the class level and, ultimately, in the construction of phylogenetic trees.

Dodge's scheme (Dodge, 1979; Fig. 8.10) favours the endosymbiotic hypothesis for the evolution of the eukaryotes; it proposes three or four separate lines derived from an ancestral prokaryote. The chromophyte and chlorophyte lines are preserved: there is an integration of the rhodophytes and cryptophytes into the chromophyte line, and a prochlorophyte-like ancestor, endosymbiotic in an ancestral flagellate, leads to the chlorophyte series. Only the Dinophyceae are isolated, and for them a 'late' endosymbiotic event is required, involving ancestors among the Prymnesiophyceae, Chrysophyceae, and an ancestral flagellate. There is no real time sequence proposed, and there are as many uncertainties in the scheme as there are in a strictly autogenous one; it does emphasize, however, how differently algal phylogeny can be portrayed, depending on the assumptions made.

The relations within the chromophyte and chlorophyte lines have been proposed in the essentially phenetic schemes of Hibberd (1979; chromophytes), and Moestrup and Ettl (1979) and Melkonian (1982a; chlorophytes). These schemes cannot comfortably accommodate the dinophytes or euglenoids. For these, the possible endosymbiotic origin of chloroplasts warrants serious consideration (Gibbs, 1978; Leedale, 1978; Whatley & Whatley, 1981). The dinophytes show no relationships with other chromophyte groups on the basis of their flagella (Moestrup, 1982) or a variety of biochemical features (Ragan & Chapman, 1978). The Cryptophyceae also pose difficulties: although Dodge (1979) was able to accommodate them in his chromophyte line (*see also* Gillot & Gibbs, 1980), they stand alone on a number of fundamental grounds, including the unique arrangement of their flagellar hairs (Moestrup, 1982).

With the exception of the euglenoids, it is generally agreed that the chlorophyte line is monophyletic. When the Chlorophyceae, Prasinophyceae, and Chlorophyceae diverged is uncertain; Moestrup (1982) has suggested that

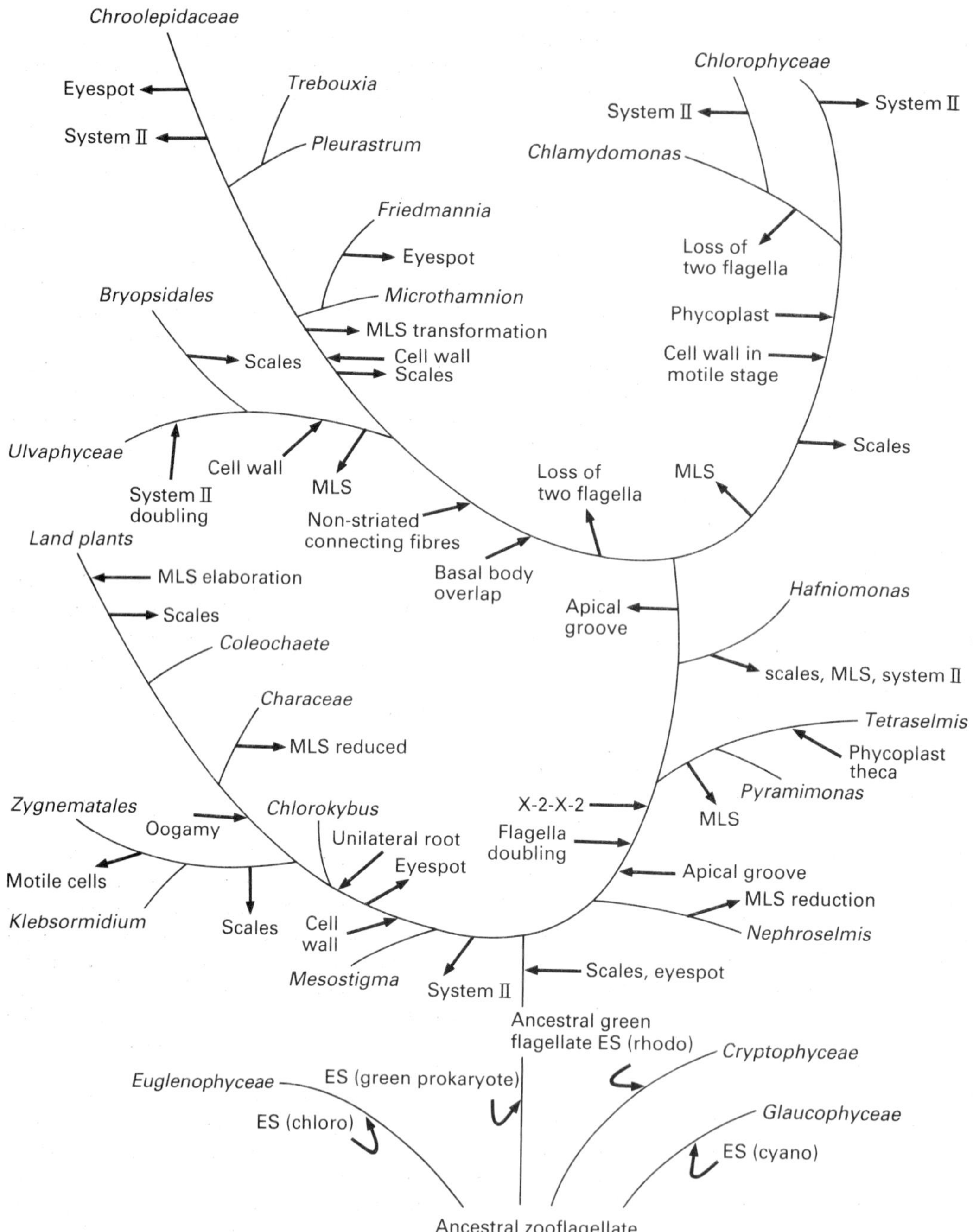

Fig. 8.13 Proposed evolution of the chlorophytes and land plants. Four endosymbiotic events are indicated (ES — curved arrows): in the evolution of the Euglenophyceae (incorporation of chloroplasts from eukaryotic green algae); the green algae (incorporation of a *Prochloron*-like green prokaryote); the Cryptophyceae (incorporation of a red alga); and the 'Glaucophyceae' (incorporation of a blue-green prokaryote). Arrows pointing towards the lines indicate a gain of a structure, those pointing away, loss of a structure. MLS, multilayered structure; system II, fibrous root, X-2-X-2; root, cruciate flagellar root system with two types of microtubular roots. From Melkonian (1982a) with permission.

members of the 'Loxophyceae' (not included in this treatment) represent a primitive component of the chlorophyte line, most related to the Prasinophyceae but possessing a number of intermediate prasinophyte/chlorophyte characteristics. He did not, however, accommodate the Loxophyceae in his proposed phlyetic scheme.

A growing body of work is stressing the value of conservative (e.g. 9 + 2 axoneme structure) versus variable (e.g. transition region) flagellar features as markers in evolutionary relationships between (Melkonian, 1982b) or among (e.g. O'Kelly & Floyd, 1984a, b) chlorophyte groups. O'Kelly and Floyd (1984) stress the importance of the absolute orientation of flagellar apparatus in green algal phylogeny: in their scheme counterclockwise (primitive) orientation occurs in all except the Chlorophyceae, which possess clockwise (advanced) orientation. In their proposed phyletic scheme three evolutionary lines (Chlorophyceae; Ulvophyceae; and *Pleurastrum*, Pleurastrophyceae) are considered to have a common ancestor resembling modern *Pyramimonas* (Prasinophyceae); the Charophyceae are thought to be of more ancient derivation. They believe that the ancestral green flagellate cell possessed the primitive characteristics of counterclockwise absolute orientation of the flagellar apparatus, and diamond-shaped body scales. Melkonian (1982b) proposed a detailed hypothetical phyletic scheme (Fig. 8.13) which interprets the currently available information on flagellar root features. Four endosymbiotic events are presumed: incorporation of a eukaryotic green alga (leading to the euglenophytes); incorporation of a *Prochloron*-like green prokaryote (leading to the Chlorophyceae); incorporation of a blue-green prokaryote (leading to the Glaucophyceae) and incorporation of a red alga (leading to the Cryptophyceae). Like all schemes, Melkonian's poses many questions and makes a number of assumptions, and there is no timescale indicated. It does not differ from the generally accepted relations among present-day groups, although it does differ in the details of interpretation.

The phyletic isolation of the euglenoids is discussed by many authors. Leedale (1978) listed the long series of distinctive features of this group, which make their placement in any scheme extremely difficult. A long-standing problem has been whether to assume a green or a colourless euglenoid line as basic to all present-day genera. Both Leedale (1978) and Dodge (1979) present phyletic schemes which favour a colourless ancestral line. Their suggestion that euglenoids only recently acquired chloroplasts through an endosymbiotic process seems to have merit when considering the probable evolutionary isolation of this group.

The position of the 'Glaucophyceae' (p.45) is discussed at length in Moestrup (1982). The phylogenetic affinities of these cyanelle-containing algae are obscure, although Moestrup and Ettl (1979) implied that they belong in the chlorophyll *b*-containing eukaryotes as a separate, monophyletic line. By contrast their treatment in Lee (1980) suggests that the glaucophytes are living examples of ancestral flagellates, and proof of the endosymbiotic origin of chloroplasts.

The fossil record

The known fossil record of classes within the chromophyte and chlorophyte lines is summarized in Fig. 8.14. The information is biased in favour of those algae with walls or other skeletal structures likely to fossilize; as a consequence a clear phylogenetic picture can be inferred for only a few of the groups.

RHODOPHYCEAE

The construction of phenetic series among the Florideophycidae is highly conjectural, as the fossil record is incomplete, and as there is a great deal of uncertainty concerning the interrelationships of the orders, especially the Nemaliales (Scagel *et al.*, 1982). There is, however, an excellent fossil record of the calcified coralline rhodophytes, described in a voluminous literature which is, regrettably, poorly known among phycologists (Johansen, 1981). Four families of calcified red algae-containing fossils are presently generally accepted (Wray, 1977), the Solenoporaceae, Gymnocodiaceae, Squamariaceae, and Corallinaceae, the first two of which are extinct. The Solenoporaceae first appeared in the Cambrian period and were abundant by the Ordovician (e.g. *Solenopora; Parachaetetes*). Two distinct lines were evident by the mid-Silurian, one of which appears to have led to the modern smaller-celled Corallinaceae (Wray, 1977; Johansen, 1981). By the end of the Mesozoic

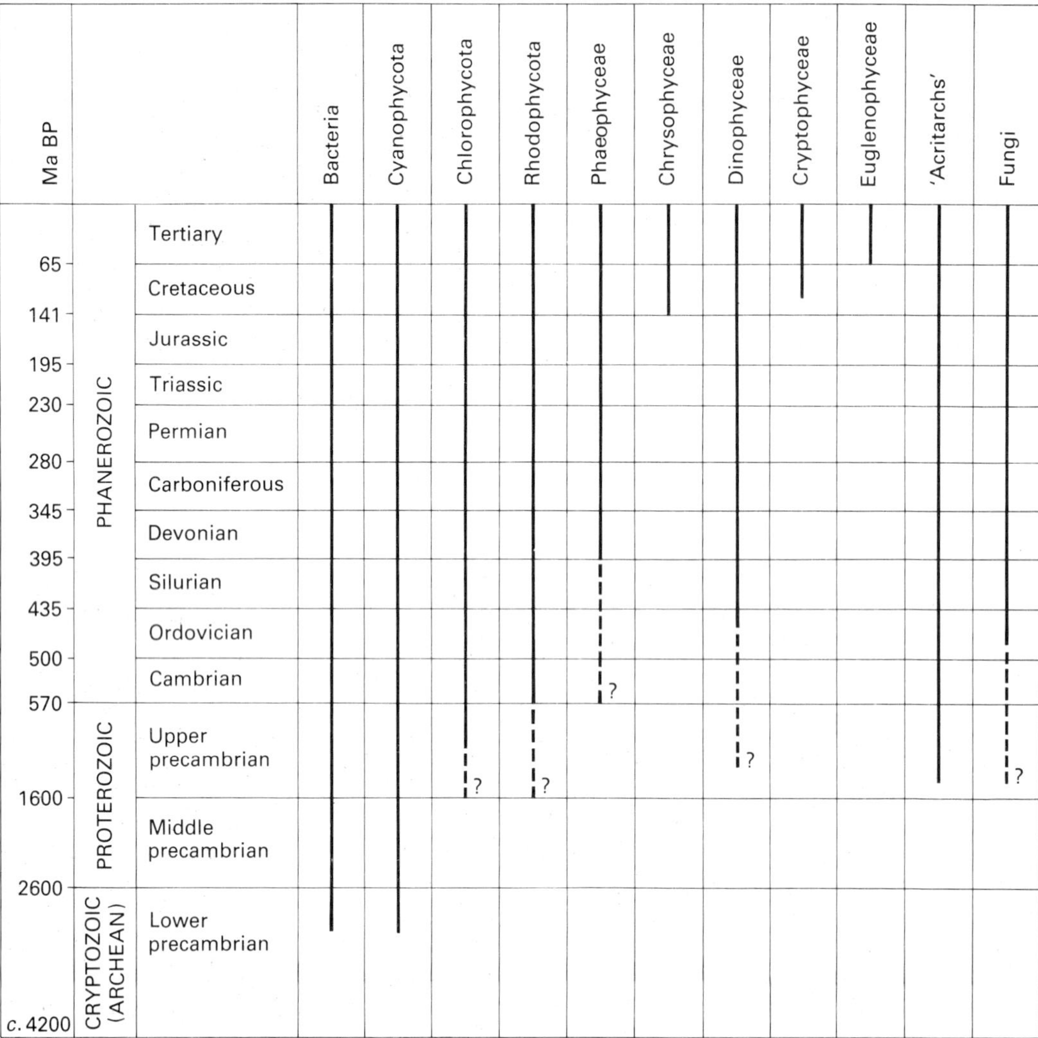

Fig. 8.14 Geological ranges of the bacteria and main algal phyla (including acritarchs). Broken lines indicate questionable or unverified records. (Modified from Scagel *et al.*, 1982.)

era, from the Cretaceous onward, many recent genera (such as *Lithothamnion; Lithophyllum*) were well established. There have been various suggestions as to how the articulated and non-articulated forms evolved; recent morphological and cultural studies suggest (Scagel *et al.*, 1982) that the segregation of modern corallines into articulated and non-articulated groups is unnatural, since phenetic data indicate that the articulated genera may have evolved from two distinct groups of early non-articulated ancestors, and not from a single line as often suggested (Johansen, 1981).

Non-articulated coralline algae occurring as free-living, nodular structures are known as rhodoliths in the fossil and living state (Fig. 8.15; = *Lithothamnion* balls, or rhodolite, but not marl; *see* Adey & MacIntyre, 1973; Johansen, 1981). They occur on the ocean floor from 50 to 200 m deep, and are widely distributed. They

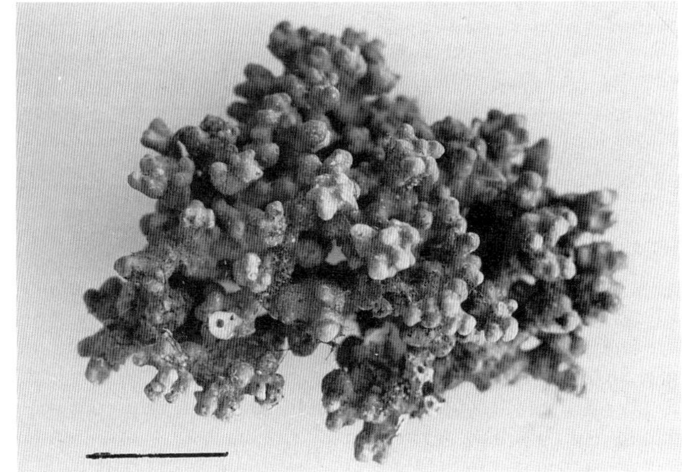

Fig. 8.15 Rhodolith, a living coralline alga (*Lithothamnion glaciale*; Rhodophycota, Corallinaceae) responsible for the laying down of rhodolith deposits in the shallow ocean. Scale = 2.0 cm.

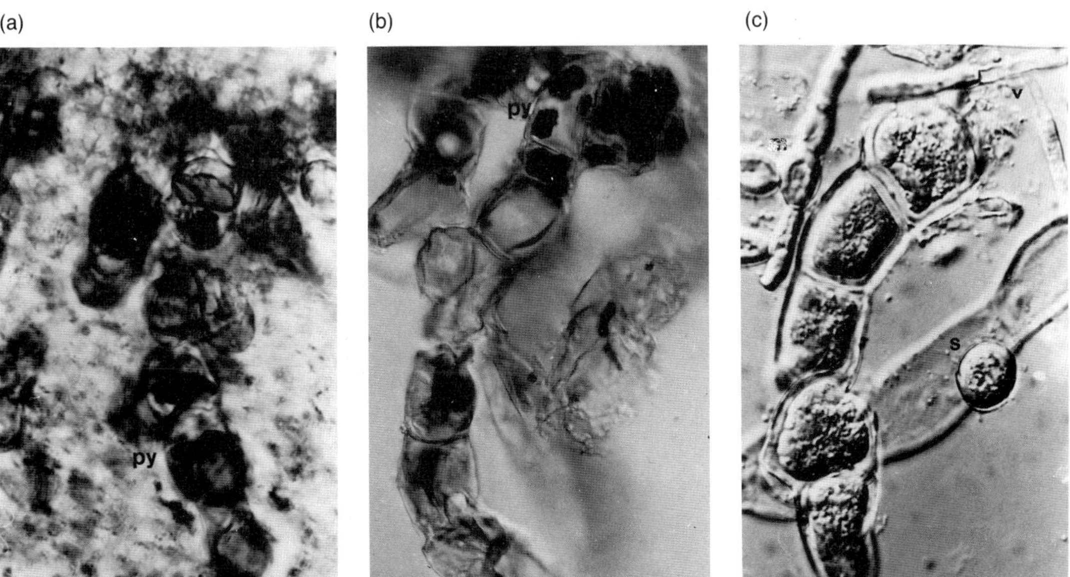

Fig. 8.16 *Palaeoconchocelis starmachii* (Rhodophycota, Bangiophycidae). This Upper Silurian (*c*. 425 Ma BP) endolithic fossil bears a striking resemblance to the living conchocelis phase of recent members of the Bangiophycidae. (a) Petrographic thin section of the *in-situ* fossil. (b) Conchosporangial branch freed from the rock. (c) Recent conchocelis of *Porphyra nereocystis*. py, pyrite within the cell; s, chochospore. From Campbell (1980) with permission.

are very slow-growing and may take 800 years or more to reach a diameter of 30 cm (Adey & MacIntyre, 1973). Rhodoliths from the Agulhas Bank off South Africa have radiocarbon-determined ages of up to 13 760 years (Siesser, 1972). Fossil rhodoliths are recorded from the Eocene, Oligocene, and Miocene.

Soft members of the Rhodophyceae are poorly represented in the fossil record, although Campbell (1980) gives a remarkable account of *Palaeoconchocelis starmachii* from the Upper Silurian of Poland (425 Ma ago). The endolithic habit of this alga accounts for its excellent preservation (Fig. 8.16). The fossilized filaments are morphologically identical with the conchocelis phase of modern *Bangia* and *Porphyra*, which Campbell describes as 'living fossils'. Tappan (1976) has attempted to place much older fossils, from the Proterozoic (as far back as 1.9 Ga ago) in the bangiophytes, thus placing the origin of this group close to that of the eukaryotic cell.

PHAEOPHYCEAE

Although a possible Precambrian origin for the Phaeophyceae has been suggested (pre 570 Ma ago), the early fossils from this group are scarce since, apart from the extant *Padina*, members of this class are uncalcified. Some fossils from the Silurian and Devonian are reportedly similar to modern members of the Fucales, Dictyotales, Chordariales, and Sphacelariales (Scagel *et al.*, 1982); and others from the Triassic (*c.* 25 Ma ago) are reportedly phaeophytes. The best fossil brown algae are from the Miocene (Parker & Dawson, 1965) and include genera assigned to the Laminariales (*Julescraneia*) and Fucales (*Paleohalidrys; Cystoseirites; Paleocystophora;* Fig. 8.17). Some very recent (*c.* 6450 years BP) non-calcareous phaeophycean fossils from northern Canada were remarkably well preserved (Illman *et al.*, 1972; Fig. 8.18).

The scant fossil record of the brown algae provides little evidence in support of phylogenetic considerations. Unicellular representatives are presently unknown. On cytological and biochemical grounds the brown algae most closely resemble the Chrysophyceae, and Christensen

Fig. 8.17 *Paleocystophora acuminata* (Phaeophyceae, Fucales). A Pliocene phaeophyte fossilized in soft white diatomite, California. Thirteen fossil species of the family Cystoseiraceae (in which *Paleocystophora* is placed) were recovered from the rich Californian deposits, indicating that the family is an old and long-diversified group (× 1.16). From Parker & Dawson (1965) with permission.

Fig. 8.18 *Sphacelaria plumosa* (Phaeophyceae, Sphacelariales). A post-glacial (*c.* 8000 years BP) fossil of an extant species recovered from deposits in the eastern Canadian Arctic, showing a remarkable state of preservation. A terminal branch with opposite, pinnate branching is shown. Scale = 0.5 mm. From Illman *et al.* (1972) with permission.

(1980) has suggested that they could be considered as clearly differentiated multicellular derivatives in a large group embracing both classes. Ragan and Chapman (1978) state a similar case on biochemical grounds, proposing that the Phaeophyceae represent the most advanced group in the chlorophyll *c* or chromophyte line. Pedersen (1984) favours a direct link between the Chrysophyceae and Phaeophyceae (= Fucophyceae), suggesting that the ancestral brown alga was derived from a chrysophycean plant with an erect, parenchymatous thallus.

BACILLARIOPHYCEAE

The siliceous walls of diatoms are easily preserved as recognizable fossils, the earliest of which are from the Lower Cretaceous (Tappan, 1980). Earlier fossils (e.g. from the Precambrian) attributed to diatoms have been dismissed as unreliable, but in an extensive review of the possible origins of diatoms Round and Crawford (1981) suggest that these data should be re-examined. The earliest forms were centric (e.g. *Stephanophyxis*). From an examination of the extensive fossil record (*see* Tappan, 1980), in combination with consideration of the known physiological and morphological features of living diatoms, Round and Crawford (1981) suggested that the prediatom ancestor to these centric forms would have possessed eukaryotic organelles and siliceous scales. They further proposed that rudimentary valves evolved from such scales, and were able to provide a hypothetical progression from a cell with siliceous scales to a true diatom with valves and girdle. Tappan (1974) has suggested that the ancestral organism was flagellate and, from the analysis of Round and Crawford (1981), it would appear likely to have possessed sexual reproduction. From a centric ancestor the subsequent evolution of the diatoms along the three classical lines (centric, araphid, and raphid) of present-day species can be readily accommodated.

All available evidence suggests that the diatoms evolved in a shallow-water benthic environment; whether this was saline or not is unclear (Round & Crawford, 1981). Centric species are prevalent in marine rather than freshwater environments, where the pennate species predominate; extensive fossil diatom deposits (as diatomaceous earth) have been laid down in both marine and freshwater habitats. All of the 60 genera known from the Upper Cretaceous to the present still exist, with essentially no change in generic character (Johnson, 1951).

The position of the diatoms in the chromophyte line is unlikely to be resolved until prediatoms are recognized in the fossil record. It appears possible, however, that they were established as a distinctive group very early in chromophyte evolution, from a eukaryote, flagellate ancestor with siliceous scales.

CHRYSOPHYCEAE

The vegetative stages of members of the Chrysophyceae are rarely represented in the fossil record, although the silicoflagellates are exceptions, being common in marine deposits. The siliceous cysts of marine and freshwater forms have also been preserved. Marine cysts are placed in the fossil Archaeomonadaceae, known from the Upper Cretaceous to the Holocene (Tappan,

1980). In the Upper Eocene of Austria they have been described as a nearly pure archaeomonadite (Tynan, 1971). Silicoflagellate remains occur in siliceous rocks, especially diatomites, where they may comprise as much as 40–60% of the rock mass (silicoflagellite; *see* Tappan, 1980). They first appear in the middle Cretaceous (the Jurassic according to Fig. 1 in Taylor, 1978), and have since undergone several fluctuations in diversity, with a peak in the Miocene (Tappan & Loeblich, 1972). Attempts to trace the evolution of silicoflagellates from their fossil record have suggested two possible trends: one towards increasing skeletal simplification, the other towards increasing complexity (Tappan, 1980). The same trends can, however, be observed in modern populations and are thus phenetically deduced. The fossil record therefore gives little indication of the origins and evolution of chrysophycean algae.

Fig. 8.19 SEM preparation of fossil coccolithophorid (Prymnesiophyceae) *Bidiscus ignotus*, showing entire coccosphere. From the Cretaceous, Spain (× 7000). From Grun & Allemann (1975) with permission.

PRYMNESIOPHYCEAE

The Prymnesiophyceae are well known from the fossil record, as coccolithophorids (p. 20; Fig. 8.19). The very extensive literature on this group is reviewed in Tappan (1980). The calcareous scales (coccoliths) preserve perfectly, and a complex and detailed terminology has been developed to describe their morphology. Coccolith fossils are known from the Mesozoic, and were abundant for most of the Jurassic. The chalky deposits of the Cretaceous are largely composed of coccoliths and coccolith debris, and the maximum diversity of the group occurred in the late Cretaceous, with many of the species lasting for 5–10 Ma (Tappan, 1980). Following this peak there was a major extinction, with new forms arising in the Tertiary, most of which have persisted to the present.

A satisfactory phylogeny of the Prymnesiophyceae has yet to be devised, as the fossil record does not provide sufficient information on which to base a phyletic series. Living species are still poorly understood and interpretation of fossils is thus rendered more difficult. It is noteworthy that as a result of investigations by Manton *et al.* (1977) a hitherto unrecognized cold-water group of prymnesiophytes has been discovered; the implications of this have yet to be assessed in palaeobiological terms. In addition, there are indications from recent studies (reported in Melkonian, 1982d) that coccoliths, once deposited, may be affected by the external environment; the taxonomic implications of this could be considerable. The Prymnesiophyceae, while superficially similar to chrysophytes, differ from other chromophyte algae in most respects (Hibberd, 1976). While there may be a common ancestry to the two classes, their placement in a phylogenetic scheme can only be conjectural at the present time.

DINOPHYCEAE

Despite their apparently ancient origins, the dinophytes (Fig. 8.20) are not well known from the fossil record until the Mesozoic and Cenozoic. Tappan (1980) suggests that they may have been unrecognized in earlier deposits, perhaps mistaken for Acritarcha (Evitt, 1963), microfossils of varied biological affinities. Alternatively, they may have had a long geological history prior to developing resistant, preservable structures. Loeblich (1974) has suggested that they were

Fig. 8.20 Fossil dinoflagellates (Dinophyceae). (a) *Phthanoperidinium amoenum* from the Oligocene, Mississippi. Dorsal view of cell (× 1280). From Drugg & Loeblich (1967) with permission. (b) *Palaeoperidinium pyrophorum* from the Paleocene, Alabama. View of whole cell (× 540). From Drugg (1969) with permission.

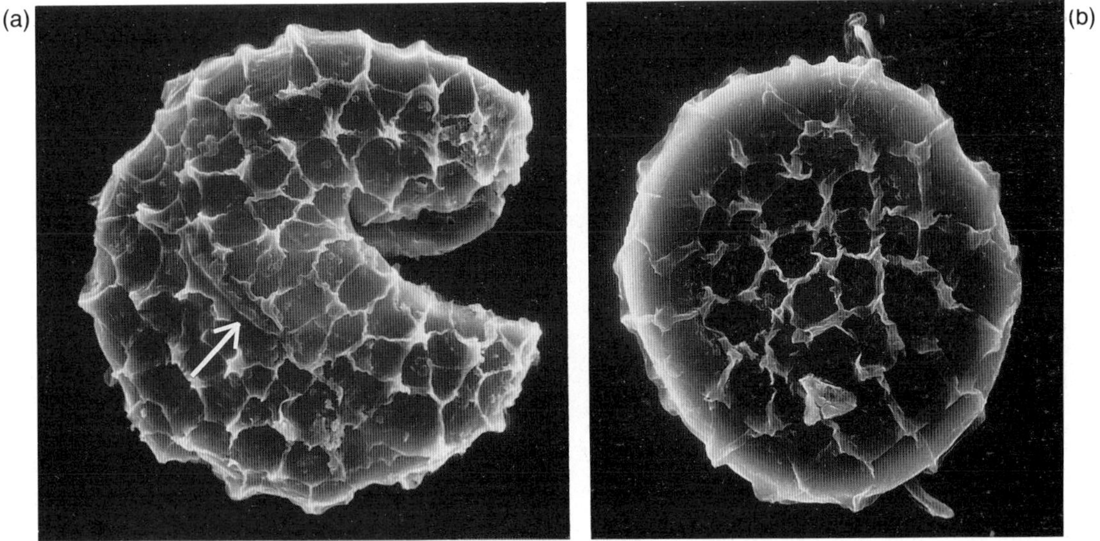

Fig. 8.21 Fossil *Cymatiosphaera reticulosa* (Prasinophyceae) from the Silurian, Missouri. (a) SEM of weakly ornamented phycoma (or cyst), showing dehiscence split (arrow). (b) SEM of a phycoma showing moderately developed ornamentation (× 2500). From Colbath (1983) with permission.

probably derived from ancestral chromophytes in the late Precambrian. A great deal of the interpretation of the evolution of this group hinges on whether the Prorocentrales are regarded as primitive or advanced (*see* Dodge [1983] for a discussion), and whether they are acritarchs, which were the dominant members of the early Palaeozoic microflora (Loeblich, 1976).

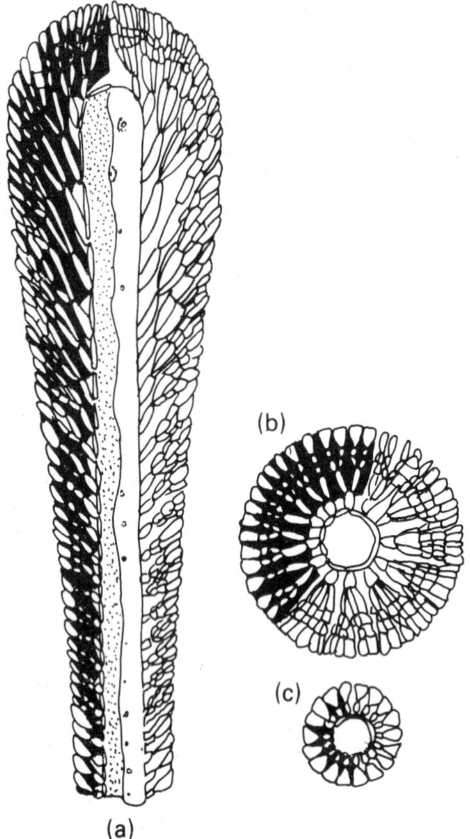

Fig. 8.22 *Palaeodasycladus mediterraneus* (Chlorophyceae, Dasycladales) from the Lower Jurassic, Italy. (a) Reconstruction of a longitudinal section showing the calcified (left) and decalcified (right, with some branches removed) thallus. (b & c) Transverse sections partly calcified and partly decalcified. There are many branches in the apical (b) and simple, unbranched laterals in the lower (c) region of the thallus (× 18). From Pia (1920).

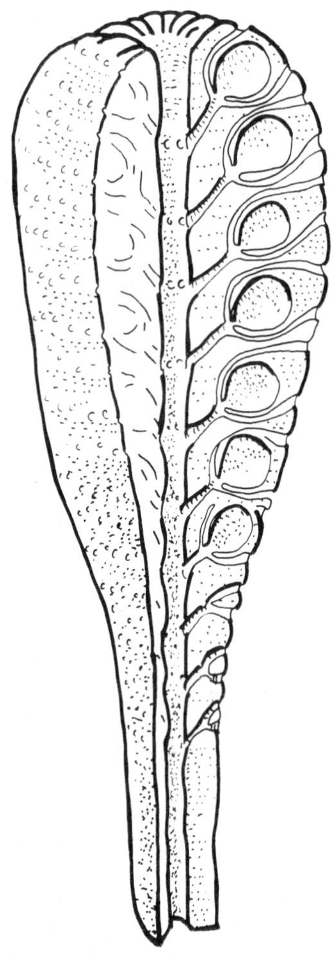

Fig. 8.23 *Cymopolia paktia* (Chlorophyceae, Dasycladales) from the Eocene, Afghanistan. Longitudinal section (× 30). From Kaever (1969) with permission.

Dodge (1983) concludes that they are an advanced group with a fairly recent history, a view that is the antithesis of that of Loeblich (1976) and Taylor (1980).

The Silurian *Arpylorus* may be the earliest dinoflagellate fossil; earlier diversification probably preceded this development of the resistant-walled cyst (Tappan, 1980; Dodge, 1983), although the phenetic evidence does not necessarily support such a notion. Dinoflagellates are excellent indicator fossils, and their record from the Permian to the present is well documented (Tappan & Loeblich, 1973). They showed a rapid diversification during the Jurassic and Cretaceous, but following the post-Cretaceous there has been a continuing decline in the cyst-forming species.

The almost entirely fossil Ebryophycidae or ebrydians (Tappan, 1980) have been variously assigned taxonomically, although they are included here with the dinoflagellates. The living cells, which lack a theca and have apically inserted flagella, superficially resemble to Desmocapsales and Prorocentrales. They are fossilized through preservation of their internal siliceous skeleton, or as thecae or cysts. They are not known from the pre-Cenozoic, but eight genera

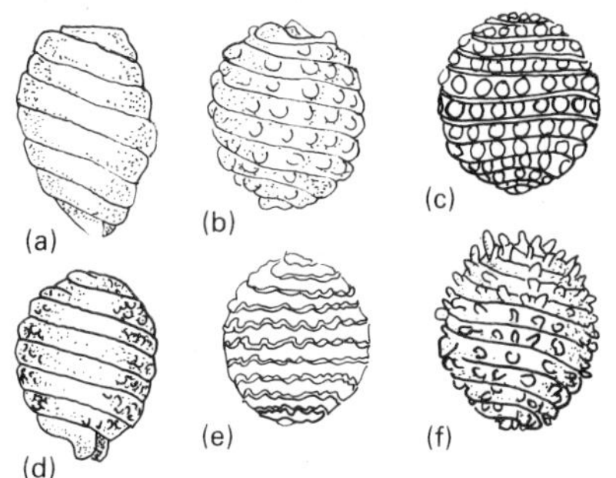

Fig. 8.24 Range of form in living (a) and fossil (b–f) charophyte (Charophyceae) gyrogonites. All are shown in lateral view. (a) *Lychnothamnus barbatus* (× 34). From Maslov (1963) with permission. (b) *Stephanochara caupta* from the lower Oligocene, England (× 23). From Gramblast (1960) with permission. (c) *Kosmogyra superba* from the Palaeocene, Italy (× 10). (After Stache, 1889.) (d) *Harrisichara vasiformis* from the Eocene, England (× 27). From Horn af Rantzien (1959) with permission. (e) *Psilochara undulata* from the Eocene, France (× 18). From Gramblast (1960) with permission. (f) *Microchara hystrix* from the Eocene, France (× 55). From Gramblast (1960) with permission.

occur in the Palaeocene, and more in the Eocene and Miocene (Tappan & Loeblich, 1972). Their relationship to the dinoflagellates is now widely accepted, although their classification is in a state of chaos because of their assignment variously to the animal or plant kingdoms. Their phylogeny is poorly understood, but in future may be of key importance in unravelling the evolution of the dinoflagellates.

THE CHLOROPHYTE SERIES

The geological record of the chlorophytes is the oldest and best known of any algae, extending from the Precambrian to the present. The euglenoids, however, are scarcely represented, and many early reports of fossils are now attributed to cysts of chrysophytes. Well-preserved *Trachelomonas* has been found in Pleistocene bituminous shales from Madagasgar (Deflandre & Lenoble, 1948), and there are reports of *Phacus*-like fossils from the Eocene (Bradley, 1929). There is otherwise little evidence preserved in the rocks telling of the past history and evolution of the euglenoids. The prasinophytes, on the other hand, are known at least from the Ordovician (*c.* 500 Ma ago), being fossilized as cyst-like phycomata (Fig. 8.21), the thick-walled tasmanites, and the thin-walled leiosphaeres (Tappan, 1980). *Tasmanites*, which occurs widely in the Palaeozoic, resembles modern *Pterosperma* (Scagel *et al.*, 1982), while some of the Precambrian and Palaeozoic leiosphaeres (e.g. *Leiosphaeridia*) resemble members of the modern Halosphaeraceae (Wall, 1962). The evolutionary conservatism of the phycoma phase of the prasinophycean life cycle has been stressed by Colbath (1983). According to Tappan (1980) the fossil prasinophytes 'appear to be "disaster species", surviving the widespread extinctions that eliminated most acritarchs in the middle Paleozoic. . . .' They became less important when dinoflagellates and other algal groups diversified. Tasmanite deposits in the Permian of Tasmania accumulated in extensive beds, evidence of the abundance of these algae in the past.

CHLOROPHYCEAE

The poor fossil record for most Chlorophyceae is overshadowed by the impressive array of the calcified, siphonous species, extending back to the Precambrian (Pia, 1920). The siphonous forms that deposit aragonite are especially significant as major rock-builders. It has been assumed that they occurred, like their modern counterparts (e.g. *Halimeda*; *Udotea; Penicillus*), in shallow, warm environments where $CaCO_3$ saturation favoured lime deposition. There have been marked fluctuations in the development and occurrence of fossil chlorophytes; they reached a maximum in the middle Cretaceous, and have declined more or less continuously since then.

The modern Dasycladaceae (Dasycladales) is represented by only a few, rather more complexly constructed remnants of a once much more prolific group (Figs 8.22 & 8.23). It is a true relict family in that about 95% of the described genera are now extinct (Tappan, 1980). Beginning in the Precambrian (1200–1400 Ma ago; e.g. *Timanella; Templuma*) the dasyclads became progressively more complex during the Palaeozoic and diversified in a major way in the Mesozoic. Pia (1920, and later studies; *see* Tappan, 1980) detailed the fossil representatives in his classical monographs, while a more recent phylogenetic account (including both living and extinct forms) has been presented by Valet (1969). While there is a rather complete record of this group, and it is possible to deduce pathways in the evolution of the thallus structure, its origins cannot be deduced from the fossil record.

The chlorophycean fossil record is by no means lacking in the remains of non-calcified species. Coccoid and filamentous fossils attributable to the Chlorococcales and Ulotrichales extend almost as far back as the calcareous forms (Tappan, 1980; Scagel *et al.*, 1982). Recognizable fossils for most other orders are also known, extending variously from the Devonian to recent times. The siphonous forms antedate the filamentous by some 230 million years, and it is uncertain whether the filamentous habit evolved earlier or later than the siphonous.

CHAROPHYCEAE

The charophyte record extends from the Silurian onwards; evolution of this group has been traced entirely from gyrogonites (Fig. 8.24), the preserved remains of their oogonia and encircling sheaths. The living charophytes are another relict group which, according to Gramblast (1974), dropped sharply in diversity at the end of the Devonian and Triassic, reached a maximum in the Lower Cretaceous, and declined in the Upper Cretaceous and Oligocene. Of about 70 recognized genera, only six now survive, and embrace a total of just over 80 species (Wood & Ihmahori, 1965). A fossil group known as the umbellinaceans has been allocated to the charophytes by some, although its affinity is questionable (Tappan, 1980).

The earliest fossil charophytes are well advanced and, as with the calcified chlorophyceans, it is impossible to speculate on their precise origins. The fossil record does, however, substantiate the view that this group was an early offshoot from the chlorophycean line.

CHAPTER 9

Algae, human affairs, and environment

Introduction

Algae have played only a minor role as disease agents, but they are significant agents of a variety of toxic and nuisance problems. They play an unquestionably central role as primary producers (Chapter 7), and over geological time have contributed to the formation of calcareous and siliceous rocks and hydrocarbon deposits. They are beneficial as a source of food, or chemical derivatives and, through the use and development of modern laboratory and cultivation techniques, offer abundant opportunities for present and future utilization (Lewin, 1983b).

Algae and human disease

The role of algae and algal metabolites in disease has been reviewed by Stein and Borden (1984). Since algae are primarily autotrophs, they are poorly adapted as pathogens: only the achlorophyllous, *Chlorella*-like *Prototheca* (Chlorophyceae; Chlorococcales; *see* Kessler, 1977, 1982) has been reported to infect humans and animals. Species of *Prototheca* are found in a variety of habitats, both marine and freshwater, and there are difficulties in relating the wild types to the disease-causing forms. They can be isolated in culture from both diseased and healthy humans. Cells reproduce asexually through the formation of up to 20 endospores (Cooke, 1968a, b).

Since the confirmation of *Prototheca* as an infective organism (Davies *et al.*, 1964) (the condition is normally described as protothecosis), there have been various reports of human infection. The disease usually involves cutaneous lesions with some subcutaneous infiltration and occasional joint involvement (Ahbel *et al.*, 1980). *Prototheca* is rarely systemically invasive, but disseminated infections do occasionally occur (Cox *et al.*, 1974). In animals protothecal infections have been reported in dogs, and in cattle as a causal agent of mastitis (Frank *et al.*, 1969). Munday and Peel (1983) implicate *Prototheca* as a possible cause of ulcerative dermatitis in the platypus (*Ornithorhynchus anatinus*), indicating that infections are potentially possible in a wide variety of animals. The treatment of protothecosis is difficult, since *Prototheca* is resistant to many antibiotics, however some may be effective, and amphotericin B, alone or with tetracycline, has been employed (Stein & Borden, 1984).

Silicosis due to the inhalation of diatomaceous earths has been reported in Sweden (Beskow, 1978). In forensic medicine, diatoms have been used in the diagnosis of drowning (Peabody, 1980), but caution must be used in interpretation, as diatoms are environmentally ubiquitous and are frequently found in normal human tissues.

Toxic algae

Toxins are released by many algae in both freshwater and marine environments and, when ingested by man or other animals, can cause detrimental or even lethal effects. The best-known problems are caused by a number of cyanophytes, a prymnesiophyte, various dinoflagellates, and some seaweeds.

TOXIC CYANOPHYTES

Cyanophyte blooms have long been associated with animal poisonings (Francis, 1878). Three species, *Microcystis aeruginosa, Anabaena flos-aquae*, and *Aphanizomenon flos-aquae* are confirmed toxin-producers. *Anabaena* blooms have the most dramatic effects, *Microcystis* is the most widespread, and *Aphanizomenon* is the least frequently reported (Carmichael, 1981).

Anabaena toxins have been termed anatoxins (Carmichael & Gorham, 1978) and appear to be of three types: alkaloids, peptides, and pteridines. The alkaloids are the most acutely toxic, producing neurological symptoms within minutes of ingestion, whereas the peptides are slower, requiring one to several hours for action and producing liver necrosis. *Microcystis* toxins are mainly peptides (Carmichael & Gorham, 1980) while *Aphanizomenon* toxins are closely related to the saxitoxins produced by the dinoflagellates

(Alam *et al.*, 1978). Not all blooms produce toxins, and even within blooms a mosaic of toxicity exists (Carmichael & Gorham, 1981). There is some evidence to suggest that the presence of plasmids in the cells of *Microcystis* is required for the production of toxins (Hauman, 1981). Acute human poisonings are generally not a problem, as obviously contaminated water is avoided. The presence of cyanophyte toxins in drinking water can, however, lead to gastroenteritis (Keleti *et al.*, 1979).

Toxins in marine *Lyngbya* spp. may be responsible for a contact dermatitis among swimmers in Hawaii (Grauer & Arnold, 1961) and Okinawa (Hashimoto *et al.*, 1976). The active toxin is the polyphenolic, debromoaplysiatoxin (Mynderse *et al.*, 1977).

PRYMNESIUM PARVUM TOXICITY

Prymnesium parvum (Prymnesiophyceae) is widespread in brackish and marine habitats and produces toxins which specifically affect gill-breathing organisms. It is capable of causing mass mortality in fish, and may be a serious nuisance in aquaculture. The toxic components are a mixture of proteo-phospholipids (Shilo, 1981). They require activation by cationic cofactors, sodium, magnesium, or calcium ions, and act by altering the permeability of the gill membranes. The toxic nature of *Prymnesium* is complex, and there is little correlation between the abundance of algal cells and fish mortality.

RED TIDES AND TOXIC DINOFLAGELLATES

Red tides are not always red, or necessarily associated with tides; they are usually monospecific blooms of dinoflagellates which produce toxins, the causal agents of paralytic shellfish poisoning (PSP) or neurotoxic shellfish poisoning (NSP) in humans, fish, and sea birds. Other non-toxic marine organisms may produce discoloured seawater, resulting in kills of marine organisms due to asphyxiation through clogging of gill surfaces (Steidinger & Haddad, 1981).

Dinoflagellate red tides occur in coastal areas and estuaries of both temperate and tropical seas. Of the more than 1000 species of dinoflagellates, approximately 20 have been shown to produce PSP toxins (Steidinger, 1983). PSP is potentially fatal in humans. Although it has been reported since the 17th century (Steidinger & Haddad, 1981), the link with toxic dinoflagellates was not established until 1937 by Sommer *et al.* Extracts of *Protogonyaulax* (= *Gonyaulax*) *catenella* produced the same symptoms in mice as the toxic extracted from toxic shellfish; non-toxic shellfish became toxic when fed on the dinoflagellates.

Protogonyaulax tamarensis (including *P. excavata*) was subsequently described as the causal agent of PSP in the North Atlantic (Needler, 1949), and *P. catenella* in British Columbia (Prakash & Taylor, 1966). Schmidt and Loeblich (1979) suggested that PSP was confined to this '*catenella*' group of species, but other genera have also been implicated (Steidinger, 1983). *Protogonyaulax* toxins may produce fish kills, the toxin being concentrated in the filter-feeding zooplankton (White, 1981).

Twelve closely related water-soluble toxins are implicated in PSP (Steidinger, 1983), of which the best known is saxitoxin. All are potent neurotoxins which inhibit nerve transmission by blocking sodium channels in the nerve axons (Schantz, 1981).

The development of *Protogonyaulax* blooms is linked to the sexual cycle. In *P. tamarensis* benthic resting cysts accumulate in offshore sediments (Lewis *et al.*, 1979). Upwelling transports them to coastal waters, where they germinate (Hartnell, 1975). Germination may be triggered by an increase in temperature (Anderson, 1980); rapid asexual planktonic growth ensues (Anderson *et al.*, 1983), and is dependent on both temperature and salinity (Watrass *et al.*, 1982).

The most spectacular red tides occur on the west coast of Florida and are caused principally by *Ptychodiscus* (= *Gymnodinium*) *brevis*, an NSP agent. Massive fish mortalities may result from these tides. The lipid-soluble NSP toxin has a different structure and action from the PSP toxins. Numbness accompanied by food poisoning symptoms may occur in humans who have eaten shellfish which have accumulated the NSP toxin, but there is no paralysis, and no fatalities have been documented (Steidinger, 1983).

It is postulated that *P. brevis* blooms start in an initiation zone (Steidinger, 1975a, b) 18–74 km offshore, with their development dependent on the seeding of surface waters with benthic cysts. Inshore red tides appear to be correlated with the intrusion of the offshore water, which carries the dinoflagellates onto the coastal shelf (Haddad & Carter, 1979).

The developmental mechanisms for red tides in estuaries are best known for *Prorocentrum minimum* (= *Exuviaella mariae-lebouriae*), implicated in poisonings resulting from consumption of short-necked clams in Japan (Nakajima, 1965). Annual blooms of *P. minimum* occur in parts of Chesapeake Bay (U.S.A.); they begin at a frontal convergence and are transported into the bay below the pycnocline in a subsurface current induced by the spring run-off. Phototaxis and the salinity boundary concentrate the algae in a frontal lens. In the summer the boundaries break down and *P. minimum* enters the surface waters of the inner part of the Bay (Tyler & Seliger, 1978, 1981).

The benthic dinoflagellate *Gambierdiscus toxicus* is reponsible for the production of ciguatera toxin (Bagnis *et al.*, 1979). Ciguatera poisoning is caused by the consumption of the visceral organs of some coral-reef fish. Symptoms include nausea, muscular weakness, and cramps, but mortality is low. Other benthic dinoflagellates, including *Prorocentrum lima*, a species of worldwide distribution, have also been implicated in ciguatera (Yasumoto *et al.*, 1980).

TOXIC SEAWEEDS

Poisonous higher plants are well known; members of many families produce secondary metabolites which act as defences against herbivores and microbial pathogens. Such compounds may be toxic to humans and many are used in medicine.

Fenical (1980) and Norris and Fenical (1982) have reviewed the occurrence of such compounds in seaweeds, and suggest they function to deter grazing invertebrates and fishes. They are found in the Chlorophyceae, Phaeophyceae, and Rhodophyceae, and include acetylenes, terpenoids, haloterpenoids, halomethanes, polyphenols, and alkaloids. The algae which produce these substances are almost totally restricted to tropical waters, a distribution that may be an evolutionary response to the high levels of grazing predation, particularly by fish, in such environments (Hiat & Strasburgh, 1960).

Several families of tropical chlorophytes contain species which are not eaten by general herbivores. *Caulerpa* spp. produce caulerpin, an indole derivative (Maiti *et al.*, 1978), as well as ichthyotoxic terpenoids (Sun & Fenical, 1979).

A general feature of the Phaeophyceae is the production of phenolic compounds (Glombitza, 1977), which are usually polymerization products of phloroglucinol (1, 3, 4-trihydroxybenzene). They are antibacterial (Conover & Sieburth, 1964) and may also cause feeding avoidance by invertebrates. A number of tropical species of the Dictyotaceae and Cystoseiraceae also produce a variety of terpenoids, which in some species of *Dictyota* may reach 5% of the dry weight.

The Rhodophyceae produce the widest variety of toxic secondary metabolites. The toxic species are primarily tropical, although the temperate Bonnemaisoniaceae also produce halogenated toxins. *Asparagopsis taxiformis*, regarded as a delicacy by native Hawaiians (Abbott & Williamson, 1974), produces the carcinogen bromoform, as well as halogenated acetates, acrylates, and ketones. *Liagora farinosa* (Nemaliales, Helminthocladiaceae) produces an acetylenic ichthyotoxic lipid. *Laurencia obtusa* (Ceramiales, Rhodomelaceae), which is avoided by most herbivores including the urchin *Diadema* (Ogden, 1976), produces brominated compounds which show exceptional toxicity to fertilized sea urchin eggs.

Toxic seaweeds are, however, eaten by a few herbivores. Some saccoglossan opisthobranchs selectively eat *Caulerpa*; they concentrate the algal toxins (Doty & Aguilar-Santos, 1970) and secrete them in an ichthyotoxic mucus (Lewin, 1970). Sea hares (*Aplysia*) feed on toxic *Laurencia* and concentrate the toxins as a defence mechanism against predators (Stallard & Faulkner, 1974).

Most seaweeds consumed by humans do not produce toxic compounds, or contain relatively low amounts which are readily metabolized (Fenical, 1980). Some, however, such as *Hizikia* sp. (Phaeophyceae, Fucales), are potentially toxic in that they concentrate arsenic (Watanabe *et al.*, 1980).

Pollution and algae

Pollution is a man-made phenomenon, arising either when the concentrations of naturally occurring substances are increased, or when unnatural synthetic compounds (xenobiotics) are released into the environment. Its effects on algae may either inhibit or stimulate growth, so that the algae themselves become the cause for concern.

ORGANIC POLLUTION

Organic substances released into the environment as a result of domestic, agricultural, and industrial activities result in organic pollution. Of major environmental concern are the effects of domestic sewage, hydrocarbons and halogenated hydrocarbons.

Domestic sewage imposes large oxygen demands, and many algae are sensitive to the oxygen status of the water. Palmer (1977) and Round (1981) described the algal assemblages likely to occur in polluted fresh waters. In the most heavily polluted areas algae may be absent, or restricted to the Cyanophyceae and a few flagellates, especially euglenoids. With reduced pollution diatoms and filamentous chlorophytes appear, and with further reduction a rich algal flora typical of eutrophic conditions develops. Much raw sewage is discharged into the marine environment, where it is usually rapidly diluted. In southern California sewage discharges have been implicated in the destruction of *Macrocystis* (Laminariales, Phaeophyceae) beds (North, 1964, 1971). Recolonization was inhibited by sediment deposition and an increase in the numbers of grazing echinoderms. Raw sewage may, however, enhance the growth of benthic marine algae (Kindig & Littler, 1980).

Hydrocarbons, including natural petroleum oil seeps (Straughan, 1982), occur naturally in the aquatic environment. Marine phytoplankton produce hydrocarbons in amounts that may reach 5 tonnes km^{-2} $annum^{-1}$ (Smith, 1954). The aromatics are more toxic than the aliphatics, and the more soluble, low molecular weight fractions are the most damaging, largely because they are present in the greatest concentrations. At equal concentrations the less soluble compounds with a higher lipid/water partition coefficient are the most acutely toxic to algae (Hutchinson *et al.*, 1978); thus naphthalene at 3 p.p.m. is more toxic than the more soluble benzene at the same concentration.

In the laboratory the general response of algal cultures to hydrocarbons is one of stimulation of growth at low concentrations and inhibition at higher concentrations (Davenport, 1982). Gordon and Prouse (1973) showed that levels of hydrocarbons below 30–50 p.p.m. stimulate photosynthesis, but that progressive inhibition occurs above 50 p.p.m. Naphthalene at 1 p.p.m. reduced the rate of photosynthesis in *Fragilaria* (Bacillariophyceae), *Monochrysis lutheri* (Chrysophyceae), and *Dunaliella* sp. (Chlorophyceae) by 40%, and at 20 p.p.m. by 80% (Vandermeulen & Ahern, 1976). Naphthalene exposure had no effect on the chlorophyll content of the algae, but did reduce the amount of cellular ATP. Naphthalene also inhibits photosynthesis and motility in *Chlamydomonas angulosa* (Chlorophyceae; Kauss *et al.*, 1973). In newly inoculated *Chlorella* cultures the lag phase was shortened by the addition of naphthalene, but the final cell concentration was also lowered. Low concentrations of hydrocarbons increase cell division in several unicellular green algae, and the degree of stimulation correlates with the carcinogenicity of the hydrocarbons. Similar stimulation occurred in sporelings of *Antithamnion plumula* and in related genera of benthic marine Rhodophyceae treated with benzanthracene; the response was dose-dependent and enhanced growth was due to increased cell division rather than to an increase in cell size (Boney & Corner, 1962; Boney, 1974).

Polychlorinated biphenyls in low concentrations enhance both photosynthesis and cell division; at higher concentrations their effect is reversed (Gotham & Rhee, 1982). Accumulation in algae is a two-part process: adsorption on the surface, followed by absorption into the cell by partitioning rather than by active transport. Desorption is slower than absorption, thus recovery is slower than contamination (Lederman & Rhee, 1981). The rate of accumulation and the effects are greater in dilute than in dense algal suspensions (Cole & Plapp, 1974). Many polycyclic hydrocarbons are broken down by sunlight, and there is evidence that breakdown is enhanced when they are bound to algal cell surfaces (Zeppa & Schlotzhauer, 1983).

Mesocosm experiments have yielded data on the effects of hydrocarbons on mixed species of phytoplankton in the presence of grazing zooplankton. A 60 000-litre enclosure in Saanich Inlet, Canada, was polluted with naphthalene-rich crude oil to a concentration of 40 p.p.m. (Lee & Takahashi, 1977). Phytoplankton production increased, and there was a decrease in the abundance of the diatom *Ceraulina bergonii* and a massive increase in the small flagellate *Chrysochromulina kappa* (Chrysophyceae), but there was no apparent effect on the zooplankton populations. Davies *et al.* (1980) used 300 000-litre enclosures in Loch Ewe, Scotland, inoculated with 100 p.p.m. North Sea crude oil. Within five weeks the oil concentrations had fallen to 25 p.p.m.; half the loss was attributed to fall-out to the sediments, the other half to microbial metabolism and evaporation. No changes in phytoplankton biomass or community structure were reported. Vargo *et al.* (1982), using 13 000-litre enclosures polluted with fuel oil, reported similar phytoplankton species composition to that of the source water, but increase in abundance attributable to decreased predation.

Oil spills in nature appear to have little effect on phytoplankton populations. Normal abundances of dinoflagellates and diatoms were reported following the wreck of the *Torrey Canyon*, though large numbers of dead prasinophyte cysts were observed (Smith, 1968). Oil from the tanker *Tsesis* in the Baltic (Johansson *et al.*, 1980) caused a transient increase in phytoplankton biomass in the immediate area, attributed to a decline in zooplankton abundance; normal levels returned, however, within five days.

The effects of oil in nature on benthic marine algae, particularly in the intertidal, have been widely studied, Diversity of the intertidal flora at Coal Point, California, where natural seepages occur, is reduced compared with similar areas which lack seeps (Nicholson & Cimberg, 1971). The grounding of the *Tampico Maru* in Baja, California, released 60 000 barrels of diesel oil in a small cove over an eight-month period (North, 1967). The immediate effect was the loss of benthic herbivores and a subsequent increase in the algal flora. Evaluation of the effects of the oil from the *Torrey Canyon* was complicated by the use of toxic detergents in the clean-up operation. A number of intertidal species were directly affected (O'Brien & Dixon, 1976), and in oiled plots surveyed prior to cleaning the biotic damage was often less than that after cleaning with emulsifiers. In areas where the weathering of the oil removes the lighter, soluble fractions prior to stranding on the shoreline, the major effects are mechanical, with injury to intertidal fucoids due to smothering, and increased loading leading to breakage by wave action (Nelson-Smith, 1973). The major effects of oil from the *Torrey Canyon* and the detergents were the removal of algal predators (Bellamy *et al.*, 1967; Smith, 1968), which together with the space made available by the removal of the larger algae allowed opportunistic species such as *Enteromorpha* (Chlorophyceae) to colonize and dominate the area. In the *Amoco Cadiz* spill in France, detergents were not widely used, and the mortality of seaweeds was correlated with the duration of their exposure to the oil. The high intertidal species *Fucus spiralis* and *Pelvetia canaliculatus* (Fucales, Phaeophyceae) were the most affected (Floc'h & Diouris, 1981). Fourteen months later oil residues could still be detected in the intertidal fucoids, but growth rates were unaffected (Topinka & Tucker, 1981).

CULTURAL EUTROPHICATION

Cultural eutrophication (Hutchinson, 1969) describes a condition in which the nutrient status of water is enriched through human activities; increases in available nitrogen, phosphorus, and carbon occur, causing an enhancement of algal production and a change in species composition. Lakes apparently do not become progressively eutrophic due to natural evolutionary processes (Beeton & Edmonston, 1972; Round, 1981). Whiteside (1983) argues that eutrophication is an inappropriate term to describe the processes of enrichment due to the ageing of lakes, and suggests that biotic interactions and/or lake morphometry could equally explain observed palaeolimnological changes. The full extent of the problem is difficult to determine, but it is estimated that in the U.S.A. 50% of the surface waters have some eutrophication (Jewell & McCarty, 1968). The principal problems arise from the massive growths of algae which physically impede waterways and increase oxygen demand during decay. Agricultural run-off is the main source of nitrogen; in England drainage from arable land contains on average 10 mg/l nitrate (Cooke & Williams, 1970), and may be an order of magnitude higher. Domestic sewage is the main source of phosphates and organic carbon, but unlike nitrogen these are usually added to waters at point sources.

Edmonston (1970, 1972, 1979) and Edmonston and Lehman (1981) have provided, for Lake

Washington, what is probably the most detailed study of the effects of sewage effluent and the subsequent recovery of a lake after treatment. Diversion of sewage effluent caused levels of phosphate to return to pre-eutrophication levels and produced a concomitant reduction in algal biomass. This reduction did not, however, result in a decrease in the nitrate or carbon levels, suggesting that phosphorus is the most important nutrient in the eutrophication process. Godfrey (1982) showed that over the previous 60 years the standing crop of algae in Cayuga Lake. U.S.A., had doubled in the summer months and showed a 30-fold increase in the spring. These changes correlated with the use of phosphate-based detergents.

Although phosphorus is clearly implicated in the eutrophication process, there has been considerable discussion of the relative importance of different nutrients. Small bottle experiments involving the artificial enrichment of lake water (Schindler, 1971) suggested that both carbon and phosphorus are implicated. Carbon is rarely limiting in nature, and in a lake experimentally fertilized with phosphorus and nitrogen (Schindler *et al.*, 1973) the carbon content due to phytoplankton growth also increased, even though the water was naturally low in carbon. Flett *et al.* (1980) showed that nitrogen-fixers among the blue-green algae could increase the nitrogen content of lake water if other nutrients were supplied in excess. This further confirms the prime role of phosphorus in eutrophication, and suggests that it is inappropriate to control nitrogen and carbon inputs on an ecosystem basis without also controlling phosphorus.

Eutrophication also affects species composition and seasonality. In unenriched temperate waters algal populations tend to peak in the spring and autumn. In highly eutrophic waters these peaks become less apparent as additional blooms occur throughout the year (Davis, 1964). In the Laurentian Great Lakes, moderate increases in phosphates caused a rapid depletion in silicate and a consequent shift from diatom-dominated assemblages to those with greater numbers of green and blue-green algae (Makarewicz & Baybutt, 1981). Still greater phosphate loading brought about a reduction in the available nitrogen and an enhancement of nitrogen-fixing cyanophytes (Bierman *et al.*, 1980).

The most striking effects of cultural eutrophication are seen in the increased growth of filamentous algae. *Cladophora glomerata* (Chlorophyceae) is the best studied of these, but other genera such as *Spirogyra, Rhizoclonium, Enteromorpha* (Chlorophyceae), *Compsopogon*, and *Thorea* (Rhodophyceae) are also affected and may create nuisances (Palmer, 1977; Whitton, 1971). The choking of rivers with long strands of *Cladophora* can be a major problem in temperate regions (Lund, 1972), and dramatic effects are seen in the more eutrophic of the Laurentian Great Lakes.

Cladophora glomerata was first identified in Lake Erie in 1847, and its growth has increased with eutrophication (Millner & Sweeney, 1982). Blooms occur in summer and autumn, and biomass exceeds 400 g dry weight/m^2 at depths between 0.5 and 3 m. The alga sloughs and may be carried ashore to foul beaches (Niell & Jackson, 1982). Graham *et al.* (1982) showed that optimum growth of *C. glomerata* occurs between 13 and 17 °C; cessation of growth occurred when water temperatures exceeded 20 °C. Reactive phosphorus was then below the level of detection in the *Cladophora* beds, and addition of phosphorus did not stimulate growth until the temperature dropped below 20 °C.

A variety of control methods has been employed to reduce growths of *C. glomerata*, including application of copper sulphate, sodium arsenite, and a variety of commercial herbicides. The successful management of the problem still lies, however, in the control of eutrophication, rather than in removal of the nuisance algae which result from it.

METAL AND ACID POLLUTION

Recent reviews of the effects of heavy metals on algae include those of Rai *et al.* (1981), Stokes (1983), and Davies (1983). Heavy metals may be used as algicides, for example copper solutions may be added to reservoirs, although some algae are remarkably tolerant of heavy metal pollution. Many studies are contradictory, and it is still not possible to generalize about the influence of metals on individual species or communities.

Studies of algal floras in mine drainage areas show that metals decrease both diversity and productivity, and that cyanophytes and diatoms are generally less tolerant than members of the Chlorophyceae. Foster (1982a) compared sites in streams polluted by copper and lead and concluded that it was the level of metal pollution rather than the metal *per se* that determined the species composition at the sites. *Ulothrix* and *Microspora* (Chlorophyceae) were especially tolerant, but members of the Zygnematales (Chlorophyceae) were typically found in areas of low metal concentration. In contrast, Gale *et al.* (1973) reported *Mougeotia, Zygnema*, and *Spirogyra* (Zygnematales) to be abundant in metal-polluted streams in Missouri (U.S.A.). Say and Whitton (1978) list species of *Netrium, Plectonema, Nitzschia, Hormidium, Ulothrix,* and *Mougeotia* as typical of the flora of zinc-contaminated streams. There is some evidence for cotolerance of metals. Foster (1982b) selected algae for copper tolerance and also found increased tolerance of nickel. In contrast, Whitton and Shehata (1982) selected strains of *Anacystis nidulans* (Cyanophyceae) for tolerance to cobalt, copper, nickel, and cadmium, and found some evidence for decreased tolerance to a second metal.

Information on the mechanisms of tolerance, which include membrane impermeability, extracellular chelation, or internal detoxification, is sparse and frequently contradictory (Stokes, 1983). At similar external levels of copper, tolerant cells of *Chlorella* sp. contained the same concentrations of copper as cells of a non-tolerant strain, suggesting the lack of an exclusion mechanism. Butler *et al.* (1980) suggested that exclusion was in part due to the complexing of heavy metals with algal extracellular products, with the resulting metal complex being more stable in tolerant strains. In *Scenedesmus*, (Chlorophyceae) however, tolerance to copper does not apparently involve the production of extracellular chelators (Mierle & Stokes, 1976); exclusion is apparently by a reduction in membrane permeability. Longer term experiments showed uptake was not the prime mechanism of tolerance (Stokes 1981). EM studies have located copper complexes in the nuclei of tolerant algae (Silverberg *et al.*, 1976). Non-tolerant algae also accumulate copper in a similar manner, but it is accompanied by extensive membrane damage. In metal-sensitive strains of *Diatoma tenue* (Bacillariophyceae) lead and copper were incorporated into polyphosphate bodies, but damage to mitochondrial and vacuolar membranes also occurred (Sicko-Goad & Stoermer, 1979; Sicko-Goad, 1982).

At many mine sites highly acidic waters are formed by the action of sulphur-oxidizing bacteria. Acidity may thus be linked with heavy metal pollution, and it is frequently difficult to separate the effects of the two pollutants. Hargreaves *et al.* (1975) investigated sites in the U.K. with pH below 3.0 and reviewed the North American literature on acid waters. They reported the occurrence of 40 species of algae under these conditions. Cassin (1974) showed that *Chlamydomonas acidophila* could grow at pH 2.0, and Hargreaves and Whitton (1976) showed that *Chlamydomonas applanata* and *Euglena mutabilis* could be grown at pH 1.3. At high pH most heavy metals occur as readily available to algae, hence they are usually more toxic under acid conditions (Rai *et al.,* 1981). Hargreaves (1976) showed, however, that the toxicity of copper and zinc to *Hormidium rivulvare* (Cyanophyceae) was less at pH 3.5 than at pH 6.0.

There are many recent reviews of the phenomenon known as acid rain, or acid precipitation (Haines, 1981). In addition to lowering the pH of lakes it may increase concentrations of aluminum and heavy metals such as manganese, zinc, and iron which are leached from lake sediments and the catchment area. Increases in concentrations of lead, nickel, copper, and cobalt have also been reported, but attributed to particulate precipitation rather than local leaching. In a study by Yan (1979), lakes at pH 4.1–4.3 with copper and nickel contamination showed a different phytoplankton composition from that of nearby uncontaminated lakes, and resembled acid lakes 80 km distant which lacked heavy metals.

Acidification of lakes causes a reduction in species number and diversity (Schindler, 1980; Hendrey, 1982; Eriksson *et al.*, 1983), though biomass and primary production need not decrease. In three Adirondak lakes (Hendrey, 1982) the smaller flagellates became dominant, this in part attributed to lower phosphate availability. Changes in composition were reported in acidified Swedish lakes (Eriksson *et al.*, 1983), with the bulk of the phytoplankton biomass composed of the acid-tolerant and oligotrophic dinoflagellates *Gymnodinium uberrimum* and

Peridinium inconspicuum. Liming of the lakes raised the pH, but the character of the phytoplankton remained acidic in the first year, due to low nutrient status. In subsequent years the phytoplankton composition came to resemble adjacent oligotrophic lakes at pH 6.0, dominated by diatoms and chrysophytes.

Increases in the biomass of benthic and epiphytic algae coupled with a decrease in species diversity has been observed in many lakes undergoing acidification (Lazarek, 1982), this a result of reduced microbial decomposition and invertebrate grazing.

While acidity is not a problem in the marine environment, local concentrations of heavy metals may occur in coastal and estuarine waters. A number of large-scale mesocosm experiments have investigated the effects of heavy metals on marine phytoplankton. Additions of copper at 10 μg/l and 50 μg/l (Thomas *et al.*, 1977a) caused photosynthetic rates to decrease after three days to 12% and 2% of the control, respectively, but gradual recovery occurred due to the development of copper-tolerant microflagellates. Similar results were obtained with mercury at 1 μg/l and 5 μg/l, with an initial decrease in productivity followed by a recovery associated with a change in species composition (Thomas *et al.*, 1977b). Thomas *et al.* (1980) also investigated the effects of mixtures of metals at relative concentrations similar to those found in moderately polluted estuaries. No significant changes were observed, but when the overall concentration was increased both photosynthetic rates were depressed. Removal of selected metals from the mixture showed that inhibitory effects were due to copper and mercury.

Marine benthic algae are also capable of taking up heavy metals. Zinc is absorbed by *Fucus vesiculosus* and may be concentrated more than 60 000-fold, with the higher concentrations in the older tissues. It is tightly bound and cannot be readily washed out of the algae (Bryan, 1969; Bryan & Hummerstone, 1973). Concentrations of copper, arsenic, lead, zinc and silver, but not of other heavy metals, in the tissues of *F. vesiculosus* (Luoma *et al.*, 1982) correlate with the concentrations of the metals in the adjacent sediments rather than the sea water. Metals may be passed through the food chain, and lead from *Egregia laevigata* has been shown to occur in grazing abalone (Stewart & Schultz-Baldes, 1976).

Much of the experimental work on heavy metal tolerance in benthic algae has been on ship-fouling and copper-tolerant strains of *Ectocarpus siliculosus* (Phaeophyceae) (Russell & Morris, 1970). Tolerance appears to be due to exclusion (Hall *et al.*, 1979; Hall, 1981) rather than detoxification, as non-tolerant strains accumulate up to seven times the amount of copper of tolerant strains. Extracellular chelators are not produced, and the copper is not bound to the cell wall, indicating that tolerance is due to differences in membrane permeability. In contrast, a number of tolerant benthic diatom species (Daniel & Chamberlain, 1981) absorb and detoxify copper by localizing it in the cell as insoluble polyphosphate–copper bodies.

THERMAL POLLUTION

Pollution by heated effluents occurs in rivers, lakes, and estuaries, and to a lesser but increasing extent in marine coastal areas. While many industrial activities return water to the environment at elevated temperatures, the prime source is from thermal electricity-generating facilities, which use water for condenser cooling and usually return it to the environment at 10–15 °C above ambient.

Phytoplankton diversity generally decreases in areas heated over 30 °C, with cyanophytes dominating where temperatures exceed this for extended periods. In the 20–30 °C range both diversity and primary productivity may increase. In marine coastal environments there is little evidence of any long-term effects on phytoplankton assemblages exposed to thermal discharges (Langford, 1983).

Mechanical damage to phytoplankton by passage through condenser systems (Kreh & Dewort, 1976) appears minimal; most studies of this entrained water report insignificant changes in cell densities or pigment levels (Langford, 1983). Ambient temperature, however, may affect photosynthetic rates of the entrained phytoplankton (Sellner *et al.*, 1984). In spring rates were higher at the outfall than at the intake, whereas the reverse was found in summer. Cooling waters are frequently chlorinated to prevent invertebrate fouling and bacterial slime formation, and this produces a reduction in the photosynthetic capacity of entrained phytoplankton (Davis & Coughlan, 1978). There is a 30% reduction at 0.4 mg/l residual chlorine and complete inhibition at 0.8 mg/l. There are also synergistic effects between temperature and chlorine (Eppley *et al.*, 1976).

Freshwater benthic and epiphytic algae show responses similar to those of the phytoplankton. In general, moderate increases in temperature produce both increased productivity and standing crop, though diversity may decrease (Langford, 1983). In a Canadian lake heated effluent kept a discharge area ice-free and enabled diatoms epiphytic on *Scirpus validus* to grow throughout winter and spring; in summer they were replaced by filamentous green algae. The thermal effluent was later diverted from the lake, and within a year standing crops in the former discharge area returned to normal (Hickman, 1982). Diatom species which had been eliminated returned, and the filamentous chlorophytes *Oedogonium, Spirogyra,* and *Cladophora* decreased in abundance.

An extreme example of thermal pollution (Tison *et al.*, 1981) involved a fluctuating discharge with temperatures to 45 °C. An algal mat community composed primarily of the cyanophytes *Fischerella* and *Phormidium* developed and remained when the temperature returned to ambient. The cyanophytes maintained an optimum growth temperature close to the upper limit to which they were exposed, a response similar to that reported from hot spring algal communities (Boylen & Brock, 1973).

Marine macroalgae are also locally affected by thermal discharges. At Turkey Point in Florida. Thorhaug *et al.* (1973) reported declines in *Halimeda* and *Penicillus* (Chlorophyceae) in areas heated 3–5 °C above winter ambient. The bare substrate was subsequently colonized by cyanophytes. Vadas *et al.* (1976) found few overall changes in the littoral macroalgae of a thermally affected Maine (U.S.A.) shore. Earlier growth of fucoids occurred in some of the moderately heated areas, but in other areas apical tips of *Ascophyllum nodosum* were not initiated the year after the discharge commenced. The basal portions survived, but were only weakly attached. In a still colder region, in Newfoundland, Canada, the effluent from a peak load station, operating only in the winter months, had no effect beyond the immediate outfall area (Whittick & Hooper, 1979). Growth of fucoids and *Chondrus* (Rhodophyceae) was enhanced, and the winter flora typified by *Monostroma, Urospora*, and *Bangia* was absent.

Mechanial nuisances

Excessive growths of freshwater algae may restrict water movement in rivers and artificial channels. Marine algae may impede the water flow through filters, and foul ships' hulls.

RESTRICTION OF WATER MOVEMENT

Filter-clogging algae are nuisances in domestic water supply systems (Palmer, 1977); the water is usually passed through sand filters to remove solid and colloidal debris. In many areas water must be pretreated by coagulants, and sedimented prior to filtration, although filters still clog and must be cleaned by back-washing. Filter-clogging algae usually have rigid walls, are filamentous or colonial, and may produce mucilage. Diatoms create the worst problems, and the most serious offenders are *Asterionella, Tabellaria,* and *Synedra* spp.

Seaweeds may impede the flow of cooling waters into thermal generating stations. At Wylfa Nuclear Station in Anglesey, U.K., *Laminaria* spp. (Phaeophyceae) detached during autumn gales have caused serious intake blockages (Langford, 1983). *Sargassum muticum* (Phaeophyceae), a Japanese species, clogs small harbours and waterways on the Pacific coast of North America, where it was introduced along with Japanese oysters (Scagel, 1956). It now occurs in the English Channel (Farnham *et al.*, 1973), where it threatens to out-compete indigenous species. *Codium fragile* (Chlorophyceae) appeared on the Atlantic coast of North America in the 1950s, again presumably introduced on oyster shells imported from Europe (Churchill & Moeller, 1972); it has now spread from Maine to New Jersey (Malinowski & Ramus, 1973). It is a nuisance in oyster beds south of Cape Cod, growing preferentially on the shells and causing their detachment.

Both of the preceding species are examples of introductions into areas in which the normal constraints of competition and predation are presumably reduced. They serve as a warning of the care that must be taken in the deliberate introduction of 'exotic' species, such as the proposal to introduce the giant kelp *Macrocystis pyrifera* to European waters (Boalch, 1981).

SHIP FOULING

On modern ships algae are the most important group of fouling organisms (Evans, 1981). An increase in roughness that decreases the normal cruising speed of a super tanker by only a fraction of a knot may increase the annual operating expenses by over U.S. $100 000 (Evans, 1981). The main fouling algae are diatoms and filamentous chlorophytes, and phaeophytes such as *Enteromorpha* (Chlorophyceae) and *Ectocarpus* (Phaeophyceae) are common fouling algae which have wide salinity and temperature tolerances. They also possess effective spore attachment mechanisms and the capacity to regenerate from basal systems following incomplete removal.

The standard antifouling paints are based on cuprous or other copper compounds and are usually formulated to leach copper ions at a rate of 10 $\mu g\ cm^{-2}\ day^{-1}$. Fouling algae vary in their tolerances to copper (Evans, 1981). Copper-resistant *Enteromorpha* spp. became a major problem on large tankers in the 1950s, but the addition of organo-tin compounds to the paints solved this problem. Other species such as *Ulothrix* and *Ectocarpus* were not so susceptible (Millner & Evans, 1980). Modern co-polymer paints in which the toxic compounds are incorporated into a water-soluble methacrylate matrix seem a promising approach, as in addition to the leaching of toxic ions the surface of the paint itself is in a constant state of erosion and is effectively self-polishing.

FIXED STRUCTURES: OIL PLATFORMS AND OFFSHORE THERMAL ENERGY CONVERSIONS

Algal growths on offshore oil production platforms are a potential nuisance (Moss *et al.*, 1981), especially in colder waters where there are adequate nutrients and low competition from sessile invertebrates. Seaweeds increase the hydrodynamic loading of structural members and impede the inspection of critical areas such as welds. They enhance the corrosion of steel, especially where thick layers promote anoxic conditions. By contrast, thin layers of photosynthesizing algae may confer some corrosion resistance (Terry & Edyvean, 1981).

One of the potential methods of generating electricity in tropical areas is by the use of offshore thermal energy conversion (OTEC), which utilizes the temperature differential between the surface and colder, deep water. A major problem in the use of such systems is the fouling of the heat-exchanger surfaces by algae (Thorhaug & Marcus, 1981).

Uses of algae

HUMAN FOOD

Marine algae have been consumed as a regular part of human diet in coastal China since 850 BC (Waaland, 1981) and today they are eaten mainly in the Orient and the Pacific Islands. Chapman and Chapman (1980) list 160 species of seaweeds eaten by humans (25 Chlorophyceae, 54 Phaeophyceae, and 81 Rhodophyceae). The principal genera consumed are:

Chlorophyceae: *Monostroma; Caulerpa; Enteromorpha; Ulva.*

Phaeophyceae: *Laminaria; Undaria; Alaria; Eisenia; Ecklonia.*

Rhodophyceae: *Porphyra; Palmaria; Gracilaria; Gelidium; Eucheuma.*

Doty (1979) estimates the annual value of food algae to be in excess of a U.S. $ 1 billion, mostly from three genera: *Porphyra* (nori, Japan; zicai, China; U.S. $500 million); *Laminaria* (kombu, Japan; haidai, China) and *Undaria* (wakame, Japan; qundaicai, China). In the North Atlantic the principal genera consumed, apart from genera such as *Gigartina* and *Chondrus* (Rhodophyceae) locally utilized in jelly-making, are *Palmaria* (dulse) in eastern Canada and Eire, and *Porphyra* (laver bread) in south Wales. Both are considered epicurean products rather than dietary staples.

Reasons for consumption include food value, flavour, colour, and texture. The structural carbohydrates of seaweeds are largely indigestible, but some soluble carbohydrates are metabolized. The protein content of many of the edible seaweeds is 20–25% dry weight. Seaweeds are an excellent source of vitamins (Kanazawa, 1963), including vitamin C at levels equivalent to citrus fruits, and vitamins A, D, B_1, B_{12}, E, riboflavin, niacin, pantothenic acid, and folic acid. Seaweeds also provide all the required trace elements for human nutrition (Yamamoto *et al.*, 1979). The attractive flavour of *Porphyra* has been attributed to the presence of isofloridiosides and free amino acids (McLachlan *et al.*, 1972).

In addition to the seaweeds, microalgae are utilized as human food. *Chlorella* is the only green microalga which forms a regular part of human diet (Soeder, 1980), but several species of cyanophytes are utilized. Terrestrial *Nostoc* is consumed in China and South America (Johnston, 1970). Farrer (1966) suggested that a staple food of the Aztecs in pre-Columbian times was a dried *Spirulina* (Cyanophyceae) cake called tecuitlatl. *Spirulina* is collected and consumed in the Lake Tchad region of North Africa (Leonhard & Compère, 1967); it has a protein content in excess of 60% and is extensively cultivated (Shelef & Soeder, 1980).

ANIMAL FOOD

Wild animals (foxes, deer, rabbits, and bears) have been reported to feed on seaweeds, and in Europe and North America cattle, sheep and horses are frequently grazed in the intertidal (Chapman & Chapman, 1980). Seaweeds may be fed directly to cattle as fodder (Hallsson, 1964), but commercially they are harvested, dried, and ground into meals which are added as supplements to prepared feed. The principal genera employed are *Laminaria, Ascophyllum*, and *Fucus* (Phaeophyceae), with an annual value estimated at U.S. $10 million (Jensen, 1979). The food value of seaweeds to animals is essentially as for humans. Trace elements are particularly important, especially to cattle grazed on marginal grassland. In poultry feeds the carotenoids produce deeply coloured yolks in the eggs.

PHYCOCOLLOIDS

Historically, seaweeds were used for the production of soda and iodine, but this source has been replaced by mineral sources and by chemical processes. Today the main chemicals commercially obtained from seaweeds are the phycocolloids.

Phycocolloids are used extensively in the food, cosmetic, and pharmaceutical industries, as emulsifiers and as gelling agents. Like their counterparts extracted from terrestrial plants, the pectins, they are cell wall components which are soluble in hot water and relatively insoluble in cold. There are a number of recent reviews of the structure, biosynthesis, and function of phycocolloids (McCandless & Craigie, 1979; Percival, 1979; Larsen, 1981; McCandless, 1981). Three major types of commercial phycocolloid are produced: the alginates, the carrageenans, and the agars. Their combined value is in excess of U.S. $200 million annually.

Alginic acid and its salts (alginates) are prepared from members of the Phaeophyceae, the principal genera utilized being *Macrocystis, Laminaria*, and *Ascophyllum*. *Ecklonia* and *Durvillaea* are also important potential sources. Total world production is 15–17 000 tonnes, with a value in excess of U.S. $100 million (Jensen, 1979).

Alginic acid is a copolymer of mannuronic and guluronic acids. Haug *et al.* (1966, 1967) showed that three types of arrangement of these monomers occur: blocks of polymannuronic acid, blocks of polyguluronic acid, and blocks of alternating mannuronic and guluronic acids. These polymers have different structural properties (McCandless, 1981) and are found in different parts of the plant. The conceptacles and growing tips of *Ascophyllum* are rich in polymannuronic alginates, while the more mature tissues show a preponderance of polyguluronic blocks (Haug *et al.*, 1974). The total alginate content varies with location in the plant, and season. In eastern Canada, Cardinal and Breton-Provencher (1977) showed that the maximum for *Laminaria longicruris* (Phaeophyceae) is in September. Chapman and Doyle (1978) investigated the genetics of alginate variation in *L. longicruris*, and found that the genetic component of the variance is small.

Laminaria and *Ascophyllum* are harvested by hand from the shore or from boats. *Macrocystis*, being a much larger plant with floating blades, is mechanically harvested by rotary or reciprocating cutters mounted on barges.

The commercial production of alginates is described by Chapman and Chapman (1980). The main steps are acid leaching, followed by alkaline digestion, washing, bleaching, drying, and chopping to produce the crude extract of calcium alginate, which in turn is converted into the free acid form or other salts. The salts formed with alkali metals are water-soluble, while those with di- and trivalent metals are relatively insoluble.

Thus, sodium and potassium alginates form viscous fluids in solution, while salts of polyvalent ions tend to form gels. The soluble salts are used in the textile and paper industry as polishes and sizes, and in the paint, food, cosmetic, and pharmaceutical industries as viscous fluids for the suspension of water-immiscible substances. Approximately half the total use in the U.S.A. is a stabilizer in the dairy products industry. The gelling alginates, particularly calcium alginate, are used in the production of gels, cellophane-like films, and gums. In dentistry they are used for the manufacture of moulds, and in medicine they are used as greaseless lubricants. Dried alginate salts can be spun into yarns, which in turn may be woven into specialist cloths. Such cloths are alkali-soluble, and sodium alginate fibres are water-soluble: they are, however, resistant to normal microbial decay and are reasonably fire-resistant.

Carrageenans are phycocolloids obtained from members of the Hypneaceae, Phyllophoraceae, Solieriaceae, and Gigartinaceae of the Rhodophyceae. Carrageenan is commercially obtained from two main sources: *Chondrus crispus* in the North Atlantic, principally Maine (U.S.A.) and the Canadian Maritime Provinces, and from various species of *Eucheuma* in the Philippines. other sources are *Gigartina, Iridaea* and *Hypnea,* while *Furcellaria* and *Phyllophora* produce structurally and functionally similar compounds. Moss (1978) estimates annual production of carrageenan at 10 000 tonnes.

Carrageenans are polymers of sulphated or pyruvated galactose and 3, 6-anhydrogalactose (McCandless, 1981). Two principal species of carrageenan have been described; solutions of κ-carrageenan gel in the presence of potassium salts, whereas the those of λ-carrageenan do not. The main chemical difference is that the gelling types posses sulphate groups at the C-4 position in the galactose units, whereas the non-gelling types do not (McCandless & Craigie, 1979). McCandless *et al.* (1973) and Chen *et al.* (1973) showed that κ-carrageenan is principally produced by the gametophytes, and λ-carrageenan is most abundant in the sporophytes. A similar alternation has been demonstrated for other members of the Gigartinaceae (McCandless, 1978), but apparently does not occur in *Eucheuma* (Doty & Santos, 1978) or *Hypnea* (McCandless, 1978).

Harvesting is by collecting of drift weed, hand raking in the intertidal, and raking from small boats in the immediate subtidal. *Eucheuma* from the Philippines is cultured rather than gathered from the wild.

Carrageenan is used in the food industry as an emulsifier, particularly in dairy products; as a size in the textile and leather industries; and as an emulsifier in the pharmaceutical industry.

Agar is a Malay word for the gelling substance extracted from *Eucheuma*, which is ironically now known to be a carrageenan. It is extracted from a number of species of *Gelidium*, and to a lesser extent from *Gracilaria, Pterocladia, Acanthopeltis*, and *Ahnfeltia* (Rhodophyceae) (Chapman & Chapman, 1980). The annual production is 10 000 tonnes with a value of approximately U.S. $50 million (Moss, 1978).

Like carrageenan, agar is basically composed of sulphated and pyruvated galactoses. Araki (1966) identified two components: a neutral gelling component, agarose, and an anionic fraction which contains the sulphate and pyruvate groups, agaropectin. There appears to be a complex spectrum of compounds, ranging from unsulphated to highly sulphated and pyruvated polygalactans (McCandless, 1981).

The highest grades of agar produce stiff gels in 1–2% aqueous solution at normal temperature, but are liquified when hot. They are principally used in the preparation of high-quality bacteriological and tissue culture media. Pure agarose is now used as a gel in electrophoretic and chromatographic studies (Pluzeck, 1981).

ALGAE AS FERTILIZERS

Chapman and Chapman (1980) have reviewed the use of seaweeds as fertilizers; drift weed has been used for this purpose in all coastal agricultural societies (Booth, 1965). Seaweeds have adequate amounts of potassium and nitrogen, but are low on phosphate. In comparison, farmyard manure has only one-third the potassium, a similar amount of nitrogen, and three times the amount of phosphates. Seaweeds are also rich in trace elements (Stephenson, 1974) and may contain hormones and growth regulators (Blunden, 1977; Wildgoose *et al.*, 1978). Seaweed manures have the advantage of being free from weeds and pathogenic fungi. In Europe and North Africa maerl deposits (calcareous algae belonging principally to the genera *Lithothamnion* and *Phymatolithon*) are applied as fertilizer to reduce soil acidity (Blundon *et al.*, 1975). Nitrogen-fixing cyanophytes are also utilized extensively in tropical rice fields in lieu of the application of nitrogenous fertilizers (Watanabe, 1975).

Several brands of liquid fertilizer based in whole or in part on seaweeds are commercially available and are used in intensive garden and greenhouse horticulture (Stephensen, 1981; Povolny, 1981), where they are reported to increase germination of seeds, and improve productivity, hardiness and disease resistance as well as shelf life.

ALGAE IN MEDICINE

Seaweeds have been extensively used in the traditional medicines of maritime nations as vermifuges, anaesthetics, and ointments, as well as for the treatment of coughs, wounds, gout, goitre, hypertension, venereal diseases, cancer, and a variety of other ills (Hoppe, 1979; Chapman & Chapman, 1980; Stein & Borden, 1984). As with most folk remedies, some are worthless while others have a substantial basis in their content of bioactive compounds. In the latter category is the use of *Digenia simplex* (Rhodophyta) which contains a potent vermifuge, kainic acid; while the iodine content of *Laminaria* (Phaeophyta) prevents goitre. The vitamin and mineral content of marine algae also are potentially important in the prevention of other dietary insufficiency diseases.

Crude extracts of many species of algae contain substances with antibiotic properties against bacteria, fungi, and viruses (Glombitza, 1979). Antiviral compounds reported in red algae include polysaccharides containing D-glycosyl groups, which are apparently active in the control of herpes viruses (Ehresmann *et al.*, 1979). They act by blocking viral attachment points on the cell membrane. Many seaweeds contain sterols and related compounds, which are antagonistic to cholesterol in mammalian systems and may reduce elevated blood pressure associated with atherosclerosis (Chapman & Capman, 1980). Weinheimer and Karns (1974) concluded that a wide range of algal extracts were 'strikingly unproductive' in anticancer tests, however Yamamoto *et al.* (1982) showed that extracts of *Sargassum* and *Laminaria* inhibited the growth of sarcoma and leukaemia cells in mice.

A number of specific compounds of algal origin have been purified and characterized and are being used experimentally in medicine. The saxitoxins produced by the PSP dinoflagellates are used in neurobiological research (Schantz, 1981). Kainic acid is also neurotoxic and causes the breakdown of nerve dendrites: it is useful for its ability to mimic the effects of Huntington's chorea (Coyle *et al.*, 1979). It is also used in studies on epilepsy (Pisa *et al.*, 1980). Marine algae produce a wide variety of haemagglutinins (Rogers *et al.*, 1980); one, obtained from *Ptilota plumosa*, is specific to human B blood group (Rogers & Blundon, 1980) and hence has some diagnostic uses.

Carrageenan may be used to induce ulceration in the stomach and intestines of animals to test anti-inflammatory compounds (Di Rosa, 1972). It is reported not to cause ulcers in humans (Grasso & Sharrat, 1973) and has been used in the treatment of peptic ulcers (MacPherson & Pfeiffer, 1976). The abilities of carrageenans and alginates to form metal salts have suggested their use as non-toxic chelating agents in the treatment of heavy metal and radionucleotide poisonings (Tanaka & Stara, 1979).

Dried *Laminaria* stipes (referred to as *Laminaria* tents) are used as cervical dilators to facilitate obstetrical procedures (Feochari, 1979; Stein & Borden, 1984). Inserted into the neck of the cervix, the dried stipe swells slowly, producing the required dilation as it absorbs the surrounding fluids.

Much of the potential of the algae in medicine has yet to be explored. The broad range of compounds found in tropical seaweeds (Norris & Fenical, 1982) offers particularly exciting prospects for the future.

Cultivation of algae

The large-scale commercial cultivation of seaweeds for human food is most advanced in the Orient, particularly in Japan and China. Recent reviews of the subject include Mathieson (1982) and Tseng (1981). Species under cultivation include the chlorophytes *Enteromorpha, Monostroma*, and *Ulva*, the phaeophytes *Laminaria, Undaria, Nemacystis, Cladosiphon*, and *Endarachne* (Saito, 1982), and numerous rhodophytes, of which *Porphyra* spp. are commercially the most important (Tseng, 1981; Kito, 1982).

Shelef and Soeder (1980) have reviewed much of the work on the mass cultivation of unicellular algae. Microalgae grown for human consumption include *Chlorella* (Chlorophyceae) and *Spirulina* (Cyanophyceae), with *Scenedesmus* being used as a feed for domestic animals. Shellish and finfish aquaculture also depends on the availability of large, constant supplies of microalgae, and the most common genera employed are *Chlorella* (Chlorophyceae), *Monochrysis*, and *Isochrysis* (Chrysophyceae), and *Chaetoceros* and *Skeletonema* (Bacillariophyceae).

CULTIVATION OF MACROALGAE

Several species of *Porphya* are grown commercially in Japan and China, the most important being *P. yezoensis* and the higher quality, but less robust, *P. tenera*. Cultivation began in the 17th century, when bundles of branches (hibi) were anchored in the intertidal, allowing the algae to attach naturally. By the early part of this century these were replaced by floating bamboo mats and nets. When Drew (1949) first showed that the filamentous shell-boring alga *Conchocelis* is the alternate phase in the life cycle of *Porphyra*, the development of commercial cultivation using artifically seeded nets became possible. *Conchocelis* production takes place in tanks which contain oyster shells; the seeded shells are then attached to nets. A recent tendency is to eliminate the shells and to seed nets directly with conchospores produced from free-living *Conchocelis* (Kito, 1982). The nets are then placed in the ocean. In intertidal cultivation they are attached to poles either at constant height or half floating to allow them to rise and fall with the tides. This prevents excessive drying at low tide yet allows the nets to remain in the surface water at high tide. The most widely utilized cultivation method is on nets attached to floating rafts which may be periodically raised above the water surface (thalli grow faster without periodic emersion, but are more susceptible to disease and are of lower quality). When the nets are covered with young *Porphyra* blades they may be removed from the sea, partially dried and stored at −20° C; the alga will resume growth when thawed. Freeze storage allows the production of multiple crops in a season, and provides insurance against crop failure. When mature the algae are mechanically removed and washed from the nets, and are then chopped, pressed into sheets and dried.

The other major macroalgae grown for direct human consumption are the kelps *Laminaria* and *Undaria* (Tseng, 1981). Although *Laminaria japonica* has been consumed in China for over 1000 years, it is not native and was imported from Japan and Korea. It was accidently introduced by shipping into the northern Yellow Sea in 1927, and further deliberate transplants established it in the marine algal flora of China. Today more than 18 000 hectares of floating rafts of *L. japonica* and over a million tonnes wet weight are produced annually.

In China gametophytes and young sporelings are produced in greenhouses using artificially cooled water at 8–10 °C. Zoospores discharged from the fertile parent fronds settle on cords attached to frames in July, and remain in the greenhouse until October, when the sea-water temperature drops below 10 °C. The frames with the young sporophytes 1–2 cm high are then sold to farmers for outplanting in the sea. After 1–2 months the sporophytes are 10–15 cm high and are removed from the frames and attached to ropes which are suspended from rafts. The *Laminaria* is harvested in 6–7 months when it is 3–4 metres in length. The growth of the outplants requires the application of fertilizer. A 5–10% solution of ammonium sulphate is sprayed in the raft areas during the period of rapid growth; aproximately 2 tonnes of fertilizer are applied per hectare each season. Fertilization has allowed the cultivation of *Laminaria japonica* in formerly infertile areas.

Undaria pinnatifida is a warmer water species of kelp. It is grown in a similar manner to *L. japonica*, and the two species are frequently intermixed. *Undaria* has a shorter growing season and is harvested in March, thus not interfering with the maturation of *L. japonica*.

The tropical rhodophyte *Eucheuma* is extensively eaten, and is now an important source of carrageenan. It is widely cultivated in the Philippines (*Eucheuma spinosum* and *E. cottonii*; Doty, 1977; Tseng, 1981) with over 1000 farms producing in excess of 10 000 tonnes of dry weight annually. Plants are propagated vegetatively and grown attached to monofilament lines or nets.

Seaweed cultivation in North America is reviewed by Mathieson (1982); here the approach is very different from that in the Orient. Most operations are at the 'pilot plant' stage; while technologically more advanced and less labour-intensive, they are not as yet commercally viable. The cultivation of the carrageenophyte *Iridaea* shows excellent potential (Mumford, 1979).

North American seaweed aquaculturists have concentrated on the phycocolloid-producing rhodophytes, principally *Chondrus crispus,* and species of *Eucheuma, Gracilaria, Hypnea, Gigartina, Gelidium,* and *Pterocladia*. The most extensive work has been conducted in Canada and New England with *Chondrus crispus* (Harvey & McLachlan, 1973). It has been successfully grown in greenhouse tank culture (Neish *et al.*, 1977), although this procedure is not considered commercially viable compared with large outdoor lagoon ponds (Mathieson, 1982). Contamination with weedy algal species is a major problem in algal cultivation; a novel approach in *Chondrus* aquaculture is the controlled use of grazers (Shacklock & Croft, 1981) which selectively predate the nuisance algae. The chemical composition of *Chondrus* is nutrient-dependent. Under high nitrogen the plants grow rapidly and have a low carrageenan content, and when transferred to unenriched sea water there is a decrease in internal nitrogen coupled with an increase in carrageenan. This observation has obvious implications for post-harvesting treatment and for future pond cultivation.

There have been limited attempts at genetic manipulation in seaweeds (van der Meer, 1983), but in the main the stock improvement has depended on the selection of superior clones from wild populations. One *Chondrus* isolate, T4 (Neish & Fox, 1971), showed sixfold increases in growth per month, but required application of fertilizer for maximum growth. Van der Meer (1983) has produced numerous mutants of the agarophyte *Gracilaria tikvahiae*, including those with faster growth rates and better agar quality than wild types (Patwary & van der Meer, 1983). Such genetic studies have obvious potential in producing commercially viable strains of marine macrophytes.

CULTIVATION OF UNICELLULAR ALGAE

The importance of algae as a food source in invertebrate and fish aquaculture is frequently overlooked because of the emphasis on the commercially cultured species (Shaw, 1982). In many instances the production of the algae is the most expensive and labour-intensive part of aquaculture systems. The most widespread use of algae is in shellfish hatcheries, where they are used to feed larvae through metamorphosis and until they are placed into the field. There are two basic methods: one utilizes cultures of single species grown under defined conditions, the other utilizes sea water that is filtered to remove larger particulates and zooplankton. Blooms of algae then develop spontaneously and are fed to the shellfish. The former method is more controlled, but the latter is considerably cheaper. A variation on the latter method has been utilized in clam culture in the Virgin Islands, where deep nutrient-rich sea water is pumped into shallow pools and inoculated with phytoplankton, the resulting blooms being utilized in clam culture.

Similar pond cultures with or without additional fertilizers are utilized in other tropical areas. In Hawaii ponds provide phytoplankton for the growth of oysters. Similar systems are utilized in shrimp and other crustacean cultures. Algae are also important components in the culture of many finfish larvae. In these cases the algae are used as food for the zooplankton on which the larval fish feed.

Unicellular algae, such as *Chlorella* and *Spirulina*, are extensively cultivated for human consumption in Mexico and in S.E. Asia for local consumption and for export, especially to the Japanese market. In Mexico *Spirulina* is cultivated in 900-hectare external ponds on the Caracol near Lake Texcoco (Durand-Chastel, 1980). *Spirulina* and *Chlorella* are also grown commercially in Taiwan. In 1977 *Chlorella* factories produced over 1000 tonnes a year, mainly for the health food market (Soong, 1980).

THE FUTURE OF ALGAL CULTIVATION

The ideal algal cultivation system will utilize organic wastes as fertilizers and produce a variety of commercial products while returning little or no waste material to the environment. Algae already play important roles in waste water treatment systems (Palmer, 1977). Experimental polyculture systems based on effluent from secondary domestic sewage treatment have been established at Woods Hole, U.S.A. (Ryther *et al.*, 1979). The effluent was mixed with sea water to grow single-celled algae in ponds; the resulting blooms of *Skeletonema costatum* in winter, and *Phaeodactylum tricornutum* in summer, were continuously fed to shellfish. The effluent was then passed to containers of the phycocolloid-producing seaweeds *Gracilaria tikvahiae* and *Agardhiella tenera*, which removed the remaining nutrients. Problems arose due to temperature fluctuations, correlation of shellfish growth with nutrient supply, and contamination of shellfish grown in domestic sewage. Such polyspecies systems are better adapted to climates with less seasonal temperature variation. Commercially successful cultivation of crustacea, fish, and *Gracilaria* is currently under way in land-based enclosures in Taiwan (Chiang, 1981).

The cultivation of algae is also being investigated for the production of biomass energy through fermentation of members of the Laminariales to produce methane (Flowers & Bryce, 1977). The deployment of large rafts seeded with *Macrocystis* has been proposed to supply raw material for this process. On a small scale, a number of algae may be grown for specific chemical products such as glycerol and β-carotene from *Dunaliella* (Ben-Amotz & Avron, 1980).

References

ABBOTT I.A. (1967) *Liagora tanakai*, a new species from southern Japan. *Bull. Jap. Soc. Phycol.* **15**, 32–7.

ABBOTT I.A. & HOLLENBERG G.J. (1976) *Marine Algae of California*. Stanford University Press, Stanford.

ABBOTT I.A. & WILLIAMSON E. (1974) *Limu: An Ethnobotanical Study of Some Edible Hawaiian Seaweeds*. Pacific Tropical Botanical Garden, Lawai, Kauai, Hawaii.

ABDEL-RAHMAN M.H. (1982) The involvement of an endogenous circadian rhythm in photoperiodic timing in *Acrochaetium asparagopsis* (Rhodophyta, Acrochaetiales). *Br. Phycol. J.* **17**, 389–92.

ABE H., UCHIYAMA M. & SATO R. (1972) Isolation and identification of native auxins in marine algae. *Agr. Biol. Chem.* **36**, 2259–60.

ABÉLARD C. & L'HARDY-HALOS M.-T. (1975) Corrélations morphogènes entre les différents éléments d'un même cladôme chez l'*Apoglossum ruscifolium* (Turner) J.Ag. (Delesseriaceae, Ceramiales). *Bull. Soc. Phycol. Fr.* **22**, 111–19.

ADEY W.H. & MACINTYRE I.G. (1973) Crustose coralline algae: a re-evaluation in the geological sciences. *Geol. Soc. Am. Bull.* **84**, 883–904.

AHBEL D.E., ALEXANDER A.H. & KLEIN M.L. (1980) Protothecal olecranon bursitis — a case and review of the literature. *J. Bone Jt Surg. Am.* **62**, 835–6.

AHMADJIAN V. (1981) Algal/fungal symbioses. *Prog. Phycol. Res.* **1**, 179–233.

AHMADJIAN V. & HALE M.E. (1973) *The Lichens*. Academic Press, New York.

AHMADJIAN V. & JACOBS J.B. (1981) Relationship between fungus and alga in the lichen *Cladonia cristatella* Tuck. *Nature* **289**, 169–72.

AJISAKI T. & UMEZAKI I. (1978) The life history of *Sphaerotrichia divaricata* (Ag.) Kylin (Phaeophyta, Chordariales) in culture. *Jap. J. Phycol.* **26**, 53–9.

ALAM M., SHIMIZU M., IKAWA M. & SASNER J.J. (1978) Re-investigation of the toxins from the blue green alga, *Aphanizomenon flos-aquae*, by a high performance chromatographic method. *J. Environ. Sci. Health* **A13**(7), 493–9.

ALLDREDGE A.L. (1981) The impact of appendicularian grazing on natural food concentration *in situ*. *Limnol. Oceanogr.* **26**, 247–57.

ALLEN M.B. (1971) High latitude phytoplankton. *Ann. Rev. Ecol. Systematics* **2**, 261–76.

ALLEN M.M. (1968) Ultrastructure of the cell wall and cell division of unicellular blue-green algae. *J. Bact.* **96**, 842–52.

ALLISON E.M. & WALSBY A.E. (1981) Potassium in the control of turgor pressure in a gas-vacuolate blue-green alga. *J. Exp. Bot.* **32**, 241–9.

ALLSOPP A. (1969) Phylogenetic relationships of the Prokaryota and the origin of the eukaryotic cell. *New Phytol.* **68**, 591–612.

AMBROSE R.F. & NELSON B.V. (1982) Inhibition of giant kelp recruitment by an introduced brown alga. *Bot. Mar.* **25**, 265–7.

AMSPOKER M.G. & CZARNECKI D.B. (1982) Bacillariophyceae: introduction and bibliography. In *Selected Papers in Phycology II* (eds Rosowski J.R. & Parker B.C.), pp.712–18. Phycological Society of America, Allen Press, Lawrence, Kansas.

ANDERSEN R.A. (1982) A light and electron microscopical investigation of *Ochromonas sphaerocystis* Matvienko (Chrysophyceae): the statopsore, vegetative cell and its peripheral vesicles. *Phycologia* **21**, 390–8.

ANDERSON D.M. (1980) Effects of temperature conditioning on development and germination of *Gonyaulax tamerensis* (Dinophyceae) hypnozygotes. *J. Phycol.* **16**, 166–72.

ANDERSON D.M., CHISHOLM S.W. & WATRASS C.J. (1983) Importance of life cycle events in the population dynamics of *Gonyaulax tamarensis*. *Mar. Biol.* **76**, 179–202.

ANDERSON J.M. & BARRETT J. (1979) Chlorophyll–protein complexes of brown algae: P700 reaction centre and light harvesting complexes. *CIBA Foundation Symposium* **61**, 81–96.

ANDERSON L.W.J. & SWEENEY B.M. (1978) Role of inorganic ions in controlling sedimentation rate of a marine centric diatom, *Ditylum brightwelli*. *J. Phycol.* **14**, 204–14.

ANDREWS H.T. (1970) Morphology and taxonomy of the euryhaline chrysophyte *Sphaeridiothrix compressa*. *J. Phycol.* **6**, 133–6.

ANTIA N.J. (1976) Effects of temperature on the darkness survival of marine microplanktonic algae. *Microb. Ecol.* **3**, 41–54.

ANTIA N.J. (1977) A critical appraisal of Lewin's prochlorophyta. *Brit. Phycol. J.* **12**, 271–6.

ANTIA N.J., BERLAND B.R., BONIN D.J. & MAESTRINI S.Y. (1975) Comparative evolution of certain organic and inorganic sources of nitrogen for phototrophic growth of marine microalgae. *J. Mar. Biol. Ass. U.K.* **55**, 519–39.

ARAKI C. (1966) Some recent studies on the polysaccharides of agarophytes. *Proc. Int. Seaweed Symp.* **5**, 3–7.

ARCHIBALD P.A. & BOLD H.C. (1970) *Phycological Studies XI.* The genus *Chlorococcum* Meneghini. *Univ. Texas Pub.* **7015**, 1–115. Austin.

ARNOLD D.E. (1971) Ingestion, assimilation, survival and reproduction by *Daphnia pulex* fed seven species of blue-green algae. *Limnol. Oceanogr.* **16**, 906–20.

ATKINSON A.W. Jr, GUNNING B.E.S. & JOHN P.C.C. (1972) Sporopollenin in the cell wall of *Chlorella* and other algae: ultrastructure, chemistry and incorporation of ^{14}C acetate, studied in synchronous cultures. *Planta* **107**, 1–32.

ATKINSON M.J. & SMITH S.V. (1983) C:N:P ratios of benthic marine plants. *Limnol. Oceanogr.* **28**, 568–74.

AUGIER H. (1976) Les hormones des algues. Etat actuel des connaissances. II. Recherche et tentatives d'identification des gibberellines, des cytokinines et de diverses autres substances de nature hormonale. *Bot. Mar.* **19**, 245–54.

AXLER R.P., GERSBERG R.M. & GOLDMAN C.R. (1980) Stimulation of nitrate uptake and photosynthesis by molybdenum in Castle Lake, California. *Can. J. Fish. Aquat. Sci.* **37**, 707–12.

BAGNIS R., HURTEL J.-M., CHANTEAU S., CHUNGYEM E., INOUE A. & YASUMOTO T. (1979) Le dinoflagellé *Gambierdiscus toxicus* Adachi et Fukuyo: agent causal probable de la ciguatera. *C. R. Acad. Sci. Paris-D* **289**, 671–4.

BAKER A.N. (1967) Algae from Lake Miers, a solar heated Antarctic lake. *N. Z. J. Bot.* **5**, 453–68.

BALAKRISHNAN M.S. & CHAGULE B.B. (1980) Cytology and life history of *Batrachospermum mahabaleshwarensis* Balakrishnan *et* Chagule. *Crypogam. Algol.* **1**, 83–97.

BARTLETT R.B. & SOUTH G.R. (1973) Observations of the life history of *Bryopsis hypnoides* Lamour from Newfoundland. *Acta Bot. Neer.* **22**, 1–15.

BAUHIN C. (1620) *Prodromus theatri botanici.* Basileae. (not seen by the authors; quoted from Prescott, 1951).

BAZIN M.J. (1968) Sexuality in a blue green alga: genetic recombination in *Anacystis nidulans*. *Nature* **218**, 282–3.

BÉ A., CARON D.A. & ANDERSON O.R. (1981) Effects of feeding frequency on life processes of the planktonic foraminifer *Globigerinoides sacculifer* in laboratory culture. *J. Mar. Biol. Ass. U.K.* **61**, 257–77.

BEAN R.C. & HASSID W.Z. (1955) Assimilation of $^{14}CO_2$ by a photosynthesizing red alga, *Iridophycus flaccidum*. *J. Biol. Chem* **212**, 411–25.

BEARDALL J.D., MUKERJI D., GLOVER H.E. & MORRIS I. (1976) The path of carbon in photosynthesis by marine phytoplankton. *J. Phycol.* **17**, 134–41.

BEETON A.M. & EDMONSTON W.T. (1972) The eutrophication problem. *J. Fish. Res. Bd. Can.* **29**, 673–82.

BELCHER J.H. (1966) *Prasinochloris sessilis* gen. et sp. nov., a coccoid member of the Prasinophyceae, with some remarks of cyst formation in *Pyramimonas*. *Br. Phycol. Bull.* **3**, 43–51.

BELCHER J.H. (1968) The fine structure of *Furcilla stigamastophora* (Skuja) Korshikov. *Arch. Mikrobiol.* **60**, 84–94.

BELCHER J.H. (1969) A morphological study of the phytoflagellate *Chrysococcus rufescens* Klebs in culture. *Brit. Phycol. J.* **4**, 105–17.

BELCHER J.H. & SWALE E.M.F. (1967) Observations on *Pteromonas tenuis* sp. nov. and *P. angusta* (Carter) Lemmermann (Chlorophyceae, Volvocales) by light and electron microscopy. *Nova Hedwiga* **13**, 353–9.

BELLAMY D.J., CLARKE P.H., JOHN D.M., JONES D., WHITTICK A. & DARKE T. (1967) Effects of pollution from the Torrey Canyon on littoral and sublittoral ecosystems. *Nature* **216**, 1170–3.

BEN-AMOTZ A. & AVRON M. (1980) Glycerol, β-carotene and dry meal production by commercial cultivation of *Dunaliella*. In *Algal Biomass: Production and Use* (eds Shelef G. & Soeder C.J.), pp. 603–10. Elsevier/North-Holland, Amsterdam.

BERGER-PERROT Y. (1981) Mise au point sur le problème concernant l'*Ulothrix speciosa* (Carm. ex Harvey) Kützing et l'*Urospora kornmannii* Berger-Perrot. Etude comparée de la reproduction et du cycle de développement des deux espèces sur les côtes de Bretagne. *Phycologia* **20**, 147–64.

BERKALOFF C. & ROUSSEAU B. (1979) Ultrastructure of male gametogenesis in *Fucus serratus* (Phaeophyceae). *J. Phycol.* **15**, 163–73.

BESKOW R. (1978) Silicosis in diatomaceous earth factory workers in Sweden. *Scand. J. Resp. Dis.* **59**, 216–21.

BIDWELL R.G.S. (1983) Carbon nutrition of plants: photosynthesis and respiration. In *Plant Physiology: A treatise. Vol. VII; Energy and Carbon Metabolism.* (ed. Steward F.C.), pp. 287–457. Academic Press, New York.

BIDWELL R.G.S. & McLACHLAN, J. (1985) Carbon nutrition of seaweeds: photosynthesis, photorespiration and respiration. *J. Exp. Mar. Biol. Ecol.* **86**, 15–46.

BIDWELL R.G.S., CRAIGIE J.S. & KROTKOV G. (1958) Photosynthesis and metabolism in marine algae. III. Distribution of photosynthesis carbon from $^{14}CO_2$ in *Fucus vesiculosus. Can. J. Bot.* **36**, 581–90.

BIEBEL P. (1973) Morphology and life cycles of saccoderm desmids in culture. *Nova Hedwiga* **42**, 39–47.

BIERMAN V.J. Jr, DOLAN D.M., STOERMER E.F., GANNON, J.E. & SMITH V.E. (1980) The development and calibration of a spatially simplified multi-class phytoplankton model for Saginaw Bay, Lake Huron. *Great Lakes Environ. Planning Study* Contribution No. 33., Great Lakes Basin Commission.

BIRD D.F. & KALFF J. (1986) Bacterial grazing by planktonic lake algae. *Science* **231**, 493–5.

BIRMINGHAM B.C. & COLMAN B. (1979) Measurement of carbon dioxide compensation points of freshwater algae. *Plant. Physiol.* **64**, 892–5.

BISALPUTRA T. (1974) Plastids. In *Algal Physiology and Biochemistry* (ed. Stewart W.D.P.), pp. 124–60. Blackwell Scientific Publications, Oxford.

BLIDING C. (1957) Studies in *Rhizoclonium*. I. Life history of two species. *Bot. Not.* **110**, 271–5.

BLIDING C. (1963) A critical survey of European taxa in Ulvales. Part I. *Capsosiphon, Percursaria, Blidingia, Enteromorpha. Opera Bot.* **8**, 1–160.

BLIDING C. (1968) A critical survey of European taxa in Ulvales. Part II. *Ulva, Ulvaria, Monostroma, Kornmannia. Bot. Not.* **121**, 535–629.

BLUNDEN G. (1977) Cytokinin activity of seaweed extracts. In *Marine Natural Products Chemistry* (eds Faulkner D.J. & Fenical W.H.), pp. 337–43. Plenum Press, New York.

BLUNDEN G., BINNS W.W. & PERKS F. (1975) Commercial collection and utilization of maerl. *Econ. Bot.* **29**, 140–5.

BOALCH G.T. (1981) Do we really need to grow *Macrocystis* in Europe? *Proc. Int. Seaweed Symp.* **10**, 657–67.

BOLD H.C. (1973) *Morphology of Plants*, 3rd Edition. Harper & Row, New York.

BOLD H.C. & WYNNE M.J. (1978) *Introduction to the Algae. Structure and Reproduction.* Prentice-Hall, Englewood Cliffs, New Jersey.

BOLD H.C. & WYNNE M.J. (1985) *Introduction to the Algae: Structure and Reproduction*, 2nd Edition. Prentice-Hall, Englewood Cliffs, New Jersey.

BOLTON J.J. (1983) Ecoclinal variation in *Ectocarpus siliculosus* (Phaeophyceae) with respect to temperature growth optima and survival limits. *Mar. Biol.* **73**, 131–8.

BOLWELL G.P., CALLOW J.A. & CALLOW M.E (1977) Cross-fertilization in fucoid seaweeds. *Nature* **268**, 626–7.

BOLWELL G.P., CALLOW J.A., CALLOW M.E. & EVANS L.V. (1979) Fertilization in brown algae. II. Evidence for lectin-sensitive complimentary receptors involved in gamete recognition in *Fucus serratus. J. Cell Sci.* **36**, 19–30.

BOLWELL G.P., CALLOW J.A. & EVANS L.V. (1980) Fertilization in brown algae. III. Preliminary characterization of putative gamete receptors from eggs and sperm of *Fucus serratus. J. Cell Sci.* **43**, 209–24.

BONEY A.D. (1974) Aromatic hydrocarbons and the growth of marine algae. *Mar. Poll. Bull.* **5**, 579–85.

BONEY A.D. & CORNER E.D.S. (1962) On the effects of some carcinogenic hydrocarbons on the growth of sporelings of marine red algae. *J. Mar. Biol. Ass. U.K.* **42,** 579–85.

BONEY A.D. & GREEN J.C. (1982). Prymnesiophyceae (Haptophyceae): introduction and bibliography. In *Selected Papers in Phycology II* (eds Rosowski J.R. & Parker B.C.), pp. 705–11. Phycological Society of America, Allen Press, Lawrence, Kansas.

BONNEAU E.R. (1978) Asexual reproduction capabilities in *Ulva lactuca* L. (Chlorophyceae). *Bot. Mar.* **21**, 117–21.

BONOTTO S. (1975) Morphogenesis in normal branched and irradiated *Acetabularia mediterranea. Pubbl. Staz. Zool. Napoli* **39**, 96–107.

BOOTH E. (1965) The manurial value of seaweed. *Bot. Mar.* **8,** 138–43.

BORDEN C.A. & STEIN J.R. (1969) Reproduction and early development in *Codium fragile* (Suringar) Hariot: Chlorophyceae. *Phycologia* **8,** 91–9.

BOROWITZKA M.J. (1982) Mechanisms in algal calcification. *Prog. Phycol. Res.* **1,** 137–78.

BOTHE H. (1982) Nitrogen fixation. In *The Biology of Cyanobacteria* (eds Carr N.G. & Whitton B.A.), pp. 87–104. Blackwell Scientific Publications, Oxford.

BOUCK G.B. (1969) Extracellular microtubules. The origin, structure, and attachment of flagellar hairs in *Fucus* and *Ascophyllum* antherozoids. *J. Cell Biol.* **40,** 446–60.

BOUCK G.B. (1970) The development and post-fertilization fate of the eyespot and the apparent photoreceptor in *Fucus* sperm. *Ann. N. Y. Acad. Sci.* **175**, 673–85.

BOUCK G.B. & SWEENEY B.M. (1966) The fine structure and ontogeny of trichocysts in marine dinoflagellates. *Protoplasma* **61**, 205–23.

BOUCK G.B., ROGALSKI A. & VALAITIS A. (1978) Surface organization and composition of *Euglena*. II. Flagellar mastigonemes. *J. Cell Biol.* **77,** 805–26.

BOURRELLY P. (1957) Recherches sur les Chrysophycées. Morphologie, phylogenie, systématique. *Rev. Algol. Mém. Hors* Ser. 1, 412 pp.

BOURRELLY P. (1970) *Les Algues d'Eau Douce. Initiation à la Systématique. III. Les Algues Bleues et Rouges, les Eugléniens, Péridiniens et Cryptomonadines.* N. Boubée et Cie, Paris.

BOURRELLY P. (1981) *Les Algues d'Eau Douce. Algues Jaunes et Brunes.* N. Boubée et Cie, Paris.

BOURRELLY P. & COUTE A. (1976) Observations en microscopie électronique à balayage des *Ceratium* d'eau douce (Dinophycées). *Phycologia* **15,** 329–38.

BOVEE E.C. (1982) Movement and locomotion of *Euglena*. In *The Biology of Euglena*. Vol. III, *Physiology* (ed. Buetow D.E.), pp.143–68. Academic Press, New York.

BOYLE J.E. & SMITH D.C. (1975) Biochemical interactions between the symbionts of *Convoluta roscoffensis. Proc. Roy. Soc. Lond. B* **189**, 121–35.

BOYLEN C.W. & BROCK T.D. (1973) Effects of thermal additions from the Yellowstone Geyser Basins on the algae of the Firehole river. *Ecology* **54,** 1283–91.

BRAARUD T., DEFLANDRE G., HALLDAL P. & KAMPTNER E.K. (1955) Terminology, nomenclature and systematics of the Coccolithophoridae. *Micropaleontology* **1,** 157–9.

BRAARUD T., GAARDER K.R. & NORDLI O. (1958) Seasonal changes in the phytoplankton at various points off the Norwegian west coast (observations at the permanent oceanographic stations, 1954–46). *Rep. Norweg. Fish. Mar. Invest.* **12** (3), 1–77.

BRADLEY W.H. (1929) Freshwater algae from the Green River Formation of Colorado. *Bull. Torrey Bot. Club* **56,** 421–8.

BRAVO L. (1965) Studies on the life history of *Prasiola meridionalis. Phycologia* **4,** 177–94.

BRAWLEY S.H. & SEARS J.R. (1980) Septal plugs in a marine green alga. *J. Cell Biol.* **87,** 62a.

BRAWLEY S. & WETHERBEE R. (1981) Cytology and ultrastructure. In *The Biology of Seaweeds* (eds Lobban C.S. & Wynne M.J.), pp.248–99. Blackwell Scientific Publications, Oxford.

BRAWLEY S., QUATRANO R.S. & WETHERBEE R. (1977) Fine structural studies of the gametes and embryos of *Fucus vesiculosus* L. (Phaeophyta). *J. Cell Sci.* **87**, 275–94.

BREEN P.A. & MANN K.H. (1976) Destructive grazing of kelp by sea urchins in Eastern Canada. *J. Fish Res. Bd Can.* **33,** 1278–83.

BROCK T.D. (1967) Life at high temperatures. *Science* **158**, 1012–19.

BROCK T.D. (1969) Microbial growth under extreme environments. *Symp. Soc. Gen. Microbiol.* **19,** 15–41.

BROCK T.D. (1973) Lower pH limit for the existence of blue green algae: Evolution and ecological implications. *Science* **179,** 480–83.

BROCK T.D. (1978) *Thermophilic Micro-organisms and Life at High Temperatures*. Springer-Verlag, New York.

BROOK A.J. (1981) *The Biology of Desmids*. Blackwell Scientific Publications, Oxford.

BROWN R.M., JOHNSON C. & BOLD H.C. (1968) Electron and phase-contrast microscopy of sexual reproduction in *Chlamydomonas moewusii. J. Phycol.* **4,** 100–20.

BRYAN G.W. (1969) The absorption of zinc and other metals by brown seaweed *Laminaria digitata. J. Mar. Biol. Ass. U.K.* **49**, 225–45.

BRYAN G.W. & HUMMERSTONE L.G. (1973) Brown seaweed as an indicator of heavy metals in estuaries in south-west England. *J. Mar. Biol. Ass. U.K.* **53,** 705–20.

BRYCESON I. & FAY P. (1981) Nitrogen fixation in *Oscillatoria (Trichodesmium) erythraea* in relation to bundle formation and trichome differentiation. *Mar. Biol.* **61,** 159–66.

BUETOW D.E. (ed.) (1968) *The Biology of Euglena, Vols I & II.* Academic Press, New York.

BUETOW D.E. (ed.) (1982) *The Biology of Euglena, Vols III & IV.* Academic Press, New York.

BUFFALOE N. (1958) A comparative cytological study of four species of *Chlamydomonas. Bull. Torrey Bot. Club.* **85**, 157–78.

BUGGELN R.G. (1974) Negative phototropism of the haptera of *Alaria esculenta* (Laminariales). *J. Phycol.* **10,** 80–2.

BUGGELN R.G. (1976) Auxin, an endogenous regulator of growth in algae? *J. Phycol.* **12**, 355–8.

BUGGELN R.G. (1978) Physiological investigations on *Alaria esculenta* (Laminariales, Phaeophyceae). IV. Inorganic and organic nitrogen in the blade. *J. Phycol.* **14,** 156–60.

BUGGELN R.G. (1981) Morphogenesis and growth regulators. In *The Biology of Seaweeds* (eds Lobban C.S. & Wynne M.J.), pp. 627–60. Blackwell Scientific Publications, Oxford.

BUGGELN (1983) Photoassimilate translocation in brown algae. *Prog. Phycol. Res.* **2,** 283–332.

BUGGELN R.G. & BAL A.K. (1977) Effects of auxins and chemically related non-auxins on photosynthesis and chloroplast ultrastructure in *Alaria esculenta* (Laminariales.) *Can. J. Bot.* **55**, 2098–105.

BUGGELN R.G., FENSOM D.S. & EMERSON C.J. (1985) Translocation of ^{11}C-photoassimilate in the blade of *Macrocystis pyrifera* (Phaeophyceae). *J. Phycol.* **21**, 35–40.

BUNT J.S. (1970) Uptake of cobalt and vitamin B_{12} by tropical marine macroalgae. *J. Phycol.* **6,** 339–43.

BUNT J.S. & WOOD E.J.C. (1963) Microbiology and antarctic sea ice. *Nature* **199,** 1254–5.

BUNT J.S., OWENS O., VAN H. & HOCH G. (1966) Exploratory studies on the physiology and ecology of a psychrophilic marine diatom. *J. Phycol.* **2**, 96–100.

BURGER-WIERSMA, T., VEENHUIS M., KORTHALS H.J., VAN DER WIEL C.C.M. & MUR L.R. (1986) A new prokaryote containing chlorophylls a and b. *Nature* **320**, 262–4.

BURR F.A. & WEST J.A. (1970) Light and electron microscope observations on the vegetative and reproductive structures of *Bryopsis hypnoides. Phycologia* **9,** 17–37.

BURR F.A. & WEST J.A. (1971) Protein bodies in *Bryopsis hypnoides*: their relationship to wound-healing and branch septum development. *J. Ultrastr. Res.* **35**, 476–98.

BURROWS E.M. (1958) Sublittoral algal population in Port Erin Bay, Isle of Man. *J. Mar. Biol. Ass. U.K.* **37,** 687–703.

BURROWS E.M., CONWAY E., LODGE S.M. & POWELL H.T. (1954) The raising of intertidal algal zones in Fair Isle. *J. Ecol.* **42,** 283–8.

BUTLER M., HASKEW A.E.J. & YOUNG M.M. (1980) Copper tolerance in the green alga, *Chlorella vulgaris. Plant Cell Environ.* **3,** 119–26.

BUTLER R.D. & ALLSOPP A. (1972) Ultrastructural investigations in the Stigonemataceae. *Arch. Mikrobiol.* **82,** 283–99.

CAMPBELL S.E. (1980) *Palaeoconchocelis starmachii*, a carbonate boring microfossil from the Upper Silurian of Poland (425 million years old): implications for the evolution of the Bangiaceae (Rhodophyta). *Phycologia* **19,** 25–36.

CANTER H.M. (1973) A new primitive protozoan devouring centric diatoms in the plankton. *J. Linn. Soc. Lond. Zool.* **52,** 63–83.

CANTER H.M. & JAWORSKI G.H.M. (1978) The isolation maintenance and host range studies of a chytrid *Rhizophidium planktonicum* Canter emend., parasitic on *Asterionella formosa* Hassall. *Ann. Bot.* **42,** 967–79.

CAPRIULO G.M. & CARPENTER E.J. (1980) Grazing by 35 to 202 mm microzooplankton in Long Island Sound. *Mar. Biol.* **56,** 319–26.

CARAM, B. (1965) Recherches sur la reproduction et le cycle sexuel de quelques Phéophycées. *Vie et Milieu* **16,** 21–226.

CARAM B. (1972) Le cycle de reproduction des Pheophycées-Pheosporées et ses modifications. *Soc. Bot. Fr. Mémoires* **1972,** 151–60.

CARDINAL A. (1964) Étude sur les Ectocarpacées de la Manche. *Beih. Nova Hedwiga* **IS**, 1–86.

CARDINAL A. & BRETON-PROVENCHER M. (1977) Variations de la teneur en acide alginique des Laminariales de l'estuaire du Saint-Laurent (Québec). *Bot. Mar.* **20,** 243–52.

CARLUCCI A.F. & BOWES P.M. (1970) Vitamin production and ultilization by phytoplankton in mixed culture. *J. Phycol.* **6,** 393–400.

CARMICHAEL W.W. (ed.) (1981) *Water Environment. Algal Toxins and Health.* Environmental Science Research Vol. 20: International Conference on Toxic Algae. Plenum Press, New York.

CARMICHAEL W.W. & GORHAM P.R. (1978) Anatoxins from clones of *Anabaena flos-aquae* isolated from lakes of western Canada. *Mitt. Int. Ver. Limnol.* **21,** 285–95.

CARMICHAEL W.W. & GORHAM P.R. (1980) Freshwater cyanophyte toxins: types and their effects on the use of micro algae biomass. In *Algal Biomass: Production and Use* (eds Shelef G. & Soeder C.J.), pp. 437–48. Elsevier/North-Holland, Amsterdam.

CARMICHAEL W.W. & GORHAM P.R. (1981) The mosaic nature of toxic blooms of Cyanobacteria. In *Water Environment. Algal Toxins and Health.* (ed. Carmichael W.W.). Environmental Science Research Vol. 20, pp. 161–72. International Conference on Toxic algae. Plenum Press, New York.

CARPENTER E.J. & PRICE C.C. (1976) Marine *Oscillatoria (Trichodesmium)*: explanation for aerobic nitrogen fixation without heterocysts. *Science* **191,** 1278–80.

CARSON J.L & BROWN R.M. (1978). Studies of Hawaiian freshwater and soil algae. II. Algal colonization and succession on a dated volcanic substrate. *J. Phycol.* **14,** 171–8.

CASSIN P.E. (1974) Isolation growth and physiology of acidophilic Chlamydomonads. *J. Phycol.* **10,** 439–47.

CASTENHOLZ R.W. (1969) Thermophilic cyanophytes of Iceland and the upper temperature limit. *J. Phycol.* **5,** 360–8.

CATTANEO A. & KALFF J. (1979) Primary production of algae growing on natural and artificial aquatic plants: a study of interactions between epiphytes and their substrate. *Limnol. Oceanogr.* **24,** 1031–7.

CAVALIER-SMITH T. (1974) Basal body and flagellar development during the vegetative cell cycle and the sexual cycle of *Chlamydomonas reinhardii. J. Cell Sci.* **16,** 529–56.

CAVALIER-SMITH T. (1975) Electron and light microscopy of gametogenesis and gamete fusion in *Chlamydomonas reinhardii. Protoplasma* **86,** 1–18.

CAVALIER-SMITH T. (1976) Electron microscopy of zygospore formation in *Chlamydomonas reinhardii. Protoplasma* **87,** 1–18.

CAVALIER-SMITH T. (1978) The evolutionary origin and phylogeny of microtubules, mitotic spindles and eucaryote flagella. *Biosystems* **10,** 93–114.

CAVALIER-SMITH T. (1983a) Endosymbiotic origin of the mitochondrial envelope. *Endocytology* **2,** 265–79.

CAVALIER-SMITH T. (1983b) A six-kingdom classification and a unified phylogeny. *Endocytology* **2,** 1027–34.

CHADEFAUD M. (1974) Possibilité d'une origine non symbiotique de la cellule des Eucaryotes. *C.R. Acad. Sci. Paris D* **278,** 3079–81.

CHADEFAUD M. (1978) Sur la notation de prochlorophytes. *Rev. Algol. N. S.* **13,** 203–6.

CHAMBERLAIN A.H.L., GORHAM J., KANE D.F. & LEWEY S.A. (1979) Laboratory growth studies on *Sargassum muticum* (Yendo) Fensholt. *Bot. Mar.* **22,** 11–19.

CHAPMAN A.R.O. (1972a) Morphological variation and its taxonomic implications in the ligulate members of the genus *Desmarestia* occurring on the west coast of North America. *Syesis* **5,** 1–20.

CHAPMAN A.R.O. (1972b) Species delimitation in the filiform, oppositely branched members of the genus *Desmarestia* Lamour. (Phaeophyceae, Desmarestiales) in the northern hemisphere. *Phycologia* **11,** 225–31.

CHAPMAN A.R.O. (1973) A critique of prevailing attitudes towards the control of seaweed zonation on the sea shore. *Bot. Mar.* **16,** 80–2.

CHAPMAN A.R.O. & BURROWS E.M. (1971) Field and culture studies of *Desmarestia aculeata. Phycologia* **10,** 63–76.

CHAPMAN A.R.O. & CRAIGIE J.S. (1977) Seasonal growth in *Laminaria longicruris*: relation with dissolved inorganic nutrients and internal reserves of nitrogen. *Mar. Biol.* **40,** 197–205.

CHAPMAN A.R.O & DOYLE R.W. (1978) Genetic analysis of alginate content in *Laminaria longicruris* (Phaeophyceae). *Proc. Int. Seaweed Symp.* **9,** 125–32.

CHAPMAN A.R.O. & LINDLEY J.E. (1980) Seasonal growth of *Laminaria solidungula* in the Canadian High Arctic in relation to irradiance and dissolved nutrient concentrations. *Mar. Biol.* **57,** 1–5.

CHAPMAN D.J. & TRENCH R.K. (1982) Prochlorophyceae: introduction and bibliography. In *Selected Papers in Phycology II* (eds Rosowski J.R. & Parker B.C.), pp. 656–8. Phycological Society of America, Allen Press, Lawrence, Kansas.

CHAPMAN V.J. (1974) *Salt Marshes and Salt Deserts of the World,* 2nd Edition. Cramer, Lehre.

CHAPMAN V.J. & CHAPMAN D.J. (1973) *The Algae*, 2nd Edition. MacMillan, London.

CHAPMAN V.J. & CHAPMAN D.J. (1976) Life forms in the algae. *Bot. Mar.* **19,** 65–74.

CHAPMAN V.J. & CHAPMAN D.J. (1980) *Seaweeds and their Uses*. Chapman & Hall, London and New York.

CHEN J.C.W. (1980) Longitudinal microfibrillar alignment in the wall of cylindrical *Nitella* rhizoidal cells — observation under polarizing and interference microscopes. *Am. J. Bot.* **67,** 859–65.

CHEN L. C.-M. & TAYLOR A.R.A (1978) Medullary tissue culture of the red alga, *Chondrus crispus. Can. J. Bot.* **56,** 883–6.

CHEN L. C.-M., McLACHLAN J., NEISH A.C. & SHACKLOCK P.F. (1973) The ratio of kappa- to lambda-carrageenan in nuclear phases of the rhodophycean algae *Chondrus crispus* and *Gigartina stellata. J. Mar. Biol. Ass. U.K.* **53,** 11–16.

CHEN L.C.-M., EDELSTEIN T. & McLACHLAN J. (1974) The life history of *Gigartina stellata* in culture. *Phycologia* **13,** 287–94.

CHENEY D.P. (1982) The determining effects of snail herbivore density on intertidal algal recruitment and composition. *Abstr. 1st Int. Phycol. Congr.*, p. 48. St Johns, Newfoundland.

CHIANG Y.-M. (1981) Cultivation of *Gracilaria* (Rhodophycophyta, Gigartinales) in Taiwan. *Proc. Int. Seaweed Symp.* **10,** 657–67.

CHIHARA M. & YOSHIZAKI M. (1972) Bonnemaisoniaceae: their gonimoblast development, life history and systematics. In *Contributions to the Systematics of Benthic Marine Algae of the North Pacific* (eds Abbott I. & Kurogi M.). Japanese Society of Phycology, Kobe, Japan.

CHISHOLM S.W. (1981) Temporal patterns of cell division in unicellular algae. *Can. Bull. Fish. Aquat. Sci.* **210,** 150–81.

CHISHOLM S.W., AZAM F. & EPPLEY R.W. (1978) Silicic acid incorporation in marine diatoms on light-cycles: use as an assay for phased cell division. *Limnol. Oceanogr.* **23**, 518–29.

CHRISTENSEN T. (1962) Alger. In *Botanik, II, Nr 2; Systematisk Botanik* (eds Böcher T.W., Lange G. & Srensen T.). Munksgaard, Copenhagen.

CHRISTENSEN T. (1964) The gross classification of the algae. In *Algae and Man* (ed. Jackson D.F.), pp. 59–64. Plenum Press, New York.

CHRISTENSEN T. (1980) *Algae, A Taxonomic Survey*. Fasc. I. AiO Tryk, Odense.

CHURCHILL A.C. & MOELLER H.W. (1972) Seasonal patterns of reproduction in New York populations of *Codium fragile* (Sur.) Hariot ssp. *tomentosoides* (van Goor) Silva. *J. Phycol.* **8,** 147–52.

CLARKE K.J. & PENNICK N.C. (1972) Flagellar scales in *Oxyrrhis marina* Dujardin. *Brit. Phycol. J.* **7,** 357–60.

CLARKE K.J. & PENNICK N.C. (1975) *Syncrypta glomerifera* sp. nov., marine member of the Chrysophyceae bearing a new form of scale. *Br. Phycol. J.* **10,** 363–70.

CLAYTON M.N. (1981) Correlated studies on seasonal changes in the sexuality, growth rate and longevity of complanate *Scytosiphon* (Scytosiphonaceae: Phaeophyta) from southern Australia, growing *in situ*. *J. Exp. Mar. Biol. Ecol.* **51,** 87–96.

CLAYTON M.N. & BEAKES G.W. (1983) Effects of fixatives on the ultrastructure of physodes in vegetative cells of *Scytosiphon lomentaria* (Scytosiphonales, Phaeophyta). *J. Phycol.* **19,** 4–16.

CLEARE, M. & PERCIVAL E. (1972) Carbohydrates of the freshwater alga *Tribonema aequale*. I. Low molecular weight and polysaccharides. *Brit. Phycol. J.* **7,** 185–93.

CLOCCHIATTI M. (1971) Sur l'existence de coccosphères portant des coccolithes de *Gephyrocapsa oceanica* et de *Emiliana huxleyi* (Coccolithophoridés). *C. R. Acad. Sci. Paris D* **273,** 318–21.

CLOUD P. (1976) Beginnings of biospheric evolution and their biogeochemical consequences. *Paleobiology* **2,** 351–87.

COESEL P.F.M. & TEXEIRA R.M.V. (1974) Notes on sexual reproduction in desmids. I. Zygospore formation in nature (with special reference to some unusual records of zygotes). *Acta Bot. Neer.* **23,** 361–8.

COHN F. (1853) Untersuchungen über die Entwicklungsgeschichte mikroskopischer Algen und Pilze. *Nov. Act. Acad. Leop. Carol.* **24,** 103–256.

COLBATH G.K. (1983) Fossil prasinophycean phycomata (Chlorophyta) from the Silurian Bainbridge Formation, Missouri, U.S.A. *Phycologia* **22,** 249–65.

COLE D.R. & PLAPP F.W. (1974) Inhibition of growth and photosynthesis in *Chlorella pyrenoidosa* by a chlorinated biphenyl and several insecticides. *Environ. Entomol.* **3,** 217–20.

COLE K. & AKINTOBE S. (1963) The life cycle of *Prasiola meridionalis* Setchel and Gardner. *Can. J. Bot.* **41,** 661–8.

COLE K. & CONWAY E. (1980) Studies in the Bangiaceae: reproductive modes. *Bot. Mar.* **23,** 545–53.

COLEMAN A.W. (1979) Sexuality in the colonial flagellates. In *Physiology and Biochemistry of the Protozoa*, Vol. 1 (eds Hunter S.H. & Levandowsky M.), pp. 307–40. Academic Press, New York.

COLEMAN A.W. (1983) The role of resting spores and akinetes in chlorophyte survival. In *Survival strategies of the Algae* (ed. Fryxell G. A.), pp. 1–21. Cambridge University Press, Cambridge.

COLMAN J.S. (1933) The nature of the intertidal zonation of plants and animals. *J. Mar. Biol. Ass. U.K.* **18,** 435–76.

COLOMBETTI G., LENCI F. & DIEHN B. (1982) Responses to photic, chemical and mechanical stimuli. In *The Biology of Euglena*. Vol. III, *Physiology* (ed. Buetow D.E.), pp. 169–95. Academic Press, New York.

CONNELL J.H. & SLAYTER R.O. (1977) Mechanisms of succession in natural communities and their role in community stability and organization. *Am. Nat.* **111.** 1119–44.

CONOVER J.T. & SIEBURTH J. McN. (1964) Effects of *Sargassum* distribution on its epibiota and antibacterial activity. *Bot. Mar.* **6,** 147–57.

COOKE G.W. & WILLIAMS R.J.B. (1970) Losses of nitrogen and phosphorus from agricultural land. *Water Treat. Exam.* **19,** 253–76.

COOKE W.B. (1968a) Studies in the genus *Prototheca*. I. Literature review. *J. Elisha Mitchell Sci. Soc.* **84,** 213–16.

COOKE W.B. (1968b) Studies in the genus *Prototheca*. II. Taxonomy. *J. Elisha Mitchell Sci. Soc.* **84,** 217–20.

COOMBS J. & GREENWOOD A.D. (1976) Compartmentation in the photosynthetic apparatus. In *The Intact Chloroplast. Topics in Photosynthesis* Vol. I (ed. Barber J.), pp. 1–51. Elsevier/North-Holland, Amsterdam.

COOMBS J. & VOLCANI B.E. (1968) Studies on the biochemistry and fine structure of silica shell formation in diatoms. Chemical changes in the wall of *Navicula pelliculosa* during its formation. *Planta* **82,** 280–92.

COOMBS J., HALICKI P.J., HOLM-HANSEN O. & VOLCANI B.E. (1967) Studies on the fine structure and silca shell formation in diatoms. *Exp. Cell Res.* **47,** 315–28.

COPELAND H.F. (1938) The kingdoms of organisms. *Quart. Rev. Biol.* **13,** 383-420.

CORDEIRO-MARINO M. & POZA A. (1981) Life history of *Gymnogrongus griffithsiae* (C.Ag.) J.Ag. (Phyllophoraceae, Gigartinales). *Proc. Int. Seaweed Symp.* **10,** 155–61.

CORTEL-BREEMAN A.M. & TEN HOOPEN A. (1978) The short day response in *Acrosymphyton purpuriferum* (J.Ag.) Sjost. (Rhodophyceae: Cryptonemiales). *Phycologia* **17,** 125–32.

CORTEL-BREEMAN A.M. & HOEK C. VAN DEN (1970) Life history studies on Rhodophyceae. I. *Acrosymphyton purpuriferum* (J. Ag.) Kyl. *Acta Bot. Neer.* **19,** 265–84.

COUGHLAN S. & TATTERSFIELD D. (1977) Photorespiration in larger littoral algae. *Bot. Mar.* **20,** 265–6.

COURT G.J. (1980). Photosynthesis and translocation studies of *Laurencia spectabilis* (Rhodophyceae). *J. Phycol.* **16,** 270–9.

COUSENS R. & HUTCHINGS M.J. (1983) The relationship between density and mean frond weight in monospecific seaweed stands. *Nature* **301,** 240–1.

COX E.R. & BOLD H.C. (1966) Taxonomic investigations of *Stigeoclonium. Univ. Tex. Publ.* 6612. 167 pp.

COX G. & DWARTE D.M. (1981) Freeze–etch ultrastructure of a *Prochloron* species — the symbiont of *Didemnum molle. New Phytol.* **88,** 427–38.

COX G.E., WILSON J.D. & BROWN P. (1974) Protothecosis: a case of disseminated algal infection. *Lancet,* **ii,** 379–82.

COYLE J.T., LONDON E.D., BIZIERE K. & ZACZEK R. (1979) Kainic acid neurotoxicity: insights into the pathophysiology of Huntington's disease. *Adv. Neurol.* **23,** 593–608.

CRAIGIE J.S. (1974) Storage products. In *Algal Physiology and Biochemistry* (ed. Stewart W.D.P.), pp. 206–35. Blackwell Scientific Publications, Oxford.

CRAWFORD R.M. (1975) The taxonomy and classification of the diatom genus *Melosira* C.Ag. The type species *M. nummuloides* C.Ag. *Br. Phycol. J.* **10**, 323–38.

CRONQUIST A. (1960) The divisions and classes of plants. *Bot. Rev. (London)* **26,** 425–82.

CRUMPTON W.G. & WETZEL R.G. (1982) Effects of differential growth and mortality in the seasonal succession of phytoplankton populations. *Ecology* **63,** 729–39.

CUBIT J. D. (1984) Herbivory and the seasonal abundance of algae on a high intertidal rocky shore. *Ecology* **65,** 1904–17.

CUSHING D.H. & VUCETIC T. (1963) Studies on a *Calanus* patch. III. The quantity of food eaten by *Calanus finmarchicus. J. Mar. Biol. Ass. U.K.* **43,** 349–71.

CZYGAN F.-C. (1970) Blutregen und Blutschnee: Stickstoffmangelzellen von *Haematococcus pluvialis* und *Chlamydomonas nivalis. Arch. Microbiol.* **74,** 69–76.

DAILY F.K. (1982) Charophyceae. In *Synopsis and Classification of Living Organisms*, Vol. I (ed. Parker S.P.), pp.161–2. McGraw-Hill, New York.

DALE B. (1983) Dinoflagellate resting cysts: 'benthic plankton'. In *Survival Strategies of the Algae* (ed. Fryxell G. A.), pp. 69–136. Cambridge University Press, Cambridge.

DANIEL G.F. & CHAMBERLAIN A.H.L. (1981) Copper immobilization in fouling diatoms. *Bot. Mar.* **24,** 229–43.

D'ANTONIO C. (1985) Epiphytes on the rocky intertidal red alga *Rhodomela larix* (Turner) C.Ag.: negative effects on the host and food for herbivores. *J. Exp. Mar. Biol. Ecol.* **86,** 197–218.

DARLEY W.M. (1982) *Algal Biology: A Physiological Approach.* Blackwell Scientific Publications, Oxford.

DARLEY W.M., SULLIVAN C.W. & VOLCANI B.E. (1976) Studies on the biochemistry and fine structure of silica shell formation in diatoms. Division cycle and chemical composition of *Navicula pelliculosa* during light dark synchronized growth. *Planta* **130,** 159–67.

DARLEY W.M., MONTAGUE C.L., PLUMLEY F.G., SAGE W.W. & PSALIDAS A.T. (1981) Factors limiting edaphic algal biomass and productivity in a Georgia salt-marsh. *J. Phycol.* **17,** 122–8.

DAVENPORT J. (1982) Oil and planktonic ecosystems. *Trans. Roy. Soc. Lond. B* **297,** 369–81.

DAVIES A.G. (1983) The effects of heavy metals upon natural marine phytoplankton populations. *Prog. Phycol. Res.* **2,** 113–45.

DAVIES J.M., BAIRD I.E., MASSIE L.C., HAY S.J. & WARD A.P. (1980) Some effects of oil-derived hydrocarbons on pelagic food webs from observations in an enriched ecosystem and considerations of their implications for monitoring. *Rapp. P-v. Réun. Cons. Int. Explor. Mer.* **179,** 201–11.

DAVIES R.R., SPENCER H. & WAKELIN P.O. (1964) A case of human protothecosis. *Trans. Roy. Soc. Trop. Med. Hyg.* **58,** 448.

DAVIS C.C. (1964) Evidence for the eutrophication of Lake Erie from phytoplankton records. *Limnol. Oceanogr.* **9,** 275–83.

DAVIS J.S. (1967) The life cycle of *Pediastrum simplex. J. Phycol.* **3,** 95–103.

DAVIS M.H. & COUGHLAN J. (1978) Response of entrained phytoplankton to low level chlorination at a coastal power station. In *Water Chlorination: Environmental Impacts and Health Effects*, Vol. 2. (ed. Jolley R.L.), pp. 369–76. Ann Arbor Science, Michigan.

DAWSON P.A. (1973) Observations on the structure of some forms of *Gomphonema parvulum* Kütz. III. Frustule formation. *J. Phycol.* **9,** 353–64.

DAYTON P.K. (1973) Dispersion, dispersal and persistence of the annual intertidal alga *Postelsia palmaeformis* Ruprecht. *Ecology* **54,** 433–8.

DAYTON P.K. (1975) Experimental evaluation of ecological dominance in a rocky intertidal algal community. *Ecol. Monogr.* **45,** 137–59.

DEASON T.R. (1984) A discussion of the classes Chlamydophyceae and Chlorophyceae and their subordinate taxa. *Pl. Syst. Evol.* **146,** 75–86.

DEASON T.R., BUTLER G.L. & RHYNE C. (1983) *Rhodella reticulata* sp. nov., a new coccoid rhodophytan alga (Porphyridiales). *J. Phycol.* **19,** 104–11.

DEBOER J.A. (1981) Nutrients. In *The Biology of Seaweeds* (eds Lobban C.S. & Wynne M.J.), pp. 356–92. Blackwell Scientific Publications, Oxford.

DECEW T.C. & WEST J.A. (1981) Life histories in the Phyllophoraceae (Rhodophyta: Gigartinales) from the Pacific coast of North America. I. *Gymnogrongus linearis* and *G. leptophyllus. J. Phycol.* **17,** 240–50.

DECEW T.C. & WEST J.A. (1982a) Investigations on the life histories of three *Farlowia* spp. (Rhodophyta, Cryponemiales, Dumontiaceae) from Pacific North America. *Phycologia* **20,** 342–51.

DECEW T.C. & WEST J.A. (1982b) A sexual life history in *Rhodophysema* (Rhodophyceae); a re-interpretation. *Phycologia* **21,** 67–74.

DEFLANDRE G. & LENOBLE A. (1948) Sur la présence d'eugleniens fossiles du genre *Trachelomonas* Ehr. dans un schiste pliocène de Madagascar. *C. R. Acad. Sci. Paris D* **226**, 509–11.

DELANEY S.F., HERDMAN M. & CARR N.G. (1976) Genetics of blue-green algae. In *The Genetics of Algae* (ed. Lewin R.A.), pp.7–28. Blackwell Scientific Publications, Oxford.

DE REAUMUR R.A. (1711) Description des fleurs et des graines des divers *Fucus*, et quelques autres observations physiques sur ces mêmes plantes. *Mém. Acad. Sci. Paris* **1711,** 383.

DE TONI G.B. (1889) *Sylloge algarum omnium hucusque cognitarum.* 1 (*Sylloge Chlorophycearum*). 1–12, I-CXXXIX, 1–1315. Patavii.

DE TONI, G.B. (1924) *Sylloge algarum omnium hucusque cognitrum.* 6 (*Sylloge Floridearum 5, Additamenta*). I-XI, 1–767. Patavii.

DEVILLY C.I. & HOUGHTON, J.A. (1977) A study of genetic transformation in *Gleocapsa alpicola. J. Gen. Microbiol.* **98,** 277–80.

DIGBY P.S.B. (1979) Reducing activity and the formation of base in the coralline algae: an electrochemical model. *J. Mar. Biol. Ass. U.K.* **59,** 455–77.

DILLENIUS J.J. (1741) *Histori Muscorum in qua circiter sexcentae veteres et novae ad sua genera relatae describuntus et inconibur genuimis·····Oxonii* (not seen by the authors; quoted from Prescott, 1951).

DILLON L.S. (1981) *Ultrastructure, Macromolecules and Evolution.* Plenum Press, New York and London.

DINSDALE M.T. & WALSBY A.E. (1972) The interrelations of cell turgor pressure, gas vacuolations, and buoyancy in a blue green alga. *J. Exp. Bot.* **23,** 561–70.

DI ROSA M. (1972) Biological properties of carrageenan. *J. Pharm. Pharmacol.* **24,** 89–102.

DIXON P.S. (1965) Perennation, vegetative propagation and algal life histories, with special reference to *Asparagopsis* and other Rhodophyta. *Bot. Gothoburg.* **3,** 67–74.

DIXON P.S. (1971) Cell enlargement in relation to the development of thallus form in Florideophyceae. *Brit. Phycol. J.* **6,** 195–205.

DIXON P.S. (1973) *The Biology of the Rhodophyta*. Oliver & Boyd, Edinburgh.

DIXON P.S. (1982) Rhodophycota. In *Synopsis and Classification of Living Organisms,* Vol. I (ed. Parker S.P.), pp. 61–79. McGraw-Hill, New York.

DODGE J.D. (1969a) A review of the fine structure of algal eyespots. *Brit. Phycol. J.* **4,** 199–210.

DODGE J.D. (1969b) The ultrastructure of *Chroomonas mesostigmatica* Butcher (Cryptophyceae). *Arch. Mikrobiol.* **69,** 266–80.

DODGE J.D. (1971) A dinoflagellate with both a mesocaryotic and a eucaryotic nucleus. I. Fine structure of the nuclei. *Protoplasma* **73,** 145–57.

DODGE J.D. (1972) The ultrastructure of the dinoflagellate pusule: a unique osmo-regulatory organelle. *Protoplasma* **75,** 285–302.

DODGE J.D. (1973) *The Fine Structure of Algal Cells*. Academic Press, New York.

DODGE J.D. (1974) Fine structure and phylogeny in the algae. *Sci. Prog. (Oxford)* **61,** 257–74.

DODGE J.D. (1979) The phytoflagellates: fine structure and phylogeny. In *Biochemistry and Physiology of the Protozoa* (eds Hutner S.H. & Levandarsky M.), pp.7–57. Academic Press, New York.

DODGE J.D. (1983) Dinoflagellates: investigation and phylogenetic speculation. *Brit. Phycol. J.* **18,** 335–56.

DODGE J.D. (1984) The functional and phylogenetic significance of dinoflagellate eyespots. *Biosystems* **16,** 259–67.

DODGE J.D. & CRAWFORD R.M. (1969) Observations on the fine structure of the eyespot and associated organelles in the dinoflagellate *Glenodinium foliaceum*. *J. Cell Sci.* **5,** 479–93.

DODGE J.D. & CRAWFORD R.M. (1970) A survey of thecal fine structure in the Dinophyceae. *Bot. J. Linn. Soc.* **63,** 53–67.

DODGE J.D. & CRAWFORD R.M. (1971) Fine structure of the dinoflagellate *Oxyrrhis marina*. II. The flagellar system. *Protistologica* **7,** 399–409.

DODGE J.D. & HERMES H.B. (1981) A scanning electron microscopical study of the apical pores of marine dinoflagellates (Dinophyceae). *Phycologia* **20,** 424–30.

DODGE J.D. & STEIDINGER K.A. (1982) Dinophyceae: introduction and bibliography. In *Selected Papers in Phycology II* (eds Rosowski J.R. & Parker B.C.), pp. 691–7. Phycological Society of America, Allen Press, Lawrence, Kansas.

DOEMEL W.N. & BROCK T.D. (1970) The upper temperature limit of *Cyanidium caldareum*. *Arch. Mikrobiol.* **72,** 326–32.

DOOLITTLE W.F. & SINGER R.A. (1974) Mutational analysis of dark endogenous metabolism of the blue green bacterium *Anacystis nidulans*. *J. Bacteriol.* **119,** 677–83.

DOP A.J. (1980) The genera *Phaeothamnion* Lagerheim. *Tetrachrysis* gen. nov. and *Sphaeridiothrix* Pascher et Vlk (Chrysophyceae). *Acta Bot. Neerl.* **29,** 65–86.

DOTY M.S. (1946) Critical tide factors that are correlated with the vertical distribution of marine algae and other organisms along the Pacific coast. *Ecology* **27,** 315–28.

DOTY M.S. (1977) *Eucheuma* current marine agronomy. In *The Marine Plant Biomass of the Pacific North-West Coast. A Potential Economic Resource* (ed. Kraus R.W.), pp. 203–14. Oregon State University Press, Corvallis.

DOTY M.S. (1979) Status of marine agronomy, with special reference to the tropics. *Proc. Int. Seaweed Symp.* **9,** 35–58.

DOTY M.S. & AGUILAR-SANTOS G. (1970) Transfer of toxic algal substances in marine food chains. *Pacific Science* **24,** 351–5.

DOTY M.S. & SANTOS G. (1978) Carrageenans from tetrasporic and cystocarpic *Eucheuma* species. *Aquatic Botany* **4,** 143–9.

DOTY M.S., GILBERT W.J. & ABBOTT I.A. (1974) Hawaiian marine algae from seaward of the algal ridge. *Phycologia* **13,** 345–57.

DREBES G. (1966) On the life history of the marine plankton diatom *Stephanopyxis palmeriana* (Grev.) Grunow. *Helgol. Wiss. Meeresunters.* **13,** 101–14.

DREBES G. (1977) Sexuality. In *The Biology of Diatoms* (ed. Werner D.). University of California Press, Berkeley.

DREBES G. (1981) Possible resting spores of *Dissodinium pseudolunula* Dinophyta and their relation to other taxa. *Brit. Phycol. J.* **16,** 207–15.

DREW K.M. (1937) *Spermothamnion snyderae* Farlow, a floridean alga bearing polysporangia. *Ann. Bot. N.S.* **1,** 463–76.

DREW K.M. (1939) An investigation of *Plumaria elegans* (Bonnem.) Schmitz with special reference to triploid plants bearing parasporangia. *Ann. Bot. N.S.* **3,** 347–67.

DREW K.M. (1949) *Conchocelis* phase in the life history of *Porphyra umbilicalis* (L.) Kütz. *Nature* **164,** 748–9.

DREWS G. & WECKESSER J. (1982) Function, structure and composition of cell walls and external layers. In *The Biology of Cyanobacteria* (eds Carr, N.G. & Whitton B.A.), pp. 333–57. Blackwell Scientific Publications, Oxford.

DRING M.J. (1967) Effects of daylength on growth and reproduction of the *Conchocelis*-phase of *Porphyra tenera. J. Mar. Biol. Ass. U.K.* **47,** 501–10.

DRING M.J. (1981) Chromatic adaptation of photosynthesis in benthic marine algae: an examination of its ecological significance using a theoretical model. *Limnol. Oceanogr.* **26,** 271–84.

DRING M.J. (1982) *The Biology of Marine Plants.* Edward Arnold, London.

DRING M.J. (1984) Photoperiodism and phycology. *Prog. Phycol. Res.* **3,** 159–92.

DRING M.J. & BROWN F.A. (1982) Photosynthesis of intertidal brown algae during and after periods of emersion: a renewed search for physiological causes of zonation. *Mar. Ecol. Prog. Ser.* **8,** 301–8.

DRING M.J. & LÜNING K. (1975) A photoperiodic effect mediated by blue light in the brown alga *Scytosiphon lomentaria. Planta* **125,** 25–32.

DROOP M.R. (1968) Vitamin B_{12} and marine ecology. IV. The kinetics of uptake, growth and inhibition in *Monochrysis lutheri. J. Mar. Biol. Assoc. U.K.* **48,** 689–733.

DROOP M.R. (1974a) Heterotrophy of carbon. In *Algal Physiology and Biochemistry* (ed. Stewart W.D.P.), pp. 530–59. Blackwell Scientific Publications, Oxford.

DROOP M.R. (1974b) The nutrient status of algal cells in continuous culture. *J. Mar. Biol. Ass. U.K.* **54,** 825–55.

DROUET F. (1968) Revision of the classification of Oscillatoriaceae. *Monogr. Acad. Nat. Sci. Philad.* **15,** 370 pp.

DROUET F. (1973) *Revision of the Nostocaceae with Cylindrical Trichomes.* Hafner Press, New York.

DROUET F. & DAILY W. (1956) Revision of the coccoid Myxophyceae. *Butler Univ. Bot. Stud.* **12,** 1–218.

DRUEHL L.D. (1978) The distribution of *Macrocystis integrifolia* in British Columbia as related to environmental parameters. *Can. J. Bot.* **56**, 69–79.

DRUEHL L.D. (1981) Geographical distribution. In *The Biology of Seaweeds* (eds Lobban C.S. & Wynne M.J.), pp. 306–25. Blackwell Scientific Publications, Oxford.

DRUEHL L.D. & GREEN J.M. (1982) Vertical distribution of intertidal seaweeds as related to patterns of submersion and emersion. *Mar. Ecol. Prog. Ser.* **9,** 163–70.

DRUGG W.S. (1969) Some new genera, species and combinations of phytoplankton from the Lower Tertiary of the Gulf Coast, U.S.A. in ultra microplankton. *Proc. N. Am. Paleontol. Convention, Chicago*, pp. 809–43. Allen Press, Lawrence, Kansas.

DRUM R.W. & HOPKINS J.T. (1966) Diatom locomotion: an explanation. *Protoplasma* **62,** 1–33.

DRUM R.W. & PANKRATZ H.S. (1963) Fine structure of a diatom centrosome. *Science (N.Y.)* **142,** 61–3.

DUBE M.A. (1967) On the life history of *Monostroma fuscum* (Postels et Rupr.) Wittrock. *J. Phycol.* **3,** 64–73.

DUCREUX P. (1977) Etude expérimentale des corrélations et des possibilités de regénération au niveau de l'apex de *Sphacelaria cirrhosa* Ag. *Ann. Sci. Nat. Bot.* **18,** 163–84.

DUFFIELD E.C.S., WAALAND S.D. & CLELAND R. (1972) Morphogenesis in the red alga *Griffithsia pacifica*: regeneration from single cells. *Planta* **105,** 185–95.

DUGDALE R.C. (1967) Nutrient limitation in the sea: dynamics, identification, and significance. *Limnol. Oceanogr.* **12,** 685–95.

DUGDALE R.C. & GOERING J.J. (1967) Uptake of new and regenerated forms of nitrogen in primary productivity. *Limnol. Oceanogr.* **12,** 685–95.

DUGGINS D.O. (1980) Kelp beds and sea otters: an experimental approach *Ecology* **61,** 447–53.

DUNSTAN P. (1979) Distribution of zooxanthellae and photosynthetic chloroplast pigments of the reef building coral *Motastrea annularis* Ellis and Solander in relation to depth on a West Indian coral reef. *Bull. Mar. Sci.* **29,** 79–95.

DURAND-CHASTEL H. (1980) Production and use of *Spirulina* in Mexico. In *Algal Biomass: Production and Use* (eds Shelef G. & Soeder C.J.), pp. 51–64. Elsevier/North-Holland, Amsterdam.

EATON J.W., BROWN J.G. & ROUND F.E. (1966) Some observations on polarity and regeneration in *Enteromorpha. Brit. Phycol. J.* **3,** 53–62.

ECHLIN P. & MORRIS I. (1965) The relationship between blue-green algae and bacteria. *Biol. Rev.* **40,** 143–87.

EDELSTEIN T. & McLACHLAN J. (1971) Further observations on *Gloiosiphonia capillaris* (Hudson) Carmichael in culture. *Phycologia* **10,** 215–19.

EDELSTEIN T., CHEN L. & McLACHLAN J. (1970) The life cycle of *Ralfsia clavata* and *R. bornetii. Can. J. Bot.* **48,** 527–31.

EDMONSTON W.T. (1970) Phosphorus, nitrogen and algae in Lake Washington after diversion of sewage. *Science* **169,** 690–1.

EDMONSTON W.T. (1972) Nutrients and phytoplankton in Lake Washington. *Limnol. Oceanogr.*, Special Symp. I., 172–93.

EDMONSTON W.T. (1979) Lake Washington and the predictability of limnological events. *Arch. Hydrobiol.* **13,** 234–41.

EDMONSTON W.T. & LEHMAN J.T. (1981) The effect of changes in the nutrient income on the condition of Lake Washington. *Limnol. Oceanogr.* **26,** 1–29.

EDWARDS P. (1969) Field and cultural studies on the season periodicity of growth and reproduction of selected Texas benthic marine algae. *Contrib. Mar. Sci.* **14,** 59–114.

EDWARDS P. (1973) Life history studies of selected British *Ceramium* species. *J. Phycol.* **9,** 181–4.

EDWARDS P. (1976) A classification of plants into higher taxa based on cytological and biochemical criteria. *Taxon* **25,** 529–42.

EGEROD L.E. (1952) An analysis of the siphonous Chlorophycophyta. *Univ. Calif. Publ. Bot.* **25**, 325–454.

EHRESMANN D.W., DEIG E.F. & HATCH M.T. (1979) Antiviral properties of algal polysaccharides and related compounds. In *Marine Algae in Pharmaceutical Science* (eds Hoppe H.A., Levring T. & Tanaka Y.), pp. 294–302. Walter de Gruyter, Berlin.

ELLIS R.J. (1976) Protein and nucleic acid synthesis by chloroplasts. In *The Intact Chloroplast* (ed. Barber J.), pp. 335–64. Topics in Photosynthesis, Vol. I. Elsevier/North-Holland, Amsterdam.

EMERSON C.J., BUGGELN R.G. & BAL A.K. (1982) Translocation in *Saccorhiza dermatodea* (Laminariales, Phaeophyceae): anatomy and physiology. *Can. J. Bot.* **60,** 2164–84.

ENGELMANN T.W. (1893) Farbe und Assimilation. *Botanische Zeitung* **41,** 1–29.

EPPLEY R.W. & COATSWORTH J.L. (1968) Nitrate and nitrite uptake by *Ditylum brightwellii*. Kinetics and mechanisms. *J. Phycol.* **4,** 151–6.

EPPLEY R.W. & PETERSON B.J. (1979) Particulate organic matter flux and planktonic new production in the deep ocean. *Nature* **282,** 677–80.

EPPLEY R.W. & RENGER E.H. (1974) Nitrogen assimilation of an oceanic diatom in nitrogen-limited continuous culture. *J. Phycol.* **10,** 15–23.

EPPLEY R.W., RENGER E.H. & WILLIAMS P.M. (1976) Chlorine reactions with seawater constituents and the inhibition of photosyntheis of natural marine phytoplankton. *Est. Coast. Mar. Sci.* **4,** 147–61.

ERIKSSON F., HORNSTROM E., MOSSBERG P. & NYBERG P. (1983) Ecological treatment of acidified lakes and rivers in Sweden. *Hydrobiologia* **101,** 145–64.

ERWIN J. & BLOCH K. (1963) Polyunsaturated fatty acids in some photosynthetic organisms. *Biochem. Z.* **388,** 496–511.

ESSER K. (1982) *Cryptogams: Cyanobacteria, Algae, Fungi, Lichens.* Cambridge University Press, Cambridge.

ESTES J.A. & PALMISANO J.F. (1974) Sea otters: their role in structuring nearshore communities. *Science* **185,** 1058–60.

ETTL H. (1978) Xanthophyceae. In *Süsswasserflora von Mitteleuropa*, Bd 3, 1 Teil. (eds Ettl H., Gerloff J. & Heynig H.). Gustav Fischer, Stuttgart.

ETTL H. (1981) Die neue Klasse Chlamydophyceae, eine natürliche Gruppe der Grünalgen (Chlorophyta). *Pl. Syst. Evol.* **137,** 107–26.

ETTL H. & KOMAREK J. (1982) Was versteht man unter dem Begriff 'coccale Grünalgen'? *Arch. Hydrobiol. (Suppl.)* **60,** 345–74.

EVENS L.V. (1965) Cytological studies in the Laminariales. *Ann. Bot. N.S.* **29,** 541–62.

EVANS L.V. (1981) Marine algae and fouling: a review with particular reference to ship fouling. *Bot. Mar.* **24,** 167–71.

EVANS L.V., CALLOW J.A. & CALLOW M.E. (1973) Structural and physiological studies of the parasitic red alga *Holmsella. New Phytol.* **72,** 393–402.

EVANS L.V., CALLOW J.A. & CALLOW M.E. (1978) Parasitic red algae: an appraisal. In *Modern approaches to the Taxonomy of Red and Brown algae* (eds Irvine D.E.G. & Price J.H.), pp. 87–140. Academic Press, New York.

EVANS L.V., CALLOW J.A. & CALLOW M.E. (1982) The biology and biochemistry of early development in *Fucus. Prog. Phycol. Res.* **1,** 67–110.

EVITT W.R. (1963) A discussion and proposals concerning fossil dinoflagellates, hystrichospheres and acritarchs. *Proc. Natl Acad. Sci. U.S.A.* **49,** 158–64.

FAIRCHILD E. & SHERIDAN R.P. (1971) A physiological investigation of the hot spring diatom *Achnanthes exigua* Grun. *J. Phycol.* **10,** 1–4.

FALKENBERG P. (1879) Die Befruchtung und der Generationswechsel von *Cutleria. Mitt. Zool. Stat. Neapel.* **1,** 420–47.

FALKOWSKI P.J. (1981) Light–shade adaptation and assimilation numbers *J. Plankton Res.* **3,** 203–16.

FARNHAM W.F., FLETCHER R.L. & IRVINE L.M. (1973) Attached *Sargassum* found in Britain. *Nature* **243,** 231–2.

FARR E.R., LEUSSINK J.A. & STAFLEU F.A. (1979) (eds) *Index Nominum Genericorum (Plantarum).* Bohn, Scheltema & Holkema, Utrecht.

FARRAR J.F. (1978) Symbiosis between fungi and algae. In *CRC Handbook Series in Nutrition and Food* (ed. Reichigl M.), pp. 121–39. CRC Press, Cleveland.

FARRAR J.F. & SMITH D.C. (1976) Ecological physiology of the lichen *Hypogymnia physodes*. III. The importance of the rewetting phase. *New Phytol.* **77,** 115–25.

FARRER W.V. (1966) Tecuitlatl, a glimpse of Aztec food technology. *Nature* **211,** 341–2.

FEHER D. (1948) Researches on the geographical distribution of soil microflora. Part II. The geographical distribution of soil algae. *Communications of the Botanical Institute of the Hungarian University of Technical and Economic Sciences*, Sopron (Hungary) **21,** 1–37.

FELDMANN J. (1951) Ecology of marine algae. In *Manual of Phycology — an Introduction to the Algae and their Biology* (ed. Smith G.M.), pp.313–34. Waltham, Massachusetts.

FENCHEL T. (1980) Suspension feeding in ciliated protozoa: feeding rates and ecological significance. *Microb. Ecol.* **6,** 13–25.

FENICAL W. (1980) Distributional and taxonomic features of toxin producing macroalgae. In *Pacific Seaweed Aquaculture* (eds Abbott I.A., Foster S. & Eklund L.F.), pp. 144–51. Institute of Marine Resources, University of California, La Jolla.

FEOCHARI C. (1979) Use of *Laminaria* tent in obstetrical practice. In *Marine Algae in Pharmaceutical Science* (eds Hoppe H.A., Levring T. & Tanaka Y.), pp. 663–73. Walter de Gruyter, Berlin.

FJELD A. & LOVLIE A. (1976) Genetics of multicellular algae. In *Genetics of Algae* (ed. Lewin R.A.), pp. 219–35. University of California Press, Berkeley.

FIKSDAHL A., WITHERS N., GUILLARD R.R.L. & LIAAEN-JENSEN S. (1984) Carotenoids of the Raphidophyceae — a chemosynthetic contribution. *Comp. Biochem. Physiol.* **78** B(1), 265–71.

FISHER C.R. & TRENCH R.K. (1980) *In vitro* carbon fixation by *Prochloron* sp. isolated from *Diplosoma virens. Biol. Bull.* **159,** 636–48.

FITT W.K. & PARDY R.L. (1981) Effects of starvation, and light and dark on the energy metabolism of symbiotic and aposymbiotic sea anemones, *Anthopleura elegantissima. Mar. Biol.* **61,** 199–205.

FLETCHER R.L. (1975) Heteroantagonism observed in mixed algal cultures. *Nature* **253**, 534–5.

FLETT R.J., SCHINDLER D.W., HAMILTON R.D. & CAMPBELL N.E.R. (1980) Nitrogen fixation in Canadian pre-cambrian Shield lakes. *Can. J. Fish. Aquat. Sci.* **37,** 494–505.

FLOC'H J.Y. & DIOURIS M. (1981) Impact du pétrole de l' Amoco Cadiz sur les algues de Portsall: suivi écologique dans une anse très polluée. In *Amoco Cadiz: Conséquences d'une pollution accidentelle par les hydrocarbures*, pp. 381–91. Actes du Colloque International Publ. Centre National pour l'Exploration des Océans, Paris.

FLOWERS A. & BRYCE A.J. (1977) Energy from marine biomass. *Sea Technol.* Oct. 1977, 18–21.

FLOYD G.L. & O'KELLY C.J. (1984) Motile cell ultrastructure and the circumscription of the orders Ulotrichales and Ulvales (Ulvophyceae, Chlorophyta). *Am. J. Bot.* **71,** 111–20.

FLOYD G.L., HOOPS H.J. & SWANSON J.A. (1980) Fine structure of the zoospore of *Ulothrix belkae* with emphasis on the flagellar apparatus. *Protoplasma* **104,** 17–31.

FOGG G.E. (1967) Observation on the snow algae of the South Orkney Islands. *Phil. Trans. Roy. Soc. Lond. B* **252,** 279–87.

FOGG G.E. (1982) Marine plankton. In *The Biology of the Cyanobacteria* (eds Whitton B.A. & Carr N.G.), pp. 491–513. Blackwell Scientific Publications, Oxford.

FOGG G.E., STEWART W.D.P., FAY P. & WALSBY A.E. (1973) *The Blue-Green Algae*. Academic Press, New York.

FORD T.W. (1984) A comparative ultrastructural study of *Cyanidium caldarium* and the unicellular red alga *Rhodosorus marinus*. *Ann. Bot.* **53,** 285–94.

FOSTER M.S. (1975) Algal succession in a *Macrocystis pyrifera* forest. *Mar. Biol.* **32,** 313–29.

FOSTER P.L. (1982a) Species associations and metal contents of algae from rivers polluted by heavy metals. *Freshwater Biol.* **12,** 17–39.

FOSTER P.L. (1982b) Metal resistances of Chlorophyta from rivers polluted by heavy metals. *Freshwater Biol.* **12,** 41–6.

FOTT B. & NOVAKOVA M. (1969) A monograph of the genus *Chlorella*. The fresh water species. In *Studies in Phycology* (ed. Fott B.), pp.10–70. Academia, Prague.

FOX G.E., STACKEBRANDT E., HESPELL R.B., GIBSON J., MANILOFF J., DYER T.A., WOLFE R.S., BALCH W.E., TANNER R.S., MAGRUM L.J., ZABLEN L.B., BLAKEMORE R., GUPTA R., BONEN L., LEWIS B.J., STAHL D.A., LUENRSEN K.R., CHEN K.N. & WOESE C.R. (1980) The phylogeny of prokaryotes. *Science* **209,** 457–63.

FOYN B. (1934) Lebenszyklus und Sexualität der Chlorophycee *Ulva lactuca* L. *Arch. Protistenk.* **83,** 154–77.

FRALICK R.A. & MATHIESON A.C. (1972) Winter fragmentation of *Codium fragile* (Suringar) Hariot spp. *tomentosoides* (van Goor) Silva (Chlorophyceae, Siphonales) in New England. *Phycologia* **11,** 67–70.

FRANCESCHI V.R. & LUCAS W.J. (1980) Structure and possible function(s) of charasome; complex plasmalemma — cell wall elaborations present in some Characean species. *Protoplasma* **104,** 253–72.

FRANCIS D. (1967) On the eyespot of the dinoflagellate *Nematodinum*. *J. Exp. Biol.* **47**, 495–502.

FRANCIS G. (1878) Poisonous Australian lake. *Nature* **18,** 11–12.

FRANK N., FERGUSON L.C., CROSS R.F. & REDMAN D.R. (1969) *Prototheca*, a cause of bovine mastitis. *Am. J. Vet. Res.* **30,** 1785–94.

FREI E. & PRESTON R.D. (1961) Variants in the structural polysaccharides of algal cell walls. *Nature* **192,** 939–43.

FREI E. & PRESTON, R.D. (1964) Non-cellulosic structural polysaccharides in algal cell walls. II. Association of xylan mannan in *Porphyra umbilicalis*. *Proc. Roy. Soc. Lond. B* **160**, 314–27.

FRIEDMANN I. (1959) Structure, life history and sex determination of *Prasiola stipitata* Suhr. *Ann. Bot. N.S.* **23,** 571–94.

FRIEDMANN E.I. (1982) Cyanophycota. In *Synopsis and Classification of Living Organisms*, Vol. 1 (ed. Parker, S.P.), pp. 45–52. McGraw-Hill, New York.

FRIES L. (1966) Influence of iodine and bromine on growth of some red algae in axenic culture. *Physiol. Plant.* **19,** 800–8.

FRIES L. (1969) The sporophyte of *Nemalion multifidum* (Weber & Mohr) J.Ag. found on the Swedish West Coast. *Svensk Bot. Tidskr.* **61,** 457–62.

FRIES L. (1973) Requirements for organic substances in seaweeds. *Bot. Mar.* **16,** 19–31.

FRIES L. (1975) Requirement of bromine in a red alga. *Z. Pflanzenphysiol.* **76**, 366–8.

FRIES N. (1979) Physiological characteristic of *Mycosphaerella ascophylli*, a fungal endophyte of the marine brown alga *Ascophyllum nodosum*. *Physiol. Plant.* **45,** 117–21.

FRITSCH F.E. (1935) *Structure and Reproduction of the Algae*, Vol. I. Cambridge University Press, Cambridge.

FRITSCH F.E. (1945) *Structure and Reproduction of the Algae*, Vol. II. Cambridge University Press, Cambridge.

FROST B.W. (1980) Grazing. In *The Physiological Ecology of Phytoplankton* (ed. Morris I.), pp. 465–91. University of California Press, Berkeley.

FRYXELL G.A. (ed.) (1983) *Survival Strategies of the Algae*. Cambridge University Press, Cambridge.

FRYXELL G.A. & HASLE G.R. (1979) The genus *Thalassiosira*: species with internal extensions of the strutted processes. *Phycologia* **18,** 378–93.

FRYXELL G.A. & HASLE G.R. (1980) The marine diatom *Thalassiosira oestrupii*: structure, taxonomy and distribution. *Am. J. Bot.* **67,** 804–14.

FUHRMAN J.A., CHISHOLM S.W. & GUILLARD R.R.L. (1978) Marine algae *Platymonas sp.* accumalates silicon without apparent requirement. *Nature* **272,** 244–6.

FUJIYAMA T. (1955) On the life history of *Prasiola japonica* Yatabe. *J. Fac. Fish. Hiroshima Univ.* **1,** 15–45.

GAGNE J.A., MANN K.H. & CHAPMAN A.R.O. (1982) Seasonal patterns of growth and storage in *Laminaria longicruris* in relation to patterns of availability of nitrogen in the water. *Mar. Biol.* **69**, 91–101.

GAILLARD J. (1972) Quelques remarques sur le cycle reproducteur des Dictyotales et sur ses variations. *Soc. Bot. Fr. Mém.* **1972**, 145–50.

GALE N.L., HARDIE M.G., JENNET J.C. & ALETI A. (1973) Transport of trace pollutants in lead mining waters, In *Trace Substances and Environmental Health*, Vol. VI (ed. Hemphell D.D.), pp. 95–106. University of Missouri, Columbia, Missouri.

GALLOP A. (1974) Evidence for the presence of a 'factor' in *Elysia viridis* which stimulates photosynthate release from its symbiotic chloroplasts. *New Phytol.* **73**, 1111–17.

GALLOP A., BARTROP J. & SMITH D.C. (1980) The biology of chloroplast acquisition by *Elysia viridis*. *Proc. Roy. Soc. Lond. B* **207**, 335–49.

GANF G.G. (1974) Phytoplankton biomass and distribution in a shallow eutrophic lake (Lake George, Uganda). *Oecologia* **16**, 9–29.

GANTT E. (1971) Micromorphology of the periplast of *Chroomonas* sp. (Cyanophyceae). *J. Phycol.* **7**, 177–84.

GANTT E. (1981) Phycobilisomes. *Ann. Rev. Plant Physiol.* **32**, 327–47.

GANTT E. & CONTI S.F. (1965) The ultrastructure of *Porphyridium cruentum*. *J. Cell Biol.* **26**, 365–81.

GANTT E., EDWARDS M.R. & CONTI S.F. (1968) Ultrastructure of *Porphyridium aeruginosum*, a blue-green colored Rhodophytan. *J. Phycol.* **4**, 65–71.

GANTT E., EDWARDS M.R. & PROVASOLI L. (1971) Chloroplast structure of the Cryptophyceae. Evidence for phycobilisomes within intrathylakoidal space. *J. Cell Biol.* **48**, 280–90.

GARBARY D.J. (1976) Life forms of algae and their distribution. *Bot. Mar.* **19**, 97–106.

GAYRAL P. & FRESNEL-MORANGE J. (1971) Résultats préliminaires sur la structure et la biologie de la coccolithacée *Ochrosphaera neopolitana* Schussnig. *C. R. Acad. Sci. Paris D* **273**, 1683–6.

GEIDER R.J., OSBORNE B.A. & RAVEN J.A. (1986) Growth photosynthesis and maintenance metabolic cost in the diatom *Phaeodactylum tricornutum* at very low light levels. *J. Phycol.* **22**, 39–48.

GEITLER L. (1959) Syncyanosen. In *Handbuch der Pflanzenphysiologie II*, pp. 530–6. Springer-Verlag, Berlin.

GEITLER L. & SCHIMAN-CZEIKA H. (1970) Ueber das sogennante Palmellastadium von *Phaeothamnion confervicola*. *Oesterr. Bot. Z.* **118**, 293–6.

GIBBS S.P. (1962a) The ultrastructure of the pyrenoids of algae, exclusive of green algae. *J. Ultrastr. Res.* **7**, 247–61.

GIBBS S.P. (1962b) The ultrastructure of the chloroplasts of algae. *J. Ultrastr. Res.* **7**, 418–35.

GIBBS S.P. (1978) The chloroplasts of *Euglena* may have evolved from symbiotic green algae. *Can. J. Bot.* **56**, 2883–9.

GIBBS S.P. (1981) The chloroplasts of some algal groups may have evolved from endosymbiotic eukaryotic algae. *Ann. N.Y. Acad. Sci.*, **361**, 193–207.

GIDDINGS T.R. Jr, WITHERS N.W. & STAEHELIN, L.A. (1980) Supramolecular structure of stacked and unstacked regions of the photosynthetic membranes of *Prochloron* sp., a prokaryote. *Proc. Natl Acad. Sci. U.S.A.* **77**, 352–6.

GILLOTT M.A. & GIBBS S.P. (1980) The cryptomonad nucleomorph: its ultrastructure and evolutionary significance. *J. Phycol.* **16**, 558–68.

GIRAUD G. & CABIOCH J. (1979) Ultrastructure and elaboration of calcified cell walls in the coralline algae (Rhodophyta, Cryptonemiales). *Biol. Cell.* **36**, 81–6.

GLIDER W.V. & PARDY R.L. (1982) Algal endozoic symbioses: an introduction and bibliography. In *Selected Papers in Phycology II* (eds Rosowski J.R. & Parker B.C.), pp. 761–72. Phycological Society of America, Lawrence, Kansas.

GLOMBITZA K.-W. (1977) Highly hydroxylated phenols of the Phaeophyceae. In *Marine Natural Products Chemistry* (eds Faulkner D.J. & Fenical W.H.), pp.191–204. Plenum Press, New York.

GLOMBITZA K.-W. (1979) Antibiotics from algae. In *Marine Algae and Pharmaceutical Science* (eds Hoppe H.A., Levring T. & Tanaka Y.), pp. 303–42. Walter de Gruyter, Berlin.

GLOOSCHENKO W.A. & CURL H. Jr (1971) Influence of nutrient enrichment on photosynthesis and assimilation ratios in natural North Pacific phytoplankton communities. *J. Fish. Res. Bd Can.* **28**, 790–3.

GMELIN S.G. (1768) *Historia Fucorum*. Petropol.

GNILOVSKAYA M.B. (1971) Ancient aquatic plants of the Vendian from the Russian Platform (latest Precambrian) (in Russian). *Paleontol. J.* **3**, 101–7.

GODWARD M.B.E. (1961) Meiosis in *Spirogyra crassa*. *Heredity* **16,** 53–62.

GODWARD M.B.E. (1966) *The Chromosomes of the Algae*. Edward Arnold, London.

GODFREY P.J. (1982) The eutrophication of Cayuga Lake: a historical analysis of the phytoplankton's response to phosphate detergents. *Freshwater Biol.* **12,** 149–66.

GOFF L.J. (1979) The biology of *Harveyella mirabilis* (Cryptonemiales, Rhodophyceae). VI. Translocation of photoassimilate ^{14}C. *J. Phycol.* **15,** 82–7.

GOFF L.J. (1982a) Symbiosis and parasitism: another viewpoint. *BioScience.* **32,** 255–6.

GOFF L.J. (1982b) The biology of parasitic red algae. *Prog. Phycol. Res.* **1,** 289–369.

GOFF L.J. (ed.) (1983) *Algal Symbiosis: A Continuum of Interaction Strategies*. Cambridge University Press, Cambridge.

GOFF L.J & COLEMAN A.W. (1984) Elucidation of fertilization and development in a red alga by quantitative DNA microspectrophotometry. *Dev. Biol.* **102,** 173–94.

GOFF L.J. & COLEMAN A.W. (1985) The role of secondary pit connections in red algal parasitism. *J. Phycol.* **21,** 483–508.

GOLDMAN J.C., McCARTHY J.J. & PEAVEY D.G. (1979) Growth rate influence on the chemical composition of phytoplankton in oceanic waters. *Nature* **279,** 210–15.

GOLDSTEIN M. (1964) Speciation and mating behaviour in *Eudorina*. *J. Protozool.* **11,** 317–44.

GOLDSTEIN M. & MORRALL S. (1970) Gametogenesis and fertilization in *Caulerpa*. *Ann. N.Y. Acad. Sci.* **175,** 660–72.

GOLTERMAN H.L., CLYMO R.S. & OHMSTAD M.A.M. (eds) (1978) *Methods for Physical and Chemical Analyses of Freshwaters*, 2nd Edition. IBP Handbook No. 8. Blackwell Scientific Publications, Oxford.

GOODENOUGH U.W. & WEISS R.L. (1975) Gametic differentiation in *Chlamydomonas reinhardtii*. III. Cell wall lysis and microfilament-associated mating structure activation in wild-type and mutant strains. *J. Cell Biol.* **67,** 623–37.

GORDON D.C. JR & PROUSE N.J. (1973) The effects of three oils on marine phytoplankton photosynthesis. *Mar. Biol.* **22,** 329–33.

GORHAM J. (1977) Plant growth regulators in *Sargassum muticum*. *J. Phycol.* **13** (suppl.), 25.

GOTHAM I.J. & RHEE G-Y. (1982) Effects of hexachlorobiphenyl and pentachlorophenol on growth and photosynthesis of phytoplankton. *J. Great Lakes Res.* **8,** 328–35.

GOVINDJEE & BRAUN B.Z. (1974) Light absorption, emission and photosynthesis. In *Algal Physiology and Biochemistry* (ed. Stewart W.D.P.), pp. 346–90. Blackwell Scientific Publications, Oxford.

GRAHAM J.M., AUER M.T., CANALE R.P. & HOFFMAN J.P. (1982) Ecological studies and mathematical modelling of *Cladophora* in lake Huron. 4. Photosynthesis and respiration as functions of light and temperature. *J. Great Lakes Res.* **8,** 100–11.

GRAHAM L.E. (1985) The origin of the life cycle of land plants. *Am. Sci.* **73,** 178–86.

GRAHAM L.E. & McBRIDE G.E. (1979) The occurrence and phylogenetic significance of a multilayered structure in *Coleochaete* spermatozoids. *Am. J. Bot.* **66,** 887–924.

GRAMBLAST L.J. (1960) *Extension Chronologique des genres chez les Charoideae*. Soc. Edit. Technip., Paris.

GRAMBLAST L.J. (1974) Phylogeny of the Charophyta. *Taxon* **23,** 463–81.

GRANT M.C. & SAWA T (1982) Charophyceae: introduction and bibliography. In *Selected Papers in Phycology II* (eds Rosowski J.R. & Parker B.C.), pp.754–9. Phycological Society of America, Lawrence, Kansas.

GRASSO P. & SHARRAT M. (1973) Studies on carrageenan and large bowel ulceration in mammals. *Food Cosmet. Toxicol.* **11,** 555–64.

GRAUER F.H. & ARNOLD H.L. JR (1961) Seaweed dermatitis: first report of a dermatitis-producing marine alga. *Arch. Dermatol.* **84,** 720–32.

GREEN J.C. (1980) The fine structure of *Pavlova pinguis* Green and a preliminary survey of the order Pavlovales (Prymnesiophyceae). *Brit. Phycol. J.* **15,** 151–91.

GREEN J.C. & HIBBERD D.J. (1977) The ultrastructure and taxonomy of *Diacronema vlkianum* (Prymnesiophyceae) with special reference to the haptonema and flagellar apparatus. *J. Mar. Biol. Ass. U.K.* **57,** 1125–36.

GREEN J.C. & PIENAAR R.N. (1977) The taxonomy of the order Isochrysidales (Prymnesiophyceae) with special reference to the genera *Isochrysis* Parke, *Dicrateria* Parke, and *Imantonia* Reynolds. *J. Mar. Biol. Ass. U.K.* **57,** 7–17.

GRETZ M.R., ARONSON J.M. & SOMMERFELD M.R. (1980) Cellulose in the cell walls of the Bangiophyceae (Rhodophyta). *Science* **207,** 779–80.

GREUET C. (1972) La nature trichocystaire du cnidoplaste dans le complexe cnidoplaste-nématocyste de *Polykrikos schwartzi* Butschli. *C. R. Acad. Sci. Paris D* **275,** 1239–42.

GREUET C. (1977) Evolution structurale et ultrastructurale de l'ocelloïde d'*Erythropsidinium pavillardi* Kofoid et Swezy (Péridinien Warnowiidae Lindemann) au cours des divisions binaire et palintomiques. *Protistologica* **13,** 127–43.

GREVILLE R.K. (1830) *Algae Britannicae.* Edinburgh.

GROOVER R.D. & BOLD, H.C. (1969) Phycological studies. VIII. The taxonomy and comparative physiology of the Chlorosarcinales and certain other edaphic algae. *Univ. Texas. Publ. 6907,* 1–165.

GRÜN W. & ALLEMANN F. (1975) The lower cretaceous of Caravaca (Spain): Berriasian calcareous nannoplankton of the Mirauetes section (Subbetic Zone, Prov. of Murcia). *Ecologae Geol. Helv.* **68,** 147–211.

GUIRY M.D. (1977) The importance of sporangia in the classification of the Florideophyceae. In *Modern Approches to the Taxonomy of Red and Brown Algae* (eds Irvine D.E.G. & Price J.H.), pp.111–44. Academic Press, London.

GUIRY M.D., WEST J.A., KIM D.H. & MASUDA M. (1984) Re-instatement of the genus *Mastocarpus* Kützing (Rhodophyta). *Taxon* **33,** 53–63.

GUNNILL F.C. (1980) Demography of the intertidal brown alga *Pelvetia fastigiata* in southern California, U.S.A. *Mar. Biol.* **59,** 169–79.

HADDAD K. & CARTER K. (1979) Oceanic intrusion: one posible initiation mechanism of red tide blooms on the west coast of Florida. In *Toxic Dinoflagellate Blooms* (eds Taylor D.L. & Seliger H.H.), pp.269–74. Developments in Marine Biology 1. Elsevier/North-Holland, New York and Amsterdam.

HAECKEL E. (1896) *Systematische Phylogenie. Entwurf eines natürlichen Systems der Organismen auf Grund ihrer Stammengeschichte.* Zweiter Theil. *Systematische Phylogenie der Wirbellosen Thiere (Invertebrata).* Verlag V. Georg Reiser, Berlin, 8 vols.

HAINES T. (1981) Acid precipitation and its consequences for aquatic ecosystems: a review. *Trans. Am. Fish. Soc.* **110,** 669–707.

HALL A. (1981) Copper accumulation in copper tolerant and non-tolerant species populations of the marine fouling alga, *Ectocarpus siliculosus* (Dillw.) Lyngbye. *Bot. Mar.* **24,** 223–8.

HALL A., FIELDING A.H. & BUTLER A.H. (1979) Mechanisms of copper tolerance in the marine fouling alga *Ectocarpus siliculosus* — evidence for an exclusion mechanism. *Mar. Biol.* **54,** 195–9.

HALL J.B. (1971) Evolution of the Prokaryotes. *J. Theor. Biol.* **30,** 429–54.

HÄLLFORS G. & THOMSEN H.A. (1979) Further observations on *Chrysochromulina birgeri* (Prymnesiophyceae) from the Tvärminne archipelago, SW coast of Finland. *Acta Bot. Fennica* **110,** 41–6.

HALLSSON S.V. (1964) The uses of seaweed in Iceland. *Proc. Int. Seaweed Symp.* **4,** 398–405.

HAMM D. & HUMM H.J. (1976) Benthic algae of the Anclote Estuary. II. Bottom-dwelling species. *Florida Sci.* **39,** 209–29.

HANIC L.A. & CRAIGIE J.S. (1969) Studies on the algal cuticle. *J. Phycol.* **5,** 89–102.

HANISAK M.D. (1979) Effect of indole-3-acetic acid on growth of *Codium fragile* subsp. *tomentosoides* (Chlorophyceae) in culture. *J. Phycol.* **15,** 124–7.

HANSEN G.I. (1980) A morphological study of *Fimbriofolium*, a new genus in the Cystocloniaceae (Gigartinales, Rhodophyta). *J. Phycol.* **16,** 207–17.

HAPPEY-WOOD C.M. (1976) Vertical migration patterns in phytoplankton of mixed species composition. *Brit. Phycol. J.* **11,** 355–69.

HARDIN G. (1960) The competitive exclusion principle. *Science* **131,** 1292–8.

HARGRAVES P.E. & FRENCH F.W. (1983) Diatom resting spores: significance and strategies. In *Survival strategies of the Algae* (ed. Fryxell G.A.), pp. 49–68. Cambridge University Press, Cambridge.

HARGREAVES J.W. (1976) Effects of pH on tolerance of *Hormidium rivulare* to zinc and copper. *Oecologia* **26,** 235–43.

HARGREAVES J.W., LLOYD E.J.H. & WHITTON B.A. (1975) Chemistry and vegetation of highly acidic streams. *Freshwater Biol.* **5,** 563–76.

HARGREAVES J.W. & WHITTON B.A. (1976) Effect of pH on growth of acid stream algae. *Brit. Phycol. J.* **11,** 215–23.

HARLIN M.M. (1980) Seagrass epiphytes. In *Handbook of Seagrass Biology. An Ecosystem Perspective* (ed. McRoy C.P.), pp. 117–52. Garland STPM Press, New York.

HARLIN M.M. & CRAIGIE J.S. (1975) The distribution of photosynthate in *Ascophyllum nodosum* as it relates to epiphytic *Polysiphonia lanosa. J. Phycol.* **11,** 109–13.

HAROLD F.M. (1966) Inorganic polyphosphates in biology: structure metabolism and function. *Bacteriol. Rev.* **30,** 772–94.

HARRIS G.P. (1980) The measurement of photosynthesis in natural populations of phytoplankton. In *The Physiological Ecology of Phytoplankton* (ed. Morris I), pp. 129–87. Blackwell Scientific Publications, Oxford.

HARRISON P.J., CONWAY H.L., HOLMES R.W. & DAVIS C.O. (1977) Marine diatoms grown in chemostat under silicate or ammonium limitation. III. Cellular chemical composition and morphology of *Chaetoceros debilis, Skeletonema costatum,* and *Thalassiosira gravida. Mar. Biol.* **43,** 19–31.

HARTNELL A.D. (1975) Hydrographic factors affecting the distribution and movement of toxic dinoflagellates in the western Gulf of Maine. In *Proceedings of the First Conference on Toxic Dinoflagellate Blooms.* (ed. LoCicero V.R.), pp. 47–68. Massachusetts Scientific and Technical Foundation, Wakefield, Mass.

HARTSHORNE J.N. (1953) The function of the eyespot in *Chlamydomonas. New Phytol.* **52,** 292–7.

HARVEY M.J. & McLACHLAN J. (eds) (1973) *Chondrus crispus.* Nova Scotia Institute of Science, Halifax, Nova Scotia.

HARVEY W.H. (1836) Algae. In *Flora Hibernica.* Mackay J.T., Dublin.

HARVEY W.H. (1846–51) *Phycologia Britannica*, 4 Vols. London.

HARVEY W.H. (1852–8) Nereis Boreali-Americana. Pt I. Melanospermae. *Smithson. Contrib. Knowl.* **3** (Art. 4), 1–150, 12 pls. Pt II. Rhodospermae. ibid. **4** (Art. 5), 1–258, 24 pls. Pt III. Chlorospermae. ibid. **10** (Art. 2), 1–140, 14 pls.

HASHIMOTO Y., KAMIYA H., YAMAZATO K. & NOZAWA K. (1975) Occurrence of a toxic blue green alga inducing skin dermatitis in Okinawa. *Taxicon* **13**, 95–6.

HASLE G.R. (1973) The 'mucilage pore' of pennate diatoms. *Nova Hedwiga* **45,** 167–94.

HASLE G.R. (1978) Some freshwater and brackish water species of the diatom genus *Thalassiosira* Cleve. *Phycologia* **17,** 263–92.

HASELKORN R. (1978) Heterocysts. *Ann. Rev. Plant Physiol.* **29,** 319–44.

HATCH M.D. & SLACK C.R. (1966) Photosynthesis by sugar cane leaves. *Biochem. J.* **101,** 103–11.

HATCHER B.G., CHAPMAN A.R.O. & MANN K.H. (1977) An annual carbon budget for the kelp *Laminaria longicruris. Mar. Biol.* **44,** 85–96.

HAUG A., LARSEN B. & SMIDSROD O. (1966) A study of the constitution of alginic acids by partial acid hydrolysis. *Acta Chem. Scand.* **20,** 183–90.

HAUG A., LARSEN, B. & SMIDSROD O. (1967) Studies on the sequence of uronic acid residues in alginic acid. *Acta Chem. Scand.* **21,** 691–704.

HAUG A., LARSEN B. & BAARDSETH E. (1969) Comparison of the constitution of alginates from different sources. *Proc. Int. Seaweed Symp.* **6,** 443–51.

HAUG A., LARSEN B. & SMIDSROD O. (1974) Uronic acid sequence in alginate from different sources. *Carbohyd. Res.* **32,** 217–25.

HAUMAN J.H. (1981) Is a plasmid(s) involved in the toxicity of *Microcystis aeruginosa*? In *Algal Toxins and Health.* Environmental Science Research, Vol. 20 (ed. Carmichael W.W.), pp. 97–102. Plenum Press, New York.

HAUPT W. (1983) Movement of chloroplasts under the control of light. *Prog. Phycol. Res.* **2,** 227–82.

HAWKES M.W. (1978) Sexual reproduction in *Porphyra gardneri* (Smith et Hollenberg) Hawkes (Bangiales, Rhodophyta). *Phycologia* **17,** 326–50.

HAWKES M.W. (1983) *Hummbrella hydra* Earle (Rhodophyta, Gigartinales): seasonality, distribution and development in laboratory culture. *Phycologia* **22,** 403–13.

HAWKINS S.J. & HARKIN E. (1985) Preliminary canopy removal experiments in algal dominated communities low on the shore and in the shallow subtidal on the Isle of Man. *Bot. Mar.* **28,** 223–30.

HAWKINS S.J. & HARTNOLL R.G. (1985) Factors determining the upper limits of intertidal canopy forming algae. *Mar. Ecol. Prog. Ser.* **20,** 265–71.

HEALEY F.P. (1973) Inorganic nutrient uptake and deficiency in algae. *Crit. Rev. Microbiol.* **3,** 69–113.

HEATH I.B. (1980) Variant mitoses in lower eukaryotes: indicators of the evolution of mitosis. *Int. Rev. Cytol.* **64,** 1–80.

HECKY R.E. & KILHAM P. (1974) Environmental control of phytoplankton cell size. *Limnol. Oceanogr.* **19,** 361–6.

HECKY R.F., MOPPER K., KILHAM P. & DEGENS E.T. (1973) The amino acid and sugar composition of diatom cell walls. *Mar. Biol.* **19,** 323–31.

HEEREBOUT G.R. (1968) Studies on the Erythropeltidaceae (Rhodophyceae, Bangiophycidae). *Blumea* **16,** 139–57.

HEIMDAL B.R. & GAARDER K.R. (1980) Coccolithophorids from the northern part of the eastern central Atlantic I. Holococcolithophorids. *'Meteor' Forsch.- Ergebnisse D* **32,** 1–14.

HEIMDAL B.R. & GAARDER K.R. (1981) Coccolithophorids from the northern part of the eastern central Atlantic II. Heterococcolithophorids. *'Meteor' Forsch.- Ergebnisse D* **33,** 37–69.

HELLEBUST J.A. (1974) Extracellular products. In *Algal Physiology and Biochemistry* (ed. Stewart W.D.P.), pp. 838–63. Blackwell Scientific Publications, Oxford.

HENDEY N.I. (1964) An introductory account of the smaller algae of British waters. Part V. Bacillariophyceae (Diatoms). Ministry of Agriculture, Fisheries and Food *Fisheries Investigation Series IV*, pp. xii + 317. HMSO, London.

HENDEY N.I. (1971) Electronmicroscope studies and the classification of diatoms. In *The Micropalaeontology of Oceans* (eds Funnell B.M. & Riede R.W.), pp. 625–31. Cambridge University Press, Cambridge.

HENDREY G.R. (1982) Acid precipitation effects on algal productivity and biomass in Adirondak lakes. NTIS publ. PB83–173203.

HENRY E.C. (1984) Syringodermatales ord. nov. and *Syringoderma floridana* sp. nov. (Phaeophyceae). *Phycologia* **23,** 419–26.

HENRY E.C. & COLE K.M. (1982a) Ultrastructure of swarmers in the Laminariales (Phaeophyceae). I. Zoospores. *J. Phycol.* **18,** 550–69.

HENRY E.C. & COLE K.M. (1982b) Ultrastructure of swarmers in the Laminariales (Phaeophyceae). II. Sperm. *J. Phycol.* **18,** 570–9.

HENRY E.C. & MÜLLER D.G. (1983) Studies on the life history of *Syringoderma phinneyi* sp. nov. (Phaeophyceae). *Phycologia* **22,** 387–93.

HERDMAN M. & STANIER R.V. (1977) The cyanelle: chloroplast of endosymbiotic prokaryote. *FEMS Letters* **1,** 7–12.

HEYWOOD P. (1982) Raphidophyceae (Chloromonadophyceae). In *Selected Papers in Phycology* II (eds Rosowski J.R. & Parker B.C.), pp. 719–22. Phycological Society of America, Allen Press, Lawrence, Kansas.

HIAT R.W. & STRASBURG D.W. (1960) Ecological relationships of the fish fauna of the Marshall Islands. *Ecol. Monogr.* **30,** 67–127.

HIBBERD D.J. (1970) Observations on the cytology and ultrastructure of *Ochromonas tuberculatus* sp. nov. (Chrysophyceae), with special reference to the discobolocysts. *Brit. Phycol. J.* **5,** 119–43.

HIBBERD D.J. (1971) Observations on the cytology and ultrastructure of *Chrysamoeba radians* Klebs (Chrysophyceae). *Br. Phycol. J.* **6,** 207–23.

HIBBERD D.J. (1976) The ultrastructure and taxonomy of the Chrysophyceae and Prymnesiophyceae (Haptophyceae), a survey with some new observations on the ultrastructure of the Chrysophyceae. *Bot. J. Linn. Soc.* **72,** 55–80.

HIBBERD D.J. (1979) The structure and phylogenetic significance of the flagellar transition region in the chlorophyll *c*-containing algae. *Biosystems* **11,** 243–61.

HIBBERD D.J. (1980a) Xanthophytes. In *Phytoflagellates* (ed. Cox E.R.), pp. 243–72. Elsevier/North-Holland, New York.

HIBBERD D.J. (1980b) Eumastigiophytes. In *Phytoflagellates* (ed. Cox E.R.), pp. 319–34. Elsevier/North-Holland, New York.

HIBBERD D.J. (1980c) Prymnesiophytes (=Haptophytes). In *Phytoflagellates* (ed. Cox E.R.), pp. 273–318. Elsevier/North-Holland, New York.

HIBBERD D.J. (1981) Notes on the taxonomy and nomenclature of the algal class Eustigmatophyceae and Tribophyceae (synonym Xanthophyceae). *Bot. J. Linn. Soc.* **82,** 93–119.

HIBBERD D.J. (1982a) Xanthophyceae. In *Synopsis and Classification of Living Organisms*, Vol. 1 (ed. Parker S.P.), pp. 91–4. McGraw-Hill, New York.

HIBBERD D.J. (1982b) Eustigmatophyceae: introduction and bibliography. In *Selected Papers in Phycology* II (eds. Rosowski J.R. & Parker B.C.), pp. 728–30. Phycological Society of America, Allen Press, Lawrence, Kansas

HIBBERD D.J. (1982c) Eustigmatophytes. In *Synopsis and Classification of Living Organisms,* Vol. 1 (ed. Parker S.P.), p. 95. McGraw-Hill, New York.

HIBBERD D.J. & LEEDALE G.F. (1971) Cytology and ultrastructure of the Xanthophyceae. II. The zoospore and vegetative cells of coccoid forms, with special reference to *Ophiocytium majus* Naegeli. *Brit. Phycol. J.* **6,** 1–23.

HIBBERD D.J., GREENWOOD A.D. & GRIFFITHS H.B. (1971) Observations on the ultrastructure of the flagella and periplast in Cryptophyceae. *Brit. Phycol. J.* **6,** 61–72.

HICKMAN M. (1982) The removal of a heated discharge from a lake and the effect upon an epiphytic algal community. *Hydrobiologia* **87,** 21–32.

HILL D.J. (1976) The physiology of lichen symbiosis. In *Lichenology (Progress and Problems)* (eds Brown D.H., Hawksworth D.L. & Bailey R.H.), pp. 457–96. Academic Press, New York.

HILL G.J.C. & MACHLIS L. (1968) An ultrastructural study of vegetative cell division in *Oedogonium bonsianum. J. Phycol.* **4,** 261–71.

HILL R. & BENDALL F. (1960) Function of two cytochrome components in chloroplasts: a working hypothesis. *Nature* **186,** 136–7.

HINDE R. (1983) Retention of algal chloroplasts by molluscs. In *Algal Symbiosis: A Continuum of Interaction Strategies* (ed. Goff L.J.), pp. 97–107. Cambridge University Press, Cambridge.

HINDE R. & SMITH D.C. (1974) Chloroplast 'symbiosis' and the extent to which it occurs in *Sacoglossa* (Gastropoda, Mollusca). *Biol. J. Linn. Soc.* **6,** 349–56.

HITCH C.J.B. & MILLBANK J.W. (1975) Nitrogen metabolism in lichens. VII. Nitrogenase activity and heterocyst frequency in lichens with blue-green phycobionts. *New Phytol.* **75,** 239–44.

HODGSON L.M. (1980) Control of the intertidal distribution of *Gastroclonium coulteri* (Harvey) Kylin, in Monterey Bay, California. *Mar. Biol.* **57,** 121–6.

HOEK C. VAN DEN (1975) Phytogeographic provinces along the coasts of the northern North Allantic Ocean. *Phycologia* **14,** 317–30.

HOEK C. VAN DEN (1981) Chlorophyta: morphology and classification. In *The Biology of Seaweeds* (eds Lobban C.S. & Wynne M.J.), pp. 86–132. Blackwell Scientific Publications, Oxford.

HOEK C. VAN DEN (1982a) *A Taxonomic Revision of the American Species of Cladophora (Chlorophyceae) in the North Atlantic Ocean and their Geographic Distribution.* North-Holland, Amsterdam.

HOEK C. VAN DEN (1982b) Phytogeographic distribution of benthic marine algae in the North Atlantic Ocean. A review of experimental evidence from life history studies. *Helgol. Wiss. Meeresunters.* **35,** 153–214.

HOEK C. VAN DEN (1982c) The distribution of benthic marine algae in relation to the temperature regulation of their life histories. *Biol. J. Linn. Soc.* **18,** 81–144.

HOEK C. VAN DEN & FLINTERMAN A. (1968) The life history of *Sphacelaria furcigera* Kütz (Phaeophyceae). *Blumea* **16,** 193–242.

HOEK C. VAN DEN & JAHNS H.M. (1978) *Einführung in die Phykologie.* Georg Thieme Verlag, Stuttgart.

HOEK C. VAN DEN, COLIJN F., CORTEL-BREEMAN A.M. & WANDERS J.B.W. (1972). Algal vegetation-type along the shores of inner bays and lagoons of Curacao and the lagoon Lac (Bonaire) Netherlands Antilles. *Ver. K. Ned. Akad. Wet.* **61,** 1–72.

HOFFMAN L.R. (1965) Cytological studies of *Oedogonium.* I. Oospore germination in *O. foveolatum. Am J. Bot.* **52,** 173–81.

HOFFMAN L.R. & MANTON I. (1962) Observations on the fine structure of the zoospore of *Oedogonium cardiacum* with special reference to the flagellar apparatus. *J. Exp. Bot.* **13,** 443–9.

HOFFMANN H.J. (1976) Precambrian microflora, Belcher Islands, Canada: significance and systematics. *J. Paleontol.* **50,** 1040–73.

HOFFMANN H.J. & AITKEN J.P. (1979) Precambrian biota from the Little Dal Group, McKenzie Mountains, northwestern Canada. *Can. J. Earth Sci.* **16,** 150–66.

HOHAM R.W. (1975a) Optimum temperatures and temperature ranges for growth of snow algae. *Arct. Alp. Res.* **7,** 13–24.

HOHAM R.W. (1975b) The life history and ecology of the snow alga *Chloromonas pichinchae* (Chlorophyta, Volvocales). *Phycologia* **16,** 53–68.

HOHAM R.W. (1980) Unicellular chlorophytes — snow algae. In *Phytoflagellates* (ed. Cox E.R.), pp. 61–84. Elsevier/North-Holland, New York.

HOHAM R.W. & BLINN D.W. (1979) The distribution of algae in an arid region. *Phycologia* **18,** 133–45.

HOHAM R.W., ROEMER S.C. & MULLET J.E. (1979) The life history and ecology of the snow alga *Chloromonas brevispina* comb. nov. (Chlorophyta, Volvocales). *Phycologia* **18,** 55–70.

HOLCOMB G.E. (1975) Hosts of the alga *Cephaleuros virescens* in Louisiana. *Proc. Am. Phytopath. Soc.* **2,** 134.

HOLLANDE A. & ENGUINET M. (1955) Sur l'évolution et la systématique des Labyrinthulidae; Etude de *Labyrinthula algenensis* nov. sp. *Ann. Sci. Nat. Zool.* II sér. **17,** 357–68.

HOLLENBERG G.J. (1935) A study of *Halicystis ovalis.* I. Morphology and reproduction. *Am. J. Bot.* **22,** 782–812.

HOLLIGAN P.M. & GOODAY G.W. (1975) Symbiosis in *Convoluta roscoffensis*. *Symp. Soc. Exp. Biol.* **29,** 205–27.

HOLMES R.W. (1956) The annual cycle of phytoplankton in the Labrador Sea, 1950–1951. *Bull. Bingham Oceanogr. Col.* **16,** 1–74.

HOLMES R.W., CRAWFORD R.M. & ROUND F.E. (1982) Variability in the structure of the genus *Cocconeis* Ehr. (Bacillariophyta) with special reference to the cingulum. *Phycologia* **21,** 370–81.

HOOPER R. & SOUTH G.R. (1977) Additions to the benthic algal flora of Newfoundland III, with observations on species new to eastern Canada and North America. *Naturaliste Can. (Qué.)* **104,** 383–94.

HOOPER R.G., SOUTH G.R. & WHITTICK A. (1980) Ecological and phenological aspects of the marine phytobenthos of the Island of Newfoundland. In *The Shore environment.* **Vol. 2;** *Ecosystems* (eds Price J.H., Irvine D.E.G. & Farnham W.F.), pp. 395–423. Systematics Association Special Volume 17b Academic Press, London and New York.

HOPKINS C.R. (1978) *Structure and Function of Cells.* W.B. Saunders, London.

HOPPE H.A. (1979) Marine algae and their products and constituents in pharmacy. In *Marine algae in Pharmaceutical Science* (eds Hoppe H.A., Levring T. & Tanaka Y.), pp. 25–119. Walter de Gruyter, Berlin.

HORI T., NORRIS R.E. & CHIHARA M. (1982) Studies on the ultrastructure and taxonomy of the genus *Tetraselmis* (Prasinophyceae). I. Subgenus *Tetraselmis. Bot. Mag. Tokyo* **95,** 49–61.

HORIGUCHI T. & CHIHARA M. (1983) *Stylodinium littorale*, a new marine dinococcalean alga (Pyrrhophyta). *Phycologia* **22,** 23–8.

HORN AF RANTZIEN H. (1959) Morphological types and organ-genera of Tertiary charophyte fructifications. *Stockh. Contr. Geol.* **4**, 45–197.

HORNER R. & ALEXANDER V. (1972) Algal populations in arctic sea ice. An investigation of heterotrophy. *Limnol. Oceanogr* **17,** 454–8.

HOSHAW R.W. (1965) A cultural study of sexuality in *Sirogonium melanosporum. J. Phycol.* **1,** 134–8.

HOVASSE R., MIGNOT J.P. & LYON L. (1967) Nouvelles observations sur les trichocystes des Crypotomonadines et les 'R boches' des particules Kappa de *Paramecium aurelia* Killer. *Protistologica* **3,** 241–55.

HOXMARK R.C. & NORDBY O. (1974) Haploid meiosis as a regular phenomenon in the life cycle of *Ulva mutabilis. Hereditas* **76,** 239–50.

HSIAO S. (1969) Life history and iodine nutrition of the marine brown alga *Petalonia fascia* (O.F. Müll.) O. Kuntze. *Can. J. Bot.* **47,** 1611–16.

HUBER M.E. & LEWIN R.A. (1986) An electrophoretic survey of the genus *Tetraselmis* (Chlorophyta, Prasinophyceae). *Phycologia* **25,** 205–9.

HUIZING H.J. & RIETEMA H. (1975) Xylan and mannan as cell wall constituents of different stages in the life histories of some siphoneous green algae. *Brit. Phycol. J.* **10,** 13–16.

HULBERT E.M., RYTHER J.H. & GUILLARD R.R.L. (1960) The phytoplankton of the Sargasso Sea off Bermuda. *J. Cons. Int. Explor. Mer.* **15,** 115–18.

HUNT M.E., FLOYD G.C. & STOUT B.B. (1979) Soil algae in field and forest environments. *Ecology* **60,** 362–75.

HUNTSMAN S.A. & SUNDA W.G. (1980) The role of trace metals in regulating phytoplankton growth. In *The Physiological Ecology of Phytoplankton* (ed. Morris I.), pp. 285–328. Blackwell Scientific Publications, Oxford.

HUTCHINS L.W. (1947) The basis for temperature zonation in geographical distribution. *Ecol. Mongr.* **17,** 81–148.

HUTCHINSON G.E. (1961) The paradox of the plankton. *Am. Nat.* **95,** 137–45.

HUTCHINSON G.E. (1969) Eutrophication past and present. In *Eutrophication: Causes, Consequences, Correctives*, pp. 17–26. Academy of Science, Washington D.C.

HUTCHINSON T.C., HELLEBUST J.A., TAM D., MACKAY D., MASCARENHAS R.A. & SHIU W.Y. (1978) The correlation of the toxicity to algae of hydrocarbons and halogenated hydrocarbons with their physical–chemical properties. In *Hydrocarbons and Halogenated Hydrocarbons in the Aquatic Environment* (eds Afghan B.K. & Mackay D.), pp. 577–86. Environmental Science Research, Vol. 16. Plenum Press. New York.

HUTH K. (1981) Der Generationswechsel von *Lemanea fluviatilis* C.Ag. in Kultur. *Nova Hedwiga.* **34,** 177–89.

ILLMAN W.I., McLACHLAN J. & EDELSTEIN T. (1972) Two assemblages of marine algae from post-glacial deposits in the eastern Canadian arctic. *Can.J. Earth Sci.* **9,** 109–15.

ILMAVIRTA K. & KOTIMAA A.L. (1974) Spatial and seasonal variations in phytoplanktonic primary production and biomass in the oligotrophic lake Paajarvi, southern Finland. *Ann. Bot. Fennici* **11,** 112–20.

INCOLL L.D., LONG S.P. & ASHMORE M.R. (1977) SI units in publications in plant science. *Curr. Adv. Plant Sci.* **28,** 331–43.

INGLE R.K. & COLMAN B. (1976) The relationship between carbonic anhydrase activity and glycolate excretion in *Coccochloris peniocystis. Planta* **128,** 217–23.

INNES D.J. & YARISH C. (1984) Genetic evidence for the occurrence of asexual reproduction in populations of *Enteromorpha linza* (L.) J.Ag. (Chorophyta, Ulvales). *Phycologia* **23,** 311–20.

INOUYE I., HORI T. & CHIHARA M. (1983) Ultrastructure and taxonomy of *Pyramimonas lunata*, a new marine species of the class Prasinophyceae. *Jap. J. Phycol.* **31**, 238–49.

IRVINE D.E.G. & JOHN D.M. (eds) (1984) *Systematics of Green Algae.* Academic Press, London.

ISHIZAWA K. & WADA S. (1979a) Growth and phototropic bending in *Boergesenia* rhizoids. *Plant Cell Physiol.* **20,** 973–82.

ISHIZAWA K. & WADA S. (1979b) Action spectrum of negative phototropism in *Boergesenia forbesii. Plant Cell Physiol.* **20,** 983–7.

JACKSON J.B.C. (1977) Competition on marine hard substrata: the adaptive significance of solitary and colonial structures. *Am. Nat.* **111,** 743–67.

JACOBS W.P. (1964) Rhizoid production and regeneration of *Caulerpa prolifera. Pubbl. Staz. Zool. Napoli* **34,** 185–96.

JACOBS W.P. & DAVIS W. (1983) Effects of gibberellic acid on the rhizome and rhizoids of the algal coenocyte *Caulerpa prolifera* in culture. *Ann. Bot.* **52,** 39–41.

JACOBS W.P., FALKENSTEIN K. & HAMILTON R.H. (1985) Nature and amount of auxin in algae. *Plant Physiol.* **78,** 844–8.

JAFFE L.F. (1958) Tropistic responses of zygotes of the Fucaceae to polarized light. *Exp. Cell Res.* **15,** 282–99.

JEFFREY C. (1962) The origin and differentiation of the archegoniate landplants. *Bot. Not.* **115,** 446–54.

JENKIN P.M. (1942) Seasonal changes in the temperature of Lake Windermere (English Lake District). *J. Animal Ecol.* **11,** 248–69.

JENSEN A. (1979) Industrial utilization of seaweeds in the past, present and future. *Proc. Int. Seaweed Symp.* **9,** 17–34.

JENSEN J.B. (1974) Morphological studies in Cystoseiraceae and Sargassaceae (Phaeophyceae) with special reference to apical organization. *Univ. Calif. Publ. Bot.* **68,** 1–61.

JEWELL W.J. & McCARTY D.L. (1968) Aerobic decomposition of algae and nutrient regeneration. *Tech. Rep. Fed. Water Pollut. Control Admin. USA.*, No. 91.

JOHANSEN H.W. (1981) *Coralline Algae, First Synthesis.* CRC Press, Boca Raton, Florida.

JOHANSSON S., LARSSON U. & BOEHM P. (1980) The Tsesis oil spill. Impact on the pelagic ecosystem. *Mar. Poll. Bull.* **11,** 284–93.

JOHNSON C.H., ROEBER J.F. & HASTINGS J.W. (1984) Circadian changes in enzyme concentration account for rhythm of enzyme activity in *Gonyaulax. Science* **223,** 1428–30.

JOHNSON J.H. (1951) Fossil algae. In *Manual of Phycology. An Introduction to the Algae and their Biology.* (ed. Smith G.M.), pp. 193–202. Ronald Press Co., New York.

JOHNSTON C.S., JONES R.G. & HUNT R.T. (1977) A seasonal carbon budget for a laminarian population in a Scottish sea loch. *Helgol. Wiss. Meeresunters.* **30,** 527–45.

JOHNSTON H.W. (1970) The biological and economic importance of algae. 3. Edible algae of fresh and brackish waters. *Tuatara* **18,** 19–35.

JOLLIFFE E.A. & TREGUNNA E.B. (1970) Studies on HCO_3 ion uptake during photosynthesis in benthic marine algae. *Phycologia* **9,** 293–303.

JONES N.S. & KAIN J.M. (1967) Subtidal algal colonization following the removal of *Echinus. Helgol. Wiss. Meeresunters.* **30,** 611–21.

JÓNSSON S. & CHESNOY L. (1974) Etude ultrastructurale de l'incorporation des axonèmes flagellaires dans les zygotes du *Monostroma grevillei* (Thuret) Wittr. Chlorophycée marine. *C.R. Acad. Sci. Paris D* **278,** 1557–60.

JOSE G. & CHOWDRAY Y.B.K. (1978) Effect of some growth substances on *Trentepohlia effusa* (Kremp) Hariot. *Indian J. Plant Physiol.* **21,** 6–33.

JOSHI G.V., KAREKAR M.D., GOWDA C.A. & BHOSALE L. (1974) Photosynthetic carbon metabolism and carboxylating enzymes in algae and mangrove under saline conditions. *Photosynthetica* **8,** 51–2.

JOUBERT J.J., RIJKENBERG F.H.J. & STEYN P.L. (1975). Studies on the physiology of a parasitic green alga *Cephaleuros* sp. *Phytopath. Z.* **84,** 147–52.

KAEVER M. (1969) Neue Dasycladaceen — *Afghanopolia fragilis* Nogen., Nosp. aus dem Mittel — Eozän von Ost-Afghanistan. *Argumenta Paleobotanica* **3,** 15–42.

KAHN N. & SWIFT E. (1978) Positive buoyancy through ionic control in the non-motile marine dinoflagellate *Pyrocystis notiluca* Murray ex Schuett. *Limnol. Oceanogr.* **23,** 649–58.

KAIN J.M. (1979) A view of the genus *Laminaria. Oceanogr. Mar. Biol. Ann. Rev.* **17,** 101–61.

KALFF J. & KNOECHEL R. (1978) Phytoplankton and their dynamics in oligotrophic and trophic lakes. *Ann. Rev. Ecol. System* **9,** 475–95.

KANAZAWA A. (1963) Vitamins in algae. *Bull. Jap. Soc. Sci. Fish.* **29,** 713–31.

KAPRAUN D.F. (1970) Field and cultural studies of *Ulva* and *Enteromorpha* in the vicinity of Port Aransas, Texas. *Contrib. Mar. Sci.* **15,** 205–83.

KAPRAUN D.F. (1977) The genus *Polysiphonia* in North Carolina, U.S.A. *Bot. Mar.* **20,** 313–31.

KARN R.C., STARR R.C. & HUDOCK G.A. (1974) Sexual and asexual differentiation in *Volvox obversus* (Shaw) Printz, Strains WD3 and WD7. *Arch. Protistenk* **116,** 142–8.

KATAOKA H. (1975a) Phototropism in *Vaucheria geminata.* I. The action spectrum. *Plant Cell Physiol.* **16,** 427–37.

KATAOKA H. (1975b) Phototropism in *Vaucheria geminata.* II. The mechanism of bending and branching. *Plant Cell Physiol.* **16,** 439–48.

KATAOKA H. (1977) Phototropic sensitivity in *Vaucheria geminata* regulated by 3′, 5′ cyclic AMP. *Plant Cell Physiol.* **18,** 431–40.

KATOH S. (1960) A new copper protein from *Chlorella ellipsoidea. Nature* **186,** 533–4.

KATSAROS C., GALATIS B. & MITRAKOS K. (1983) Fine structural studies on the interphase and dividing apical cells of *Sphacelaria tribuloides* (Phaeophyta). *J. Phycol.* **19,** 16–30.

KAUSS H. (1978) Osmotic regulation in algae. *Prog. Phytochem.* **5,** 1–28.

KAUSS P., HUTCHINSON C.S., HELLEBUST J. & GRIFFITHS M. (1973) The toxicity of crude oil and its components to fresh water algae. In *Proceedings of the 1973 Conference on Prevention and Control of Oil Spills*, pp. 703–14. American Petroleum Institute, Washington D.C.

KELETI G., SYKORA J.L., LIPPY E.C. & SHAPIRO M.A. (1979) Composition and biological properties of lipopolysaccharides isolated from *Schizothrix calciola* (Ag.) Gomont (Cyanobacteria). *Appl. Environ. Microbiol.* **38,** 471–7.

KERSHAW K.A. (1985) *Physiological Ecology of Lichens* (Cambridge Studies in Ecology). Cambridge University Press, Cambridge.

KESSLER E. (1977) Physiological and biochemical contributions to the taxonomy of the genus *Prototheca.* I. Hydrogenase, acid tolerance, salt tolerance, thermophily, and liquefaction of gelatin. *Arch. Mikrobiol.* **113,** 139–44.

KESSLER E. (1982) Physiological and biochemical contributions to the taxonomy of the genus *Prototheca.* III. Utilization of organic carbon and nitrogen compounds. *Arch. Mikrobiol.* **132,** 103–6.

KIES L. (1976) Untersuchungen zur Feinstruktur und taxonomischen Einordnung von *Gloeochaete wittrockiana*, einer apoplastidialen capsalen Alge mit blaugrünen Endosymbioten (Cyanellen). *Protoplasma* **87,** 419–46.

KIES L. (1980) Morphology and systematic position of some endocyanomes. In *Endocytology. Endosymbiosis and Cell Biology*, Vol. 1 (eds Schwemmler W. & Schenk H.E.A.), pp. 7–10. Walter de Gruyter & Co., Berlin & New York.

KINDIG A.C. & LITTLER M.M. (1980) Growth and primary productivity of marine macrophytes exposed to domestic sewage effluents. *Mar. Environ. Res.* **3,** 81–100.

KING J.M. & WARD C.H. (1977). Distribution of edaphic algae as related to land usage. *Phycologia* **16,** 23–30.

KITO H. (1982) Recent problems of nori (*Porphyra*) culture in Japan. In *Proceedings of the 6th U.S.–Japan Meeting on Aquaculture* (ed. Sinderman C.J.), pp. 7–12. NOAA Technical Report NMFS Circ. 442.

KIVIC P.A. & WALNE P.L. (1983) Algal photosensory apparatus probably represents multiple parallel evolutions. *Biosystems* **16,** 31–8.

KJELLMAN F.R. (1897) Phaeophyceae (Fucoideae). In *Die natürlichen Pflanzenfamilien*, Vol. 1, Ab. 2 (eds Engler A. & Prantl K.A.E.), 121 pp., 62 figs. Leipzig.

KLAVENESS D. (1973) The microanatomy of *Calyptrosphaera sphaeroidea*, with some supplementary observations on the motile stage of *Coccolithus pelagica, Norw. J. Bot.* **20,** 151–62.

KLAVENESS D. & PAASCHE E. (1979) Physiology of coccolithophorids. In *Biochemistry and Physiology of Protozoa,* Vol. 1, 2nd Edition (eds Levandowsky M. & Hunter S.H.), pp. 191–215. Academic Press, New York.

KLEIN R.M. & CRONQUIST A. (1967) A consideration of the evolutionary and taxonomic significance of some biochemical, micromorphological, and physiological characters in the thallophytes. *Quart. Rev. Biol.* **42,** 105–296.

KLINTWORTH G.K., FETTER B.F. & NIELSEN H.S. (1968) Protothecosis, an algal infection: report of a case in man. *J. Med. Microbiol.* **1,** 211–16.

KNAGGS F. (1969) A review of Florideophycidean life histories and the culture techniques employed in their investigation. *Nova Hedwiga* **18,** 293–330.

KNIGHT M. (1923) Studies in the Ectocarpaceae. I. The life history and cytology of *Pylaiella littoralis* (L.) Kjellm. Trans. Roy. Soc. Edinb. **53,** 343–61.

KNIGHT M. (1929) Studies in the Ectocarpaceae. II. The life history and cytology of *Ectocarpus siliculosus* Dillw. *Trans. Roy. Soc. Edinb.* **56,** 307–32.

KNOECHEL R. & KALFF J. (1976) Track autoradiography, a method for the determination of phytoplankton species productivity. *Limnol. Oceanogr.* **21,** 590–6.

KOEMAN R.P.T. & CORTEL-BREEMAN A.M. (1976) Observations on the life history of *Elachista fucicola* (Vell.) Aresch. (Phaeophyceae) in culture. *Phycologia* **15,** 107–17.

KOFOID C.A. & SKOGSBERG T. (1928) The Dinoflagellata: the Dinophysoidae. *Mem. Mus. Comp. Zool. Harv.* **51,** 1–766.

KOHLMEYER J. & KOHLMEYER E. (1972) Is *Ascophyllum* lichenized? *Bot. Mar.* **15,** 109–12.

KOHLMEYER J. & KOHLMEYER E. (1979) *Marine Mycology. The Higher Fungi.* Academic Press, New York.

KOOP H.-U. (1979) The life cycle of *Acetabularia* (Dasycladales, Chlorophyceae): a compilation of evidence for meiosis in the primary nucleus. *Protoplasma* **100,** 353–66.

KORNMANN P. (1956) Zur Morphologie und Entwicklung von *Percursaria percursa. Helgol. Wiss. Meeresunters.* **5,** 259–72.

KORNMANN P. (1960) Die heterogene Gattung *Gomontia* II. Der fädige Anteil, *Eugomonita sacculata* nov. gen. et nov. spec. *Helgol. Wiss. Meeresunters.* **7,** 59–71.

KORNMANN P. (1961) Über *Codiolum* and *Urospora. Helgol. Wiss. Meeresunters.* **8,** 42–57.

KORNMANN P. (1962) Der Lebenszyklus von *Desmarestia viridis. Helgol. Wiss. Meeresunters.* **8,** 287–92.

KORNMANN P. (1964) Der Lebenszyklus von *Acrosiphonia arcta. Helgol. Wiss. Meeresunters.* **11,** 110–17.

KORNMANN P. (1970a) Mutation in the siphonaceous green alga *Derbesia marina. Helgol. Wiss. Meeresunters.* **21,** 1–8.

KORNMANN P. (1970b) Phylogenetische Beziehungen in der Grünalgengattung *Acrosiphonia. Helgol. Wiss. Meeresunters.* **21,** 292–304.

KORNMANN P. (1972) Ein Beitrag zur Taxonomie der Gattung *Chaetomorpha* (Cladophorales, Chlorophyta). *Helgol. Wiss. Meeresunters.* **23,** 1–31.

KORNMANN P. (1973) Codiolophyceae, a new class of Chlorophyta. *Helgol. Wiss. Meeresunters.* **25,** 1–13.

KORNMANN P. & SAHLING P.-H. (1962) Zur Taxonomie und Entwicklung der *Monostroma*-Arten von Helgoland. *Helgol. Wiss. Meeresunters.* **8,** 302–20.

KORNMANN P. & SAHLING P.H. (1974) Prasiolales (Chlorophyta) von Helgoland. *Helgol. Wiss. Meeresunters.* **8,** 320.

KRAFT G.T. (1976) The morphology of *Beckerella scalaramosa*, a new species of Gelidiales (Rhodophyta) from the Philippines. *Phycologia* **15,** 85–91.

KRAFT G.T. (1978) Studies on marine algae in the lesser-known families of the Gigartinales (Rhodophyta). III. The Mychodeaceae and Mychodeophyllaceae. *Aust. J. Bot.* **26,** 515–610.

KRAFT G.T. (1981) Rhodophyta: morphology and classification. In *The Biology of Seaweeds* (eds Lobban C.S. & Wynne M.J.), pp. 6–51. Blackwell Scientific Publications, Oxford.

KREGER D.R. & VAN DER VEER J. (1971) Paramylon in a Chrysophyte. *Acta Bot. Neer.* **19,** 401–2.

KREH T.V. & DEWORT J.E. (1976) Effects of entrainment through Oconee Nuclear Station on carbon-14 assimilation rates of phytoplankton. In *Thermal Ecology II* (eds Esch G.W. & McFarlane R.W.), pp. 331–5. ERDA. Symposium Series (Conf. 750425), Springfield, Mass., U.S.A.

KREMER B.P. (1981a) Carbon metabolism. In *The Biology of Seaweeds* (eds Lobban C.S. & Wynne M.J.), pp. 493–533, Blackwell Scientific Publications, Oxford.

KREMER B.P. (1981b) Dark reactions of photosynthesis. *Can. Bull. Fish. Aquat. Sci.* **210,** 55–82.

KREMER B.P. (1982) *Cyanidium caldarium*: a discussion of biochemical features and taxonomic problems. *Brit. Phycol. J.* **17,** 51–61.

KREMER B.P. (1983) Carbon economy and nutrition of the alloparasitic red alga *Harveyella mirabilis. Mar. Biol.* **76,** 231–9.

KREMER B.P. & BERKS R. (1978) Photosynthesis and carbon metabolism in marine and freshwater diatoms. *Z. Pflanzenphysiol.* **87**, 149–67.

KREMER B.P. & KÜPPERS U. (1977) Carboxylating enzymes and pathway of photosynthetic carbon assimilation in different marine algae — evidence for the C_4 pathway? *Planta* **133**, 191–6.

KREMER B.P., KIES L. & ROSTAMI-RABET A. (1979) Photosynthetic performance of cyanelles in the endogamones *Cyanophora, Glaucosphaera, Gloeochaete*, and *Glaucocystis. Z. Pflanzenphysiol.* **92**, 303–17.

KREMER B.P., SCHMALJOHANN R. & ROTTGER R. (1980) Features and nutritional significance of photosynthetates produced by unicellular algae symbiotic with larger foraminifera. *Mar. Ecol. Prog. Ser.* **2,** 225–8.

KRISTIANSEN J. (1974) The fine structure of the zoospore of *Urospora penicilliformis*, with special reference to the flagellar apparatus. *Brit. Phycol. J.* **9**, 201–13.

KRISTIANSEN J. (1982) Chromophycota and Chrysophyceae. In *Synopsis and Classification of Living Organisms*, Vol. 1 (ed. Parker S.P.), pp. 81–6. McGraw-Hill, New York.

KRISTIANSEN J. & TAKAHASHI E. (1982) Chrysophyceae: introduction and bibliography. In *Selected Papers in Phycology II* (eds Rosowski J.R. & Parker B.C.), pp. 698–704. Phycological Society of America, Allen Press, Lawrence, Kansas.

KRISTIANSEN J. & WALNE P.L. (1976) Structural connections between flagellar base and stigma in *Dinobryon. Protoplasma* **99,** 371–4.

KRISTIANSEN J. & WALNE P.L. (1977) Fine structure of photokinetic systems in *Dinobryon cylindricum* var. *alpinum* (Chrysophyceae). *Brit. Phycol. J.* **12,** 329–41.

KUCKUCK P. (1912) Die Fortpflanzung der Phaeosporeen. *Wiss. Meeresunters. N.F. (Abt. Helgoland)* **5,** 153–86.

KUENZLER E.J. (1965) Glucose-6-phosphate utilization by marine algae, *J. Phycol* **1,** 156–64.

KUGRENS P. & WEST, J.A. (1972) Ultrastructure of tetrasporogenesis in the parasitic red alga *Levringiella gardneri* (Setchell) Kylin. *J. Phycol.* **8,** 370–83.

KUGRENS P. & WEST J.A. (1973) The ultrastructure of an alloparasitic red alga *Choreocolax polysiphoniae. Phycologia* **12,** 175–86.

KUHLENKAMP R. & MÜLLER D.G. (1985) Culture studies of the life history of *Haplospora globosa* and *Tilopoteris mertensii* (Tilopteridales, Phaeophyceae). *Brit. Phycol. J.* **20,** 301–12.

KULAEV I.S. (1975) Biochemistry of inorganic polyphosphates. *Rev. Physiol. Biochem. Pharmacol.* **73,** 133–57.

KULLBERG R.G. (1971) Algal distribution in six thermal stream effluents. *Trans. Am. Microscop. Soc.* **90,** 412–34.

KÜTZING F.T. (1843) *Phycologia generalis*. Leipzig.

KÜTZING F.T. (1849) *Species algarum*. Leipzig.

KUWABARA J.S. (1982) Micronutrients and kelp cultures: evidence for cobalt deficiency in southern California deep seawater. *Science* **216,** 1219–21.

KYLIN H. (1914) Studien über die Entwicklungsgeschichte von *Rhodomela virgata* Kjellm. *Svensk. Bot. Tidsskr.* **8,** 33–70.

KYLIN H. (1956) *Die Gattungen der Rhodophyceen*. C.W.K. Gleerup, Lund.

LAMBEIN F. & WOLK C.P. (1973) Structural studies on the glycolipids from the envelope of the heterocyst of *Anabaena cylindrica. Biochemistry* **12,** 791–8.

LAMOUROUX J.V.F. (1813) Essai sur les genres de la famille des thalassiophytes non articulées. *Mém. Mus. Nat. Hist. Paris* **20,** 21–47, 115–39, 267–93.

LANG N.J. & WALSBY A.E. (1982) Cyanophyceae: introduction and bibliography. In *Selected Papers in Phycology II* (eds Rosowski J.R. & Parker B.C.), pp. 647–55. Phycological Society of America, Allen Press, Lawrence, Kansas.

LANG N.J., SIMON R.D. & WOLK C.P. (1972) Correspondence of cyanophycean granules with structured granules in *Anabaena cylindrica. Arch. Mikrobiol.* **83,** 313–20.

LANGMUIR I. (1938) Surface motion of water induced by wind. *Science* **87,** 119–23.

LANGFORD T.E. (1983) *Electricity Generation and the Ecology of Natural Waters*. Liverpool University Press, Liverpool.

LARSEN B. (1981) Biosynthesis of alginate. *Proc. Int. Seaweed Symp.* **10,** 7–34.

LAWRENCE J.M. (1975) On the relationship between marine plants and sea-urchins. *Oceanogr. Mar. Biol. Ann. Rev.* **13,** 213–86.

LAWREY J.D. (1984) *Biology of Lichenized Fungi*. Praeger, New York.

LAWSON G.W. (1978) The distribution of seaweeds in the tropical and subtropical Atlantic Ocean: a quantitative approach. *Bot. J. Linn. Soc.* **76,** 177–93.

LAZAREK S. (1982) Structure and productivity of epiphytic algal communities on *Lobelia dortmanna* in acidified and limed lakes. *Water Air Soil Pollut.* **18,** 333–42.

LEADBEATER B.S.C. (1970) Preliminary observations on differences of scale morphology at various stages in the life cycle of '*Apistonema-Syracosphaera*' *sensu* von Stosch. *Brit. Phycol. J.* **5,** 57–69.

LEADBEATER B.S.C. (1972) Fine structural observations on six new species of *Chrysochromulina* (Haptophyceae) from Norway, with preliminary observations on scale production in *C. microcylindrica* sp. nov. *Sarsia* **49,** 65–80.

LEADBEATER B.S.C. & DODGE J.D. (1967) An electron microscope study of dinoflagellate flagella. *J. Gen. Microbiol.* **46,** 305–14.

LEADBEATER B.S.C. & MANTON I. (1974) Preliminary observations on the chemistry and biology of the lorica in a collared flagellate (*Stephanoeca diplocostata*). *J. Mar. Biol. Ass. U.K.* **54,** 269–79.

LEDERMAN T.C. & RHEE G-Y. (1981) Bioconcentration of hexachorobiphenyl in Great Lakes planktonic algae. *Can. J. Fish. Aquat. Sci.* **39,** 380–7.

LEE J.J (1980) Nutrition and physiology of the foraminifera. In *Biochemistry and Physiology of the Protozoa*, Vol. 3 (eds M.Levandowsky & S.H. Hunter), pp. 43–66. Academic Press, New York.

LEE J.J & McENERY M.E. (1983) Symbiosis in foraminifera. In *Algal symbiosis: A Continuum of Interaction Strategies* (ed. L.J. Goff), pp. 37–68. Cambridge University Press, Cambridge.

LEE K.W. & BOLD H.C. (1973) *Pseudocharaciopsis texensis* gen et sp. nov., a new member of the Eustigmatophyceae. *Brit. Phycol. J.* **8,** 31–7.

LEE R.E. (1980) *Phycology*. Cambridge University Press, Cambridge.

LEE R.F. & TAKAHASHI M. (1977) The fate and effect of petroleum in controlled ecosystem enclosures. *Rapp. P.-v. Réun. Cons. Int. Explor. Mer.* **171,** 150–6.

LEEDALE G.F. (1967) *Euglenoid Flagellates*. Prentice-Hall Inc., Englewood Cliffs, New Jersey.

LEEDALE G.F. (1968) The nucleus in *Euglena*. In *The Biology of Euglena*, Vol. 1 (ed. Buetow D.E.), pp. 185–242. Academic Press, New York.

LEEDALE G.F. (1974) How many are the kingdoms of organisms? *Taxon* **23,** 261–70.

LEEDALE G.F. (1978) Phylogenetic criteria in euglenoid flagellates. *Biosystems* **10,** 183–7.

LEEDALE G.F. (1982a) Euglenophycota. In *Synopsis and Classification of Living Organisms*, Vol. 1 (ed. Parker S.P.), pp. 129–31. McGraw-Hill, New York.

LEEDALE G.F. (1982b) Ultrastructure. In *The Biology of Euglena* Vol. III, *Physiology* (ed. Buetow D.E.), pp. 1–27. Academic Press, New York.

LEEDALE G.F., SCHIFF J.A. & BUETOW D.E. (1982) Euglenophyceae: introduction and bibliography. In *Selected Papers in Phycology II* (eds Rosowski J.R. & Parker B.C.), pp. 687–90. Phycological Society of America, Allen Press, Lawrence, Kansas.

LEMBI C.A. (1980) Unicellular chlorophytes. In *Phytoflagellates* (ed. Cox E.R.), pp. 5–60. Elsevier/North-Holland, New York.

LEMBI C.A. & WALNE P.L. (1971) Ultrastructure of pseudocilia in *Tetraspora lubrica* (Roth) Ag. *J. Cell Sci.* **9,** 569–79.

LEMIEUX C., TURMEL M. & LEE R.W. (1981) Physical evidence for recombination of chloroplast DNA in hybrid progeny of *Chlamydomonas eugametos* and *C. moewusii*. *Current Genetics* **3,** 97–103.

LEONHARD J. & COMPÈRE P. (1967) *Spirulina platensis* (Gom.) Geitl., algue bleue de grande valeur alimentaire par sa richesse en protéines. *Bull. Jard. Bot. Bruxelles* **37,** Suppl. 1.

LEWIN J.C. (1962) Silicification. In *Physiology and Biochemistry of the Algae* (ed. Lewin R.A.), pp. 445–55. Academic Press, New York.

LEWIN J.C. (1966) Silicon metabolism in diatoms. V. Germanium dioxide a specific inhibitor of diatom growth. *Phycologia* **6,** 1–12.

LEWIN J.C. (1976) Effects of boron deficiency on chemical composition of a marine diatom. *J. Exp. Bot.* **27,** 916–21.

LEWIN J.C. & CHEN C.-H. (1968) Silicon metabolism in diatoms. VI. Silicic acid uptake by a colourless marine diatom, *Nitzschia alba* Lewin & Lewin. *J. Phycol.* **4,** 161–8.

LEWIN J.C., LEWIN R.A. & PHILPOTT D.E. (1958) Observations on *Phaeodactylum tricornutum*. *J. Gen. Microbiol.* **18,** 418–26.

LEWIN R.A. (1970) Toxin secretion and tail autonomy by irritated *Oxynoe panamensis*. *Pacific Sci.* **24,** 356–9.

LEWIN R.A. (1976a) *The Genetics of Algae*. Blackwell Scientific Publications, Oxford.

LEWIN R.A. (1976b) Prochlorophyta as a proposed new division of algae. *Nature* **261**, 697–8.

LEWIN, R.A. (1977) *Prochloron*-type genus of the Prochlorophyta. *Phycologia* **16**, 217.

LEWIN R.A. (1981a) *Prochloron* and the theory of symbiosis. *Ann. N. Y. Acad. Sci.* **361**, 325–9.

LEWIN R.A. (1981b) The prochlorophytes. In *The Prokaryotes. A Handbook on Habitat, Isolation, and Identification of Bacteria* (eds Starr M.P., Stolp H., Truper H.G., Balows A. & Schlegel H.G.), pp. 257–66. Springer-Verlag, Berlin.

LEWIN R.A. (1983a) The problems of *Prochloron. Ann. Microbiol. (Inst. Pasteur)* **134B**, 37–41.

LEWIN R.A. (1983b) Phycotechnology — how microbial geneticists might help. *BioScience* **33**, 177–9.

LEWIN R.A. (1984) *Prochloron* — a status report. *Phycologia* **23**, 203–8.

LEWIN R.A. & GIBBS S.P. (1982) Algae of uncertain position: introduction and bibliography. In *Selected Papers in Phycology II* (eds Rosowski J.R. & Parker B.C.), pp. 659–62. Phycological Society of America, Allen Press, Lawrence, Kansas.

LEWIN R.A. & ROBERTSON J.A. (1971) Influence of salinity on the form of *Asterocytis* in pure culture. *J. Phycol.* **7**, 236–8.

LEWIS C.M., YENTSCH C.M. & DALE B. (1979) Distribution of *Gonyaulax excavata* resting cysts in the sediments of the Gulf of Maine. In *Toxic Dinoflagellate Blooms. Proceedings of the 2nd International Conference on Toxic Dinoflagellate Blooms* (eds Taylor D.L. & Seliger H.H.), pp. 235–8. Developments in Marine Biology 1. Elsevier/North-Holland, Amsterdam.

LEWIS J.R. (1961) The littoral zone on rocky shores — a biological or physical entity? *Oikos* **12**, 280–1.

LEWIS J.R. (1964) *The Ecology of Rocky Shores*. English Universities Press, London.

LEWIS W.M. (1978) A compositional phytogeographical and elementary structural analysis of the phytoplankton in a tropical lake, Lake Lanao, Philippines. *J. Ecol.* **66**, 213–26.

L'HARDY-HALOS M.-TH. (1971a) Manifestation d'une dominance apicale chez les algues à structure cladomienne du genre *Antithamnion* (Rhodophycées, Céramiales). *C.R. Acad. Sci. Paris D* **272**, 2301–4.

L'HARDY-HALOS M.-TH. (1971b) Recherches sur les Ceramiacées (Rhodophycées-Céramiales) et leur morphogénèse. III. Observations et recherches expérimentales sur la polarité cellulaire et sur la hiérarchisation des éléments de la fronde. *Rev. Gén. Bot.* **78**, 407–91.

LIN H., SOMMERFELD M.R. & SWAFFORD J.R. (1975) Light and electron microscope observations on motile cells of *Porphyridium purpureum* (Rhodophyta). *J. Phycol.* **11**, 452–7.

LINNAEUS C. (1753) *Species plantarum*, Ed. 1. Holmiae.

LINNAEUS C. (1754) *Genera plantarum.*, Ed. 5. Holmiae.

LIST H. (1930) Die Entwicklungsgeschichte von *Cladophora glomerata* Kütz. *Arch. Protistenk.* **72**, 453–81.

LITTLER M.M. & ARNOLD K.E. (1982) Primary productivity of marine macroalgal functional form groups from south-western North America. *J. Phycol.* **18**, 307–11.

LITTLER M.M. & LITTLER D.S. (1980) The evolution of thallus form and survival strategies in benthic marine macroalgae: field and laboratory tests of a functional form model. *Am. Nat.* **116**, 25–42.

LITTLER M.M., LITTLER D.S., BLAIR S.M. & NORRIS J.N. (1985) Deepest known plant life discovered on an uncharted seamount. *Science* **227**, 57–9.

LLOYD D. (1974) Dark respiration. In *Algal Physiology and Biochemistry* (ed. Stewart W.D.P.), pp. 505–29. Blackwell Scientific Publications, Oxford.

LLOYD N.D.H., CANVIN D.T. & CULVER D.A. (1977) Photosynthesis and photorespiration in algae. *Plant Physiol.* **59**, 936–40.

LOBBAN C.S. & WYNNE M.J. (eds) (1981) *The Biology of Seaweeds*. Blackwell Scientific Publications, Oxford.

LODGE S.M. (1948) Algal growth in the absence of *Patella* on an experimental strip of foreshore, Port St. Mary, Isle of man. *Proc. Trans. Liverpool Biol. Soc.* **56**, 78–83.

LOEBLICH A.R. JR (1974) Protistan phylogeny as indicated by the fossil record. *Taxon* **23**, 277–90.

LOEBLICH A.R. III (1970) The amphiesma or dinoflagellate cell covering. *North American Paleontological Convention, Chicago*, 1969. Proc. G, 867–929.

LOEBLICH A.R. III (1976) Dinoflagellate evolution: speculation and evidence. *J. Protozool.* **23**, 13–28.

LOEBLICH A.R. III (1982) Dinophyceae. In *Synopsis and Clasification of Living Organisms*, Vol. 1 (ed. Parker S.P.), pp. 101–15. McGraw-Hill, New York.

LOEBLICH A.R. III & LOEBLILCH L.A. (1978) Division Eustigmatophyta. In *CRC Handbook of Microbiology*, 2nd Edition. Vol. II, *Fungi, Algae, Protozoa and Viruses* (eds Laskin A.I. & Lechevalier H.A.), pp. 481–7. CRC Press, West Palm Beach.

LOEBLICH A.R. III & SHERLEY J.L. (1979) Observations on the theca of the motile phase of freeliving and symbiotic isolates of *Zooxanthella microadriatica* (Freudenthal) comb. nov. *J. Mar. Biol. Ass. U.K.* **59,** 195–205.

LOEBLICH A.R. JR & TAPPAN H. (1966) Annotated index and bibliography of the calcareous nanoplankton. *Phycologia* **5,** 81–216.

LOGAN B.W. (1961) *Cryptozoon* and associated stromatolities from the Recent, Shark Bay, Western Australia. *J. Geol.* **69,** 517–33.

LOISEAUX S. (1970a) Notes on several Myrionemataceae from California using culture studies. *J. Phycol.* **6,** 248–60.

LOISEAUX S. (1970b) *Streblonema anomalum* S.et G. and *Compsonema sporangiiferum* S.et G. stages in the life history of a minute *Scytosiphon*. *Phycologia* **9,** 185–91.

LOKHORST G.M. (1978) Taxonomic studies on the marine and brackish-water species of *Ulothrix* (Ulotrichales, Chlorophyceae) in western Europe. *Blumea* **24,** 191–299.

LOKHORST G.M. & STAR W. (1983) Fine structure of mitosis and cytokinesis in *Urospora* (Acrosiphoniales, Chlorophyta). *Protoplasma* **117,** 142–53.

LOMBARD E.H. & CAPON B. (1971) *Peridinium gregarium*, a new species of dinoflagellate. *J. Phycol.* **7,** 184–7.

LOVLIE A. & BRYHNI E. (1978) On the relation between sexual and parthenogenic reproduction in haplo-diplontic algae. *Bot. Mar.* **21,** 155–63.

LUBCHENCO J. (1978) Plant species diversity in a marine intertidal community: importance of herbivore food preference and algal competitive ability. *Am. Nat.* **112,** 23–39.

LUBCHENCO J. (1980) Algal zonation in the New England rocky intertidal community: an experimental analysis. *Ecology* **61,** 333–44.

LUBCHENCO J. & CUBIT J. (1980) Heteromorphic life histories of certain marine algae as adaptations to variations in herbivory. *Ecology* **61,** 676–87.

LUBCHENCO J. & GAINES S.D. (1981) A unified approach to marine plant–herbivore interactions. I. Populations and communities. *Ann. Rev. Ecol. Syst.* **12,** 405–37.

LUBCHENCO J. & MENGE B.A. (1978) Community development and persistence in a low rocky interitdal zone. *Ecol. Mongr.* **48,** 67–94.

LUND J.W.G. (1954) The seasonal cycle of the planktonic diatom *Melosira italica* subsp. *subarctica* Müll. *J. Ecol.* **42,** 151–79.

LUND J.W.G. (1972) Eutrophication. *Proc. Roy. Soc. Lond. B* **180,** 371–82.

LUND S. (1966) On a sporangia-bearing microthallus of *Scytosiphon lomentaria* from nature. *Phycologia* **6,** 67–78.

LÜNING K. (1979) Growth strategies of three *Laminaria* species (Phaeophyceae) inhabiting different depth zones in the sublittoral region of Helgoland (North Sea). *Mar. Ecol. Prog. Ser.* **1,** 195–207.

LÜNING K. (1980) Control of life histories by daylength and temperature. In *The Shore Environment*. Vol. 2: *Ecosystems* (eds Price J.H., Irvine D.E.G. & Farnham W.F.), pp. 915–45, Systematics Association Special Volume 17b. Academic Press, London and New York.

LÜNING K. (1981) Light. In *Biology of Seaweeds* (eds Lobban C.S. & Wynne M.J.), pp. 326–55. Blackwell Scientific Publications, Oxford.

LÜNING K. & DRING M.J. (1975) Reproduction, growth and photosynthesis of gametophytes of *Laminaria saccharina* grown in blue and red light. *Mar. Biol.* **29,** 195–200.

LÜNING K., CHAPMAN A.R.O. & MANN K.H. (1978) Crossing experiments in the non-digitate complex of *Laminaria* from both sides of the Atlantic. *Phycologia* **17,** 293–8.

LUOMA S.N., BRYAN G.W. & LANGSTON W.J. (1982) Scavenging of heavy metals from particulates by brown seaweed. *Mar. Poll. Bull.* **13,** 394–6.

LYNGBYE H.C. (1819) *Tentamen Hydrophytologiae Danicae*. Copenhagen.

McCANDLESS E.L. (1978) The importance of cell wall constituents in algal taxonomy. In *Modern Approaches to the Taxonomy of Red and Brown Algae* (eds Irvine D.E.G. & PRICE J.H.), pp. 63–85. Academic Press, London and New York.

McCANDLESS E.L. (1981) Polysaccharides of seaweeds. In *The Biology of Seaweeds* (eds Lobban C.S. & Wynne M.J.), pp. 559–88. Blackwell Scientific Publications., Oxford.

McCANDLESS E.L. & CRAIGIE J.S. (1979) Sulphated polysaccharides in red and brown algae. *Ann. Rev. Plant Physiol.* **30,** 41–53.

McCANDLESS E.L., CRAIGIE J.S. & WALTER J.A. (1973) Carrageenans in the gametophytic and sporophytic stages of *Chondrus crispus*. *Can. J. Bot.* **55,** 2053–64.

McCANDLESS E.L. & VOLLMER C.M. (1984) The nemathecium of *Gymnogrongus chiton* (Rhodophyceae, Gigartinales): immunochemical evidence of meiosis. *Phycologia* **23,** 119–23.

McCARTHY J.J. (1981) The kinetics of nutrient utilization. *Can Bull. Fish. Aquat. Sci.* **210,** 211–33.

McCARTHY J.J. & GOLDMAN J.C. (1979) Nitrogenous nutrition of marine phytoplankton in nutrient depleted waters. *Science* **203,** 670–2.

McCULLY M.E. (1966) Histological studies on the genus *Fucus*. I. Light microscopy of the mature vegetative plant. *Protoplasma* **62,** 287–305.

McDONALD K. (1972) The ultrastructure of mitosis in the marine red alga *Membranoptera platyphylla. J. Phycol.* **8,** 156–66.

McFADDEN G.I., MOESTRUP Ø. & WETHERBEE R. (1982) *Pyramimonas gelidicola* sp. nov. (Prasinophyceae), a new species isolated from Antarctic sea ice. *Phycologia* **21,** 103–11.

MACHLIS L., HILL G.G.C., STEINBECK K.E. & REED W. (1974) Some characteristics of the sperm attractant in *Oedogonium cardiacum. J. Phycol.* **10,** 199–204.

MACKIE I.M. & PERCIVAL E. (1959) The constitution of xylan from the green seaweed *Caulerpa filiformis. J. Chem. Soc.* **30,** 10–15.

MACKIE W. & PRESTON R.D. (1974) Cell wall and intercellular region polysaccharides. In *Algal Physiology and Biochemistry* (ed. Stewart W.D.P.), pp. 40–85. Blackwell Scientific Publications, Oxford.

McLACHLAN J. (1977) Effects of nutrients on growth and development of embryos of *Fucus edentatus* Pyl. (Phaeophyceae; Fucales). *Phycologia* **16,** 329–38.

McLACHLAN J. & BIDWELL R.G.S. (1983) Effects of colored light on the growth and metabolism of *Fucus* embryos and apices in culture. *Can. J. Bot.* **61,** 1993–2003.

McLACHLAN J. & CHEN L.C. (1972) Formation of adventive embryos from rhizoidal filaments in sporelings of four species of *Fucus. Can. J. Bot.* **50,** 1841–4.

McLACHLAN J., CHEN L. C-M, & EDELSTEIN T. (1971) The culture of four species of *Fucus* under laboratory conditions. *Can. J. Bot.* **49,** 1463–9.

McLACHLAN J. & CRAIGIE J.S. (1964) Algal inhibition by yellow ultraviolet absorbing substances from *Fucus vesiculosus. Can. J. Bot.* **42,** 287–92.

McLACHLAN J., CRAIGIE J.S., CHEN L.C-M & OGETZE E. (1972) *Porphyra linearis* Grev.: an edible species of nori from Nova Scotia. *Proc. Int. Seaweed Symp.* **7,** 473–6.

McLEAN R.J., LAURENDI C.J. & BROWN R.M. Jr (1974) The relationship of gamone to the mating reaction in *Chlamydomonas moewusii. Proc. Natl Acad. Sci.* **71,** 2610–13.

MACPHERSON B. & PFEIFFER C.J. (1976) Experimental colitis. *Digestion* **14,** 424–52.

McQUADE A.B. (1983) Origins of the nucleate organisms. II. *Biosystems* **16,** 39–55.

MAESTRINI S.Y. & BONIN D.J. (1981a) Allelopathic relationships between phytoplankton species. *Can. Bull. Fish. Aquat. Sci.* **210,** 323–46.

MAESTRINI S.Y. & BONIN D.J. (1981b). Competition among phytoplankton based on inorganic macronutrients. *Can. Bull. Fish. Aquat. Sci.* **210,** 264–78.

MAITI B.C., THOMSON R.H. & MAHENDRAN M. (1978) The structure of caulerpin, a pigment from *Caulerpa* algae. *J. Chem. Res.* **4,** 126–7.

MAGGS C.A. & GUIRY M.D. (1982) Morphology, phenology and photoperiodism in *Halymenia latifolia* Kutz (Rhodophyta) from Ireland. *Bot. Mar.* **25,** 589–99.

MAGNE F. (1967) Sur le déroulement et le lieu de la méiose chez Lemanéacées (Rhodophycées, Nemalionales). *C.R. Acad. Sci. Paris D* **265,** 670–3.

MAIER I. (1982) New aspects of pheromone-triggered spermatozoid release in *Laminaria digitata* (Phaeophyta). *Protoplasma* **113,** 137–43.

MAIER I. & MÜLLER D.G. (1982) Antheridium fine structure and spermatozoid release in *Laminaria digitata* (Phaeophyceae). *Phycologia* **21,** 1–8.

MAIER I. & MÜLLER D.G. (1986) Sexual pheromones in algae. *Biol. Bull.* **170,** 145–75.

MAKAREWICZ J.C. & BAYBUTT R.I. (1981) Longterm (1927–1978) changes in the phytoplankton community of Lake Michigan at Chicago. *Bull. Torrey Bot. Club* **108,** 240–54.

MALINOWSKI K.C. & RAMUS J. (1973) Growth of the green alga *Codium fragile*, in a Connecticut estuary. *J. Phycol.* **9,** 102–10.

MANN D.G. (1982) Structure, life history and systematics of *Rhoicosphenia* (Bacillariophyta). II. Auxospore formation and perizonium structure of *Rh. curvata. J. Phycol.* **18,** 264–74.

MANNY B.A. (1972) Seasonal changes in organic nitrogen content of net- and nanoplankton in two hardwater lakes. *Arch. Hydrobiol.* **71,** 103–23.

MANTON I. (1959) Observations on the internal structure of the spermatozoid of *Dictyota*. *J. Exp. Bot.* **10,** 448–61.

MANTON I. (1964a) Further observations on the fine structure of the haptonema in *Prymnesium parvum*. *Arch. Mikrobiol.* **49,** 315–30.

MANTON I. (1964b) Observations with the electron microscope on the division cycle in the flagellate *Prymnesium parvum* Carter. *J. Roy. Microsc. Soc. Ser. 3,* **83,** 317–25.

MANTON, I (1965) Some phyletic implications of flagellar structure in plants. In *Advances in Botanical Research*, Vol. 2. (ed. Preston R.D.), pp. 1–34. Academic Press, New York.

MANTON I. (1977) *Dolichomastix* (Prasinophyceae) from arctic Canada, Alaska, and South Africa: a new genus of flagellates with scaly flagella. *Phycologia* **16**, 427–38.

MANTON I. & PARKE M. (1960) Further observation on small green flagellates with special reference to possible relatives of *Chromulina pusilla* Butcher. *J. Mar. Biol. Ass. U.K.* **39,** 275–98.

MANTON I., KOWALLIK K. & STOSCH H. A. von (1969) Observations on the fine structure and development of the spindle at mitosis and meiosis in a marine centric diatom (*Lithodesmium undulatum*). I. Preliminary survey of mitosis in spermatogonia. *J. Microscop.* **89,** 295–320.

MANTON I., SUTHERLAND J. & OATES K. (1977) Arctic coccolithophorids: *Wigwamma arctica* gen. et sp. nov. from Greenland and Arctic Canada, *W. annulifera* sp. nov. from South Africa and S. Alaska and *Calciarus alaskensis* gen. et sp. nov. from S. Alaska. *Proc. Roy. Soc. Lond. B* **197,** 145–68.

MANTON I., OATES K. & SUTHERLAND J. (1981) Cylinder-scales in marine flagellates from the genus *Chrysochromulina* (Haptophyceae=Prymnesiophyceae): further observations from *C. microcylindrica* Leadbeater and *C. cyathophora* Thomsen. *J. Mar. Biol. Ass. U.K.* **61,** 27–33.

MARCHANT H.J. (1974) Mitosis, cytokinesis and colony formation in *Pediastrum boryanum*. *Ann. Bot.* **38,** 883–8.

MARGALEF R. (1978) Life forms of phytoplankton as survival alternatives in an unstable environment. *Oceanologica Acta* **1,** 493–509.

MARGULIS L. (1970) *Origin of Eukaryotic Cells.* Yale University Press, New Haven.

MARGULIS L. (1981) *Symbiosis in Cell Evolution. Life and its Environment on the Early Earth.* W.H. Freeman & Co., San Francisco.

MARGULIS L. (1982) *Phyla of the Five Kingdoms. An Illustrated Guide to the Kinds of Life on Earth.* W.H. Freeman & Co., San Francisco.

MARGULIS L. & SCHWARTZ K.V. (1982) *Five Kingdoms. An Illustrated Guide to the Phyla of Life on Earth.* W.H. Freeman & Co., San Francisco.

MARKEY D.R. & WILCE R.T. (1975) The ultrastructure of reproduction in the brown alga *Pylaiella littoralis*. I. Mitosis and cytokinesis in the plurilocular gametangia. *Protoplasma* **85,** 219–41.

MARKHAM J.W. & HAGMEIER E. (1982) Observations on the effects of germanium dioxide on the growth of macroalgae and diatoms. *Phycologia,* **21,** 125–30.

MARTIN M.T. (1969) A review of life histories in the Nemalionales and some allied genera. *Brit. Phycol. J.* **4,** 145–58.

MARTIN T.C. & WYATT J.T. (1974) Extracellular investments in blue-green algae with particular emphasis on the genus *Nostoc*. *J. Phycol.* **10,** 204–10.

MASLOV V.P. (1963) *Introduction to the Study of Fossil Charophytes.* Acad. Science U.S.S.R. Geol. Inst. Transactions, Vol. 82, 104 pp. (in Russian).

MASON T.R. & VAN BRUNN V. (1977) 3-Gyr-old stromatolites from South Africa. *Nature* **266,** 47–9.

MATHIESON A.C. (1967) Morphology and life history of *Phaeostrophion irregulare* S. et G. *Nova Hedwiga* **13,** 293–318.

MATHIESON A.C. (1982) Seaweed cultivation: a review. In *Proceedings of the 6th U.S.–Japan Meeting on Aquaculture* (ed. Sinderman C.J.), pp. 25–66. NOAA Technical Report NMFS Circ. 442.

MATILSKY M.B. & JACOBS W.P. (1983) Accumulation of amyloplasts on the bottom of normal and inverted rhizome tips of *Caulerpa prolifera* (Forsskal) Lamouroux. *Planta* **159,** 189–92.

MEHTA S.C., VENKATARAMAN G.S. & DAS S.C. (1961) The fine structure and the cell wall nature of *Diatoma hiemale* var. *mesodon* (Ehr.) Grun. *Rev. Algol. N. S.* **6,** 49–52.

MEINESZ A. (1980) Connaissances actuelles et contribution à l' étude de la reproduction et du cycle des Udotéacées (Caulerpales, Chlorophytes). *Phycologia* **19,** 110–38.

MELKONIAN M. (1980) Ultrastructural aspects of basal body associated fibrous structures in green algae: a critical review. *Biosystems* **12,** 85–104.

MELKONIAN M. (1981a) Structure and significance of cruciate flagellar root systems in green algae: female gametes of *Bryopsis lyngbyei* (Bryopsidales). *Helgol. Wiss. Meeresunters.* **34,** 355–69.

MELKONIAN M. (1981b) The flagellar apparatus of the scaly green flagellate *Pyramimonas obovata*: absolute configuration. *Protoplasma* **108,** 341–55.

MELKONIAN M. (1982a) Structural and evolutionary aspects of the flagellar apparatus in green algae and land plants. *Taxon* **31,** 255–65.

MELKONIAN M. (1982b) The functional analysis of the flagellar apparatus in green algae. In *Prokaryotic and Eukaryotic Flagella* (eds Amos W.B. & Duckett J.G.). Cambridge University Press, Cambridge.

MELKONIAN M. (1982c) Effect of divalent cations on flagellar scales in the green flagellate *Tetraselmis cordiformis*. *Protoplasma* **111,** 221–33.

MELKONIAN M. (1982d) Systematics and evolution in the algae. *Prog. Bot.* **44,** 315–44.

MELKONIAN M. & BERNS B. (1983) Zoospore ultrastructure in the green alga *Friedmannia israelensis*: an absolute configuration analysis. *Protoplasma* **114,** 67–84.

MELKONIAN M. & ICHIMURA T. (1982) Chlorophyceae: introduction and bibliography. In *Selected Papers in Phycology II* (eds Rosowski J.R. & Parker B.C.), pp.747–53. Phycological Society of America, Allen Press, Lawrence, Kansas.

MERESCHOWSKY C. (1905) Über Natur und Ursprung der Chromatophoren im Pflanzenreiche. *Biol. Centr.* **25,** 593–604.

METTING B. (1981) The systematics and ecology of soil algae. *Bot. Rev.* **47,** 195–312.

MIERLE G.M. & STOKES P.M. (1976) Heavy metal tolerance and metal accumulation by planktonic algae. In *Trace substances and Environmental Health* Vol. XI (ed. Hemphill D.D.), pp. 113–22. University of Missouri, Columbia, Missouri.

MIGNOT J.-P. (1967) Structure et ultrastructure de quelques Chloromonadines. *Protistologica* **3,** 5–23.

MIGNOT J.-P. (1976) Compléments à l'étude des Chloromonadines. Ultrastructure de *Chattonella subsalsa* Biecheler. Flagellé d'eaux saumâtres. *Protistologica* **12,** 279–93.

MILLER M.M. & LANG N.J. (1968) The fine structure of akinete formation and germination in *Cylindrospermum*. *Arch. Mikrobiol.* **60,** 303–13.

MILLER R.J. (1985) Seaweeds, sea-urchins and lobsters: a re-appraisal. *Can. J. Fish. Aquat. Sci.* **42,** 2061–72.

MILLER W.I. III & COLLIER A. (1978) Ultrastructure of the frustule of *Triceratium favus* (Bacillariophyceae). *J. Phycol.* **14,** 56–62.

MILLNER G.C. & SWEENEY R.A. (1982) Lake Erie *Cladophora* in perspective. *J. Great Lakes Res.* **8,** 27–9.

MILLNER P.A. & EVANS L.V. (1980) The effects of triphenyltin chloride on respiration and photosynthesis of the green algae *Enteromorpha intestinalis* and *Ulothrix flacca*. *Plant Cell Environ.* **3,** 339–48.

MILTON D.J. (1966) Drifting organisms in the Precambrian sea. *Science* **153,** 293–4.

MIX M. (1975) Die Feinstruktur der Zellwände der Conjugaten und ihre systematische Bedeutung. *Nova Hedwiga* **42,** 179–94.

MOE R.L. & HENRY E.C. (1982) Reproduction and early development of *Ascoseira mirabilis* Skottsberg (Phaeophyta), with notes on Ascoseirales Petrov. *Phycologia* **21,** 55–66.

MOE R.L. & SILVA P.C. (1977) Antarctic marine flora: uniquely devoid of kelps. *Science* **196,** 1206–8.

MOEBUS K., JOHNSON K.M. & SIEBURTH J. McN. (1974) Re-hydration of desiccated intertidal brown algae: release of dissolved organic carbon and water uptake. *Mar. Biol.* **26,** 127–34.

MOESTRUP Ø. (1970) On the fine structure of the spermatozoids of *Vaucheria sescuplicaria* and on the later stages in spermatogenesis. *J. Mar. Biol. Ass. U.K.* **50,** 513–23.

MOESTRUP Ø. (1972) Observations on the fine structure of spermatozoids and vegetative cells of the green alga *Golenkinia*. *Brit. Phycol. J.* **7,** 169–83.

MOESTRUP Ø. (1974) Ultrastructure of the scale-covered zoospores of the green alga *Chaetosphaeridium*, a possible ancestor of the higher plants or bryophytes. *Biol. J. Linn. Soc.* **6,** 111–25.

MOESTRUP Ø. (1978) On the phylogenetic validity of the flagellar apparatus in green algae and other chlorophyll a and b containing plants. *Biosystems,* **10,** 117–44.

MOESTRUP Ø. (1982) Flagellar structure in algae: a review, with new observations particularly on the Chrysophyceae, Phaeophyceae (Fucophyceae), Euglenophyceae, and *Reckertia*. *Phycologia* **21,** 427–528.

MOESTRUP Ø. & ETTL H. (1979) A light and electron microscopical study of the flagellate *Nephroselmis olivacea* (Prasinophyceae). *Opera Bot.* **49,** 1–39.

MOESTRUP Ø. & THOMSEN H.A. (1974) An ultrastructural study of the flagellate *Pyramimonas orientalis* with particular emphasis on Golgi apparatus activity and the flagellar apparatus. *Protoplasma* **81,** 247–69.

MONOD J. (1942) *Recherches sur la Croissance des Cultures Bactériennes*. Hermann & Cie, Paris.

MORIARTY D.J.W. (1979) Muramic acid in the cell walls of *Prochloron*. *Arch. Mikrobiol.* **210,** 191–4.

MORIARTY D.J.W. & MORIARTY C.M. (1973) The assimilation of carbon from phytoplankton by two herbivorous fishes: *Tilapia nilotica* and *Haplochromis nigripinnis*. *J. Zool.* **171,** 41–55.

MORNIN L. & FRANCIS D. (1967) The fine structure of *Nematodinium armatum*, a naked dinoflagellate. *J. Microscop.* **6,** 759–72.

MORRIL L.C. & LOEBLICH A.R. III (1983) Formation and release of body scales in the dinoflagellate genus *Heterocapsa*. *J. Mar. Biol. Ass. U.K.* **63,** 905–13.

MORRIS I. (ed.) (1980) *The Physiological Ecology of Phytoplankton*. Blackwell Scientific Publications, Oxford.

MOSS B. (1977) Adaptations of epipelic and episammic freshwater algae. *Oecologia* **28,** 103–8.

MOSS B.L. (1964) Growth and regeneration of *Fucus vesiculosus* in culture. *Brit. Phycol. Bull.* **2,** 377–80.

MOSS B.L. (1965) Apical dominance in *Fucus vesiculosus*. *New Phytol.* **64,** 387–92.

MOSS B.L. (1966) Polarity and apical dominance in *Fucus vesiculosus*. *Brit. Phycol. Bull.* **3,** 209–12.

MOSS B.L. (1970) Meristems and growth control in *Ascophyllum nodosum* (L.) Le Jolis. *New Phytol.* **69,** 253–60.

MOSS B.L. (1982) The control of epiphytes on *Halidrys siliquosa* (L.) Lyngb. (Phaeophyta, Cystoseiraceae). *Phycologia* **21,** 185–91.

MOSS B.L., TOVEY D. & COURT P. (1981) Kelps as fouling organisms on North Sea oil platforms. *Bot. Mar.* **24,** 207.

MOSS J.R. (1978) Essential considerations for establishing seaweed extraction factories. In *The Marine Plant Biomass of the Pacific North-west Coast* (ed. Kraus R.W.), pp. 301–14. Oregon state University Press, Corvallis.

MUELLER-HAEKEL A. (1973) Experimente zum Bewegungsverhalten von einzelligen Fliesswasseralgen. *Hydrobiology* **41,** 221–46.

MUKAI L.S., CRAIGIE J.S. & BROWN R.G. (1981) Chemical composition and structure of the cell walls of the *Conchocelis* and thallus phases of *Porphyra tenera* (Rhodophyceae). *J. Phycol.* **17,** 192–8.

MÜLLER D.G. (1962) Über jahres- und lunarperiodische Erscheinungen bei einigen Braunalgen. *Bot. Mar.* **4,** 140–55.

MÜLLER D.G. (1967) Generationswechsel, Kernphasenwechsel und Sexualität der Braunalge *Ectocarpus* im Kulturversuch. *Planta* **75,** 39–54.

MÜLLER D.G. (1972) Life cycle of the brown alga *Ectocarpus fasciculatus* var. *refractus* (Kütz) Ardis. (Phaeophyceae, Ectocarpales) in culture. *Phycologia* **11,** 11–13.

MÜLLER, D.G. (1975) Experimental evidence against sexual fusion of spores from unilocular sporangia of *Ectocarpus siliculosus* (Phaeophyta). *Brit. Phycol. J.* **10,** 315–20.

MÜLLER D.G. (1981) Sexuality and sex attraction. In *The Biology of Seaweeds* (eds Lobban C.S. & Wynne M.J.), pp. 661–74. Blackwell Scientific Publications, Oxford.

MÜLLER D.G. & FALK H. (1973) Flagellar structure of the gametes of *Ectocarpus siliculosus* (Phaeophyta) as revealed by negative staining. *Arch. Mikrobiol.* **91,** 313–22.

MÜLLER D.G. & MEEL H. (1982) Culture studies on the life history of *Arthrocladia villosa* (Desmarestiales, Phaeophyceae). *Brit. Phycol. J.* **4,** 419–25.

MUMFORD T.F. (1979) Field and laboratory experiments with *Iridaea cordata* (Florideophyceae) grown on nylon netting. *Proc. Int. Seaweed Symp.* **9,** 515–23.

MUNDAY B.L. & PEEL B.F. (1983) Severe ulcerative dermatitis in Platypus (*Ornithorhynchus anatinus*). *J. Wildlife Diseases* **19,** 363–5.

MURRAY S., DIXON P.S. & SCOTT J.L. (1972) The life history of *Porphyropsis coccininea* var. *dawsonii* in culture. *Brit. Phycol. J.* **7,** 323–33.

MUSCATINE L. (1973) Nutrition of corals. In *Biology and Geology of Coral Reefs. Vol. II, Biology I* (eds Jones O.A. & Endean R.), pp. 77–115. Academic Press, New York.

MUSCATINE L. (1980) Productivity of zooxanthellae. In *Primary Productivity in the Sea* (ed Falkowski P.G.), pp. 381–402, Plenum Press, New York.

MUSCATINE L. & McAULEY P.J. (1982) Transmission of symbiotic algae to eggs of green hydra. *Cytobios* **33,** 111–23.

MUSCATINE L. & PORTER J.W. (1977) Reef corals: mutualistic symbioses adapted to nutrient-poor environments. *BioScience* **27,** 454–60.

MYNDERSE J.S., MOORE R.E., KASHIWAGI M. & NORTON T.R. (1977) Antileukemia activity in the oscillatoriaceae, isolation of debromoaplysiatoxin from *Lyngbya*. *Science* **196,** 538–40.

NAKAHARA H. & NAKAMURA Y. (1973) Parthenogenesis, apogamy and apospory in *Alaria crassifolia* (Laminariales). *Mar. Biol.* **18,** 327–32.

NAKAJIMA M. (1965) Studies on the source of shellfish poison in Lake Hamana. III. Poisonous effects of shellfishes feeding on *Prorocentrum* sp. *Bull. Jap. Soc. Sci. Fish.* **31,** 282–5.

NAKAMURA K., OGAWA T. & SHIBATA K. (1976) Chlorophyll and peptide compositions in the two photosystems of marine green algae. *Biochim. Biophys. Acta* **423,** 227–36.

NAKAMURA Y. & TATEWAKI M. (1975) The life history of some species of Scytosiphonales. *Sci. Pap. Inst. Algol. Res. Hokkaido Univ.* **6,** 57–93.

NAYLOR M. (1954) A note on *Xiphophora chondrophylla* var. *maxima* J.Ag. *New Phytol.* **53,** 155–9.

NEEDLER A.B. (1949) Paralytic shell fish poisoning and *Gonyaulax tamarensis*. *J. Fish. Res. Bd Can.* **7,** 490–504.

NEILSON A.H. & LARSSON T. (1980) The utilization of organic nitrogen for growth of algae: physiological aspects. *Physiol. Plant.* **48,** 542–53.

NEISH A.C. & FOX C.H. (1971) Greenhouse experiments on the vegetative propagation of *Chondrus crispus* Irish Moss. *Natl Res. Coun. Can. Atl. Reg. Lab. Tech. Rep.* **14,** 25p.

NEISH A.C., SHACKLOCK P.F., FOX C.H. & SIMPSON F.J. (1977) The cultivation of *Chondrus crispus*. Factors affecting growth under greenhouse conditions. *Can. J. Bot.* **55,** 2263–71.

NELSON D.M., GOERING J.J., KILHAM S.S. & GUILLARD R.R.L. (1976) Kinetics of silicic acid uptake and rates of silica dissolution in the marine diatom *Thalassiosira pseudonana*. *J. Phycol.* **12,** 246–52.

NELSON-SMITH A. (1973) *Oil Pollution and Marine Ecology*. Plenum Press, New York.

NEWROTH P. (1971) Studies on life histories in the Phyllophoraceae. I. *Phyllophora truncata*. (Rhodophyceae, Gigartinales). *Phycologia* **10,** 345–54.

NEWROTH P. (1972) Studies on life histories in the Phyllophoraceae II. *Phyllophora pseudoceranoides* and notes on *P. crispa* and *P. heridia* (Rhodophyta, Gigartinales). *Phycologia* **11,** 99–107.

NEWTON J.W. & HERMAN A.I. (1979) Isolation of cyanobacteria from the aquatic fern *Azolla*. *Arch. Mikrobiol.* **120,** 161–5.

NICHOLS K.H. (1981) *Chrysococcus furcatus* (Dolg.) comb. nov.: a new name for *Chrysastrella furcata* (Dolg.) Defl. based on the discovery of the vegetative stage. *Phycologia* **20,** 16–21.

NICHOLS H.W. (1965) Culture and development of *Hildenbrandia rivularis* from Denmark and North America. *Am. J. Bot.* **51,** 180–8.

NICHOLS H.W. & LISSANT E.K. (1967) Developmental studies of *Erythrocladia* Rosenvinge in culture. *J. Phycol.* **3,** 6–18.

NICHOLS J.M. & ADAMS D.G. (1982) Akinetes. In *The Biology of Cyanobacteria* (eds Whitton B.A. & Carr N.G.), pp. 389–412. Blackwell Scientific Publications, Oxford.

NICHOLSON N.L. & CIMBERG R.L. (1971) The Santa Barbara oil spills of 1969: a post spill survey of the rocky intertidal. In *Biological and Oceanographic Survey of the Santa Barbara Channel Oil Spills of 1969–1970*, Vol. 1, pp. 325–400. University of Southern California.

NIELL J.H. & JACKSON M.B. (1982) Monitoring *Cladophora* growth conditions and the effect of phosphorus additions at a shoreline site in North-eastern Lake Erie. *J. Great Lakes Res.* **8,** 30–4.

NIENHUIS P.H. (1974) Variability in the life cycle of *Rhizoclonium riparium* (Roth) Harv. (Chlorophyceae: Cladophorales) under Dutch estuarine conditions. *Hydrobiol. Bull.* **8,** 172–8.

NIZAMUDDIN M. (1962) Classification and distribution of the Fucales. *Bot. Mar.* **4,** 191–203.

NODA H. & HORIGUCHI Y. (1971) The significance of zinc as a nutrient for the red alga *Porphyra tenera*. *Proc. Int. Seaweed Symp.* **7,** 368–72.

NONOMURA A.M. & WEST J.A. (1981) Host specificity of *Janczewskia* (Ceramiales, Rhodophyta). *Phycologia* **20,** 251–8.

NORRIS J.N. & FENICAL W. (1982) Chemical defense in tropical marine algae. In *The Atlantic Barrier Reef Ecosystem at Carrie Bow Cay, Belize, I. Structure and Communities* (eds Rutzler K. & Macintyre I.S.), pp. 417–32. Smithsonian Institute Press, Washington.

NORRIS J.N. & KUGRENS P. (1982) Marine Rhodophyceae: introduction and bibliography. In *Selected Papers in Phycology* II (eds Rosowski J.R. & Parker B.C.), pp. 663–70. Phycological Society of America, Allen Press, Lawrence, Kansas.

NORRIS R.E. (1977) Flagellate cells in the life-history of *Stichogloea* (Chrysophyceae). *Phycologia* **16,** 75–8.

NORRIS R.E. (1980) Prasinophytes. In *Phytoflagellates* (ed. Cox E.R.), pp. 85–146. Elsevier/North-Holland, New York.

NORRIS R.E. (1982a) Prymnesiophyceae. In *Synopsis and Classification of Living Organisms*, Vol. 1 (ed. Parker S.P.), pp. 86–91. McGraw-Hill, New York.

NORRIS R.E. (1982b) Prasinophyceae. In *Synopsis and Classification of Living Organisms*, Vol. 1 (ed Parker S.P.), pp. 162–4. McGraw-Hill, New York.

NORRIS R.E. (1982c) Prasinophyceae: introduction and bibliography. In *Selected Papers in Phycology* II (eds Rosowski J.R. & Parker B.C.), pp. 740–6. Phycological Society of America, Allen Press, Lawrence, Kansas.

NORRIS R.E., HORI T. & CHIHARA M. (1980) Revision of the genus *Tetraselmis* (Class Prasinophyceae). *Bot. Mag. Tokyo* **93,** 317–39.

NORTH W.J. (1964) Ecology of the rocky near shore environment in southern California and possible influence of discharged wastes. *Adv. Water Poll. Res.* **3,** 247–74.

NORTH W.J. (1967) Tampico: a study of destruction and restoration. *Sea Frontiers* **13,** 212–17.

NORTH W.J. (ed.) (1971) The biology of giant kelp beds in California. *Nova Hedwiga* **32,** 1–600.

NORTON T.A. & MATHIESON A.C. (1983) The biology of unattached seaweeds. *Prog. Phycol. Res.* **2,** 333–86.

NORTON T.A., MATHIESON A.C. & NEUSHUL M. (1981) Morphology and environment. In *The Biology of Seaweeds* (eds Lobban C.S. & Wynne M.J.), pp. 421–51. Blackwell Scientific Publications, Oxford.

NOZAKI H. (1981) The life history of Japanese *Pandorina unicocca* (Chlorophyta, Volvocales). *J. Jap. Bot.* **56,** 65–72.

NYGREN S. (1975) Life history of some Phaeophyceae from Sweden. *Bot. Mar.* **18,** 131–41.

OAKLEY B.R. & SANTORE U.J. (1982) Cryptophyceae: introduction and bibliography. In *Selected Papers in Phycology II* (eds Rosowski J.R. & Parker B.C.), pp. 682–6. Phycological Society of America, Allen Press, Lawrence, Kansas.

O'BRIEN P.Y. & DIXON P.S. (1976) The effects of oils and oil components on algae: a review. *Brit. Phycol. J.* **11,** 115–42.

OGDEN J.C. (1976) Some aspects of herbivore–plant relationships on Caribbean reefs and seagrass beds. *Aquatic Botany* **2,** 103–6.

O'HEOCHA C. (1971) Pigments of the red algae. *Oceanogr. Mar. Biol. Ann. Rev.* **9,** 61–82.

OKAZAKI M., IKAWA T., FURUYA K., NISIZAWA K. & MIWA T. (1970) Studies on calcium carbonate deposition of a calcareous red alga *Serraticardia maxina*. *Bot. Mag. Tokyo* **83,** 193–201.

O'KELLY C.J. & FLOYD G.L. (1983) The flagellar apparatus of *Entocladia viridis* motile cells, and the taxonomic position of the resurrected family Ulvellaceae (Ulvales, Chlorophyta). *J. Phycol.* **19,** 153–64.

O'KELLY, C.J. & FLOYD, G.L. (1984a) The absolute configuration of the flagellar apparatus in zoospores from two species of Laminariales (Phaeophyceae). *Protoplasma* **123,** 18–25.

O'KELLY C.J. & FLOYD G.L. (1984b) Flagellar apparatus absolute orientations and the phylogeny of the green algae. *Biosystems* **16,** 227–51.

O'KELLY C.J. & YARISH C. (1981) On the circumscription of the genus *Entocladia* Reinke. *Phycologia* **20,** 32–45.

O'KELLY J.C. (1974) Inorganic nutrients. In *Algal Physiology and Biochemistry* (ed. Stewart W.D.P.), pp. 610–34. Blackwell Scientific Publications, Oxford.

OKUDA K. & TATEWAKI M. (1982) A circadian rhythm of gametangium formation in *Pseudobryopsis* sp. (Chlorophyta, Codiales). *Jap. Soc. Phycol.* **30,** 171–80.

OLTMANNS F. (1904) *Morphologie und Biologie der Algen*, Vols I & II. Jena.

OLTMANNS F. (1922) *Morphologie und Biologie der Algen*, Vols I & II, 2nd Edition. Jena.

OTT D.W. (1982) Tribophyceae (Xanthophyceae): introduction and bibliography. In *Selected Papers in Phycology* II (eds Rosowski J.R. & Parker B.C.), pp. 723–7. Phycological Society of America, Allen Press, Lawrence, Kansas.

OTT D.W. & BROWN R.M. JR (1972) Light and electon microscopical observations on mitosis in *Vaucheria litorea* Hofman ex C. Agardh. *Brit. Phycol. J.* **7,** 361–74.

OTT D.W. & BROWN R.M. Jr (1974a) Developmental cytology of the genus *Vaucheria*. II. Sporogenesis in *V. fontinalis* (L.) Christensen. *Brit. Phycol. J.* **9,** 333–51.

OTT D.W. & BROWN R.M. (1974b) Developmental cytology of the genus *Vaucheria*. I. Organization of the vegetative filament. *Brit. Phycol. J.* **9,** 111–26.

OTT D.W. & HOMMERSAND M.H. (1974) Vaucheriae of North Carolina. I. Marine and brackish water species. *J. Phycol.* **10,** 373–85.

OTT F.D. & SOMMERFELD M.R. (1982) Freshwater Rhodophyceae: introduction and bibliography. In *Selected Papers in Phycology II* (eds Rosowski J.R. & Parker B.C.), pp. 671–81. Phycological Society of America, Allen Press, Lawrence, Kansas.

PAASCHE E. (1980) Silicon. In *The Physiological Ecology of Phytoplankton* (ed. Morris I.), pp. 259–84. Blackwell Scientific Publications, Oxford.

PAASCHE E. & KLAVENESS D. (1970) A physiological comparison of coccolith forming and naked cells of *Coccolithus huxleyi. Arch. Mikrobiol.* **73,** 143–52.

PADAN E. & COHEN Y. (1982) Anoxogenic photosynthesis. In *The Biology of Cyanobacteria* (eds Carr N.G. & Whitton B.A.), pp. 215–36. Blackwell Scientific Publications, Oxford.

PAERL H.W., TUCKER J. & BLAND P.T. (1983) Carotenoid enhancement and its role in maintaining blue green algal (*Microcystis aeruginosa*) surface blooms. *Limnol. Oceanogr.* **28,** 847–57.

PAINE R.T. (1969) A note on trophic complexity and community stability. *Am. Nat.* **103,** 91–3.

PAINE R.T. (1974) Intertidal community structure: experimental studies on the relationship between a dominant competitor and its principal predator. *Oecologia* **15,** 93–120.

PAINE R.T. & VADAS R.L. (1969) The effects of grazing by sea urchins *Strongylocentrotus* spp. on benthic algal populations. *Limnol. Oceanogr.* **16,** 86–98.

PALMER C.M. (1977) *Algae and Water Pollution: An Illustrated Manual on the Identification, Significance and Control of Algae in Water Supplies and Polluted Water.* EPA-600/9-77-036. US Environmental Protection Agency, Cincinnati, U.S.A.

PALMER J.D. & ROUND F.E. (1965) Persistent, vertical migration rhythms in benthic microflora. I. The effect of light and temperature on rhythmic behaviour of *Euglena obtusa. J. Mar. Biol. Ass. U.K.* **45,** 567–82.

PALMER J.D. & ROUND F.E. (1967) Persistent, vertical migration rhythms in benthic microflora. VI. The tidal and diurnal nature of the rhythm in the diatom *Hantzschia virgata. Biol. Bull.* **132,** 44–55.

PALMISANO A.C., KOTTMEIER S.T., MOE R.L. & SULLIVAN C.W. (1985) Sea ice microbial communities. IV. The effect of light perturbation on microalgae at the ice seawater interface in McMurdo Sound, Antarctica. *Mar. Ecol. Prog. Ser.* **21,** 37–45.

PALMISANO A.C. & SULLIVAN C.W. (1983) Sea ice microbial communities (SIMCOs). I. Distribution, abundance and primary production of ice microalgae in McMurdo Sound in 1980. *Polar Biol.* **2,** 171–7.

PARDY R.L. (1981) Cell size distribution of green symbionts from *Hydra viridis. Cytobios* **32,** 71–7.

PARDY R.L., LEWIN R.A. & LEE K. (1983) The *Prochloron* symbiosis. In *Algal Symbiosis: A Continuum of Interaction Strategies* (ed. L.J. Goff), pp. 91–6. Cambridge University Press, Cambridge.

PARKE M. & ADAMS I. (1960) The motile (*Crystallolithus hyalinus* Gaarder and Markali) and non motile phases in the life history of *Coccolithus pelagicus* (Wallich) Schiller. *J. Mar. Biol. Ass. U.K.* **39,** 263–74.

PARKE M. & ADAMS I. (1961) The *Pyramimonas*-like motile stage of *Halosphaera viridis* Schmitz. *Bull. Res. Coun. Israel* **10D,** 94–100.

PARKE M. & GREEN J.C. (1976) Chlorophyta, Prasinophyceae. In Parke M. & Dixon P.S. Check-list of British marine algae — third revision, 564–6. *J. Mar. Biol. Ass. U.K.* **56,** 527–94.

PARKER B.C. (1969) The occurrence of silica in brown and green algae. *Can. J. Bot.* **47,** 537–40.

PARKER B.C. & DAWSON E.Y. (1965) Non-calcareous marine algae from California Miocene deposits. *Nova Hedwiga* **10,** 273–95.

PARKER B.C., PRESTON R.D. & FOGG G.E. (1963) Studies on the structure and chemical composition of the cell walls of Vaucheriaceae and Saprolegniaceae., *Proc. Roy. Soc. Lond. B* **158,** 435–45.

PARKER S.P. (ed.) (1982) *Synopsis and Classification of living Organisms*, Vols 1 & 2. McGraw-Hill, New York.

PARSONS T.R. & TAKAHASHI M. (1973) Environmental control of phytoplankton cell size. *Limnol. Oceanogr.* **18,** 511–15.

PARSONS T.R., MAITA Y. & LALLI C.M. (1984) *A Manual of Chemical and Biological Methods for Seawater Analysis.* Pergamon Press, Oxford.

PASCHER A. (1917) Von der merkwürdigen Bewegungsweise einiger Flagellaten. *Biol. Centralbl.* **37,** 421–9.

PATEL R.J. (1971) Cytotaxonomical studies of British species of *Chaetomorpha* I. *Chaetomorpha linum* Kütz. and *Chaetomorpha aerea* Kütz. *Phykos* **10,** 127–36.

PATEL R.J. (1972) Cytotaxonomical studies of British species of *Chaetomorpha* II. *Chaetomorpha melagonium* Kütz. *Phykos* **11,** 17–22.

PATRIQUIN D.G. & McCLUNG C.R. (1978) Nitrogen accretion, and the nature and possible significance of N_2 fixation (acetylene reduction) in a Nova Scotian *Spartina alterniflora* stand. *Mar. Biol.* **16,** 49–58.

PATTERSON G.M.L. & WITHERS N.W. (1982) Laboratory cultivation of *Prochloron*, a trypophan auxotroph. *Science* **217,** 1934–5.

PATWARY M.U. & VAN DER MEER J.P. (1983) Genetics of *Gracilaria tikvahiae* (Rhodophyceae). IX. Some properties of agar extracted from morphological mutants. *Bot. Mar.* **26,** 295–9.

PEABODY A.J. (1980) Diatoms and drowning — a review. *Med. Sci. Law* **20,** 254–64.

PEARLMUTTER N.L. & LEMBI C.A. (1978) Localization of chitin in algal and fungal cell walls by light and electron microscopy. *J. Histochem. Cytochem.* **26,** 782–91.

PEARSE V.B. (1972) Radioisotope study of calcification in the articulated coralline alga *Bossiella orbigniana. J. Phycol.* **8,** 88–97.

PEARSON B.R. & NORRIS R.E. (1975) Fine structure of cell division in *Pyramimonas parkeae* Norris and Pearson (Chlorophyta, Prasinophyceae). *J. Phycol.* **11,** 113–24.

PEDERSEN M. (1968) *Ectocarpus fasciculatus*: marine brown alga requiring kinetin. *Nature* **218,** 776.

PEDERSEN M. (1969) The demand for iodine and bromine of three marine brown algae grown in bacteria-free cultures. *Physiol. Plant.* **22,** 680–5.

PEDERSEN M. (1973) Identification of a cytokinin, 6-(3-methyl-2-butenylamino) purine, in seawater and the effects of cytokinins on brown algae. *Physiol. Plant.* **28,** 101–5.

PEDERSEN P.M. (1981) Phaeophyta: Life histories. In *The Biology of Seaweeds* (eds Lobban C.S. & Wynne M.J.), pp. 194–217. Blackwell Scientific Publications, Oxford.

PEDERSEN P.M. (1984) Studies on primitive brown algae (Fucophyceae). *Opera Bot.* **74,** 1–76.

PELROY R.A. & BASSHAM J.A. (1972) Photosynthetic and dark carbon metabolism in unicellular blue-green algae. *Arch. Mikrobiol.* **86,** 25–38.

PERCIVAL E. (1979) The polysaccharides of green, red, and brown seaweeds: their basic structure, biosynthesis and function. *Brit. Phycol. J.* **14,** 103–17.

PERCIVAL E. & McDOWELL R.H. (1967) *Chemistry and Enzymology of Marine Algal Polysaccharides.* Academic Press, New York.

PERKINS E.J. (1960) The diurnal rhythm of the littoral diatoms of the River Eden estuary. *J. Ecol.* **48,** 103–8.

PERRONE G. & FELICINI G.P. (1972) Sur les bourgeons adventifs de *Petroglossum nicaeese* (Duby) Schotter (Rhodophycées, Gigartinales) en culture. *Phycologia* **11,** 87–95.

PERRY M.J. (1972) Alkaline phosphatase activity in subtropical central North Pacific waters using a sensitive fluorometric method. *Mar. Biol.* **15,** 113–19.

PERRY M.J. (1976) Phosphate utilization by an oceanic diatom in phosphorus- limited chemostat culture and in oligotrophic waters of the central North Pacific. *Limnol. Oceanogr.* **21,** 88–107.

PERRY M.J., TOLBERT M.C. & ALBERTE R.S. (1981) Photoadaptation in marine phytoplankton: response of the photosynthetic unit. *Mar. Biol.* **62,** 91–101.

PETERS A.F. (1984) Observations on the life history of *Papenfussiella callitricha* (Phaeophyceae, Chordariales) in culture. *J. Phycol.* **20,** 409–14.

PETERS G.A. (1978) Blue-green algae and algal associations. *BioScience* **28,** 580–5.

PETERS G.A. & CALVERT H.E. (1983) The *Azolla–Anabaena azollae* symbiosis. In *Algal symbiosis: A Continuum of Interaction Strategies* (ed. Goff L.J.), pp. 109–46. Cambridge University Press, Cambridge.

PFIESTER L.A. (1977) Sexual reproduction of *Peridinium gatunense* (Dinophyceae). *J. Phycol.* **13,** 92–5.

PFIESTER L.A. & SKVARLA J.J. (1980) Comparative ultrastructure of vegetative and sexual thecae of *Peridinium limbatum* and *Peridinium cinctum* (Dinophyceae). *Am. J. Bot.* **67,** 955–8.

PHILLIPS D. & SCOTT J. (1981) Ultrastructure of cell division and reproductive differentiation of male plants in the Florideophycidae (Rhodophyta). Mitosis in *Dasya baillouviana. Protoplasma* **106,** 329–41.

PIA J. (1920) Die *Siphoneae verticillatae* vom Karbon bis zur Kreide. *Abh. Zool. Bot. Ges. Wien* **11**(2), 1–263.

PICCINNI E., ALBERGONI V. & COPPELLOTTI O. (1975) ATP-ase activity in flagella from *Euglena gracilis.* Localization of the enzyme and effects of detergents. *J. Protozool.* **22,** 331–5.

PICKETT-HEAPS J.D. (1975) *Green Algae.* Sinauer Association, Sunderland, Mass.

PICKETT-HEAPS J.D. (1976) Cell division in eukaryotic algae. *BioScience* **26,** 445–50.

PICKETT-HEAPS J.D. (1979) Electron microscopy and the phylogeny of green algae and land plants. *Am. Zool.* **19,** 545–54.

PICKETT-HEAPS J. & KOWALSKI S.E. (1981) Valve morphogenesis and the microtubule center of the diatom *Hantzschia amphioxys. Eur. J. Cell Biol.* **25,** 150–70.

PICKETT-HEAPS J. & MARCHANT H.J. (1972) The phylogeny of the green algae: a new proposal. *Cytobios* **6,** 255–64.

PICKETT-HEAPS J.D. & STAEHELIN L.A. (1975) The ultrastructure of *Scenedesmus* (Chlorophyceae). II. Cell division and colony formation. *J. Phycol.* **11,** 186–202.

PICKETT-HEAPS J.D. & WEIK K. (1977) Cell division in *Euglena* and *Phacus.* I. Mitosis. In *Mechanisms of Control of Cell Division* (eds Post L. & Gifford E.M.), pp. 308–36. Dowden, Hutchinson & Ross, Stroudsburg, Pennsylvania.

PIELOU E.C. (1977) The latitudinal spans of seaweed species and their patterns of overlap. *J. Biogeogr.* **4,** 299–311.

PIELOU E.C. (1978) Latitudinal overlap of seaweed species: evidence of quasisympatric speciation. *J. Biogeogr.* **5,** 227–38.

PIENAAR R.N. (1980) Chrysophytes. In *Phytoflagellates* (ed. Cox E.R.), pp. 213–42. Elsevier/North-Holland, New York.

PINTNER I.J. & ALTMYER V.L. (1979) Vitamin B_{12} binder and other algal inhibitors. *J. Phycol.* **15,** 391–8.

PISA M., SANBERG P.R., CORCORAN M.E. & FIBIGER H.C. (1980) Spontaneously recurrent seizures after intracerebral injections of kainic acid in rat: a possible model of human temporal lobe epilepsy. *Brain Res.* **200,** 481–7.

PLATT T. (ed.) (1981) Physiological bases of phytoplankton ecology. *Can. Bull. Fish. Aquat. Sci.* **210,** 1–346.

PLATT T. & DENHAM K. (1980) Patchiness in phytoplankton distribution. In *The Physiological Ecology of Phytoplankton* (ed. Morris I.), pp. 413–31. Blackwell Scientific Publications, Oxford.

PLUZECK K. (1981) Applications of agarose in biomedical techniques in relation to chemical and physical properties. *Proc. Int. Seaweed Symp.* **10,** 711–18.

POLANSHEK A.R. & WEST J.A. (1977) Culture and hybridization studies on *Gigartina papillata. J. Phycol.* **13,** 141–9.

PORCELLA R.A. & WALNE P.L. (1980) Microarchitecture and envelope development in *Dysmorphococcus globosus* (Phacotaceae, Chlorophyceae). *J. Phycol.* **16,** 280–90.

PORTER K.G. (1973) Selective grazing and differential digestion of algae by zooplankton. *Nature* **244,** 179–80.

PORTER K.G. (1976) Enhancement of algal growth and productivity by grazing zooplankton. *Science* **192,** 1332–4.

PORTERFIELD W.M. (1922) References to the algae in the Chinese classics. *Bull. Torrey Bot. Club.* **49,** 297–300.

POTTS M. (1980) Blue-green algae (Cyanophyta) in marine coastal environments of the Sinai Peninsula; distribution, zonation, stratification and taxonomic diversity. *Phycologia* **19,** 60–73.

POVOLNY M. (1981) The effect of steeping of peat–cellulose flower pots (Jiffypots) in extracts of seaweeds on the quality of tomato seedlings. *Proc. Int. Seaweed Symp.* **8,** 730–3.

PRAKASH A. & TAYLOR F.J.R. (1966) A red water bloom of *Gonyaulax acatenella* in the strait of Georgia and its relation to paralytic shell fish toxicity. *J. Fish. Res. Bd Can.* **23,** 1256–70.

PRASK J.A. & PLOCKE D.J. (1971) A role for zinc in the structural integrity of the cytoplasmic ribosomes of *Euglena gracilis. Plant. Physiol. Lancaster* **37,** 428–33.

PRESCOTT G.W. (1951) History of phycology. In *Manual of Phycology* (ed. Smith G.M.), pp. 1–11. The Ronald Press Company, New York.

PREZLIN B.B. (1981) Light reactions in photosynthesis. *Can. Bull. Fish. Aquat. Sci.* **210,** 1–43.

PREZLIN B.B. & ALBERTE R.S. (1978) Photosynthetic characteristics and organization of chlorophyll in marine dinoflagellates. *Proc. Natl Acad. Sci.* **75,** 1801–4.

PREZLIN B.B. & SWEENEY B.M. (1978) Photoadaptation of photosynthesis in *Gonyaulax polyhedra. Mar. Biol.* **48,** 27–35.

PRICE I.A. (1972) Zygote development in *Caulerpa* (Chlorophyta, Caulerpales). *Phycologia* **11,** 217–18.

PRICE I.R. & DUCKER S.C. (1966) The life history of the brown alga *Splachnidium rugosum. Phycologia* **5,** 261–73.

PRINGSHEIM E.G. (1968) Kleine Mitteilungen über Flagellaten und Algen. XV. Zur Kenntnis der Gattung *Porphyridium. Arch. Mikrobiol.* **61,** 169–80.

PROCTOR V.W. (1971) Taxonomic significance of monoecism and dioecism in the genus *Chara. Phycologia* **10,** 299–307.

PROVASOLI L. & CARLUCCI A.F. (1974) Vitamins and growth regulators. In *Algal Physiology and Biochemistry* (ed. Stewart W.D.P.), pp. 741–87. Blackwell Scientific Publications, Oxford.

PROVASOLI L., McLAUGHLIN J.J.A. & DROOP M.R. (1957) The development of artificial media for marine algae. *Arch. Mikrobiol.* **25,** 392–428.

PROVASOLI L., PINTNER I.J. & SAMPATHKUMAR S. (1977) Morphogenetic substances for *Monostroma oxyspermum* from marine bacteria. *J. Phycol.* **13** (suppl.), 56.

PUESCHEL, C.M. (1977) A freeze–etch study of the ultrastructure of red algal pit plugs. *Protoplasma* **91,** 15–30.

PUESCHEL C.M. (1980) A re-appraisal of the cytochemical properties of rhodophycean pit plugs. *Phycologia* **19,** 210–17.

QUATRANO R.S. & STEVENS P.T. (1976) Cell wall assembly in *Fucus* zygotes. *Plant Physiol.* **58,** 224–31.

RAGAN M.A. (1981) Chemical constituents of seaweeds. In *The Biology of Seaweeds* (eds Lobban C.S. & Wynne M.J.), pp. 589–626. Blackwell Scientific Publications, Oxford.

RAGAN M.A. & CHAPMAN D.J. (1978) *A Biochemical Phylogeny of the Protists*. Academic Press, New York.

RAI L.C., GAUR J.P. & KUMAR H.D. (1981) Phycology and heavy metal pollution. *Biol. Rev.* **56,** 99–151.

RAMUS J. (1969) The developmental sequence of the marine red alga *Pseudogloiophloea* in culture. *Univ. Calif. Publ. Bot.* **52,** 1–42.

RAMUS J. (1971) Properties of septal plugs from the red alga *Griffithsia pacifica*. *Phycologia* **10,** 99–103.

RAMUS J. (1972) The production of extracellular polysaccharide by the unicellular red alga *Porphyridium aerugineum*. *J. Phycol.* **8,** 97–111.

RAMUS J. (1981) The capture and transduction of light energy. In *The Biology of Seaweeds* (eds Lobban C.S. & Wynne M.J.), pp. 458–92. Blackwell Scientific Publications, Oxford.

RAMUS J., BEALE S.I., MAUZERALL D. & HOWARD K.L. (1976) Changes in photosynthetic pigment concentration in seaweeds as a function of water depth. *Mar. Biol.* **37,** 223–9.

RAMUS J., LEMONS F. & ZIMMERMAN C. (1977) Adaptation of light-harvesting pigments to downwelling light and consequent photosynthetic performance of the eulittoral rockweeds *Ascophyllum nodosum* and *Fucus vesiculosus*. *Mar. Biol.* **42,** 293–303.

RANWELL D.S. (1972) *Ecology of Salt Marshes and Sand Dunes*. Chapman & Hall, London.

RAUNKIAER C. (1934) *The Life Forms of Plants and Statistical Plant Geography*. Clarendon Press, Oxford.

RAVEN J.A. (1970) Exogenous inorganic carbon sources in plant photosynthesis. *Biol. Rev.* **45,** 167–221.

RAVEN J.A. (1974) Carbon dioxide fixation. In *Algal Physiology and Biochemistry* (ed. Stewart W.D.P.), pp. 434–55. Blackwell Scientific Publications, Oxford.

RAVEN J.A. & BEARDALL J. (1981) Respiration and photorespiration. *Can. Bull. Fish. Aquat. Sci.* **210,** 55–82.

RAWITSCHER-KUNKEL E. & MACHLIS L. (1962) The hormonal integration of sexual reproduction in *Oedogonium*. *Am. J. Bot.* **49,** 177–83.

RAWLENCE D.J. (1972) An ultrastructural study of the relationship between the rhizoids of *Polysiphonia lanosa* (L.) Tandy (Rhodophyceae) and the tissue of *Ascophyllum nodosum* (L.) Le Jolis (Phaeophyceae). *Phycologia* **11,** 279–90.

REDFIELD A.C., KETCHUM B.H. & RICHARDS F.A. (1963) The influence of organisms on the composition of seawater. In *The Sea* (ed. Hill M.N.), pp. 26–77. Wiley–Interscience, New York.

REED M.L. & GRAHAM D. (1981) Carbonic anhydrase in plants: distribution, properties and possible physiological functions. *Prog. Phytochem.* **7,** 47–96.

REED R.H. (1983) Measurement and osmotic significance of β-dimethylsulphonioproponate in marine macroalgae. *Mar. Biol. Lett.* **4,** 173–81.

REED R.H., DAVISION I.R., CHUDEK J.A. & FOSTER R. (1985) The osmotic role of mannitol in the Phaeophyta: an appraisal. *Phycologia* **24,** 35–47.

REICHARDT W., OVERBECK J. & STENBING L. (1967) Free dissolved enzymes in lake waters. *Nature* **216,** 1345–7.

REIMANN B.E.F., LEWIN J.C. & VOLCANI B.E. (1966) Studies on the biochemistry and fine structure of silica shell formation in diatoms. II. The structure of the cell wall of *Navicula pelliculosa* (Breb.) Hilse. *J. Phycol.* **2,** 74–84.

REINKE J. (1878) Entwicklungsgeschichtliche Untersuchungen über die Cutleriaceen des Golfs von Neapel. *Nova Acta K. Leop.-Carol. Deutsch. Akad. Naturforsch.* **40,** 59–96.

RENTSCHLER H.-G. (1967) Photoperiodische Induktion der Monosporenbildung bei *Porphyra tenera* Kjellm. (Rhodophyta-Bangiophyceae). *Planta* **76,** 65–74.

REYNOLDS C.S. (1984) *The Ecology of Freshwater Phytoplankton*. Cambridge University Press, Cambridge.

REYNOLDS C.S. & WALSBY A.E. (1975) Water blooms. *Biol. Rev.* **50,** 437–81.

REYNOLDS R.C. Jr (1978) Polyphenol inhibition of calcite precipitation in Lake Powel. *Limnol. Oceanogr.* **23,** 585–97.

REYSSAC J. & ROUX M. (1972) Communautés phytoplanctoniques dans les eaux de la Côte d'Ivoire. Groupes d' espèces associeés. *Mar. Biol.* **13,** 14–33.

RHEE G-Y. (1980) Continuous culture in phytoplankton ecology. *Advances Aquatic Microbiol.* **2,** 151–203.

RICE E.L. & CHAPMAN A.R.O. (1982) Net productivity of two cohorts of *Chordaria flagelliformis* (Phaeophyta) in Nova Scotia, Canada. *Mar. Biol.* **71,** 107–11.

RICHARDSON J.L. (1968) Diatoms and lake typology in East and Cental Africa. *Int. Rev. Ges. Hydrobiol. Hydrogr.* **53,** 299–338.

RICHARDSON N. (1970) Studies on the photobiology of *Bangia fuscopurpurea*. *J. Phycol.* **6,** 215–19.

RICHARDSON N. & DIXON P.S. (1968) Life history of *Bangia fuscopurpurea* (Dillw.) Lynb. in culture. *Nature* **218,** 496–7.

RICHARDSON N. & DIXON P. (1969) The Conchocelis phase of *Smithora naiadum* (Anders.) Hollenb. *Brit. Phycol. J.* **7,** 49–51.

RICHERSON P., ARMSTRONG R. & GOLDMAN C.R. (1970) Contempoeraneous disequilibrium, a new hypothesis to explain the 'paradox of the plankton'. *Proc. Natl Acad. Sci.* **67,** 1710–14.

RIETEMA H. (1970) Life histories of *Bryopsis plumosa,* Caulerpales, Chlorophyceae from European coasts. *Acta Bot. Neer.* **19,** 859–66.

RIETEMA H. (1971) Life history studies in the genus *Bryopsis*, Chlorophyceae. Part 4, life histories in *Bryopsis hypnoides*, from different parts of the European coasts. *Acta Bot. Neer.* **20,** 291–8.

RIETEMA H. (1972) A morphological, developmental and caryological study on the life history of *Bryopsis halymeniae* (Chlorophyceae). *Neth. J. Sea Res.* **5,** 445–57.

RILEY G.A. (1967) The plankton of estuaries. In *Estuaries* (ed. Lauff G.H.), pp. 316–26. A.A.A.S. Publ. 83, Washington, D.C.

RINGO D.L. (1967) Flagellar motion and fine structure of the flagellar apparatus in *Chlamydomonas. J. Cell Biol.* **33,** 543–71.

RIPPKA R., DERUELLES J., WATERBURY J.B., HERDMAN, M. & STANIER R.Y. (1979) Generic assignments, strain histories and properties of pure cultures of Cyanobacteria. *J. Gen. Microbiol.* **111,** 1–61.

ROBENEK H. & MELKONIAN M. (1983) Structural specialization of the paraflagellar body membranes of *Euglena. Protoplasma* **117,** 154–7.

ROBERTS K., GURNEY-SMITH M. & HILLS G.J. (1972) Structure, composition and morphogenesis of the cell wall of *Chlamydomonas reinhardii.* I. Ultrastructure and preliminary chemical analyses. *J. Ultrastr. Res.* **40,** 599–613.

ROBINSON D.G. & PRESTON R.D. (1971) Studies on the fine structure of *Glaucocystis nostochinearum* Itzigs. II. Membrane morphology and taxonomy. *Brit. Phycol. J.* **6,** 113–28.

ROELOFSEN P.A. (1965) Ultrastructure of the wall in growing cells and its relation to the direction of growth. *Adv. Bot. Res.* **2,** 69–149.

ROGERS D.J. & BLUNDEN G. (1980) Structural properties of the anti-B lectin from the red alga *Ptilota plumosa* (Huds.) C.Ag. *Bot. Mar.* **23**, 459–62.

ROGERS D.J., BLUNDEN G., TOPLISS J.A. & GUIRY M.D. (1980) A survey of some marine organisms for haemagglutinins. *Bot. Mar.* **23,** 569–77.

ROSENBAUM J.L. & CHILD F.M. (1967) Flagellar regeneration in protozoan flagellates. *J. Cell Biol.* **34,** 345–64.

ROSENBAUM J.L., MOULDER J.E. & RINGO D.L. (1969) Flagellar elongation and shortening in *Chlamydomonas.* The use of cycloheximide and colchicine to study the synthesis and assembly of flagellar proteins. *J. Cell Biol.* **41,** 600–19.

ROSENBERG G. & RAMUS J. (1982) Ecological growth strategies in the seaweeds *Gracilaria foliifera* (Rhodophyceae) and *Ulva* sp. (Chlorophyceae): soluble nitrogen and reserve carbohydrates. *Mar. Biol.* **66,** 251–9.

ROSENVINGE L.K. (1924) The marine algae of Denmark, contributions to their natural history. Pt III. Rhodophyceae (Ceramiales). *Det. Kgl. Vidensk. Selsk. Skrift.* **7(3),** 287–486.

ROSOWSKI J.R. (1977) Development of mucilaginous surfaces in euglenoids. II. Flagellated, creeping and palmelloid cells of *Euglena. J. Phycol.* **13,** 323–8.

ROSS R. (1982) Bacillariophyceae. In *Synopsis and Classification of Living Organisms*, Vol. 1 (ed. Parker S.P.), pp. 95–101. McGraw-Hill, New York.

RÖTTGER R., IRWAN A., SCHMALJOHANN R. & FRANZISKET L. (1980) Growth of the endosymbiont-bearing foraminifer *Amphistegina lessonii* d'Orbigny and *Heterostegina depressa* d'Orbigny (Protozoa). In *Endocytobiology*, Vol. 1 (eds Schwemmler W. & Schenk H.E.A.), pp. 155–62. Walter de Gruyter, Berlin.

ROUND F.E. (1957) The late-glacial and post-glacial diatom succession in the Kentmere valley deposit. I. Introduction, methods and flora. *New Phytol.* **58,** 98–126.

ROUND F.E. (1971) The taxonomy of the Chlorophyta. II. *Brit. Phycol. J.* **6,** 235–64.

ROUND F.E. (1973) *The Biology of the Algae*, 2nd Edition. St Martin's Press, New York.

ROUND F.E. (1981) *The Ecology of the Algae.* Cambridge University Press, Cambridge.

ROUND F.E. & CRAWFORD R.M. (1981) The lines of evolution of the Bacillariophyceae. I. Origin. *Proc. Roy. Soc. Lond. B* **211,** 237–60.

RUDDAT M. (1961) Versuche zur Beeinflussung und Auslosung der Endogen tagesrhythmik bei *Oedogonium cardiaceum* Wittr. *Z. Bot.* **49,** 23–46.

RUENESS J. (1973) Culture and field observations on growth and reproduction of *Ceramium strictum* Harv. from the Oslofjord, Norway. *Norw. J. Bot.* **20,** 61–5.

RUENESS J. (1978) A note on development and reproduction in *Gigartina stellata* (Rhodophyceae, Gigartinales) From Norway. *Brit. Phycol. J.* **13,** 87–90.

RUSSELL G. (1964) *Laminariocolax tomentosoides* on the Isle of Man. *J. Mar. Biol. Ass. U.K.* **44,** 601–12.

RUSSELL G. (1977) Vegetation on rocky shores at some north Irish sea sites. *J. Ecol.* **65,** 485–95.

RUSSELL G. (1979) Heavy receptacles in estuarine *Fucus vesiculosus* L. *Est. Cstl Mar. Sci.* **9,** 659–61.

RUSSELL G. & FIELDING A.H. (1974) The competitive properties of marine algae in culture. *J. Ecol.* **62,** 689–98.

RUSSELL G. & MORRIS O.P. (1970) Copper tolerance in the marine fouling alga *Ectocarpus siliculosus. Nature* **228,** 288–89.

RUSSELL G. & VELTKAMP C.J. (1982) Epiphytes and anti-fouling characteristics of *Himanthalia* (brown algae). *Br. Phycol. J.* **17**, 239.

RYTHER J.H. & DUNSTAN W.M. (1971) Nitrogen, phosphorus and eutrophication in the coastal marine environment. *Science* **171,** 1008–13.

RYTHER J.H., DEBOER J.A. & LAPOINTE B.E. (1979) Cultivation of seaweeds for hydrocolloids, waste treatment and energy conversion. *Proc. Int. Seaweed Symp.* **9,** 1–16.

SACHS J. (1874) *Lehrbuch der Botanik*, 4th Edition. Leipzig.

SAFFERMAN R.S. (1973) Phycoviruses. In *The Biology of the Blue Green Algae* (eds Carr N.G. & Whitton B.A.), pp. 214–37. Blackwell Scientific Publications, Oxford.

SAGA N., UCHIDA T. & SAKAI Y. (1978) Clone *Laminaria* from single isolated cell. *Jap. Soc. Sci. Fish.* **44,** 87.

SAGAN L. (MARGULIS L.) (1967) On the origin of mitosing cells. *J. Theor. Biol.* **14,** 225–75.

SAITO Y. (1982) Information on the culture of phytoplankton for aquacultural needs in Japan. In *Proceedings of the 6th U.S.–Japan Meeting on Aquaculture* (ed. Sinderman C.J.), pp. 1–6. NOAA Technical Report NMFS Circ. 442.

SAMPAIO M.J.A.M., RAI A.N., ROWELL P. & STEWART W.D.P. (1979) Occurrence, synthesis and activity of glutamine synthetase in N_2 fixing lichens. *FEMS Microbiol. Lett.* **6,** 107–10.

SANBONSUGA Y. & NEUSHUL M. (1978) Hybridization of *Macrocystis* (Phaeophyta) with other float bearing kelps. *J. Phycol.* **14,** 214–24.

SANDGREN C.D. (1980) An ultrastructural investigation of resting cyst formation in *Dinobryon cylindricum* Imhoff (Chrysophyceae, Chrysophycota). *Protistologica* **16,** 259–76.

SANDGREN C.D. (1981) Characteristics of sexual and asexual resting cyst (statospore) formation in *Dinobryon cylindricum. J. Phycol.* **17,** 199–210.

SANDGREN C.D. (1983) Survival strategies of chrysophycean flagellates: reproduction and the formation of resistant resting cysts. In *Survival Strategies of the Algae* (ed. Fryxell G.A.) pp. 23–49. Cambridge University Press, Cambridge.

SANTORE U.J. (1977) Scanning electron microscopy and comparative micromorphology of the periplast of *Hemiselmis rufescens, Chroomonas* sp. and members of the genus *Cryptomonas* (Cryptophyceae). *Brit. Phycol. J.* **12,** 255–70.

SANTORE U.J. (1982a) Comparative ultrastructure of two members of the Cryptophyceae assigned to the genus *Chroomonas* — with comments on their taxonomy. *Arch. Protistenk.* **125,** 5–29.

SANTORE U.J. (1982b) The ultrastructure of *Hemiselmis brunnescens* and *Hemiselmis virescens* with additional observations on *Hemiselmis rufescens* and comments about the Hemiselmidaceae as a natural group of the Cryptophyceae. *Brit. Phycol. J.* **17,** 81–99.

SARJEANT N.A.S. (1974) *Fossil and Living Dinoflagellates.* Academic Press, New York.

SARMA Y.S.R.K. & SHYAM R. (1974) A new member of colonial Volvocales (Chlorophyceae), *Pyrobotrys acuminata* sp. nov. from India. *Phycologia* **13,** 121–24.

SAUVAGEAU C. (1899) Les Cultériacées et leur alternance de générations. *Ann. Sci. Nat. Bot. Sér. 8* **10,** 265–362.

SAUVAGEAU C. (1915) Sur la sexualité hétérogamique d'une Laminaire (*Saccorhiza bulbosa*). *C. R. Acad. Sci. Paris D* **161,** 769–99.

SAWA T. (1965) Cytotaxonomy of the Characeae: karyotype analysis of *Nitella opaca* and *Nitella flexilis. Am. J. Bot.* **52,** 962–70.

SAWA T. & FRAME P.W. (1974) Comparative anatomy of Charophyta. I. Oogonia and oospores of *Tolypella* with special reference to the sterile oogonial cell. *Bull. Torrey Bot. Club.* **101,** 136–44.

SAY P.J & WHITTON B.A. (1978) Chemistry and benthic algae of a zinc polluted stream in the northern Pennines. *Brit. Phycol. J.* **13,** 206.

SCAGEL R.F. (1956) Introduction of a Japanese alga, *Sargassum muticum*, into the North-west Pacific. *Fish. Res. Pap. Wash. Dept Fish* **1,** 49–58.

SCAGEL R.F., BANDONI R.J., MAZE J.R., ROUSE G.E., SCHOFIELD W.G. & STEIN J.R. (1982) *Nonvascular Plants. An Evolutionary Survey.* Wadsworth, Belmont.

SCHAEDE R. (1951) Über die Blaualgensymbiose von *Gunnera. Planta* **39,** 154–70.

SCHANTZ E.J. (1981) Poisons produced by dinoflagellates — a review. In *The Water Environment: Algal Toxins and Health* (ed. Carmichael W.W.), pp. 25–36. Plenum Press, New York.

SCHINDLER D.W. (1971) Carbon, nitrogen and phosphorus and the eutrophication of freshwater lakes. *J. Phycol.* **7,** 321–9.

SCHINDLER D.W. (1974) Eutrophication and recovery in experimental lakes: implications for lake management. *Science* **184,** 897–8.

SCHINDLER D.W. (1977) Evolution of phosphorus limitation in lakes. *Science* **195,** 260–7.

SCHINDLER D.W. (1980) Experimental acidification of a whole lake, a test of the oligotrophication hypothesis. In *Ecological Impact of Acid Precipitation* (eds Drablos B. & Tollan A.), pp. 370–4. SNSF, Oslo.

SCHINDLER D.W., KLING H., SCHMIDT R.V., PROKOPOWICH J., FROST V.E., REID R.L. & CAPEL M. (1973) Eutrophication of Lake 227 by addition of phosphate and nitrate: the second, third and forth years of enrichment 1970, 1971, 1972. *J. Fish. Res. Bd Can.* **30,** 1415–40.

SCHLÖSSER U. (1976) Enzymatisch gesteuerte Freisetzung von Zoosporen bei *Chlamydomonas reinhardtii* Dangeard in Synchronkultur. *Arch. Mikrobiol.* **54,** 129–54.

SCHMIDT R.J. & LOEBLICH A.R. III (1979) Distribution of paralytic shell fish poison among Pyrrophyta. *J. Mar. Biol. Ass. U.K.* **59,** 479–87.

SCHMITZ Fr. (1883) Untersuchungen über die Befruchtung der Florideen. *Sitzbr. Akad. Wiss. Berlin* **883,** 215–58.

SCHMITZ Fr. (1889) Systematische Übersicht der bisher bekannten Gattungen der Florideen. *Flora (Jena)* **72,** 435–56.

SCHMITZ K. (1981) Translocation. In *The Biology of Seaweeds* (eds Lobban C.S. & Wynne M.J.), pp. 534–558. Blackwell Scientific Publications, Oxford.

SCHMITZ K. & SRIVASTAVA L.M. (1975) On the fine structure of sieve tubes and the physiology of assimilate transport in *Alaria marginata. Can. J. Bot.* **53,** 861–76.

SCHNEIDER C.W. (1976) Spatial and temporal distributions of benthic marine algae on the continental shelf of the Carolinas. *Bull. Mar. Sci.* **26,** 133–51.

SCHOENBERG D.A. & TRENCH R.K. (1980a) Genetic variation in *Symbiodinium* (=*Gymnodinium*) *microadriaticum* Freudenthal, and specificity in its symbiosis with marine invertebrates. I. Isoenzyme and soluble protein patterns of axenic cultures of *Symbiodinium microadriaticum. Proc. Roy. Soc. Lond. B* **207,** 405–27.

SCHOENBERG D.A. & TRENCH R.K. (1980b) Genetic variation in *Symbiodinium* (=*Gymnodinium*) *microadriaticum* Freudenthal, and specificity in its symbiosis with marine invertebrates. II. Morphological variation in *Symbiodinium microadriaticum. Proc. Roy. Soc. Lond. B* **207,** 429–44.

SCHOENBERG D.A. & TRENCH R.K. (1980c) Genetic variation in *Symbiodinium* (=*Gymnodinium*) *microadriaticum* Freudenthal, and specificity in its symbiosis with marine invertebrates. III. Specificity and infectivity of *Symbiodinium microadriaticum. Proc. Roy. Soc. Lond. B* **207,** 445–60.

SCHONBECK M. & NORTON T.A. (1978) Factors controlling the upper limit of fucoid algae on the shore. *J. Exp. Mar. Biol. Ecol.* **31,** 303–30.

SCHONBECK M. & NORTON T.A. (1980) Factors controlling the lower limits of fucoid algae on the shore. *J. Exp. Mar. Biol. Ecol.* **43,** 131–50.

SCHOPF J.W. (1970) Precambrian micro-organisms and evolutionary events prior to the origin of vascular plants. *Biol. Rev.* **45,** 319–52.

SCHOPF J.W. (1974a) Paleobiology of the Precambrian: the age of blue-green algae. In *Evolutionary Biology* (eds Dobzhansky T., Hecht M.K. & Steere W.C.), pp. 1–43. Plenum Press, New York.

SCHOPF J.W. (1974b) The development and diversification of Precambrian life. *Origins of Life* **5,** 119–35.

SCHOPF J.W. (1976) Are the oldest 'fossils' fossils? *Origins of Life* **7,** 19–36.

SCHOPF J.W. (1978) The evolution of the earliest cells. *Sci. Am.,* **239,** 110–34.

SCHOPF J.W. & WALTER M.R. (1982a) Origin and early evolution of Cyanobacteria: the geological evidence. In *The Biology of Cyanobacteria* (eds Carr N.G. & Whitton B.A.), pp. 543–64. Blackwell Scientific Publications, Oxford.

SCHOPF J.W. & WALTER M.R. (1982b) Archean microfossils: new evidence of ancient microbes. In *Origin and Evolution of Earth's Earliest Biosphere: An Interdisciplinary Study* (ed. Schopf J.W.). Princeton University Press, Princeton, New Jersey.

SCHREIBER E. (1930) Untersuchungen über Parthenogenesis Geschlechtsbestimmung und Bastardierungs vermögen bei Laminarien. *Planta* **12,** 331–51.

SCHULTZ M.E. & TRAINOR F.R. (1968) Production of male gametes and auxospores in the centric diatoms *Cyclotella meneghiniana* and *C. crypta. J. Phycol.* **4,** 85–8.

SCOFFIN T.P. (1970) The trapping and binding of subtidal carbonate sediments by marine vegetation in Bimini Lagoon, Bahama. *J. Sedim. Petrol.* **40,** 249–73.

SCOTT J. & BULLOCK K.W. (1976) Ultrastructure of cell division in *Cladophora.* Pregametangial cell division in the haploid generation of *Cladophora flexuosa. Can. J. Bot.* **54**, 1546–60.

SCOTT J.L. & DIXON P.S. (1971) The life history of *Pikea californica* Harv. *J. Phycol.* **7,** 295–300.

SCOTT J.L., BOSCO C., SCHORNSTEIN K. & THOMAS J. (1980) Ultrastructure of cell division and reproductive differentiation of male plants in the Florideophyceae (Rhodophyta): cell division in *Polysiphonia. J. Phycol.* **16,** 507–24.

SEAPY R.R. & LITTLER M.M. (1978) The distribution, abundance, community structure and primary productivity of macro organisms from two central California rocky intertidal habitats. *Pacific Sci.* **32,** 293–314.

SEARS, J.R. & WILCE R.T. (1970) Reproduction and systematics of the marine alga *Derbesia* (Chlorophyceae) in New England. *J. Phycol.* **6,** 381–92.

SEARS J.R. & WILCE R.T. (1975) Sublittoral, benthic algae of southern Cape Cod and adjacent islands: seasonal periodicity, associations, diversity and floristic composition. *Ecol. Monogr.* **45,** 337–65.

SELLNER K.G., KACHUR M.E. & LYONS L. (1984) Alterations in carbon fixation during power plant entrainment of estuarine phytoplankton. *Water Air Soil Poll.* **21,** 359–74.

SETCHELL W.A. (1915) The law of temperature connected with the distribution of the marine algae. *Ann. Mo. Bot. Gard.* **2,** 287–305.

SHACKLOCK P.F. & CROFT G.B. (1981) Effect of grazers on *Chondrus crispus* in culture. *Aquaculture* **22,** 331–42.

SHARABI N.E. & PRAMER D. (1973) A spectrophotofluorometric method for studying algae in the soil. *Bull. Ecol. Res. Comm. (Stockholm)* **17,** 77–84.

SHAW W.N. (1982) The use of phytoplankton for aquaculture needs — a status report. In *Proceedings of the 6th U.S.–Japan Meeting on Aquaculture* (ed. Sinderman C.J.), pp. 19–24. NOAA Technical Report NMFS Circ. 442.

SHEATH R.G. (1984) The biology of freshwater red algae. *Prog. Phycol. Res.* **3,** 89–157.

SHEATH R.G. & HELLEBUST J.A. (1974) Glucose transport systems and growth characteristics of *Bracteococcus minor. J. Phycol.* **10,** 43–41.

SHELEF G. & SOEDER C.J. (eds) (1980) *Algal Biomass: Production and Use.* Elsevier/North-Holland, Amsterdam.

SHEN E.Y.F. (1967) Amitosis in *Chara. Cytologia* **32,** 481–8.

SHEN E.Y.F. (1967) Microspectrophotometric analysis of nuclear DNA in *Chara zeylanica. J. Cell Biol.* **35,** 377–84.

SHERIDAN R.P. (1979) Seasonal variation in sun–shade ecotypes of *Plectonema notatum* (Cyanophyta). *J. Phycol.* **15,** 223–6.

SHIELDS L.M. & DURRELL L.W. (1964) Algae in relation to soil fertility. *Bot. Rev.* **30,** 92–128.

SHIHARA I. & KRAUSS R.W. (1964) *Chlorella: Physiology and Taxonomy of Forty-one Isolates.* University of Maryland, College Park.

SHILO M. (1971) Biological agents which cause lysis of blue-green algae. *Mitt. Int. Verein. Limnol.* **19,** 206–13.

SHILO M. (1981) The toxic priciples of *Prymnesium parvum.* In *Water Environment. Algal Toxins and Health.* Environment Science Research Vol. 20 (ed. Carmichael W.W.), pp. 34–47. Plenum Press, New York.

SICKO-GOAD L. (1982) A morphometric analysis of algal response to low dose, short term heavy metal exposure. *Protoplasma* **110,** 75–86.

SICKO-GOAD L. & STOERMER E.F. (1979) A morphometric study of lead and copper effects on *Diatoma tenue* var. *elongatum* (Bacillariophyta). *J. Phycol.* **15,** 316–21.

SIDDEQUE M. & FARIDI M.A.F. (1977) The life history of *Cladophora crispata* (Roth) Ag. *Pak. J. Bot.* **9,** 159–62.

SIEGEL B.T. & SIEGEL S.M. (1973) The chemical composition of algal cell walls. *Crit. Rev. Microbiol.* **3,** 1–26.

SIESSER W.G. (1972) Relief algal nodules (rhodolites) from the South African continental shelf. *J. Geol.* **80,** 611.

SIEVERS A., HEINEMANN B. & RODRIGUEZ-GARCIA M.I. (1979) Nachweis des subapikalen differentiellen Flankenwachstums im *Chara*-Rhizoid während der Graviresponse. *Z. Pflanzenphysiol.* **91,** 435–42.

SIKES C.S., ROER R.D. & WILBUR K.M. (1980) Photosynthesis and coccolith formation: inorganic carbon sources and net inorganic reaction of deposition. *Limnol. Oceanogr.* **25,** 248–61.

SILVA P.C. (1951) The genus *Codium* in California with observations on the structure of the walls of the utricles. *Univ. Calif. Pub. Bot.* **25,** 79–114.

SILVA P.C. (1979) Review of the taxonomic history and nomenclature of the yellow-green algae. *Arch. Protistenkd.* **121,** 20–63.

SILVA P.C. (1980) Names and classes of living algae. *Regnum Vegetabile* **103,** 1–156.

SILVA P.C. (1982) Thallobionta. In *Synopsis and Classification of Living Organisms*, Vol. 1 (ed. Parker S.P.), pp. 59–60. McGraw-Hill, New York.

SILVER M.W. & BRULAND K.W. (1981) Differential feeding and fecal pellet composition of salps and pteropods, and the possible origin of the deep water flora and olive green cells. *Mar. Biol.* **62,** 263–73.

SILVERBERG B.A., STOKES P.M. & FERSTENBERG L.B. (1976) Intranuclear complexes in a copper-tolerant green alga. *J. Cell Biol.* **69,** 210–14.

SIMON R.D. (1973) The effect of chloramphenicol on the production of cyanophycin granule polypeptide in the blue green alga *Anabaena cylindrica. Arch. Mikrobiol.* **92,** 115–22.

SKUJA H. (1954) Glaucophyta. In Engler A., *Syllabus der Pflanzenfamilien,* 12, I (eds Melchoir H. & Nerdermann E.), pp. 56–7. Borntraeger, Berlin.

SLANKIS T. & GIBBS S.P. (1972) The fine structure of mitosis and cell division in the Chrysophycean alga *Ochromonas danica. J. Phycol.* **8,** 243–56.

SLOCUM C.J. (1980) Differential susceptibility to grazers in two phases of an intertidal alga: advantages of heteromorphic generations. *J. Exp. Mar. Biol. Ecol.* **46,** 99–110.

SLOCUM R.D., AHMADJIAN V. & HILDRETH K.C. (1980) Zoosporogenesis in *Trebouxia gelatinosa*: ultrastructure, potential for zoospore release and implications for the lichen association. *Lichenologist* **12,** 173–87.

SLUIMAN H.J., ROBERTS K.R., STEWART K.D. & MATTOX K.R. (1980) Comparative cytology and taxonomy of the Ulvaphyceae. I. The zoospore of *Ulotrhix zonata* (Chlorophyta). *J. Phycol.* **16,** 537–45.

SLUIMAN H.J., ROBERTS K.R., STEWART K.D., MATTOX K.R. & LOKHORST G.M. (1981) A reinvestigation of the flagellar apparatus of the zoospore of *Urospora* (Chlorophyta). *Brit. Phycol. J.* **16,** 140.

SLUIMAN H.J., ROBERTS K.R., STEWART K.D. & MATTOX K.R. (1982) The flagellar apparatus of the zoospore of *Urospora penicilliformis* (Chlorophyta). *J. Phycol.* **18,** 1–12.

SMAYDA T.J. (1970) The suspension and sinking of phytoplankton in the sea. *Ann. Rev. Oceanogr. Mar. Biol.* **8,** 353–414.

SMAYDA T.J. (1980) Phytoplankton species succession. In *The Physiological Ecology of Phytoplankton* (ed. Morris I.), pp. 493–570. Blackwell Scientific Publications, Oxford.

SMITH D.C. (1978) What can lichens tell us about real fungi? *Mycologia* **70,** 915–34.

SMITH D.C. (1980) Mechanisms of nutrient movements between lichen symbionts. In *Cellular Interactions in Symbiosis and Parasitism* (eds Cook C.B., Pappas P.W. & Rudolph E.D.), pp. 197–227. Ohio State University Press, Columbus.

SMITH D. & WIEBE W. (1977) Rates of carbon fixation, organic carbon release and translocation in a reef building foraminifer *Maringopora vertebralis. Aust. J. Mar. Freshwater Res.* **28,** 311–19.

SMITH G.M. (ed.) (1951) *Manual of Phycology. An Introduction to the Algae and their Biology.* The Ronald Press Company, New York.

SMITH G.M. (1955) *Cryptogamic Botany. Vol. I, Algae and Fungi,* 2nd Edition. McGraw-Hill, New York.

SMITH J.E. (ed.) (1968) *'Torrey Canyon' Pollution and Marine Life.* Cambridge University Press, Cambridge.

SMITH P.V. (1954) Studies on the origin of petroleum; occurrence of hydrocarbons in recent sediments. *Bull. Am. Ass. Petrol. Geol.* **38,** 377–404.

SOEDER C.J. (1980) The scope of microalgae for food and feed. In *Algal Biomass: Production and use* (eds Shelef G. & Soeder C.J.), pp. 9–20. Elsevier/North-Holland, Amsterdam.

SOMMER H., WHEDON W.F., KOFOID C.A. & STOHLER R. (1937) Relation of paralytic shellfish poison to certain plankton organisms of the genus *Gonyaulax. A.M.A. Arch. Pathol.* **24,** 537–46.

SOMMER U. (1983) Nutrient competition between phytoplankton species in multispecies chemostat experiments. *Arch. Hydrobiol.* **96,** 399–416.

SOMMER U. (1984) The paradox of the plankton: fluctuations of phosphorus availability maintain diversity of phytoplankton in flow-through cultures. *Limnol. Oceanogr.* **29,** 633–36.

SOMMERFELD M.R. & NICHOLS H.W. (1970) Comparative studies in the genus *Porphyridium* Naeg. *J. Phycol.* **6,** 67–78.

SOONG P. (1980) Production and development of *Chlorella* and *Spirulina* in Taiwan. In *Algal Biomass: Production and Use* (eds Shelef G. & Soeder C.J.), pp. 97–114. Elsevier/North-Holland, Amsterdam.

SOURNIA A. (1976) Primary production of sands in the lagoon of an atoll and the role of foraminiferan symbionts. *Mar. Biol.* **37,** 29–32.

SOURNIA A. (ed.) (1978) *Phytoplankton Manual.* UNESCO, Paris.

SOURNIA A. (1981) Morphological bases of competition and succession. *Can. Bull. Fish. Aquatic Sci.* **210,** 339–46.

SOURNIA A. (1982) Is there a shade flora in the marine plankton? *J. Plankton Res.* **4,** 391–9.

SOUSA W.P., SCHROETER S.C. & GAINES S.D. (1981) Latitudinal variation in intertidal algal community structure: the influence of grazing and vegetative propagation. *Oecologia* **48,** 297–307.

SOUTH G.R. (1975) Contribution to the flora of marine algae of eastern Canada. III. Tilopteridales. *Naturaliste Can. (Qué)* **102,** 693–702.

SOUTH G.R. & ADAMS N.M. (1976) *Erythrotrichia foliiformis* sp. nov. (Rhodophyta, Erythropeltidaceae) from New Zealand. *J. Roy. Soc. N.Z.* **6,** 399–405.

SOUTH G.R. & HILL R.D. (1970) Studies on marine algae of Newfoundland. 1. Occurrence and distribution of free-living *Ascophyllum nodosum* in Newfoundland. *Can. J. Bot.* **48,** 1697–701.

SPJELDNAES N. (1963) A new fossil (*Papillomembrana* sp.) from the upper pre-Cambrian of Norway. *Nature* **200,** 63–5.

STACKEBRANDT E., SEEWALDT E., FOWLER V.J. & SCHLEIFER K.-H. (1982) The relatedness of *Prochloron* sp. isolated from different didemnid ascidian hosts. *Arch. Mikrobiol.* **132,** 216–17.

STACKHOUSE J. (1801) *Nereis Britannica.* Bath.

STACHE K.H.H.G. (1889) Die Liburnische Stufe und deren Grenz-Horizonte. Eine Studie über die Schichtenfolgen der Cretacisch-Eucänen oder Protocänen Landbildung speriode im Bereiche der Küstenlander von Oesterreich-Ungarn. Wien, *Geol. Abh.* **13** (1), 1–170.

STALLARD M.O. & FAULKNER D.J. (1974) Chemical constituents of the digestive gland of *Aplysia californica*, I. Importance of diet. *Comp. Biochem. Physiol.* **49B,** 25–36.

STANIER R.Y. (1961) La place des bactéries dans le monde vivant. *Annl Inst. Pasteur Paris* **101,** 297–312.

STANIER R.Y. (1974) The origins of photosynthesis in eukaryotes. *Symp. Soc. Gen. Microbiol.* **24,** 219–40.

STANIER R.Y. & COHEN-BAZIRE G. (1977) Phototrophic prokaryotes: cyanobacteria. *Ann. Rev. Microbiol.,* **31,** 225–70.

STANIER R.Y., KUNISAWA R., MANDEL M. & COHEN-BAZIRE G. (1971) Purification and properties of unicellular blue-green algae (order Chroococcales). *Bacteriol. Rev.* **35,** 171–205.

STANIER R.Y., SISTROM W.R., HANSEN T.A., WHITTON B.A., CASTENHOLZ R.W., PFENNIG B.A., GORLENKO V.N., KONDRATIEVA E.N., EIMHJELLEN K.E., WHITTENBURY R., GHERNA R.L. & TRUPER H.G. (1978) Proposal to place the nomenclature of the cyanobacteria (blue-green algae) under the rules of the international code of nomenclature of bacteria. *Int. J. Syst. Bacteriol.* **28,** 335–6.

STARKS T.L., SHUBERT L.E. & TRAINOR F.R. (1981) Ecology of soil algae: a review. *Phycologia* **20,** 65–80.

STARMACH K. (1972) *Chrysosphaera stigmatica* n. sp. (Chrysophyceae). *Bull. Acad. Polon. Sci. Ser. Biol.* **20,** 577–9.

STARR R.C. (1955) Isolation of sexual strains of placoderm desmids. *Bull. Torrey Bot. Club* **82,** 261–5.

STARR R.C. (1969) Structure reproduction and diffferentiation in *Volvox carteri* f. *nagarensis.* Iyengar, strains HK9 and 10. *Arch. Protistenk.* **111,** 204–22.

STARR R.C. (1975) Meiosis in *Volvox carteri* f. *nagariensis Arch. Protistenk.* **117,** 187–91.

STARR R.C. (1980) Colonial chlorophytes. In *Phytoflagellates* (ed. Cox E.R.), pp. 147–64, Elsevier/North-Holland, New York.

STARR R.C. & RAYBURN W.R. (1964) Sexual reproduction in *Mesotaeneum kramstai. Phycologia* **4,** 23–46.

STARR R.C., O'NEIL R.M. & MILLER III C.E. (1980) L-glutamic acid as a mediator of morphogenesis in *Volvox capensis. Proc. Natl Acad. Sci.* **77,** 1025–8.

STARR M.P., STOLP H., TRUPER H.G., BALOWS A. & SCHLEGEL H.G. (eds) (1981) *The Prokaryotes,* Vols 1 & 2. Springer-Verlag, Berlin.

STEIDINGER K.A. (1975a) Implications of dinoflagellate life cycles on initiation of *Gymnodinium breve* red tides. *Environ. Lett.* **9,** 129–39.

STEIDINGER K.A. (1975b) Basic factors influencing red tides. In *Proceedings of the 1st International Conference on Toxic Dinoflagellate Blooms* (ed. LoCicero V.R), pp. 153–62. Massachusetts Science and Technical Foundation, Wakefield, Mass.

STEIDINGER K.A. (1983) A re-evaluation of toxic dinoflagellate biology and ecology. *Prog. Phycol. Res.* **2,** 147–88.

STEIDINGER K.A. & HADDAD K. (1981) Biologic and hydrographic aspects of red tides. *BioScience* **31,** 814–19.

STEIN J. & BORDEN C.A. (1984) Causative and beneficial algae in human disease conditions: a review. *Phycologia* **23,** 485–501.

STENECK R.S. & WATLING L. (1982) Feeding capabilities and limitation of herbivorous molluscs: a functional approach. *Mar. Biol.* **68,** 299–312.

STEPHENSEN J.W. (1981) The effects of a seaweed extract on the yield of a variety of field and glasshouse crops. *Proc. Int. Seaweed Symp.* **8,** 740–4.

STEPHENSON T.A. & STEPHENSON A. (1949) The universal features of zonation between tide marks on rocky coasts. *J. Ecol.* **37,** 289–305.

STEPHENSON T.A. & STEPHENSON A. (1972) *Life Between Tide Marks on Rocky Shores.* W.H. Freeman, San Francisco.

STEPHENSON W.A. (1974) *Seaweed in Agriculture and Horticulture*, 3rd. Edition. Rateaver Publ. Valley, California.

STEVENS S.E. & PORTER R.D. (1980) Transformation in *Agmenellum quadruplicatum. Proc. Natl Acad. Sci. U.S.A.* **77,** 6052–6.

STEWART K.D. & MATTOX K.R. (1975) Comparative cytology, evolution and classification of the green algae with some consideration of the origin of other organisms with chlorophylls *a* and *b*. *Bot. Rev.* **41,** 104–35.

STEWART K.D. & MATTOX K.R. (1980) Phylogeny of phytoflagellates. In *Phytoflagellates* (ed. Cox E.R.), pp. 433–62. Elsevier/North-Holland, New York.

STEWART J.G. & SCHULTZ-BALDES M. (1976) Long term lead accumulation in the abalone (*Haliotus* sp.) fed on lead treated brown algae (*Egregia laevigata*.) *Mar. Biol* **36,** 19–24.

STEWART W.D.P. & RODGERS G.A. (1977) The cyanophyte hepatic symbiosis. II. Nitrogen fixation and the interchange of nitrogen and carbon. *New Phytol.* **78,** 459–71.

STEWART W.D.P. & ROWELL P. (1977) Modifications of nitrogen fixing algae in lichen symbioses. *Nature* **265,** 371–2.

STOKES P.M. (1981) Multiple metal tolerance in copper tolerant green algae. *J. Plant Nutrit.* **3,** 667–78.

STOKES P.M. (1983) Responses of freshwater algae to metals. *Prog. Phycol. Res.* **2,** 87–112.

STOSCH H.A. VON (1954) Die Oogamie von *Biddulphia mobiliensis* und die bisher bekannten Auxosporenbildungen bei den Centrales. *Int. Bot. Congr.* 8, *Rap. Comm. Sec.* **17,** 58–68.

STOSCH H.A. VON (1964) Wirkungen von Jod und Arsenit auf Meeresalgaen in Kultur. *Proc. Int. Seaweed Symp.* **4,** 151–6.

STOSCH H.A. VON (1965a) Manipulierung der Zellgrösse von Diatomeen im Experiment. *Phycologia* **5,** 21–44.

STOSCH H.A. VON (1965b) The sporophyte of *Liagora farinosa* Lamour. *Brit. Phycol. Bull.* **2,** 486–96.

STOSCH H.A. VON (1967) Haptophyceae. In *Vegetative Fortplanzung, Parthenogenese und Apogamie bei Algen.* (ed. Rohland W.), pp. 646–56. Encyclopaedia of Plant Physiology, Vol. 18.

STOSCH H.A. VON (1973) Observations on vegetative reproduction and sexual life cycles of two freshwater dinoflagellates, *Gymnodinium pseudopalustre* Schiller and *Woloszynskia apiculata* sp. nov. *Brit. Phycol. J.* **8,** 105–34.

STOSCH H.A. VON & TIEL G. (1979) A new mode of life history in the freshwater red algal genus *Batrachospermum. Am. J. Bot.* **66,** 105–7.

STOSCH H.A., VON THEIL G. & KOWALLIK K.V. (1973). Entwicklungsgeschichtliche Untersuchungen an zentrischen Diatomeen. V. Bau und Lebenszyklus von *Chaetoceros didymum* mit Beobachtungen über einige andere Arten der Gattung. *Helgol. Wiss. Meeresunters.* **25,** 384–445.

STRASBURGER E. (1906) Zur Frage eines Generationswechsels bei Phaeophyceen. *Bot. Zeit.* **64,** 1–7.

STRAUGHAN D. (1982) Observations on the effects of oil seeps in the Coal Oil Point area. *Phil. Trans. Roy. Soc. Lond. B* **297,** 269–81.

SULLIVAN C.W. (1976) Diatom mineralization of silicic acid. I. $Si(OH)_4$ transport characteristics in *Navicula pelliculosa. J. Phycol.* **12,** 390–6.

SULLIVAN C.W. (1977) Diatom mineralization of silicic acid. II. Regulation of $Si(OH)_4$ transport rates during the cell cycle of *Navicula pelliculosa*. *J. Phycol.* **13,** 86–91.

SUN H. & FENICAL W. (1979) Rhipocephenal and rhipocephalin, novel linear sesquiterpenoids from the tropical green alga *Rhipocephallus phoenix*. *Tetrahed. Lett.* **1979**(8), 84–8.

SUNDENE O. (1962) Reproduction and morphology in strains of *Antithamnion boreale* originating from Spitzbergen and Scandinavia. *Nor. Vidensk. Akad. Oslo. I. Mat.-nat. Kl. N.S.* **5,** 1–19.

SUNESON S. (1950) The cytology of bispore formation in two species of *Lithophyllum* and the significance of the bispores in the Corallinaceae. *Bot. Not.* **1950,** 429–50.

SUTHERLAND J.M., HERDMAN M. & STEWART W.D.P. (1979) Akinetes of the cyanobacterium *Nostoc* PCC 7524: macromolecular composition, structure and control of differentiation. *J. Gen. Microbiol.* **115,** 273–87.

SUTHERLAND J.P. & KARLSON R.H. (1977) Development and stability of the fouling community at Beaufort, North Carolina. *Ecol. Mongr.* **47,** 425–66.

SVEDELIUS N. (1908) Über den bau und die Entwicklung der Florideengattung *Martenisa. Vet. Akad. handl. Bd.* **43,** Stockholm.

SVEDELIUS N. (1937) The apomeiotic tetrad division in *Lomentaria rosea* in comparison with the normal development in *Lomentaria clavellosa*. *Symb. Bot. Uppsala* **2,** 1–53.

SWEENEY B.M. (1979) The bioluminescence of dinoflagellates. In *Biochemistry and Physiology of Protozoa*, 2nd Edition, Vol. I (eds Levandrowsky M. & Hunter G.H.), pp. 287–306. Academic Press, London.

SWEENEY B.M. (1983) Circadian time keeping in eukaryotic cells, models and hypotheses. *Prog. Phycol. Res.* **2,** 189–225.

SWIFT D.G. (1980) Vitamins and phytoplankton growth. In *The Physiological Ecology of Phytoplankton* (ed. Morris I.), pp. 329–68. Blackwell Scientific Publications, Oxford.

SYRETT P.J. (1981) Nitrogen metabolism of microalgae. *Can. Bull. Fish. Aquat. Sci.* **210,** 182–210.

SZCZEPANSKI A. (1966) Bermerkungen über die Primarproduction des Pelagials. *Verh. Int. Ver. Limnol.* **16,** 364–71.

TAKAHASHI E. (1975) Studies on genera *Mallomonas* and *Synura*, and other plankton in the fresh water with the electron microscope. IX. *Mallomonas harrisae* sp. nov. (Chrysophyceae). *Phycologia* **14,** 41–4.

TANAKA Y. & STARA J.F. (1979) Algal polysaccharides—their potential use to prevent chronic metal poisoning. In *Marine Algae in Pharmaceutical Science* (eds Hoppe H.A., Levring T. & Tanaka Y.), pp. 525–43. Walter de Gruyter, Berlin.

TANDEAU DE MARSAC N.T. (1977) Occurrence and nature of chromatic adaptation in cyanobacteria. *J. Bact.* **130,** 82–91.

TANNER C.E. (1981) Chlorophyta life histories. In *The Biology of Seaweeds* (eds Lobban C.S. & Wynne M.J.), pp. 218–47. Blackwell Scientific Publications, Oxford.

TAPPAN H. (1974) Protistan phylogeny: multiple working hypotheses. *Taxon* **32,** 271–6.

TAPPAN H. (1976) Possible eukaryotic algae (Bangiophycidae) among early Proterozoic microfossils. *Geol. Soc. Am. Bull.* **87,** 633–9.

TAPPAN H. (1980) *The Paleobiology of Plant Protists*. W.H. Freeman & Co., San Francisco.

TAPPAN H. & LOEBLICH A.R. Jr (1972) Fluctuating rates of protistan evolution, diversification and extinction. *XXIV Int. Geol. Congr., Sec. 7 Paleont.*, 205–13.

TAPPAN H. & LOEBLICH A.R. Jr (1973) Evolution of the oceanic plankton. *Earth-Sci. Rev.* **9,** 207–40.

TATEWAKI M. (1966) Formation of a crustose sporophyte with unilocular sporangia in *Scytosiphon lomentaria*. *Phycologia* **6,** 62–6.

TATEWAKI M. & IIMA M. (1984) Life histories of *Blidingia minima* (Chlorophyceae) especially sexual reproduction. *J. Phycol.* **20,** 368–76.

TATEWAKI M. & NAGATA N. (1970) Surviving protoplasts *in vitro* and their development in *Bryopsis*. *J. Phycol.* **6,** 401–3.

TASCH P. (1973) *Paleobiology of the Invertebrates*. John Wiley & Sons, New York.

TAYLOR D.L. (1973) The cellular interaction of algal–invertebrate symbiosis. *Adv. Mar. Biol.* **11,** 1–56.

TAYLOR D.L. (1974) Symbiotic marine algae: taxonomy and biological fitness. In *Symbiosis in the Sea* (ed. Vernberg W.B.), pp. 245–62. University of South Carolina Press, Columbia.

TAYLOR D.L. & SELIGER H.H. (eds) (1979) *Toxic Dinoflagellate Blooms*. Elsevier/North-Holland, New York.

TAYLOR F.J.R. (1974) Implications and extensions of the serial endosymbiosis theory of the origin of eukaryotes. *Taxon* **23,** 229–58.

TAYLOR F.J.R. (1976) Autogenesis theories for the origin of eukaryotes. *Taxon* **25,** 377–90.

TAYLOR F.J.R. (1978) Problems in the development of an explicit hypothetical phylogeny of the lower eukaryotes. *Biosystems* **10,** 67–89.

TAYLOR F.J.R. (1980) On dinoflagellate evolution. *Biosystems* **13,** 65–108.

TAYLOR W.R. (1969) Phycology. In *A Short History of Botany in the United States* (ed. Ewan J.), pp. 74–81. Hafner Publishing Co., New York and London.

TERRY L.A. & EDYVEAN R.G.T. (1981) Microalgae and corrosion. *Bot. Mar.* **24,** 177–83.

THACKER A. & SYRETT P.J. (1972) The assimilation of nitrate and ammonium by *Chlamydomonas reinhardii. New Phytol.* **71,** 423–33.

THINH L.V. (1979) *Prochloron* (Prochlorophyta) associated with the ascidian *Tridemnum cyclops. Phycologia* **18,** 77–82.

THOMAS R.N. & COX E.R. (1973) Observations on the symbiosis of *Peridinium balticum* and its intracellular alga. I. Ultrastructure. *J. Phycol.* **9,** 304–23.

THOMAS W.H., HOLM-HANSEN O., SEIBERT D.L.R., AZAM F., HODSON R. & TAKAHASHI M. (1977a) Effects of copper on the dominance and the diversity of algae: controlled ecosystem pollution experiment. *Bull. Mar. Sci.* **27,** 19–33.

THOMAS W.H., SEIBERT D.L.R. & TAKAHASHI M. (1977b) Controlled ecosystem pollution experiment: effect of mercury on enclosed water columns III. Phytoplankton dynamics and production. *Mar. Sci. Commun.* **3,** 331–54.

THOMAS W.H., HOLLIBAUGH J.T., SEIBERT D.L.R. & WALLACE G.T. JR (1980) Toxicity of a mixture of ten metals to phytoplankton. *Mar. Ecol. Prog. Ser.* **2,** 213–20.

THOMPSON A. (1973) The flagella of *Cyanophora paradoxa* Korsch. *J. South Afr. Bot.* **39,** 35–9.

THORHAUG A. & MARCUS J.H. (1981) The effects of temperature and light on attached forms of tropical and semi-tropical macroalgae potentially associated with OTEC (ocean thermal energy conversion) machine operation. *Bot. Mar.* **24,** 393–8.

THORHAUG A., SEGAR D. & ROESSLER M.A. (1973) Impact of a power plant on a subtropical estuarine environment. *Mar. Poll. Bull.* **4,** 166–9.

THORNBER J.P. & BARBER J. (1979) Photosynthetic pigments and models for their organization *in vivo.* In *Photosynthesis in Relation to Model Systems* (ed. Barber J.), pp. 27–70. Elsevier, New York.

TILMAN D. (1977) Resource competition between planktonic algae: an experimental and theoretical approach. *Ecology* **58,** 338–48.

TISON D.L., WILDE E.W., POPE D.H. & FOIERMANS C.B. (1981) Productivity and species composition of algal mat communities exposed to a fluctuating thermal regime. *Microbiol. Ecol.* **7,** 151–65.

TOKIDA J. (1960) Marine algae epiphytic on Laminariales plants. *Bull. Fac. Fish. Hokkaido Univ.* **11,** 73–105.

TOPINKA J.A. & TUCKER L.R. (1981) Long-term oil contamination of fucoid macroalgae following the Amoco Cadiz oil spill. In *Amoco Cadiz: Conséquences d'une Pollution Accidentelle par les Hydrocarbures*, pp. 381–91. Publ. Centre national pour l'exploration des océans, Paris.

TOTH R. (1976) The release, settlement and germination of zoospores in *Chorda tomentosa* (Phaeophyceae, Laminariales). *J. Phycol.* **12,** 222–33.

TOTH R. & MARKEY D.R. (1973) Synaptonemal complexes in brown algae. *Nature* **243,** 236–7.

TRAINOR F.R. (1978) *Introductory Phycology.* Wiley, New York.

TRAINOR F.R. (1983) Survival of algae in soil after high temperature treatment. *Phycologia* **22,** 201–2.

TRENCH R.K. (1975) Of 'leaves that crawl': functional chloroplasts in animal cells. *Symp. Soc. Exp. Biol.* **29,** 229–65.

TRENCH R.K. (1979) The cell biology of plant–animal symbiosis. *Ann. Rev. Plant Physiol.* **30,** 485–531.

TRENCH R.K. (1980) Uptake, retention and function of chloroplasts in animal cells. In *Endocytobiology: Endosymbiosis and cell Biology, a Synthesis of Recent Research* (eds. Schwemmler W. & Schenk H.E.A.), pp. 703–27. Walter de Gruyter, Berlin.

TRENCH R.K. (1982) Physiology, ultrastructure and biochemistry of cyanellae. *Prog. Phycol. Res.* **1,** 257–87.

TRENCH R.K. & SIEBENS H.B. (1978) Aspects of the relation between *Cyanophora paradoxa* (Korschikoff) and its endosymbiotic cyanelles *Cyanocyta korschikoffiana* (Hall and Claus). IV. The effects of rifampicin, chloramphenicol and cycloheximide on the synthesis of ribosomal ribonucleic acids and chlorophyll. *Proc. Roy. Soc. Lond. B* **202,** 473–83.

TRENCH R.K., BOYLE J.E. & SMITH D.C. (1973) The association between chloroplasts of *Codium fragile* and the mollusc *Elysia viridis*. II. Chloroplast ultrastructure and photosynthetic carbon fixation in *E. viridis. Proc. Roy. Soc. Lond. B* **184,** 63–81.

TRIEMER R.E. & BROWN R.M. Jr (1975) Ultrastructure of fertilization in *Chlamydomonas reinhardtii*. *Brit. Phycol. J.* **12,** 23–44.

TSCHERMAK-WOESS E. (1959) Extreme Anisogamie und ein bemerkenswerter Fall der Geschlechtsbestimmung bei einer neuen *Chlamydomonas*-Art. *Planta* **52,** 606–22.

TSCHERMAK-WOESS E. (1962) Zur Kenntnis von *Chlamydomonas suboogama*. *Planta* **59,** 68–76.

TSCHERMAK-WOESS E. (1978) *Myrmecia reticulata* as a phycobiont and free living *Trebouxia* — the problem of *Stenocybe sepatata*. *Lichenologist* **10,** 69–79.

TSENG C.K. (1981) Commercial cultivation. In *The Biology of the Seaweeds* (eds Lobban C.S. & Wynne M.J.), pp. 680–741. Blackwell Scientific Publications, Oxford.

TSUBO Y. (1961) Chemotaxis and sexual behaviour in *Chlamydomonas*. *J. Protozool.* **8,** 114–21.

TURNER C.H. & EVANS L. V. (1977). Physiological studies on the relationship between *Ascophyllum nodosum* and *Polysiphonia lanosa*. *New Phytol.* **79,** 363–71.

TURNER C.H. & EVANS L.V. (1978) Translocation of photoassimilated ^{14}C in the red alga *Polysiphonia lanosa*. *Brit. Phycol. J.* **13,** 51–5.

TURNER D. (1802) *A Synopsis of the British Fuci*. London.

TYLER M.A. & SELIGER H.H. (1978) Annual subsurface transport of a red tide dinoflagellate in its bloom area: water circulation patterns and organism distribution in the Chesapeake Bay. *Limnol. Oceanogr.* **23,** 227–46.

TYLER M.A. & SELIGER H.H. (1981) Selection for a red tide organism: physiological responses to the physical environment. *Limnol. Oceanogr.* **26,** 310–24.

TYNAN E.J. (1971) Geologic occurrence of the archaeomonads. In *Proceedings II. Planktonic Conference, Roma, 1970*, Vol. 2 (ed Farinacci A.), pp. 1225–30. Edizioni Tecnoscienza, Rome.

UNDERWOOD A.J. (1978) A refutation of critical tide levels as determinants of the structure of intertidal communities on British Shores. *J. Exp. Mar. Biol. Ecol.* **33,** 261–76.

UNDERWOOD A.J. & DENLEY E.J. (1984) Paradigms, explanations, and generalizations in models for the structure of intertidal communities on rocky shores. In *Ecological Communities: Conceptual Issues and the Evidence* (eds Strong D.R. Jr, Simberloff D., Abele L.G. & Thistle A.B.), pp. 151–80, Princeton University Press, Princeton.

VADAS R.L. (1977) Preferential feeding: an optimization strategy in sea urchins. *Ecol. Monogr.* **47,** 337–71.

VADAS R.L., KESER M. & RUSANOWSKI P.C. (1976) Influence of thermal loading on the ecology of intertidal algae. In *Thermal Ecology* II (eds Esch G.W. & McFarlane R.W.), pp. 331–5. E.R.D.A. Symposium Series, Conf. 750425 Springfield, Mass., U.S.A.

VALET G. (1969) Contribution à l'étude des Dasycladales, 2. Cytologie et reproduction. 3. Révision systématique. *Nova Hedwiga* **17,** 551–644, pls 133–57.

VALIELA I. (1984) *Marine Ecological Processes*. Springer-Verlag, New York.

VAN DEN ENDE H. (1976) *Sexual Interactions in Plants*. Academic Press, London.

VAN DER MEER J.P. (1983) The domestication of seaweeds. *BioScience* **33,** 172–6.

VAN DER MEER J.P. & TODD E.R. (1977) Genetics of *Gracilaria* sp. (Rhodophyceae, Gigartinales). IV. Mitotic recombination and its relationship to mixed phases in the life history. *Can. J. Bot.* **55,** 2810–17.

VAN DER MEER J.P. & TODD E.R. (1980) The life history of *Palmaria palmata* in culture. A new type for the Rhodophyceae. *Can. J. Bot.* **58,** 1250–6.

VANDERMEULEN J.H & AHERN T.P. (1976) Effect of petroleum hydrocarbons on algal physiology: review and progress report. In *Effects of Pollutants on Aquatic Organisms* (ed. Lockwood A.P.M.), pp. 107–26. Cambridge University Press, Cambridge.

VAN DER VEER J. (1976) *Pavlova calceolata* (Haptophyceae), a new species from the Tamar Estuary, Cornwall, England. *J. Mar. Biol. Ass. U.K.* **56,** 21–30.

VAN DER VELDE H.H. & HEMRIKA-WAGNER A.M. (1978) The detection of phytochrome in the red alga *Achrocheaetium daviesii*. *Plant Sci. Lett.* **11,** 145–9.

VARGO S.A., HUTCHINS M. & ALMQUIST G. (1982) The effect of low, chronic levels of No.2 fuel oil on natural phytoplankton assemblages in microcosms: 1. Species composition and seasonal succession. *Mar. Environ. Res.* **6,** 245–6.

VAUCHER J.P. (1803) *Histoire des Conferves d'eau douce*. Genève.

VESK M. & JEFFREY S.W. (1977) Effect of blue-green light on photosynthetic pigments and chloroplast structure in unicellular marine algae in six classes. *J. Phycol.* **13,** 280–8.

VIDAL G. (1984) The oldest eukaryotic cells. *Sci. Am.* **250,** 48–57.

VINAYAKUMAR M. & KESSLER E. (1975) Physiological and biochemical contributions to the taxonomy of the genus *Chlorella*. *Arch. Mikrobiol.* **103,** 13–19.

VINCE-PRUE D. (1975) Photomorphogenesis and flowering. In *Encyclopedia of Plant Physiology. Vol. 16B, Photomorphogenesis* (eds Shropshire W. & Mohr H.), pp. 457–90. Springer, Heidelberg.

VINCENT W.F. & GOLDMAN C.R. (1980) Evidence for algal heteroptrophy in Lake Tahoe, California–Nevada. *Limnol. Oceanogr.* **25,** 89–99.

VON ZALUZIAN Z. (1592) *Methodi herbariae libri tres.* Prague (not seen by the authors; quoted from Prescott, 1951).

VOSS V.G. (1983) *International Code of Botanical Nomenclature.* Bohn, Scheltema & Holtema, Utrecht & Antwerpen. W. Junk, The Hague and Boston.

WAALAND J.R. (1981) Commercial cultivation. In *The Biology of the Seaweeds* (eds Lobban C.S. & Wynne M.J.), pp. 726–41. Blackwell Scientific Publications, Oxford.

WAALAND J.R., WAALAND S.D. & BATES G. (1974) Chloroplast structure and pigment composition in the red alga *Griffithsia pacifica*: regulation by light intensity. *J. Phycol.* **10,** 193–8.

WAALAND S.D. (1975) Evidence for a species-specific cell fusion hormone in red algae. *Protoplasma* **86,** 253–63.

WAALAND S.D. & CLELAND R. (1972) Development in the red alga *Griffithsia pacifica*: control by internal and external factors. *Planta* **105,** 196–204.

WAALAND S.D. & WAALAND J.R. (1975) Analysis of cell elongation in red algae by fluorescent labelling. *Planta* **126,** 127–38.

WALKER D.C. & HENRY E.C. (1978) Unusual reproductive structures in *Syringoderma abyssicola* (S. & G.) Levring. *New Phytol.* **80,** 193–7.

WALKER J.C.G., KLEIN C., SCHIDLOWSKI M., SCHOPF J.W., STEVENSON D.J. & WALTER M.R. (1982) Environmental evolution of the Archean–early Proterozoic earth. In *Origin and Evolution of Earth's Earliest Biosphere: An Interdisciplinary Study* (ed. Schopf J.W.). Princeton University Press, Princeton, New Jersey.

WALKER L.M. (1984) Life histories, dispersal and survival in marine, planktonic dinoflagellates. In *Marine Plankton Life Cycle Strategies* (eds Steidinger K.A. & Walker L.M.), pp. 19–34. CRC press, Boca Raton.

WALL D. (1962) Evidence from recent plankton regarding the biological affinities of *Tasmanites* Newton 1875 and *Leiosphaeridia* Eisenack 1958. *Geol. Mag.* **99,** 353–62.

WALNE P.L. (1980) Euglenoid flagellates. In *Phytoflagellates* (ed. Cox E.R.), pp. 165–212. Elsevier/North-Holland, New York.

WALNE P.L. & ARNOTT H.J. (1967) The comparative ultrastructure and possible function of eyespots: *Euglena granulata* and *Chlamydomonas eugametos. Planta* **77,** 325–53.

WALSBY A.E. (1978) The gas vesicles of aquatic prokaryotes. In *Relations Between Structure and Function in the Prokaryotic Cell* (eds Stanier R.Y. *et al.*), pp. 327–57. Cambridge University Press, Cambridge.

WALSBY A.E. & XYPOLYTA A. (1977) The form resistance of chitin fibres attached to the cells of *Thalassiosira fluviatilis* Hustedt. *Brit. Phycol. J.* **12,** 215–23.

WALTER M.R., OEHLER J.H. & OEHLER D.Z. (1976) Megascopic algae 1300 million years old from the Belt Supergroup, Montana: a reinterpretation of Walcott's helminthoidichnites. *J. Paleontol.* **50,** 872–81.

WANDERS J.B.W (1977) The role of benthic algae in the shallow reef of Curaçao (Netherlands Antilles). III: The significance of grazing. *Aquatic Bot.* **3,** 357–90.

WATANABE A. (1975) Practical significance of algae in Japan. Nitrogen fixation by algae. In *Advance of Phycology in Japan* (eds Tokida J. & Hirose H.W.), pp. 255–72. Junk, The Hague.

WATANABE T. & KONDO N. (1976) Ethylene evolution in marine algae and a proteinaceous inhibitor of ethylene biosynthesis from a red alga. *Plant Cell Physiol* **17,** 1159–66.

WATANABE T., HIRAYAMA T., TAKAHASI T., KOKUBU T. & IKEDAM M. (1980) Toxicological evaluation of arsenic in edible seaweed *Hizikia* species. *Toxicology* **14,** 1–22.

WATRASS C.J., CHISHOLM S.W. & ANDERSON D.M. (1982) Regulation of growth in an estuarine clone of *Gonyaulax tamarensis* Lebour: salinity dependent temperature responses. *J. Exp. Mar. Biol. Ecol.* **62,** 25–37.

WATSON B.A. & WAALAND S.D. (1983) Partial purification of a glycoprotein cell fusion hormone from *Griffithsia pacifica*, a red alga. *Plant Physiol.* **71,** 327–32.

WATSON S. & KALFF J. (1981) Relationships between nannoplankton and lake trophic status. *Can. J. Fish. Aquat. Sci.* **38,** 960–7.

WEIDNER M. & KUPPERS U. (1982) Metabolic conversion of ^{14}C-aspartate, ^{14}C-malate and ^{14}C-mannitol by tissue disks of *Laminaria hyperborea*: role of phosphoenolpyruvate carboxykinase. *Z. Pflanzenphysiol.* **108,** 353–64.

WEINHEIMER A.J. & KARNS T.K.B. (1974) A search for anti-cancer and cardiovascular agents in marine organisms. In *Food–Drugs from the Sea* (eds Webber H.H. & Ruggieri R.), pp. 491–6. Marine Science Center, Puerto Rico.

WERNER D. (ed.) (1977) *The Biology of Diatoms*. Blackwell Scientific Publications, Oxford.

WEST J.A. (1968) Morphology and reproduction of the red alga *Acrochaetium pectinatum*. *J. Phycol.* **4,** 89–99.

WEST J.A. (1972) The life history of *Petrocelis franciscana*. *Brit. Phycol. J.* **7,** 299–308.

WEST J.A. (1972) Environmental regulation of reproduction in *Rhodochorton purpureum*. In *Contributions to the Systematics of the Benthic Marine Algae of the North Pacific* (eds Abbott I.A. & Kurogi M.), pp. 213–30. Japanese Society of Phycology, Kobe, Japan.

WEST J.A. & HOMMERSAND M.H. (1981) Rhodophyta: life histories. In *The Biology of Seaweeds* (eds Lobban C.S. & Wynne M.J.), pp. 133–93. Blackwell Scientific Publications, Oxford.

WEST L.K. & WALNE P.L. (1980) *Trachelomonas hispida* var. *coronata* (Euglenophyceae): II. Envelope substructure. *J. Phycol.* **16,** 498–506.

WEST L.K., WALNE P.L. & ROSOWSKI R.J. (1980a) *Trachelomona hispida* var. *coronata* (Euglenophyceae): I. Ultrastructure of cytoskeletal and flagellar systems. *J. Phycol.* **16,** 489–97.

WEST L.K., WALNE P.L. & BENTLEY J. (1980b) *Trachelomonas hispida* var. *coronata* (Euglenophyceae): III. Envelope elemental composition and mineralization. *J. Phycol.* **16,** 582–91.

WETHERBEE R. (1979) 'Transfer connection': specialized pathways for nutrient translocation in a red alga? *Science* **204,** 858–9.

WHATLEY J.M. (1977) The fine structure of *Prochloron*. *New Phytol.* **79,** 369–73.

WHATLEY J.M. (1982) Ultrastructure of plastid inheritance: green algae to angiosperms. *Biol. Rev.* **57,** 527–69.

WHATLEY J.M. & WHATLEY F.R. (1981) Chloroplast evolution. *New Phytol.* **87,** 233–47.

WHEELER P.A., NORTH B.B. & STEPHENS C.G. (1974) Aminoacid uptake by marine phytoplankters. *Limnol. Oceanogr.* **19,** 249–95.

WHITE A.W. (1981) Marine zooplankton can accumulate and retain dinoflagellate toxins and cause fish kills. *Limnol. Oceanogr.* **26,** 103–7.

WHITESIDE M.C. (1983) The mythical concept of eutrophication. *Hydrobiologia* **103,** 107–11.

WHITTAKER R.H. (1959) On the broad classification of organisms. *Quart. Rev. Biol.* **34,** 210–26.

WHITTAKER R.H. (1969) New concepts of kingdoms of organisms. *Science* **163,** 150–60.

WHITTAKER R.H. & MARGULIS L. (1978) Protist classification and the kingdoms of organisms. *Biosystems* **10,** 3–18.

WHITTICK A. (1977) The reproductive ecology of *Plumaria elegans* (Bonnem.) Schmitz (Ceramiaceae, Rhodophyta) at its northern limits in the western Atlantic. *J. Exp. Mar. Biol. Ecol.* **29,** 223–30.

WHITTICK A. (1978) The life history and phenology of *Callithamnion corymbosum* (Rhodophyta: Ceramiaceae) in Newfoundland. *Can. J. Bot.* **56,** 2497–9.

WHITTICK A. (1983) Spatial and temporal distributions of dominant epiphytes on the stipes of *Laminaria hyperborea*, (Gunn.) Fosl. (Phaeophyta: Laminariales) in S.E. Scotland. *J. Exp. Mar. Biol. Ecol.* **73,** 1–10.

WHITTICK A. & HOOPER R.G. (1979) Seasonal changes in the phytobenthos of Conception Bay, Newfoundland, in the vicinity of a thermal outfall. *Brit. Phycol. J.* **14,** 128–9.

WHITTON B.A. (1971) Filamentous algae as weeds. In *3rd European Weed Res. Council Symposium*, pp. 249–63. Oxford.

WHITTON B.A. (1973) Interactions with other organisms. In *The Biology of the Blue Green Algae* (eds Carr N.D. & Whitton B.A.), pp. 415–33. University of California Press, Berkeley & Los Angeles.

WHITTON B.A. & SHEHATA F.H.A. (1982) Influence of cobalt, nickel, copper and cadmium on the bluegreen alga *Anacytis nidulans*. *Environ. Pollut. (Ser. A)*, **27,** 275–81.

WIESE L. (1974) Nature of sex specific glycoprotein agglutinins in *Chlamydomonas*. *Ann. N.Y. Acad. Sci.* **234,** 383–95.

WIK-SJOSTEDT A. (1970) Cytogenetic investigations in *Cladophora*. *Hereditas* **66,** 233–62.

WILBUR K.M. & WATABE N. (1963) Experimental studies of calcification in molluscs and the alga *Coccolithus huxleyi*. *Ann. N.Y. Acad. Sci.* **109,** 82–112.

WILCE R.T. (1967) Heterotrophy in Arctic sublittoral seaweeds: an hypothesis. *Bot. Mar.* **10,** 185–97.

WILCE R.T., SCHNEIDER C.W., QUINLAN A.V. & VAN DEN BOSCH K. (1982) The life history and morphology of a free living *Pilayella littoralis* (L.) Kjellm. (Ectocarpaceae, Ectocarpales) in Nahant Bay, Massachusetts. *Phycologia* **21,** 336–54.

WILDGOOSE P.B., BLUNDEN G. & JEWERS K. (1978) Seasonal variations in gibberellin activity of some species of Fucacaeae and Laminariaceae. *Bot. Mar.* **21,** 63–5.

WILDMAN R.B., LOESCHER J.H. & WINGER C.L. (1975) Development and germination of akinetes of *Aphanizomenon flos-aquae*. *J. Phycol.* **11,** 96–104.

WILKINSON M. & BURROWS E.M. (1970) *Eugomontia sacculata* Kornm. in Britain and North America. *Brit. Phycol. J.* **5,** 235–8.

WILLE J.N.F. (1897–1911) Conjugatae. Chlorophyceae. Characeae. In *Die naturalichen Pflanzenfamilien*, Teil 1, Abt. 2 (1897) (eds Engler A. & Prantl K.A.E.), 175 pp; 128 figs; ibid. (1909), 134 pp., 70 figs; ibid. (1911), 136 pp., 70 figs. Leipzig.

WILLIAMS J.L. (1897) The antherozoids of *Dictyota* and *Taonia. Ann. Bot.* **11,** 545–53.

WILLIAMS J.L. (1898) Reproduction in *Dictyota dichotoma. Ann. Bot.* **12,** 559–60.

WOELKERLING W.J. (1983) The *Audouinella* (*Acrochaetium — Rhodochorton*) complex (Rhodophyta): present perspectives. *Phycologia* **22,** 59–92.

WOESE C.R. (1981) Archaebacteria. *Sci. Am.* **299,** 98–122.

WOLK C.P. (1982) Heterocysts. In *The Biology of Cyanobacteria* (eds Carr N.G. & Whitton B.A.), pp. 359–86. Blackwell Scientific Publications, Oxford.

WOLLASTON, E. (1984) Species of Ceramiaceae (Rhodophyta) recorded from the International Indian Ocean Expedition, 1962. *Phycologia* **23,** 281–99.

WOMERSLEY H.B.S. & NORRIS R.E. (1959) A free floating marine red alga. *Nature* **811,** 828.

WOOD B.J.B. (1974) Fatty acids and saponifiable lipids. In *Algal Physiology and Biochemistry* (ed. Stewart W.D.P.), pp. 236–65. Blackwell Scientific Publications, Oxford.

WOOD R.D. (1967) *Charophytes of North America.* Stella's Printing, West Kingston, Rhode Island.

WOOD R.D. & IMAHORI K. (1965) *A Revision of the Characeae. Vol. 1, Monograph of the Characeae.* Cramer, Weinheim.

WRAY J.L. (1977) Late Paleozoic calcareous red algae. In *Fossil Algae, Recent Results and Development* (ed. Flugel E.). Springer-Verlag, Berlin.

WYNNE M.J. (1969) Life history and systematic studies of some Pacific North American Phaeophyceae (brown algae). *Univ. Calif. Publ. Bot.* **50,** 1–88.

WYNNE M.J. (1972) Culture studies of Pacific coast Phaeophyceae. *Soc. Bot. Fr. Memoires* **1972,** 129–44.

WYNNE M.J. (1981) Phaeophyta: morphology and classification. In *The Biology of Seaweeds* (eds Lobban C.S. & Wynne M.J.), pp. 52–85. Blackwell Scientific Publications, Oxford.

WYNNE M.J. (1982a) Phaeophyceae. In *Synopsis and Classification of Living Organisms*, Vol. 1 (ed. Parker S.P.), pp. 115–25. McGraw-Hill, New York.

WYNNE M.J. (1982b) Phaeophyceae: introduction and bibliography. In *Selected Papers in Phycology II* (eds Rosowski J.R. & Parker B.C.), pp. 731–9. Phycological Society of America, Allen Press, Lawrence, Kansas.

WYNNE M.J. & LOISEAUX S. (1976) Recent advances in life history studies of the Phaeophyta. *Phycologia* **15,** 435–53.

YAMADA T., IKAWA T. & NISIZAWA K. (1979) Circadian rhythm of the enzymes participating in the CO_2 photoassimilation of the brown alga *Spatoglossum pacificum. Bot. Mar.* **22,** 203–9.

YAMAMOTO I., TAKAHASHI M., TAMURA E. & MARUYAMA H. (1982) Antitumor activity of crude extracts from edible marine algae against L1210 leukemia. *Bot. Mar.* **25,** 455–7.

YAMAMOTO T., YAMAOKA T., TUNO S., TOKURA R., NISHIMURA T. & HIROSE H. (1979) Microconstituents in seaweeds. *Proc. Int. Seaweed Symp.* **9,** 445–50.

YAMANOUCHI S. (1906) The life history of *Polysiphonia violacea. Bot. Gaz.* **42,** 401–49.

YAN N.D. (1979) Phytoplankton community of an acidified heavy metal-contaminated lake near Sudbury, Ontario: 1973-1977. *Water Soil Air Pollut.* **11,** 43–55.

YARISH C., BREEMAN A.M. & HOEK C. VAN DEN (1984) Temperature, light and photoperiod responses of some Northeast American and West European endemic rhodophytes in relation to their geographic distribution. *Helgol. Wiss. Merresunters.* **38,** 273–304.

YASUMOTO T., INOUE A., OCHI T., FUJIMOTO K., OSHIMA Y., FUKUYO Y., ADACHI R. & BAGNIS R. (1980) Environmental studies on a toxic dinoflagellate responsible for ciguatera. *Bull. Jap. Soc. Sci. Fish.* **46,** 1405–11.

YENTSCH C.S. (1980) Light attenuation and phytoplankton phostosynthesis. In *The Physiological Ecology of Phytoplankton* (ed. Morris I.), pp. 95–128. University of California Press, Berkeley & Los Angeles.

YENTSCH C.M., YENTSCH C.S. & STRUBE L.R. (1977) Variations in ammonium enhancement, an indication of nitrogen deficiency in New England coastal phytoplankton populations. *J. Mar. Res.* **35,** 539–55.

YENTSCH C.M. & TAYLOR D.L. (1982) *Red Tides*. Harvard University Press, Boston.

YOUNGMAN R.E., JOHNSON D. & FARLEY M.R. (1976) Factors influencing phytoplankton growth and succession in Farmoor Reservoir. *Freshwater Biol.* **6,** 253–63.

ZEITZSCHEL B. (1978) Why study phytoplankton? In *Phytoplankton Manual* (ed. Sournia A.), pp. 1–6. UNESCO, Paris.

ZEPPA R.G. & SCHLOTZHAUER P.F. (1983) Influence of algae on photolysis rates of chemicals in water. *Environ. Sci. Technol.* **17**, 426–68.

ZIESENISZ E., REISSER W. & WIESSNER W. (1981) Evidence of *de novo* synthesis of maltose excreted by the endosymbiotic *Chlorella* from *Paramecium bursaria. Planta* **153**, 481–5.

ZIMMERMANN U. (1977) Cell turgor pressure regulation and turgor pressure mediated transport processes. In *Integration of Activity in the Higher Plant* (ed Jennings D.H.), pp. 117–54. Cambridge University Press, Cambridge.

ZIMMERMANN U., STEUDLE E. & LELKES P.I. (1976) Turgor pressure regulation in *Valonia utricularis*, effect of cell wall elasticity and auxin. *Plant Physiol.* **58**, 608–13.

ZINGMARK R.G. (1970) Sexual reproduction in the dinoflagellate *Noctiluca miliaris* Suriray. *J. Phycol.* **6**, 122–6.

Index